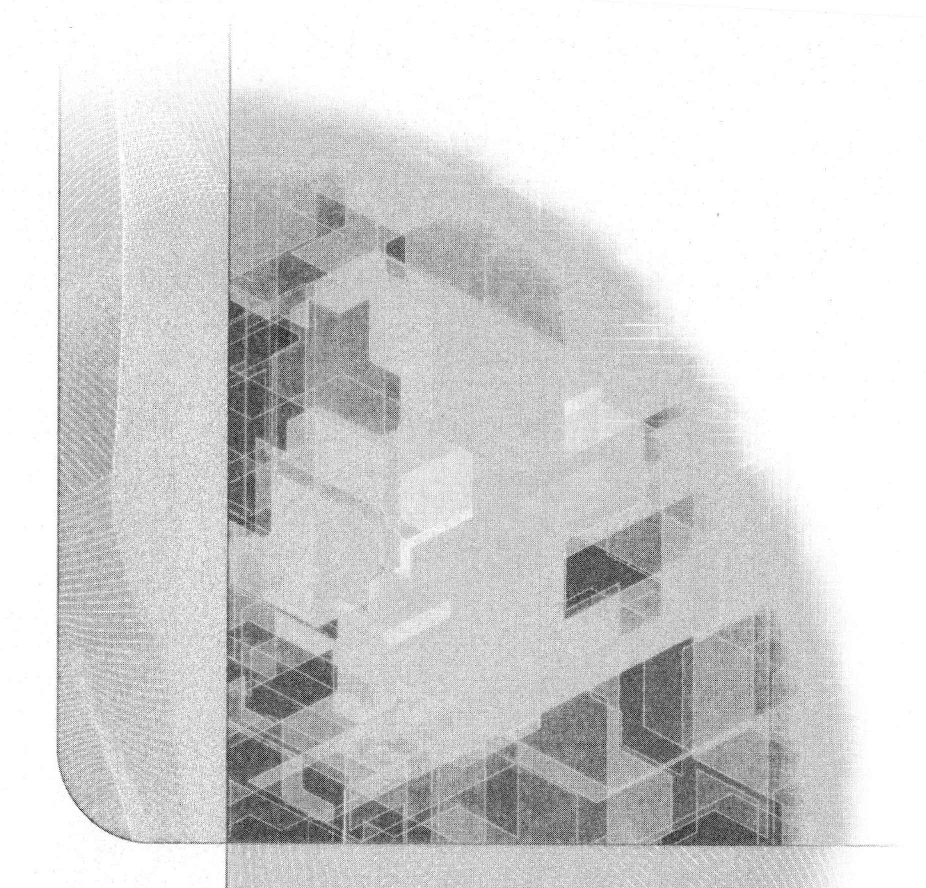

SolidWorks 2012 Tutor

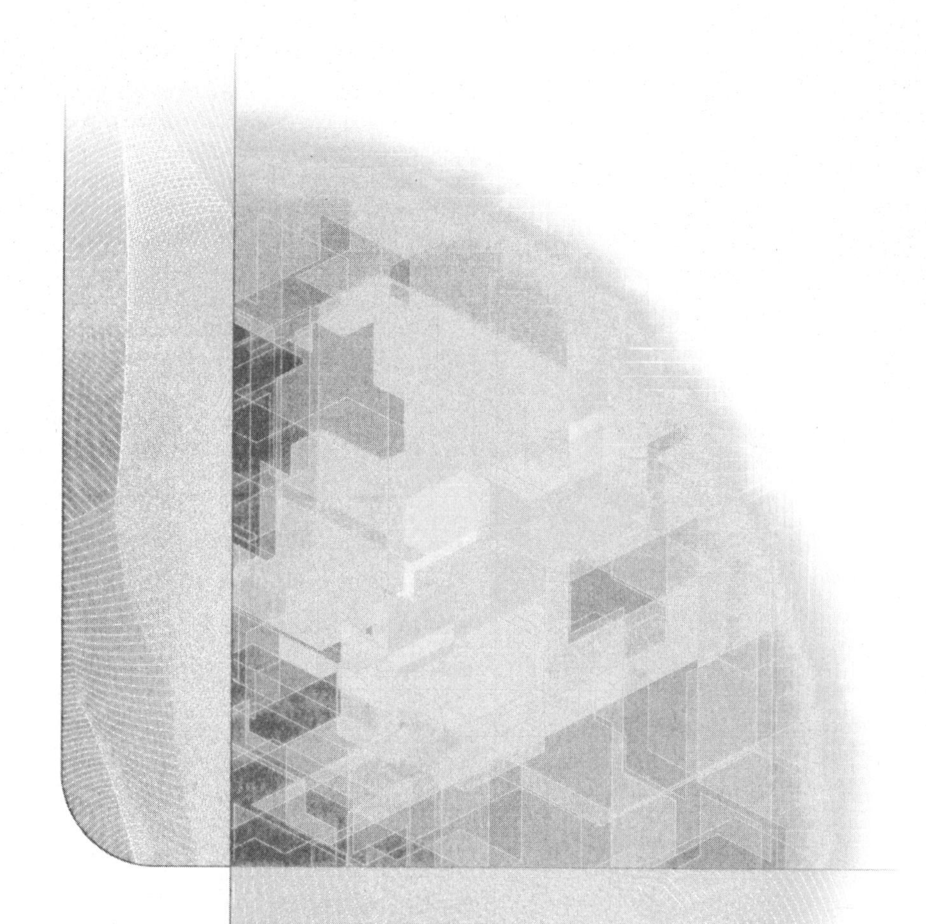

SolidWorks 2012 Tutor

ALAN J. KALAMEJA

MARK VOISINET

Program Coordinator of the Design and Drafting Program, Niagara County Community College

DELMAR CENGAGE Learning

Australia • Brazil • Japan • Korea • Mexico • Singapore • Spain • United Kingdom • United States

SolidWorks 2012 Tutor
Alan J. Kalameja and Mark Voisinet

Vice President, Editorial: Dave Garza
Director of Learning Solutions: Sandy Clark
Acquisitions Editor: Stacy Masucci
Managing Editor: Larry Main
Senior Product Manager: John Fisher
Editorial Assistant: Kaitlin Murphy
Vice President, Marketing: Jennifer Baker
Marketing Director: Deborah Yarnell
Marketing Manager: Erin Brennan
Marketing Coordinator: Jillian Borden
Production Director: Wendy Troeger
Production Manager: Mark Bernard
Content Project Management: PreMediaGlobal
Production Technology Assistant: Emily Gross
Art Direction: PreMediaGlobal
Technology Project Manager: Joe Pliss
Cover Image: courtesy of Mark Biasotti and 3DS SolidWorks

Library of Congress Control Number: 2012935641

ISBN-13: 978-1-4354-9678-1

ISBN-10: 1-4354-9678-7

Delmar
5 Maxwell Drive
Clifton Park, NY 12065-2919
USA

Cengage Learning is a leading provider of customized learning solutions with office locations around the globe, including Singapore, the United Kingdom, Australia, Mexico, Brazil, and Japan. Locate your local office at:
international.cengage.com/region

Cengage Learning products are represented in Canada by Nelson Education, Ltd.

To learn more about Delmar, visit **www.cengage.com/delmar**

Purchase any of our products at your local college store or at our preferred online store **www.cengagebrain.com**

Printed in the United States of America
1 2 3 4 5 6 7 16 15 14 13 12

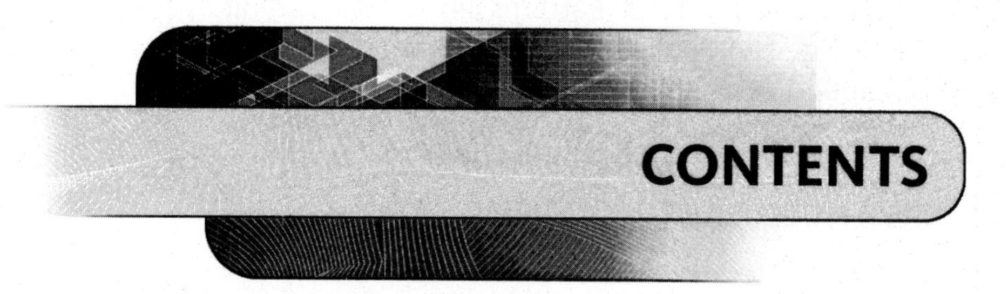

CONTENTS

CHAPTER 8 ADVANCED ASSEMBLY MODELING 325

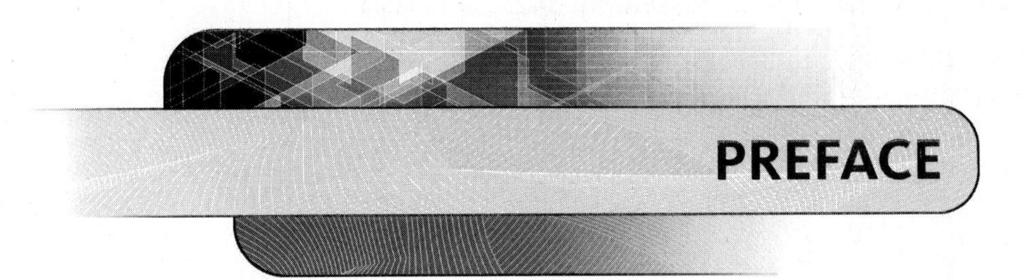

PREFACE

A long-standing phrase used to establish the need for drafting is "anything that is manufactured or built first has to have a technical drawing made of it." In the last couple of decades the field of design and drafting has evolved from hand drafting to simple 2D CAD drawings to 3D solid modeling. As will be discussed at the start of Chapter 5, because of the incorporation of technological advancements including solid modeling, the engineering design process has changed considerably. Although the initial statement is still true, time has come where we can now start using the phrase "Anything that is going to be manufactured or built first has to be virtually modeled."

INTENT

SolidWorks Tutor 2012 is written for the beginning SolidWorks user and the subject matter continues into an intermediate level. The textbook teaches the student the skill sets needed to create virtual models of complex parts and how to create technical drawings of parts and assemblies.

This text can be used to instruct the solid modeling component in a design and drafting or other technology-based curriculum. This text is designed in such a way that after completing all of the chapters the student can function efficiently using SolidWorks. The student will have been introduced thoroughly in the creation of parts and assemblies as well as drawing creation of parts and assemblies.

Future editions of this text as well as the Student Online Companion will offer more advanced topics in SolidWorks and the engineering design process.

HOW TO USE THIS TEXT

This text can be used in a number of ways. As the concepts have been ordered carefully, a beginning user can self-learn by going through the text from beginning to end. Critical concepts are incorporated into either very short assignments that are indicated by the "Try It!" icons or the longer step-by-step tutorials.

The text is also designed to be used in a classroom setting. The material covered can assist in the instruction of the solid modeling functionality component of a course.

Along with the tutorials and "Try It!" topics, additional chapter end components can be utilized as course assignments. There are chapter end review questions as well as chapter end modeling and drawing assignments.

Order matters! Following this text in the order the topics are presented will make it easier for the student to master working with SolidWorks. In order to fully comprehend the material, it is not recommended to jump around the text.

Along with the textbook, many files are available on the companion website. A number of these files are incorporated in the tutorials so the student doesn't have to start from scratch, instead they can use pre-created files that allow them to practice commands without having to create too much preliminary geometry. It is recommended that students download all of the tutorial files at one time so that they need not continually go to the website for individual files.

STUDENT ONLINE COMPANION

This text utilizes an internet companion piece. The Student Online Companion is your link to *SolidWorks Tutor 2012* on the internet. Facts and tutorials will be added to the website to accentuate the topics covered in this text. Tutorial files can be downloaded from this student companion site.

To access the Student Companion Site from CengageBrain:

1. Go to: http://www.cengagebrain.com
2. Type author, title, or ISBN in the Search window
3. Locate the desired product and click on the title
4. When you arrive at the Product page, click on the Access Now tab
5. Under Book Resources, download the drawing files for the book's tutorials and Try It! exercises.

ACKNOWLEDGMENTS

First, I would like to thank my colleagues who were willing to review this textbook prior to publication. They are Kirk Barnes from Ivy Tech Community College in Bloomington, Indiana; Steven Beyerlein from the University of Idaho in Moscow, Idaho; Pam R. Benson from Rochester Community and Technical College in Rochester, Minnesota; Yalcin Ertekin from Drexel University in Philadelphia, Pennsylvania; and Alex Lepeska from Renton Technical College in Renton, Washington.

I would also like to thank all of my students who assisted me in various ways on this project, particularly Alex Strauss and Corinne Cialfi.

A special thanks to John Fisher the Project Manager and Stacy Masucci the Acquisitons Editor for this text, particularly for their patience and understanding while I worked through my first textbook.

I'd also like to thank my family for their support throughout this project. My parents for their constant support, my wife for her patience, and my lovely daughters who had to resort to visiting the house dungeon in order to spend time with me while I was writing.

Lastly and most importantly, I'd like to honor the memory of my co-author Alan Kalameja. Alan had conceived the idea to offer a hybrid based textbook that is tutorial based as well offering chapter end assignments. Alan had partially completed this text before he passed away.

Regretfully I had never had the opportunity to meet Alan, but when I did a little research on Alan's life, I realized that we are very similar people. Alan was raised in Buffalo, NY, the city of good neighbors, only a couple of miles from where I was raised and still reside. Alan also had a strong faith in God and was active with his church, ideals in which I hold dearly also. Alan and I also both raised two wonderful daughters each.

I sense that if Alan and I had an opportunity to know each other, we would have been good friends. At least we had the opportunity to connect through our writing. God bless Alan, it is an honor being your co-author.

Mark Voisinet

INSTRUCTOR SITE

An Instruction Companion Website containing supplementary material is available.

This site contains an Instructor Guide, test bank, image gallery of text figures, and chapter presentations done in PowerPoint. Contact Delmar Cengage Learning or your local sales representative to obtain an instructor account.

To access an Instructor Companion Website site from SSO Front Door:

1. Go to: "http://login.cengage.com/" and log-in using the Instructor email address and password
2. Enter author, title, or ISBN in the Add a title to your bookshelf search box and click on Search button
3. Click Add to My Bookshelf to add Instructor Resources
4. At the Product page click on the Instructor Companion site link.

New Users

If you're new to Cengage brain and do not have a password, contact your sales representative.

SolidWorks 2012 Tutor

CHAPTER 1

Getting Started with SolidWorks 2012

This chapter discusses the basics of SolidWorks and includes the following topics:

- The SolidWorks Default Screen
- The SolidWorks Command Manager and the Feature Manager
- File types used in SolidWorks
- Manipulating part, drawing, and assembly files
- Using display tools for viewing a part or assembly
- Creating metric and inch part template files
- Walking through the creation of a sample part and drawing file

THE SOLIDWORKS DEFAULT SCREEN

The typical SolidWorks display screen consists of various screen elements designed to easily navigate through the creation of part, drawing, and assembly models. Figure 1.1 explains these elements in greater detail. The *Graphics Area* contains the part, drawing, or assembly model. *Pull-down menus* aid in the selection of most commands. Toolbars can be used to select the most popular of 3D commands. The *Command Manager* is an enhanced toolbar that displays a flyout explaining the most popular commands. The *Reference Triad* displays three axes that represent the X, Y, and Z axes that are used for determining the reference position of a 3D part or assembly. The *Feature Manager* is used for managing features as they are created in a part. The *Status Area* displays useful information on the lower right corner of the display screen. The *Orientation* pull-down displays a series of views that allow you to display your part or assembly in such standard views as Front, Top, Right, or even Isometric.

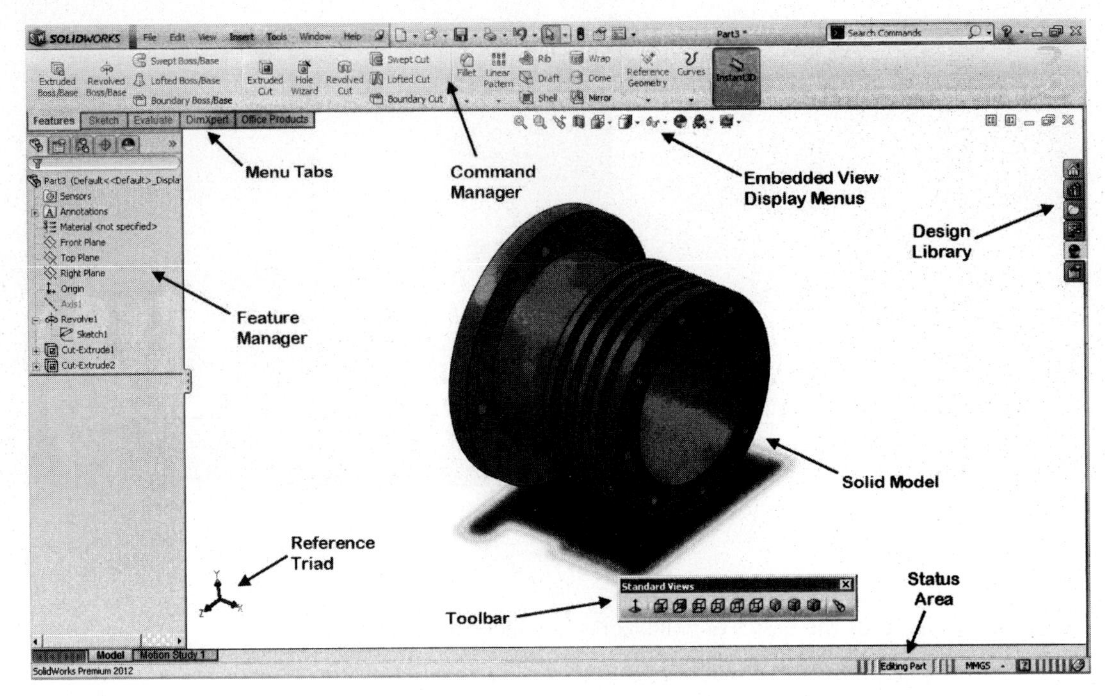

FIGURE 1.1

SELECTING COMMANDS FROM THE MENU BAR TOOLBAR

Basic commands such as New, Open, Save, Print, Undo, Select, Rebuild, and Options can be selected through a menu bar toolbar as shown in Figure 1.2. These items automatically display on the screen next to the SolidWorks banner.

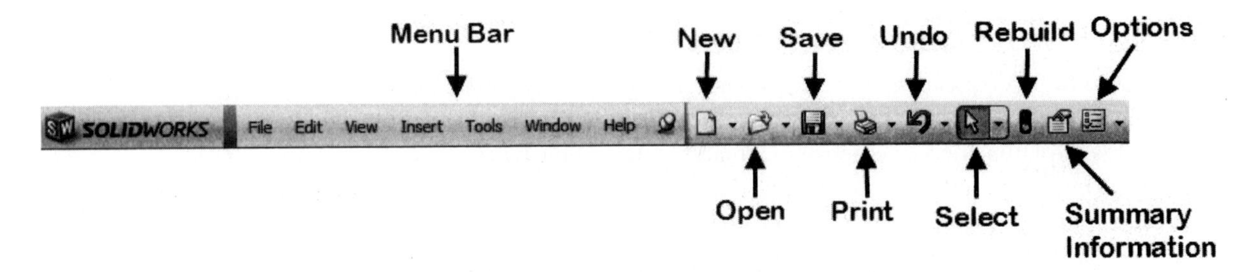

FIGURE 1.2

SELECTING COMMANDS FROM THE MENU BAR

Although most common commands are found on the Command Manager menus, commands can also be selected by utilizing any of the Menu Bar pull-down menus (Figure 1.3).

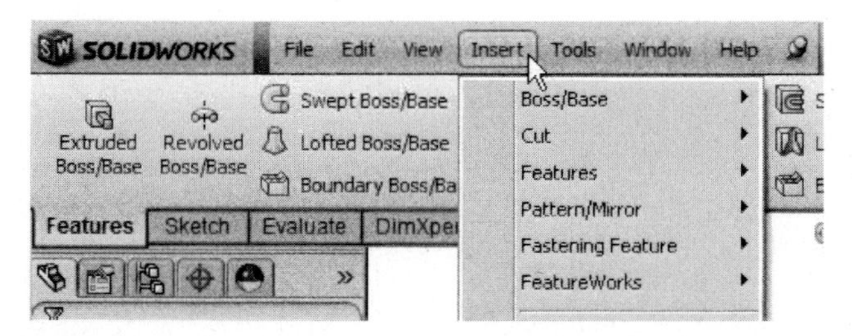

FIGURE 1.3

THE SOLIDWORKS COMMANDMANAGER

One of the more convenient methods of choosing SolidWorks commands is through the Command Manager. The more popular commands can be found in this area. Different sets of commands are available depending on the tab selected. Figure 1.4 illustrates commands located in the Sketch and Features Command Managers.

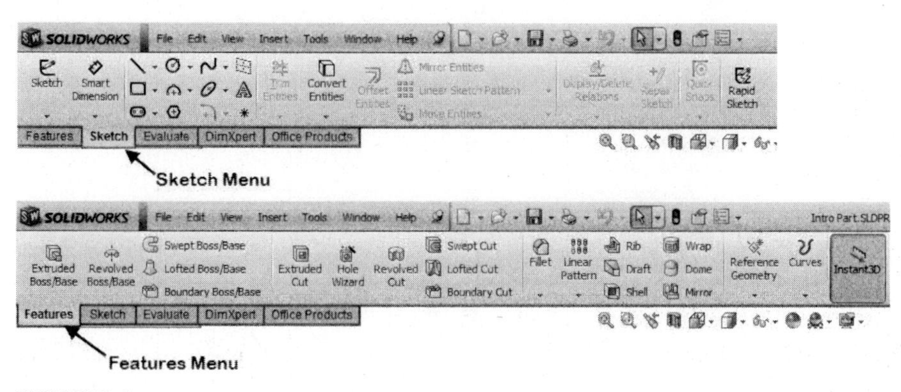

FIGURE 1.4

THE HEADS-UP VIEW TOOLBAR

A heads-up view toolbar appears at the top of the display screen as shown in Figure 1.5. It appears transparently on the screen and is used for choosing common tools for manipulating views, zooming, and hiding or showing items.

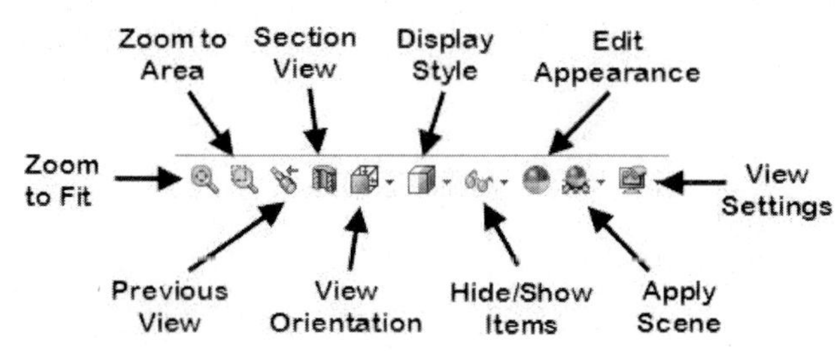

FIGURE 1.5

ACCESSING TOOLBARS IN SOLIDWORKS

While inside of the menu bar, clicking on View followed by Toolbars displays a menu that lists all toolbars available in SolidWorks. Use these toolbars to get further access to commands used in SolidWorks (Figure 1.6).

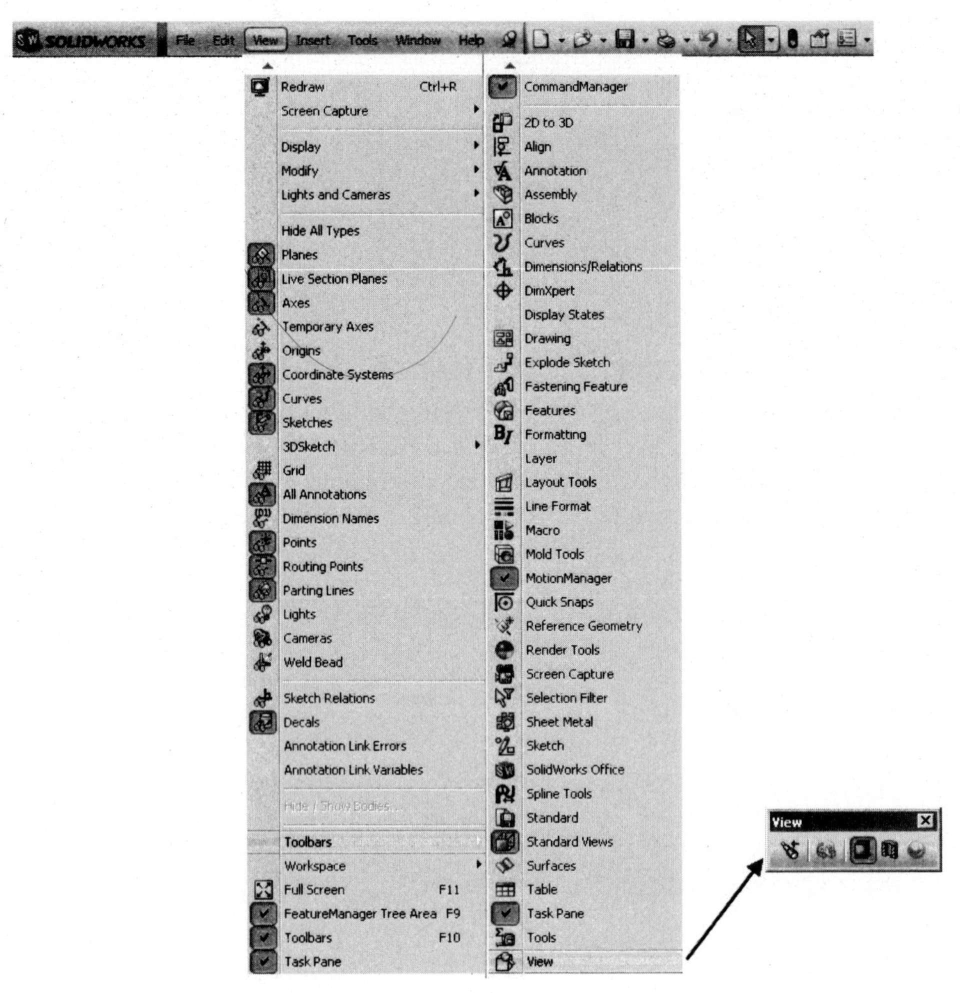

FIGURE 1.6

SCREEN MANAGER ENVIRONMENTS

The pane located on the left side of the display screen displays a number of tools used for managing various phases of the part modeling, drawing, assembly, and dimensioning environments. The top of this pane identifies four tabs used for managing these phases. The four tabs are illustrated in Figure 1.7 and will be explained in greater detail in this and other chapters.

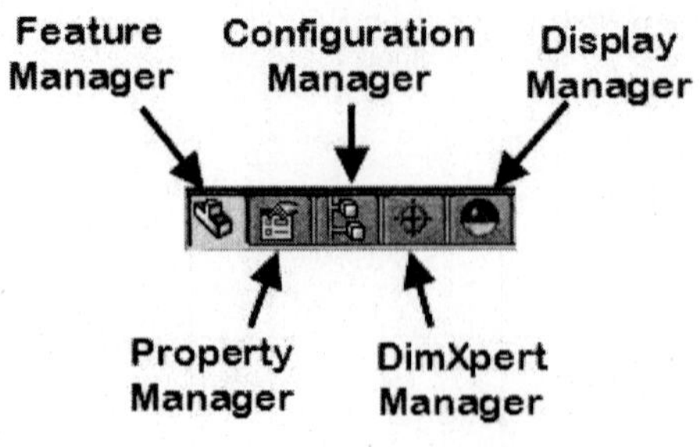

FIGURE 1.7

THE FEATURE MANAGER

The Feature Manager design tree displays the order that features, drawing sheets, or parts appear in a chronological order. With part files, for instance, as features are created, they are added to the Feature Manager as shown in Figure 1.8 on the left.

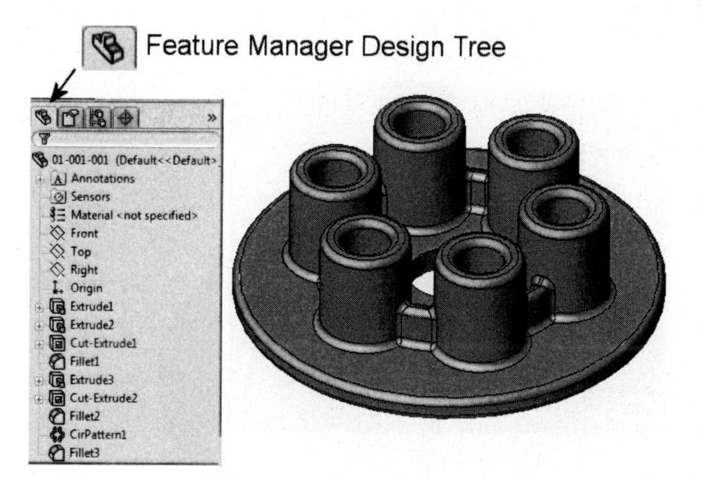

FIGURE 1.8

Dedicated Feature Managers are available in part, drawing, and assembly environments as shown in Figure 1.9. When creating a part model, the Feature Manager displays those features used for creating the part. While in drawing mode, the Feature Manager displays the various drawing sheets and the views associated with the sheet. While in Assembly mode, the Feature Manager displays the individual part files that make up the assembly model.

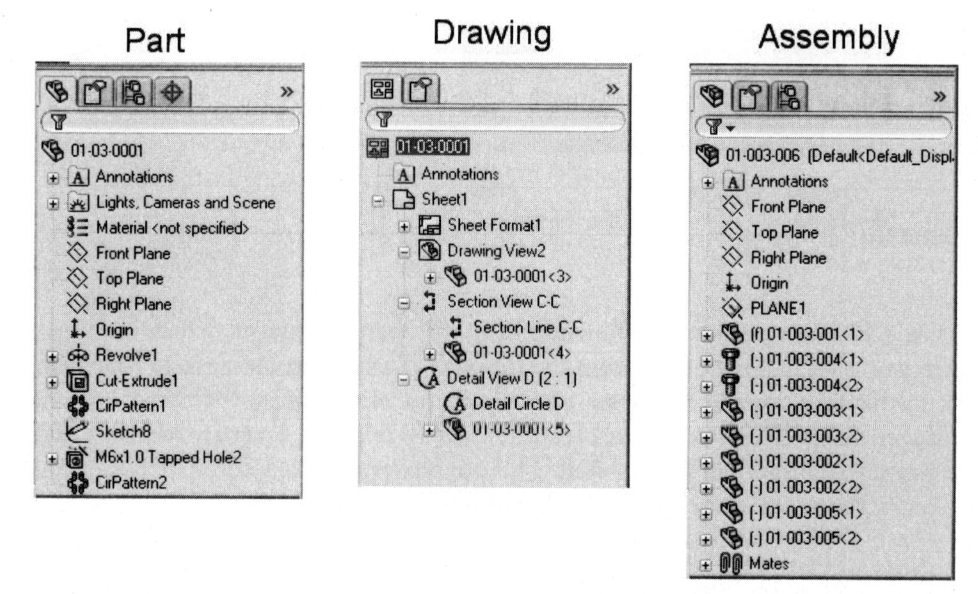

FIGURE 1.9

THE PROPERTY MANAGER

The second tab present at the top left corner of a typical SolidWorks display screen is the Property Manager as shown in Figure 1.10. In this example, the Extruded Base/Boss tool was activated and flagged for editing. The Property Manager automatically activates and displays information such as direction, flip, and distance, just to name a few. Also displayed in this figure is a preview of the extruded distance in preview mode.

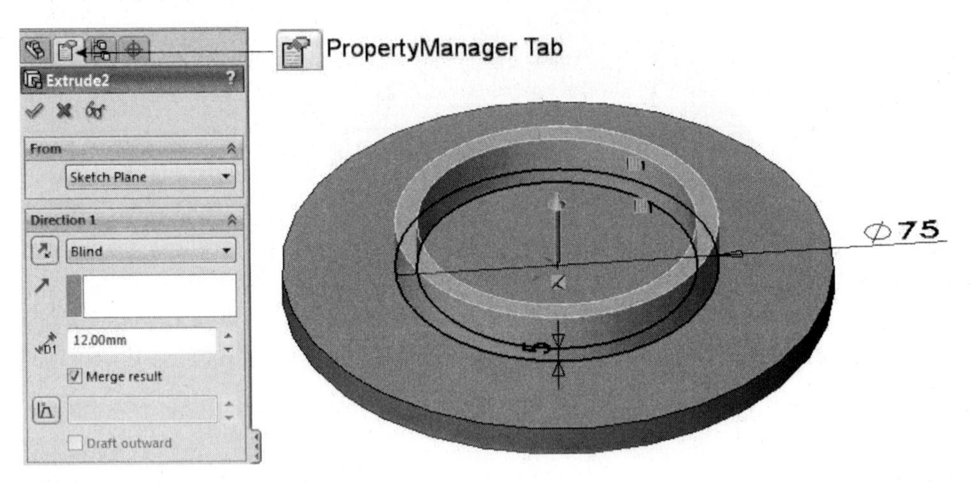

FIGURE 1.10

Also available at the top of the Property Manager title bar are a number of acknowledgment buttons used for accepting, canceling, or previewing changes as shown in Figure 1.11.

FIGURE 1.11

Figure 1.12 illustrates another function of the Property Manager. There are times when you need to switch back from the Property Manager to the Feature Manager. To do this task, expand the name of the part model to display the flyout Feature Manager design tree as shown in Figure 1.12. Now both the Property Manager and the Feature Manager can be displayed. This can be especially helpful when displaying reference planes or features while in the Property Manager.

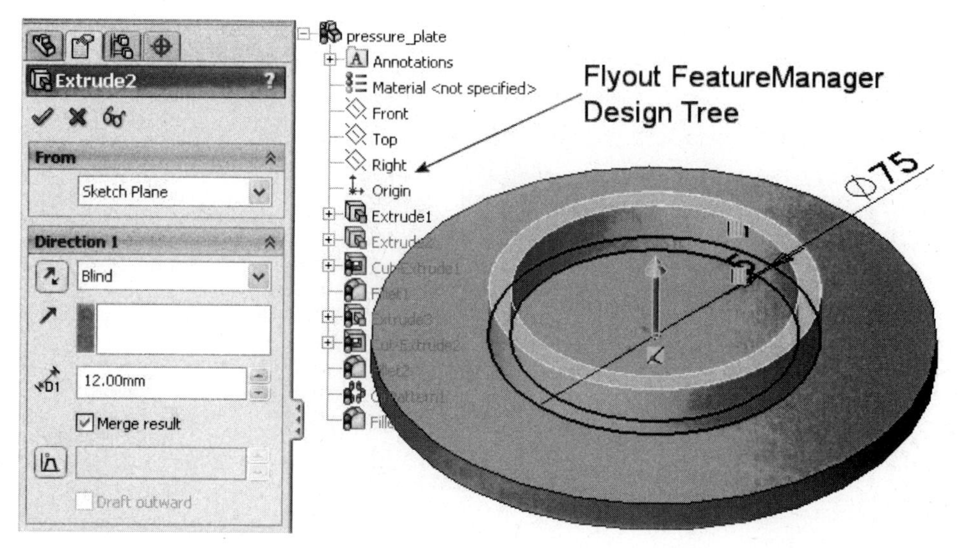

FIGURE 1.12

THE CONFIGURATION MANAGER

Another tab can also be present in the upper left corner of the SolidWorks display screen; it is called the Configuration Manager. This tool is used to create, select, and view multiple configurations of parts and assemblies in a document. Think of a configuration as a family of a part that can be controlled by a single feature. Figure 1.13 displays a structural member that has a length of 426mm. Rather than creating separate part files that have different lengths, configurations are created to group like parts of different lengths under the same file name.

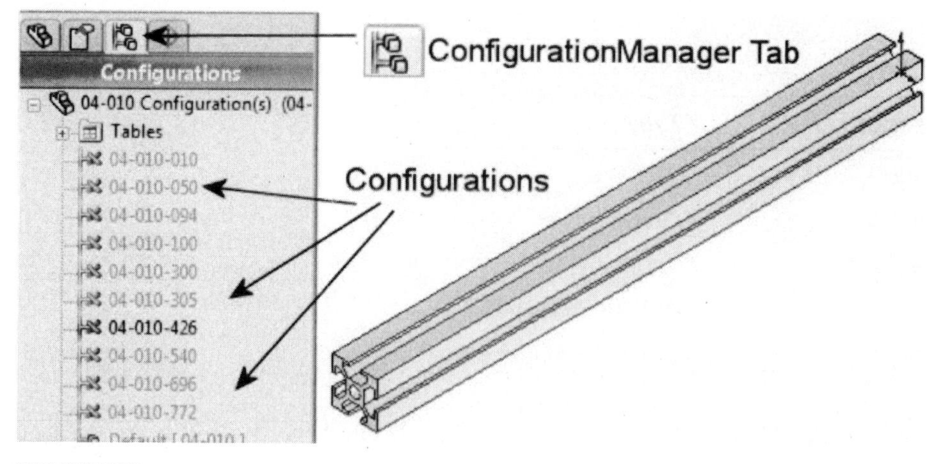

FIGURE 1.13

THE DIMXPERT MANAGER

The DimXpert Manager lists the tolerance features defined by the DimXpert for parts function as shown in Figure 1.14. It also displays DimXpert tools that are used to insert dimensions and tolerances into parts. These dimensions and tolerances can then be imported into drawings.

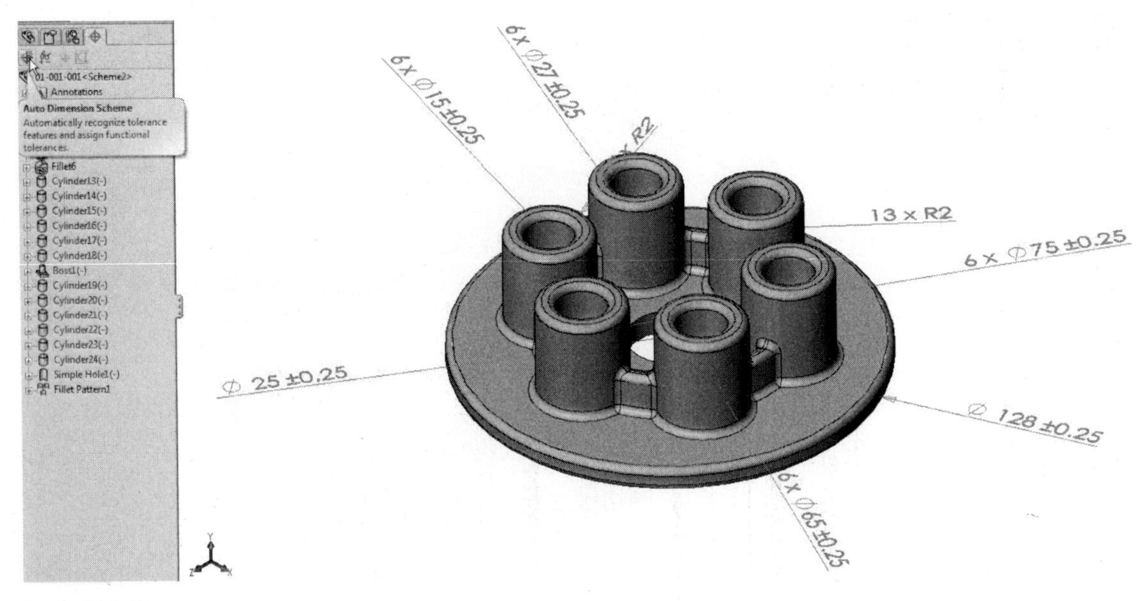

FIGURE 1.14

COMMONLY USED SOLIDWORKS FILE TYPES

When building part, drawing, and assemblies in SolidWorks, separate file extensions are used to keep these different environments separate. Generally, you will begin creating part files that have an extension of SLDPRT. After enough part files are created, you will generate drawing views and create drawing files that have an extension of SLDDRW. Finally, you will assemble a number of part files into an assembly that has an extension of SLDASM. Use the following table to identify these file and template extensions.

TABLE 1.1

Extension	Type
SLDPRT	Part File
SLDDRW	Drawing File
SLDASM	Assembly File
PRTDOT	Part Template File
DRWDOT	Drawing Template File
ASMDOT	Assembly Template File

DISPLAY TOOLS

The ability to view a part file in any orientation is a basic yet powerful function of the SolidWorks program. Display tools are available from the mouse or through various pull-down menus, and toolbars can be selected by selecting the Tools pull-down menu (select Customize) (Figure 1.15). An image of each view along with a description is given in the following table. Toolbars can be moved to suit.

FIGURE 1.15

TABLE 1.2

Tool	Description
	Normal To—Displays the view normal or perpendicular to the current sketch plane or current surface selected.
	Front View—Rotates and zooms the model to display the front view.
	Back View—Rotates and zooms the model to display the back view.
	Left View—Rotates and zooms the model to display the left view.
	Right View—Rotates and zooms the model to display the right view.
	Top View—Rotates and zooms the model to display the top view.
	Bottom View—Rotates and zooms the model to display the bottom view.
	Isometric—Rotates and zooms the model to display an isometric view.
	Trimetric—Rotates and zooms the model to display a trimetric view.
	Dimetric—Rotates and zooms the model to display a dimetric view.
	View Orientation—Displays a dialog box to select standard or user-defined views.

Display Functions Using the Mouse

The following table outlines a few of the basic display functions available with using a wheel mouse. Rolling the wheel will magnify or demagnify the image of a part, drawing, or assembly model. Pressing and holding down on the mouse wheel while moving it along the screen will rotate the part and assembly models. At times, you need to keep the same rotation orientation yet pan to a different location on the part or assembly model. To pan the model, press and hold down the CTRL key while also pressing and holding down on the wheel. Moving the mouse will pan the image to a new location. Other rotation tools include tapping the arrow keys to rotate a part or assembly horizontally or vertically.

TABLE 1.3

Mouse and Arrow Key Functions	Function Description
Roll the Wheel	Increases or decreases the magnification of a part, drawing, or assembly
Press on the Wheel	Rotates the part or assembly in 3D
Double-Click on the Wheel	Fits the part, drawing, or assembly to the current display screen
CTRL + Wheel	Press the CTRL key while pressing the wheel to pan the part or assembly without rotating or magnifying
Directional Arrow Keys	Rotate the part or assembly vertically or horizontally
SHIFT + Arrow Keys	Rotate the part or assembly vertically or horizontally in 90° increments
ALT + Left/Right Arrow Keys	Rotate the part or assembly clockwise or counterclockwise

Display Functions Using the Orientation Dialog Box

One of the more convenient methods of manipulating a part or assembly model is through the Orientation dialog box as shown in Figure 1.16 on the left. To activate this dialog box, press the space bar. Once displayed, you can select from a number of standard views which include Isometric, Trimetric, and Dimetric. All three of these examples are illustrated in the following figure. Note that while you click one of the views, the image changes and the Orientation dialog box disappears. You must press the space bar to activate it again.

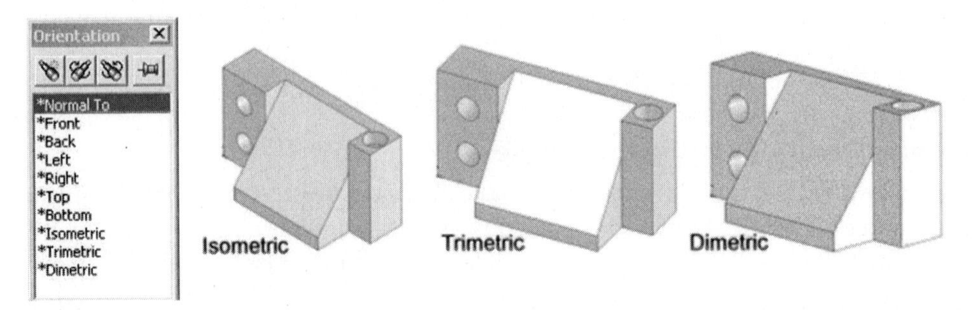

FIGURE 1.16

In addition to the three main pictorial views shown in Figure 1.16, you can view a model from one of the six main orthographic views as shown in Figure 1.17.

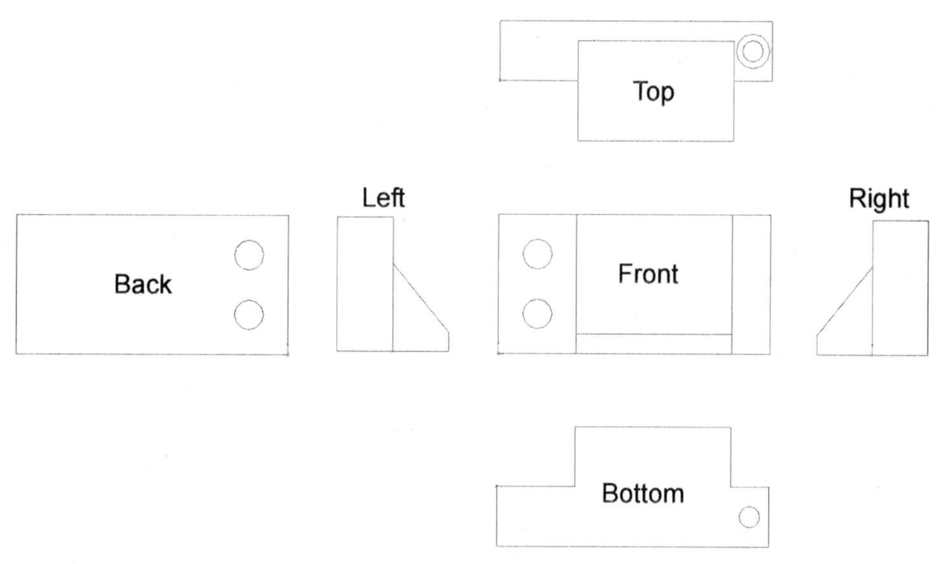

FIGURE 1.17

In addition to the Orientation dialog box, the on-screen display pull-down menu is always available to display the standard views as shown in Figure 1.18 on the left. You can even split your screen into one of the four standard viewing port positions. With the Orientation dialog box already mentioned, it is possible to display the dialog box without having to continually press the space bar. This is accomplished by clicking on the push pin icon located in the upper left corner of this dialog box to keep the dialog box displayed as shown in the following figure.

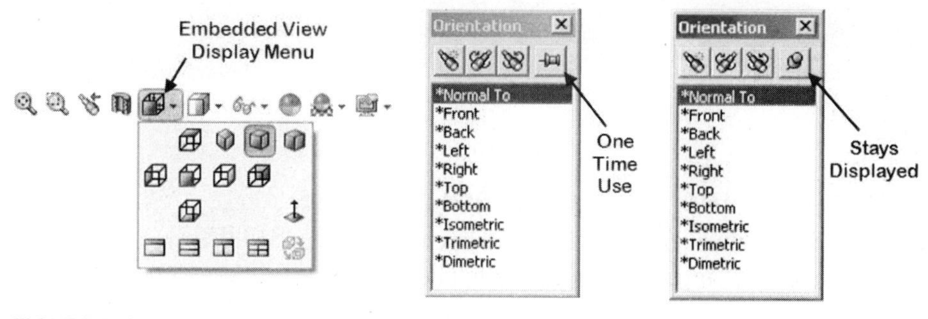

FIGURE 1.18

SHADING MODES FOR PART MODELS

As parts and assemblies are being displayed, a number of shading modes are available. Some shade modes deal specifically with the appearance of the part while other modes assist in the construction of parts and assemblies. The seven shade modes are shown in Figure 1.19.

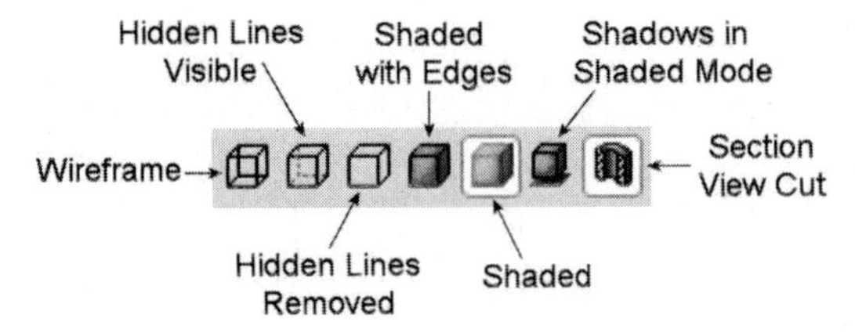

FIGURE 1.19

Shading Modes

Figure 1.20 illustrates examples of the six primary shading modes. Note that the Shadows in Shaded Mode tool are either on or off and affects only the shaded display of the part.

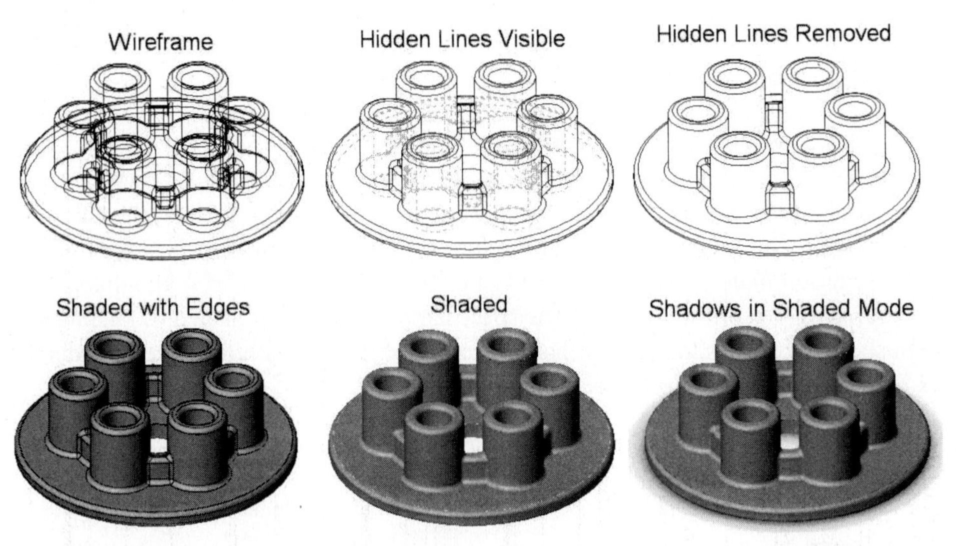

FIGURE 1.20

OPENING EXISTING MODELS IN SOLIDWORKS

Use the Open tool shown in Figure 1.21 to open up existing SolidWorks models. An Open dialog box will appear allowing you to choose the file or files to open.

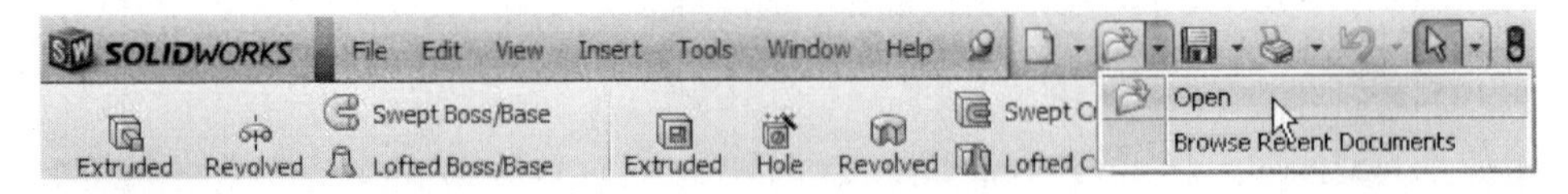

FIGURE 1.21

When working with existing models, the previous seven models appear under the File tab as shown in Figure 1.22 on the left. Moving your cursor over one of these files will display a flyout as shown in the following figure on the right. This gives you a more graphical view when opening up files.

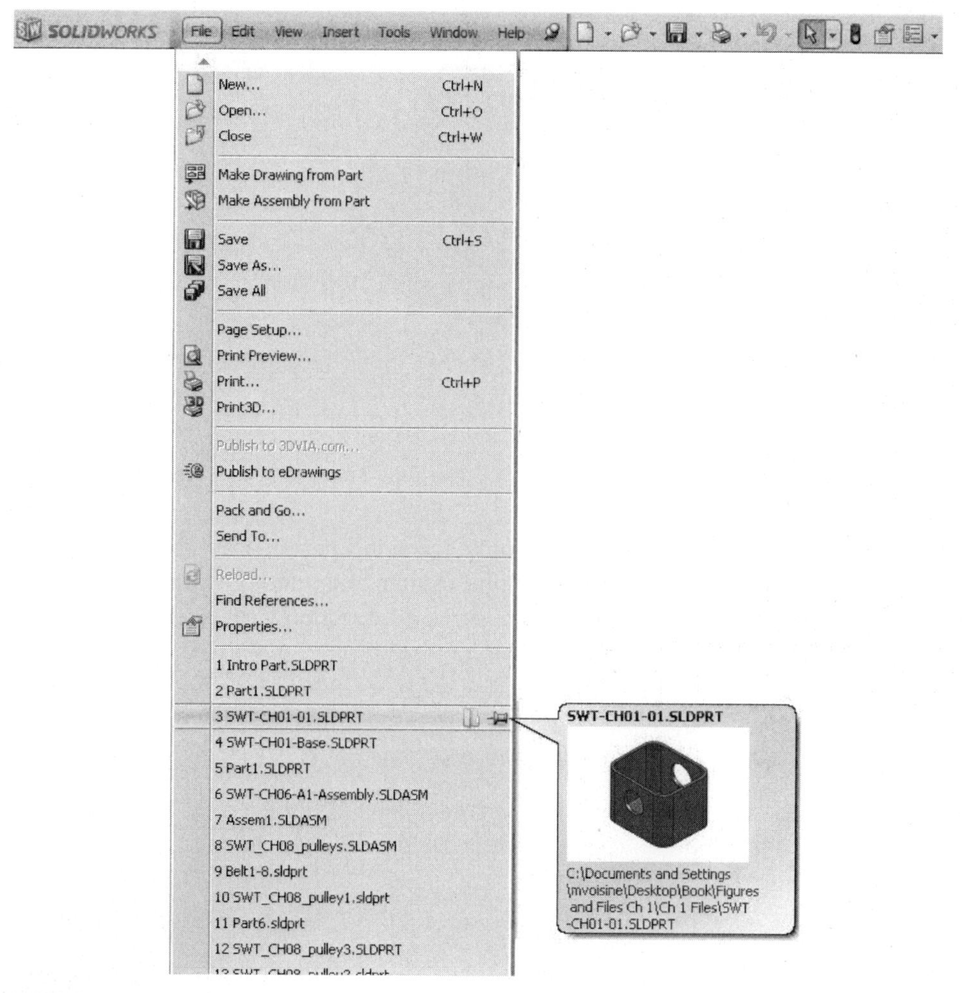

FIGURE 1.22

WORKING WITH PART FILES (SLDPRT)

Begin experimenting with the part file environment by opening the file SWT-01-001-001 as shown in Figure 1.23. This file represents an existing 3D model built in SolidWorks. The following tasks will be performed on this part file: Rotating; Generating orthographic views; Editing sketches and features; Rolling back features; and Changing the color of the model.

TRY IT!

For example, press and hold down the wheel while moving the mouse. Notice how the object rotates. Experiment rotating the part until you can see its underside.

To return the part back to its original orientation, press the space bar to display the Orientation dialog box and double-click on Isometric. This will display the part as shown in Figure 1.23. Also experiment with choosing Trimetric and Dimetric from the Orientation dialog box.

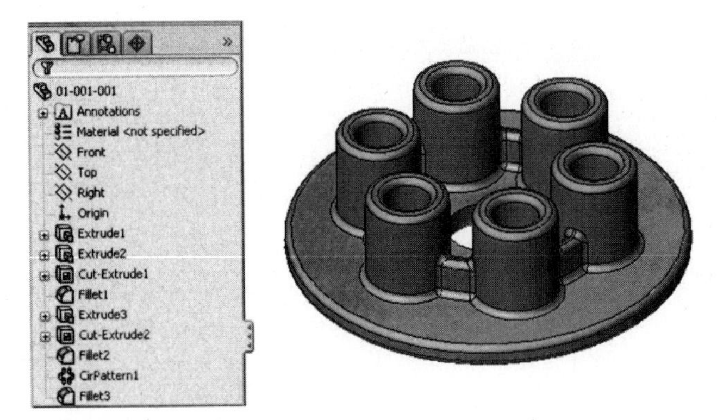

FIGURE 1.23

Section View Cuts

Another shading mode technique is to split or section a part in order to view its internal features. This mode is especially helpful during constructing features where it is important to view complicated internal features. The section view is created based on an existing plane as shown in Figure 1.24.

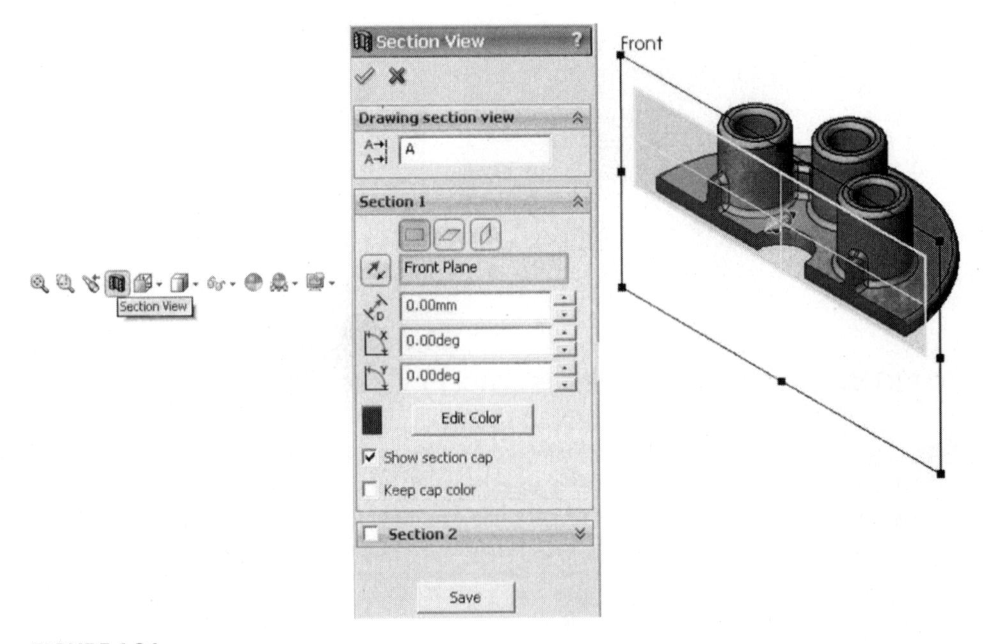

FIGURE 1.24

The results of performing the section view cut are shown in Figure 1.25. You can also reverse or flip the resulting cut as shown in the following figure.

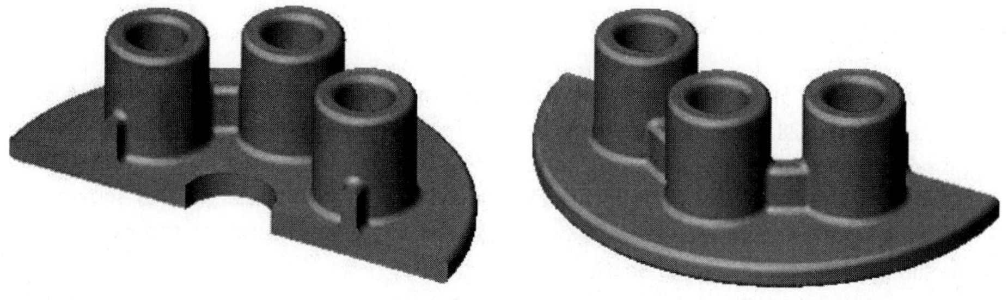

FIGURE 1.25

Rolling Back Operations

Once a part is finished, you can roll back a single feature or multiple features in order to view how a part was created. In Figure 1.26, the existing part was rolled back to its original extrusion by pressing and dragging the bottom line of the Feature Manager. When you release the mouse button under a specific feature, the remainder features disappear and display only what is above the Feature Manager rollback line. To have your part return back to its original state, press and drag the rollback bar to the bottom of the last feature.

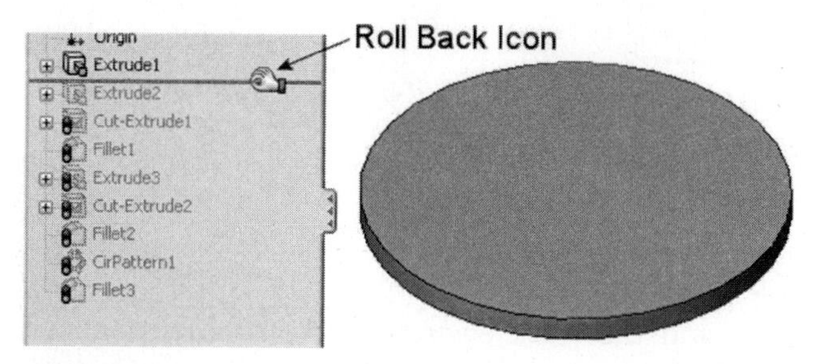

FIGURE 1.26

WORKING WITH DRAWING FILES (SLDDRW)

TRY IT!

Experiment with the drawing environment by opening the file SWT-01-001-001.SLDDRW as shown in Figure 1.27. This file represents an existing 2D drawing file that was generated from the 3D model. You will perform the following tasks on this part file: Zooming, Panning; Moving drawing views and observing how they remain justified to one another; Observing the makeup of the drawing Feature Manager as shown in the following figure on the left.

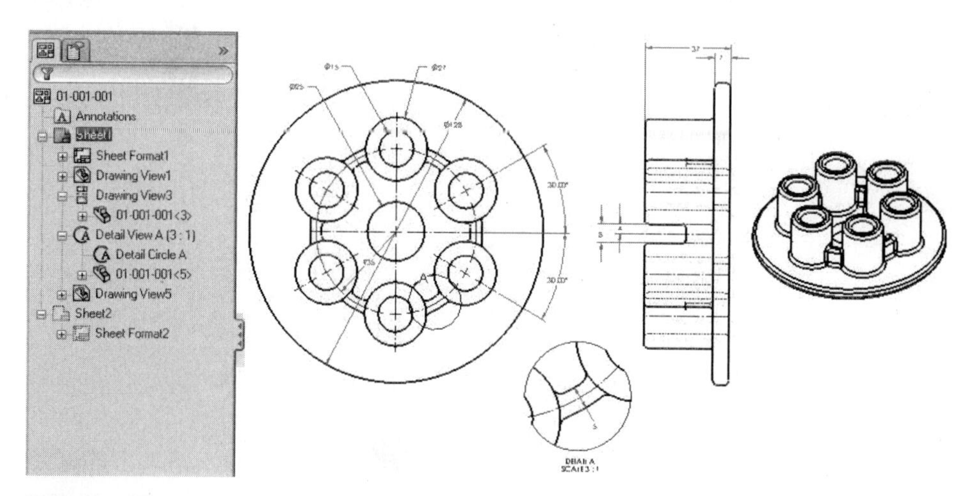

FIGURE 1.27

WORKING WITH ASSEMBLY FILES (SLDASM)

TRY IT!

Experiment with the assembly environment by opening the file SWT-01-003-006.SLDASM as shown in Figure 1.28. This file represents an existing assembly model of a gear train built with numerous parts. You will perform the following tasks on this part file: Rotating, Panning; Observing the individual parts located in the assembly Feature Manager as shown in the following figure on the left; Press and drag on the left gear and notice the right gear rotating along with the hexagonal shafts.

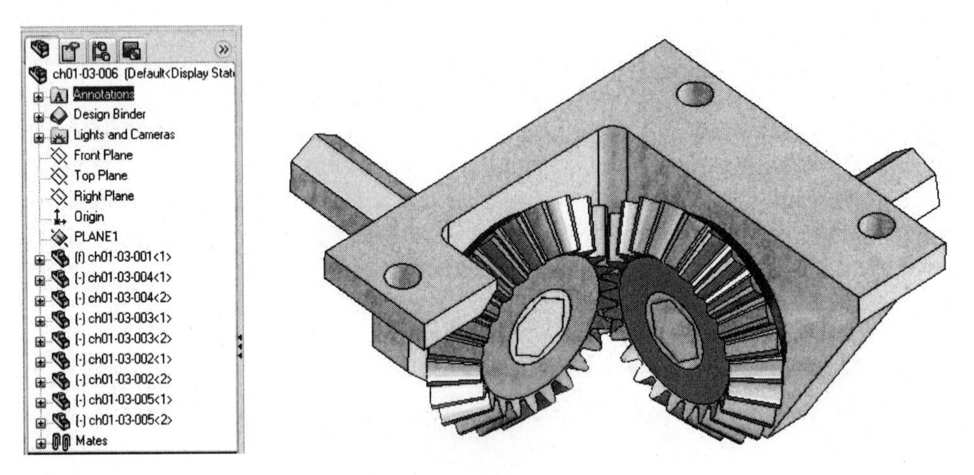

FIGURE 1.28

MANAGING DOCUMENT SETTINGS

Items such as drafting standards and the work environment can be customized through the Document Properties dialog box, which is activated by clicking Options from the Tools pull-down menu as shown in Figure 1.29 on the left. Two tabs available through this dialog box include System Options and Document Properties as shown in the following figure on the right.

With System Options, items saved under this tab affect every file that is opened in a typical SolidWorks design session. System Options allow you to customize the SolidWorks work environment by making changes to such items as the screen background.

With Document Properties tab, specific settings are saved only to the individual design file. Through this tab, you can make changes to the current design units or even change the drafting standards.

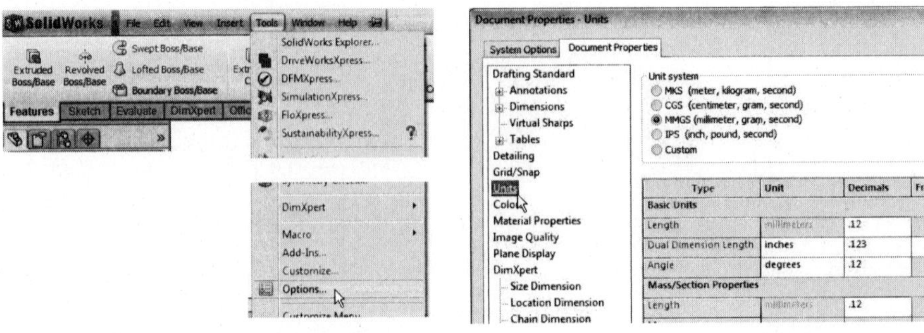

FIGURE 1.29

TEMPLATE FILES (PRTDOT, DRWDOT, ASMDOT)

Clicking the New tool launches the New SolidWorks Document dialog box as shown in Figure 1.30 on the left. These represent the standard part, assembly, and drawing templates available when SolidWorks was initially loaded. Notice the Advanced button located in the lower left corner of this dialog box; clicking on it takes you to the Templates tab that lists all available templates as shown in the following figure on the right. This list displays the same three default templates; you do have the ability to create custom templates and have them display in the Templates tab.

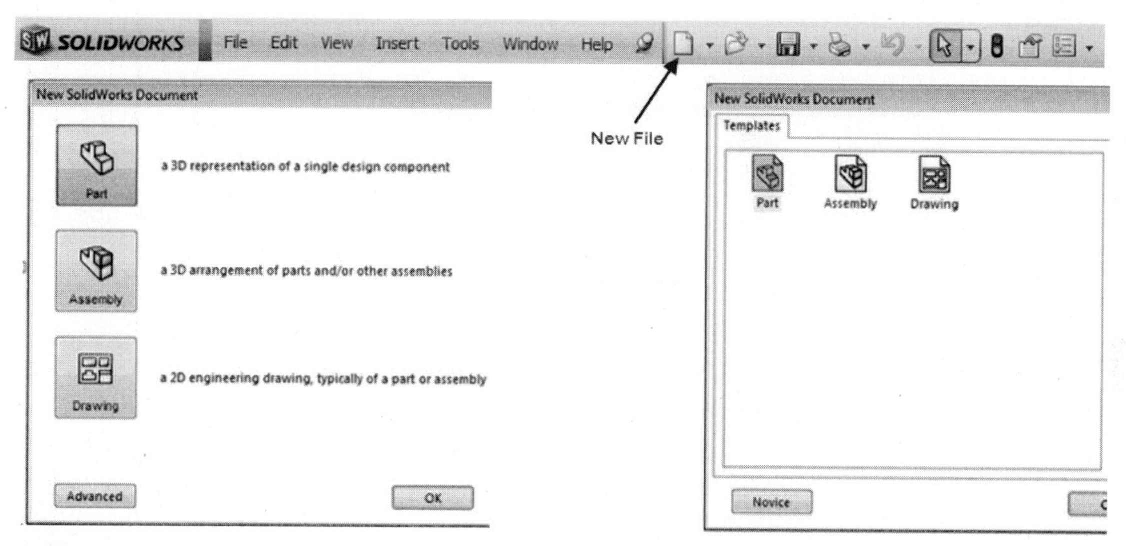

FIGURE 1.30

Creating a Part Template That Uses Millimeters as the Units of Measure

This segment deals with creating a custom template for use in creating metric part files. This template may already be available in metric units; however, we will first check the units and then save the template under a name unique to metric units. First, click the Open tool to launch the Open dialog box. Change the Files of type: to Template as shown in Figure 1.31 and click the open button. Next, click on the file Part.PRTDOT; this file extension identifies a part template.

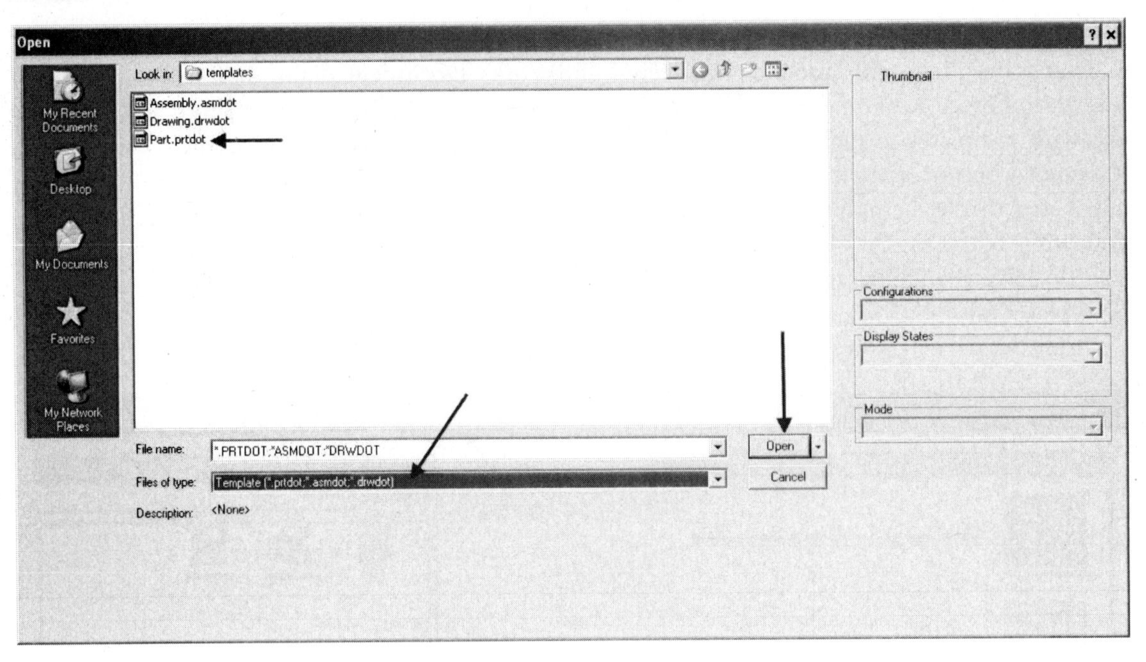

FIGURE 1.31

You have just opened the Part.PRTDOT template file. Your display screen will appear similar to any part modeling environment. You will change or modify two items, namely, the dimension standard and the units. Both items can be found by first clicking on Options under the Tools pull-down menu as shown in Figure 1.32 on the left. When the Document Properties dialog box displays, switch to the Document Properties tab and click on Drafting Standard as shown in the following figure on the right. It is here that you change to the proper dimensioning standard. Normally, you would pick ISO, DIN, or JIS, which represent one of the many metric dimensioning standards available. You could also elect to use the ANSI or English dimensioning style.

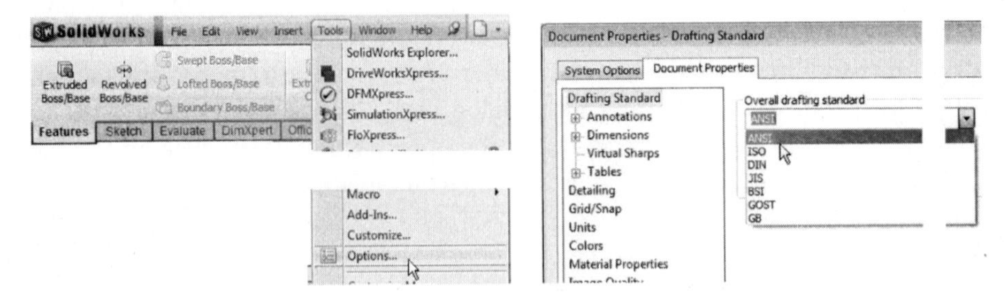

FIGURE 1.32

Next, click Units from the list on the left side of the Document Properties tab. Under the Unit System heading, change the units to MMGS (millimeters, grams, seconds) as shown in Figure 1.33. Click the OK button to save and exit this dialog box.

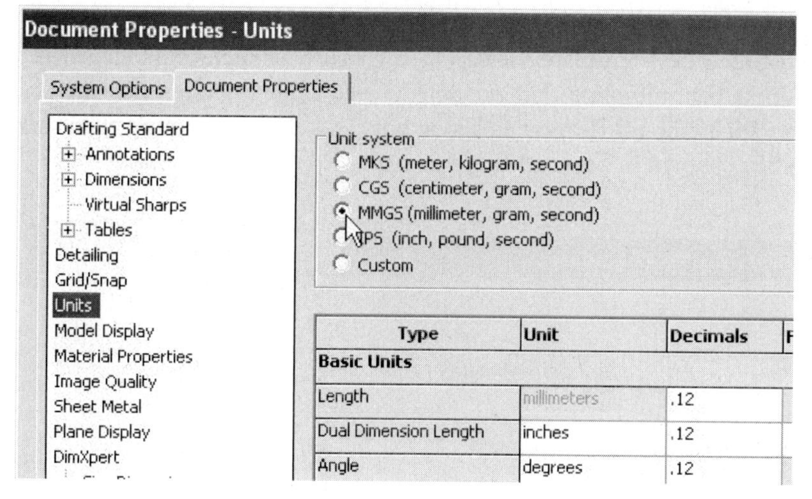

FIGURE 1.33

You will now save this template under a different name. First click the Save As button to launch the Save As dialog box. Verify that the file type is still set to Part Templates with the extension of PRTDOT as shown in Figure 1.34 on the left. Enter the new template name as Part (mm). This should place the new template in the same area as the existing templates. This completes the creation of the metric part file template.

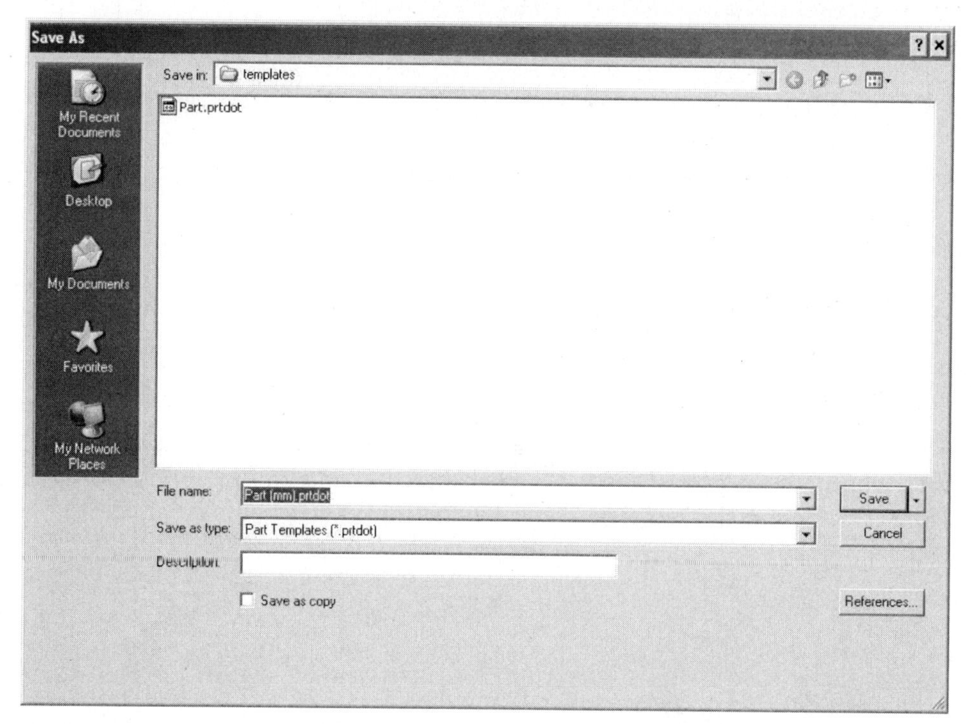

FIGURE 1.34

Creating a Part Template That Uses Inches as the Units of Measure

Continue by creating a new template for use with English or inch units. Open the existing SolidWorks part template that previously used millimeters as the units of measure, launch the Document Properties dialog box, and change the Dimensioning Standard to ANSI as shown in Figure 1.35.

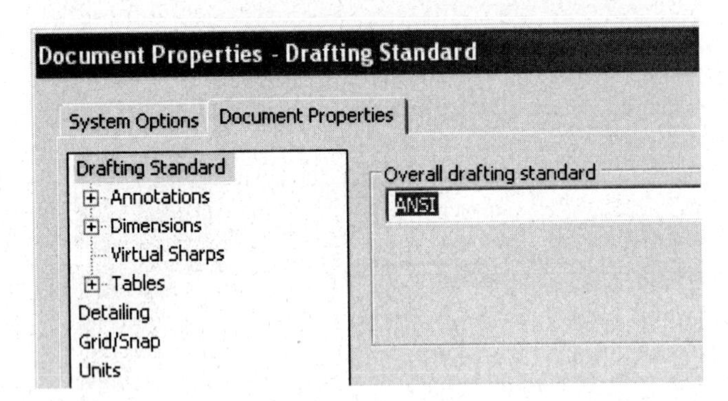

FIGURE 1.35

While still in the Document Properties tab, change to the Units category and change the Unit System to IPS (inch, pound, second) as shown in Figure 1.36. When finished, click the OK button to save the changes and exit this dialog box.

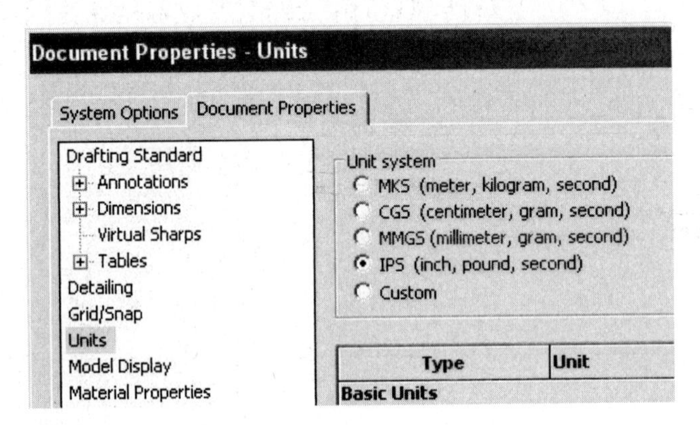

FIGURE 1.36

Use the Save As tool to save the template under a different name. Verify that the Part Template is the current file type and change file name to Part (in) as shown in Figure 1.37. Be sure this template is saved in the correct area (templates folder).

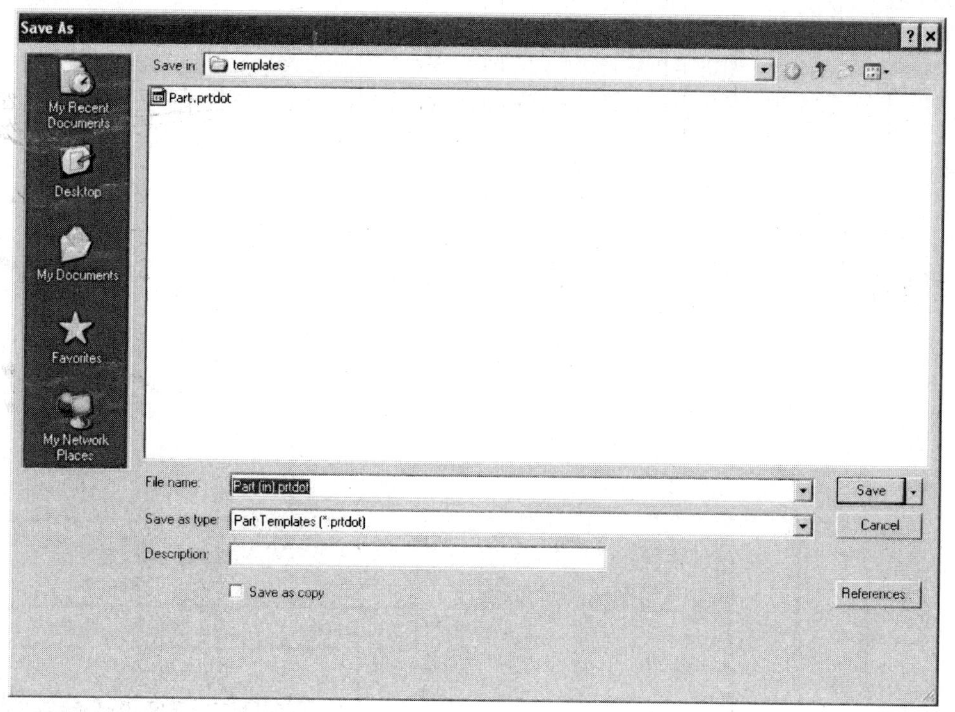

FIGURE 1.37

Test to see that both templates are available in the New SolidWorks Document dialog box. Click the New tool to display the New SolidWorks Document dialog box. Notice that you now have two new part templates, namely, Part (in) and Part (mm), that will be used for various part modeling exercises as shown in Figure 1.38.

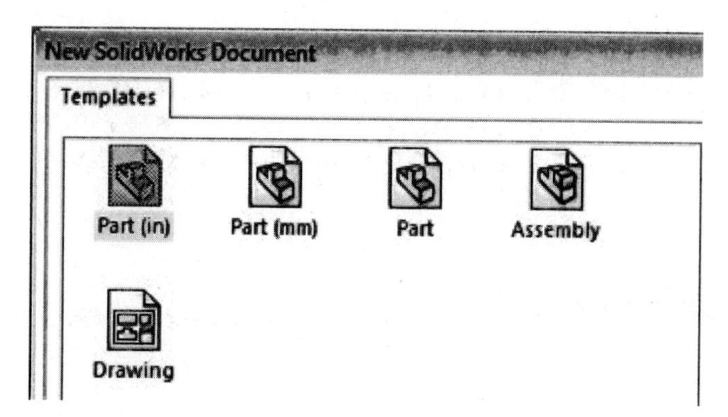

FIGURE 1.38

PART NUMBERING STANDARDS

An important topic that seems to be overlooked has nothing to do with the creation of part, drawing, or assembly files. This topic is instrumental in the managing of the data. This process starts with the assignment of a part number. Companies are required to utilize a part numbering scheme for the purpose of tracking files. If there is a change in the form, fit, or function of an existing part, a new part number is assigned. Part numbers are not only used as identifiers, they can also be used to track a part through its manufacturing cycle. Part numbers are also tied directly to a

Manufacturing Extension System (MES) where part descriptions and costs are tracked. The assigning of part numbers is an integral part of any training or production environment. In Figure 1.39, an eight-digit/character part number is displayed as an example of a part number assigned to part and assembly files. Three subsets separated by hyphens (-) will track the information in the following table, the current class (251); the week number (w01) for week 1, note the lower case "w"; the actual number of the part (01). Drawings generated from models will share the same part number. This should not be an issue since the parts (SLDPRT) and drawings (SLDDRW) have different file extensions.

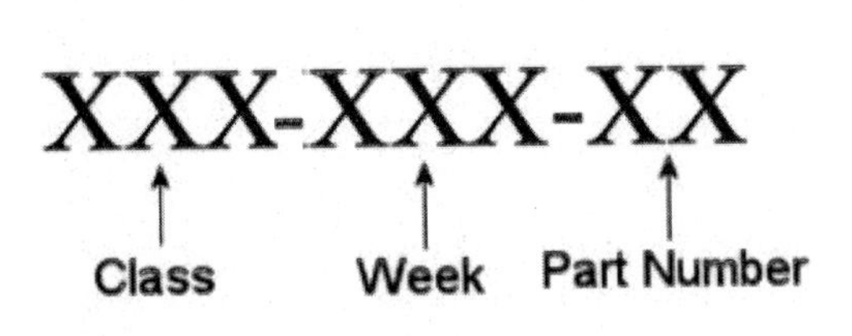

Part Numbering Example: 251-w01-01	
Code	Description
251	EGT 251
w01	Week 1
01	Part Number 01

FIGURE 1.39

BACKING UP FILES

Do not use Save As to save and copy files to your removable jump drive. If you perform this type of save on individual part files, your data will be saved and your jump drive will become the current drive. This can be considered a good thing. If however you perform this operation on a drawing or assembly file, then some files may be located safely on your jump drive while other files may still reside on your hard drive. Get into the habit of saving your work to the hard drive and then using Windows Explorer to copy these files to your jump drive.

When saving and backing up, it is best to follow these steps:

1. Work entirely from a folder located on your hard drive.
2. Whether creating part, drawing, or assembly files, save these directly to your hard drive.
3. When finished for the day, exit the SolidWorks program.
4. Right-click on the Start button and pick Explore.
5. Locate the folder where your SolidWorks data is found and copy the data to your jump drive or external hard drive.

TUTORIAL EXERCISE: SWT-CH01-01.SLDPRT

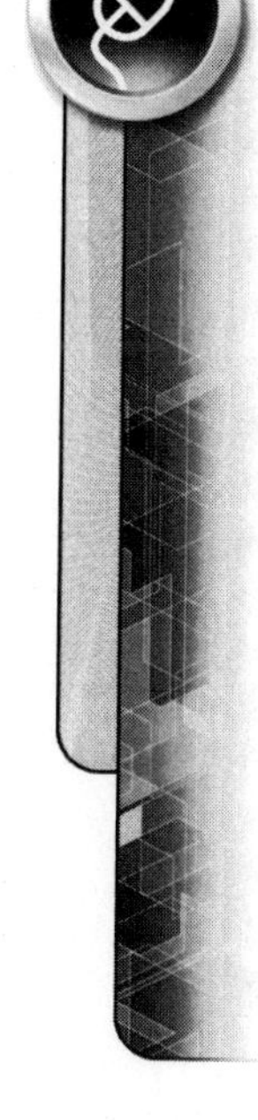

Part Number: SWT-CH01-01.SLDPRT

Description: Container

Units: Metric

This tutorial is designed as an overview to the part and drawing environments of SolidWorks. You will first create a simple part file described as a Container and then produce a three-view drawing including an isometric view as shown in Figure 1.40.

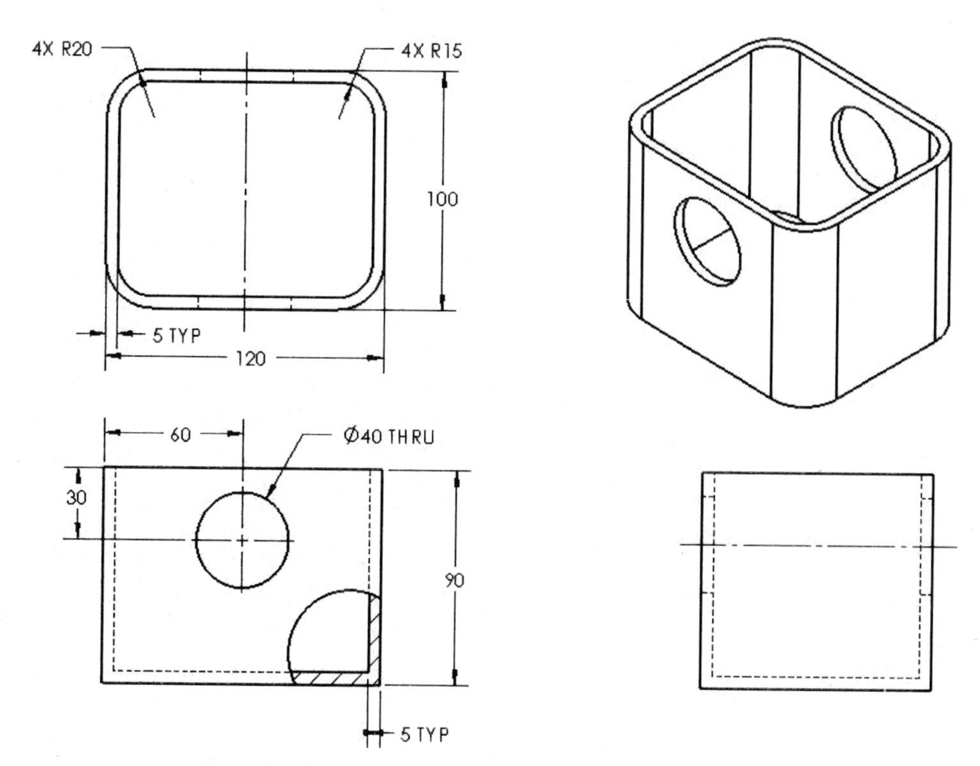

FIGURE 1.40

STEP 1

While in the SolidWorks neutral screen, click on the *New* button (Figure 1.2). When the New SolidWorks Document dialog box appears, click the *Part* template then *OK*, from either the Novice or Advanced dialog box as shown in Figure 1.41.

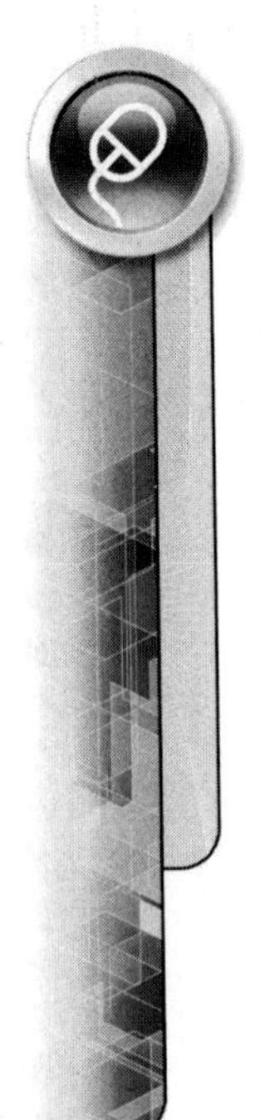

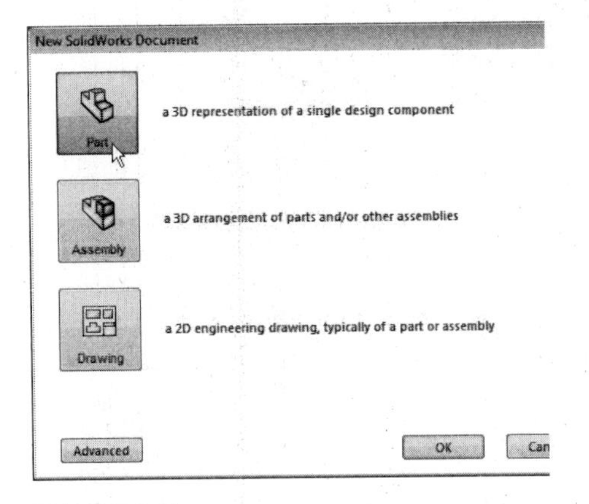

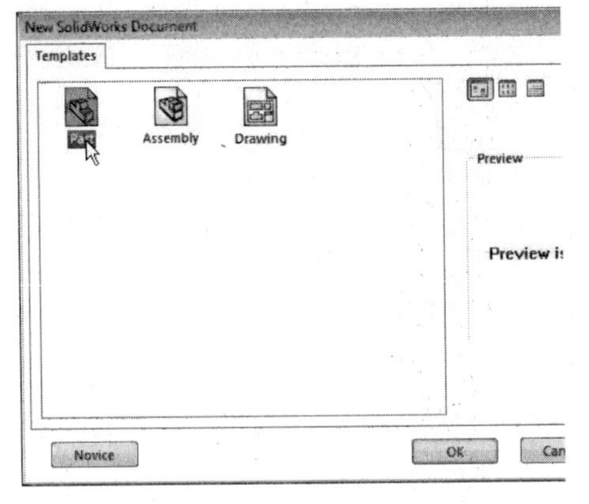

FIGURE 1.41

STEP 2

Click the *Sketch* tab followed by the *Sketch* tool as shown in Figure 1.42 on the left. When the three default planes appear, select the Top Plane as shown in the following figure on the right.

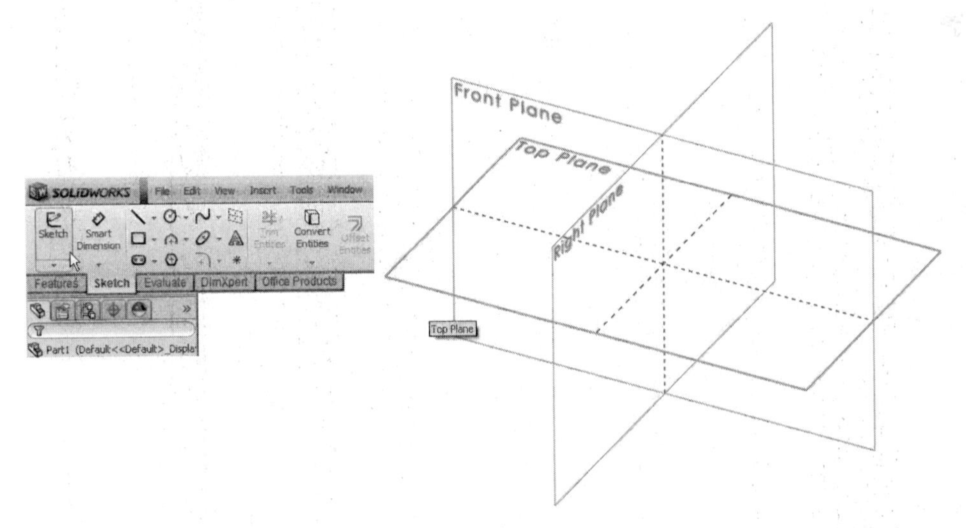

FIGURE 1.42

STEP 3

Click the *Corner Rectangle* tool to construct a rectangle entity. Begin the rectangle on the red origin and select the opposite corner of the rectangle as shown in Figure 1.43 on the left. Select the *Smart Dimension* tool. Click on the top line and click to place the dimension above it, then change the primary value to 120 mm. While still in the Smart Dimension command, click on the right line and click to locate the dimension to the right, then change the primary dimension to 100 mm.

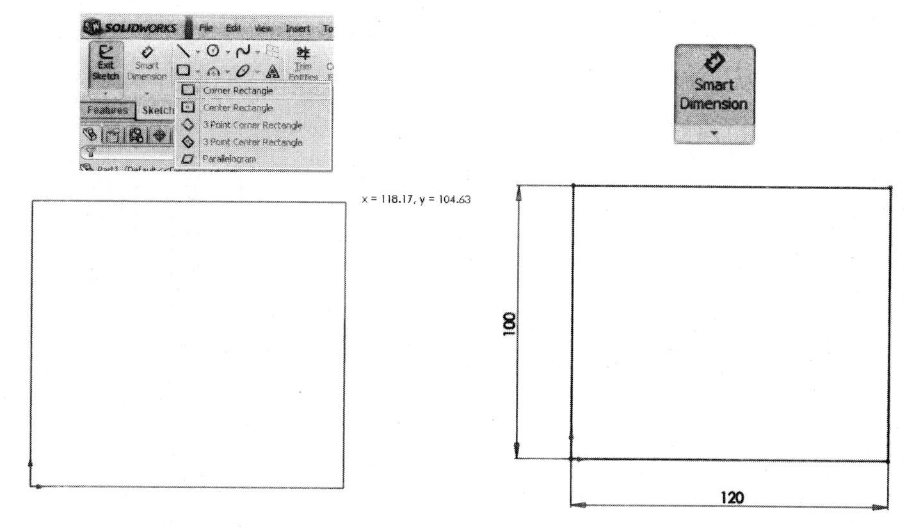

FIGURE 1.43

STEP 4

Select the Features menu tab and then the Extruded Boss/Base tool. Once inside the Feature Manager, change the depth of the extrusion from 10 to 90. This should form the solid box as shown in Figure 1.44 on the right. Accept the command by selecting the green checkmark.

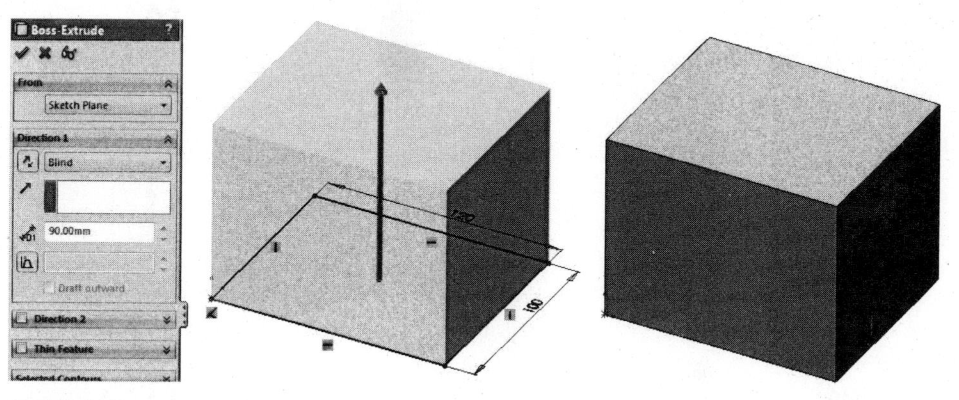

FIGURE 1.44

STEP 5

Select *Zoom to Fit* (the magnifying glass icon from the display menu). Select the Fillet tool and select the four edges as shown in Figure 1.45 in the middle. Verify that the fillet radius is set to 20 and click the green checkmark to apply the fillet. The results should be displayed in the following figure on the right.

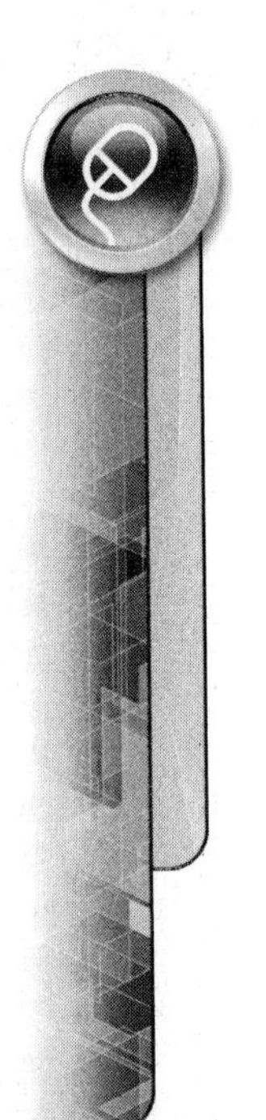

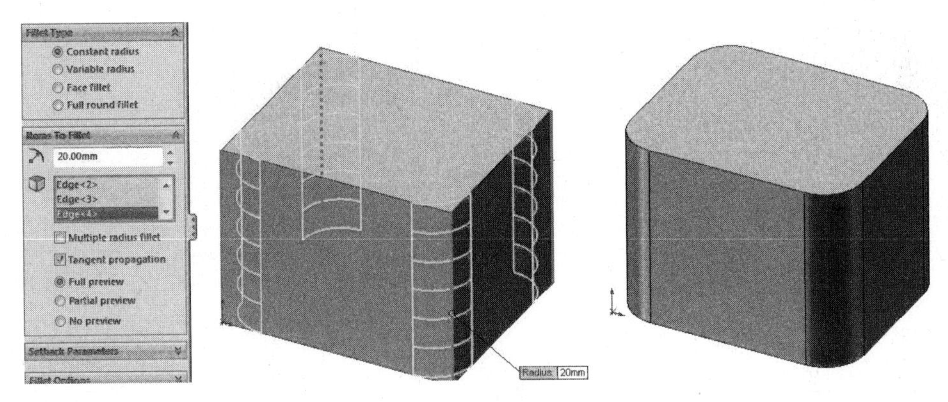

FIGURE 1.45

STEP 6

Create a thin wall of 5 units by selecting the *Shell* tool. Select on the top of the box to remove this face from the selection as shown in Figure 1.46 in the middle. Verify the shell distance is set to 5 and accept the size by clicking the green arrow. The results should be displayed in the following figure on the right. The entire object has a thin wall of 5mm except for the top face, which was removed from having the wall thickness applied to it.

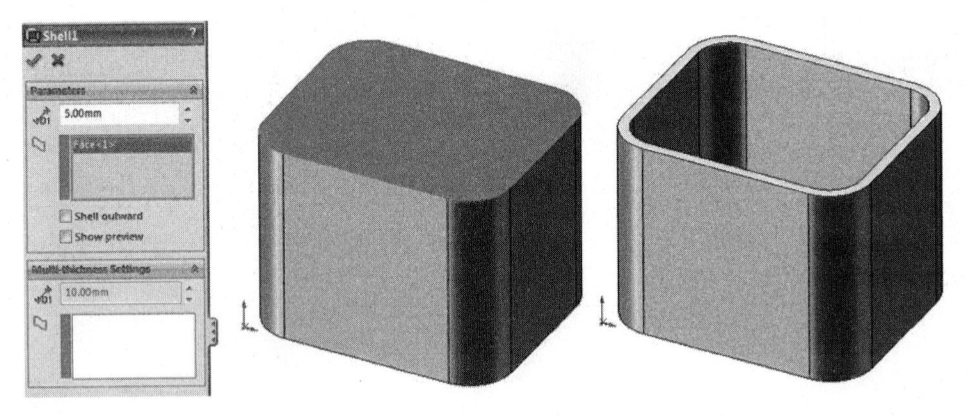

FIGURE 1.46

STEP 7

Click the *Sketch* tab and pick the front face as shown in Figure 1.47 on the left. Create a *circle* as shown in the middle figure by selecting the circle command and locating it accordingly. Add the three dimensions using the *Smart Dimension* tool as shown in the following figure on the right. The three dimensions represent the vertical circle dimension locator (55mm), the width circle dimension locator (60mm), and the circle diameter (40mm).

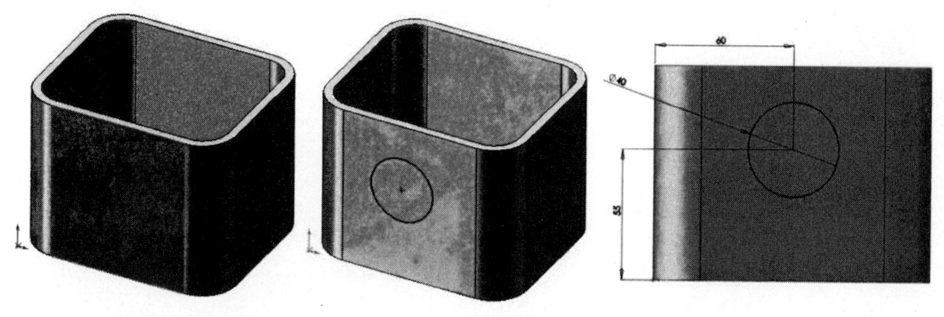

FIGURE 1.47

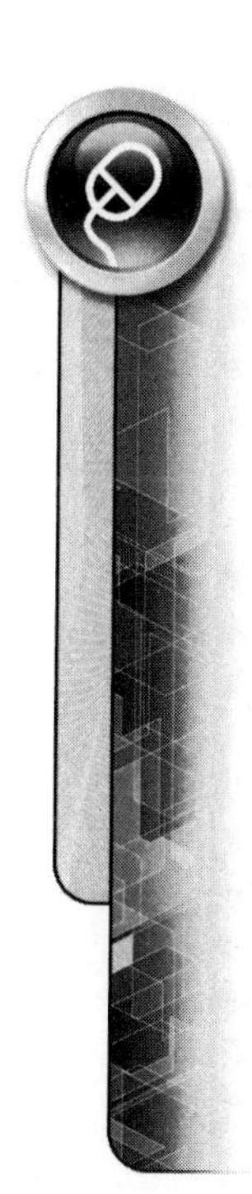

STEP 8

Select the *Features* tab and then the *Extruded Cut* tool, and change the direction from *Blind* to *Through All* as shown in Figure 1.48 on the left. When the preview appears as shown in the following figure in the middle, click the green check box to cut the hole through the object as shown in the following figure on the right. This completes the creation of the part.

FIGURE 1.48

STEP 9

Check the accuracy of the part by performing a volume calculation. Click the *Mass Properties* tool from the *Tools* pull-down menu as shown in Figure 1.49 on the left. This will automatically display the Mass Properties dialog box. A centroid will also appear on the part signifying the center of the mass.

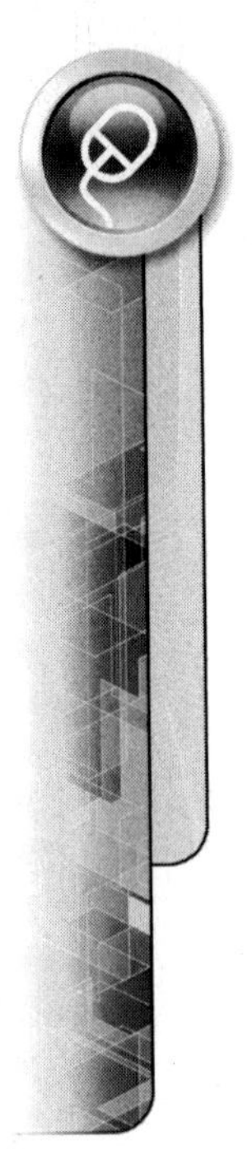

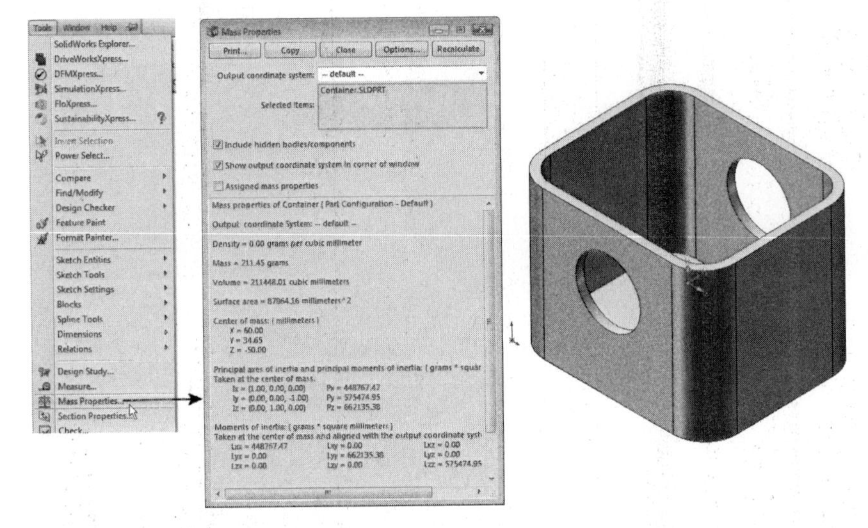

FIGURE 1.49

NOTE

Whenever a tool is mistakenly selected, it can always be exited by selecting the ESC button on the keyboard.

STEP 10

Experiment with changing the color of the model. First right select the part name (e.g., Part 1) and select the *Appearance* icon, then the *Part*, as shown in Figure 1.50 on the left. When the Color And Optics bar appears, change to a different color as shown in the following figure in the middle and then select the green checkmark. This will update the color of the model as shown in the following figure on the right. Save the part but do not close it.

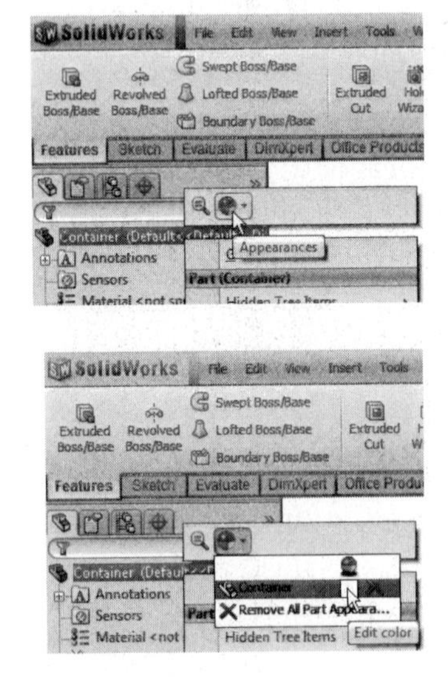

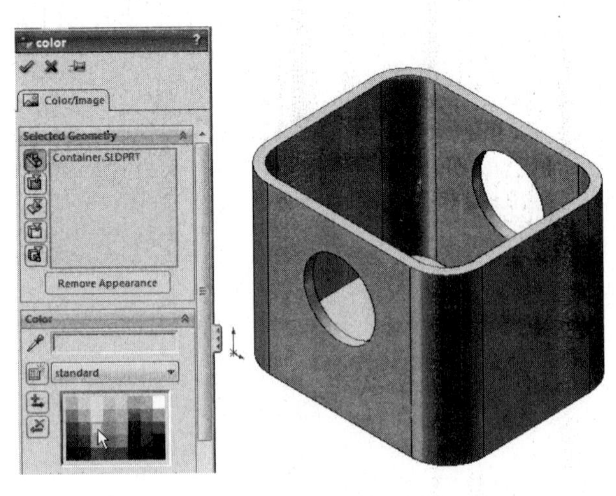

FIGURE 1.50

STEP 11

Begin a *New* file from the File pull-down menu. When the New SolidWorks Document dialog box appears, click on the default *Drawing* template as shown in Figure 1.51 on the left. When the Sheet Format/Size dialog box appears, remove the check from the box next to *Only Show Standard Formats* and click on the C-Landscape standard sheet size as shown in the following figure on the right and then click OK.

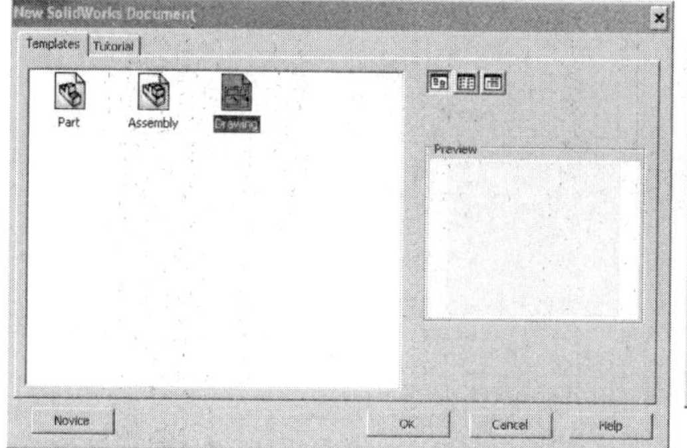

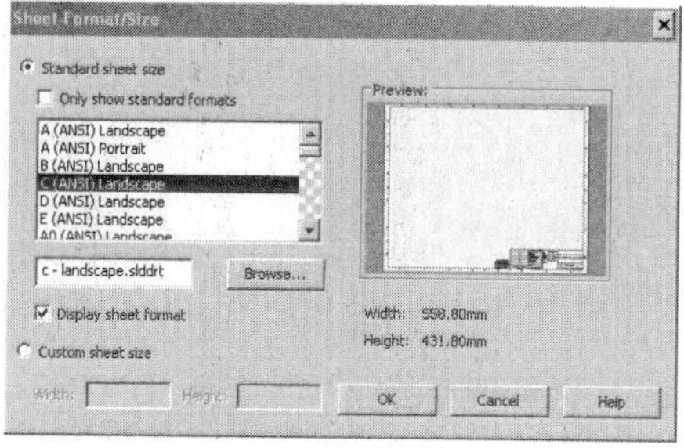

FIGURE 1.51

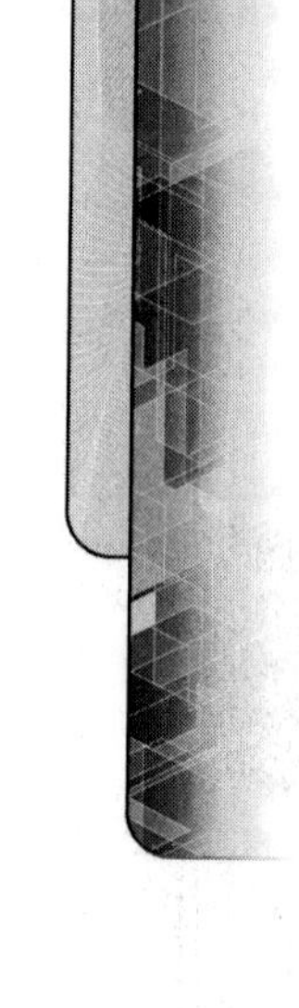

STEP 12

When the drawing sheet appears, right-click on *Sheet1* in the Feature Manager and pick on *Properties* from the menu as shown in Figure 1.52 on the left. When the Sheet Properties dialog box appears, notice the two types of projection present, namely, First angle and Third angle. The First angle method of projecting views is designed for metric use. Third angle projection is used primarily in the United States. Change First angle to Third angle as shown in the following figure on the right.

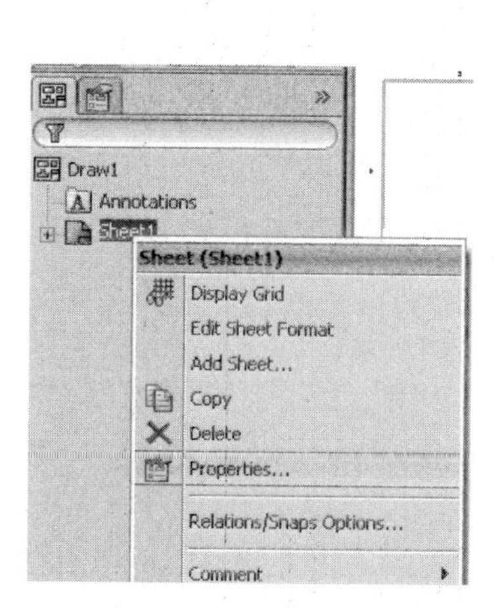

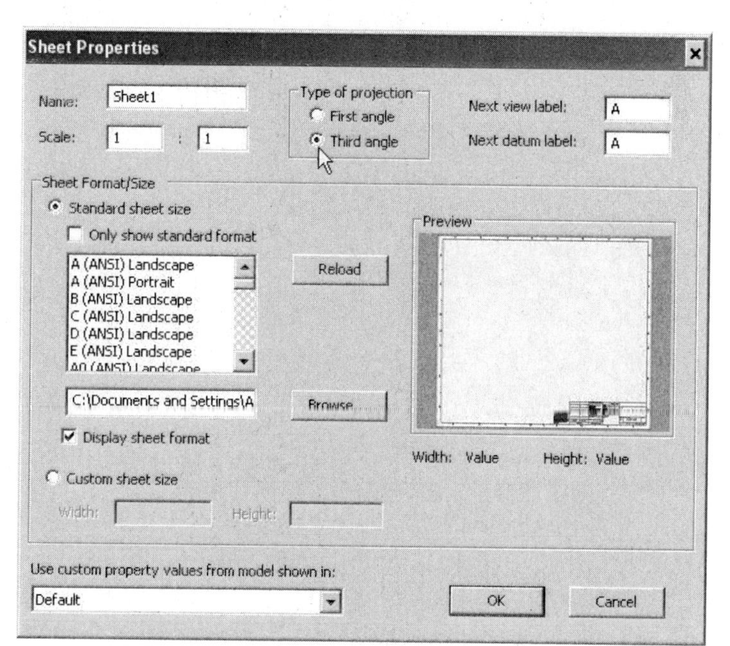

FIGURE 1.52

STEP 13

Select the *View Layout* tab and click the *Standard 3 View* button as shown in Figure 1.53 on the left. Click the green checkbox as shown in the following figure in the middle to display the three standard views, namely, front, top, and right side views. The results should appear similar in the following figure on the right. You could also move the views to better locations by selecting the red dotted box that surrounds each view and dragging this box to move the view.

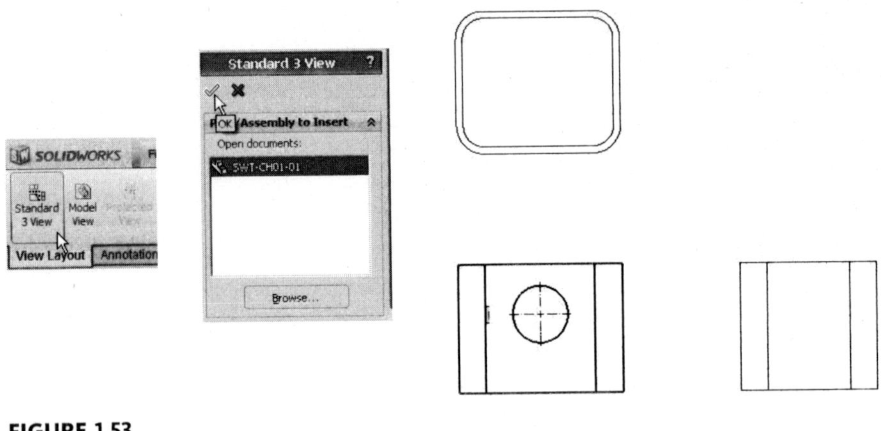

FIGURE 1.53

STEP 14

Click the *View Layout* tab and then the *Projected View* tool as shown in Figure 1.54 on the left. Click the front view and move your cursor in an upward right direction. This should generate the isometric view as shown in the following figure on the right; click to place.

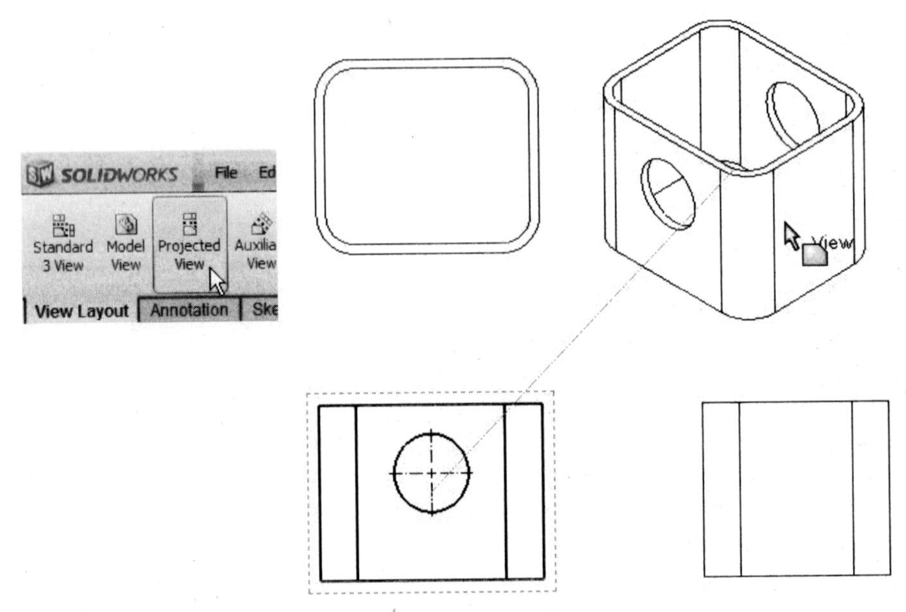

FIGURE 1.54

STEP 15

If hidden lines are not automatically generated, click the front view. Locate the Display Style heading in the drawing Property Manager and select the icon for *Hidden Lines Visible* as shown in Figure 1.55 on the left. All views should display hidden lines as shown in the following figure on the right. Click the green checkmark to accept.

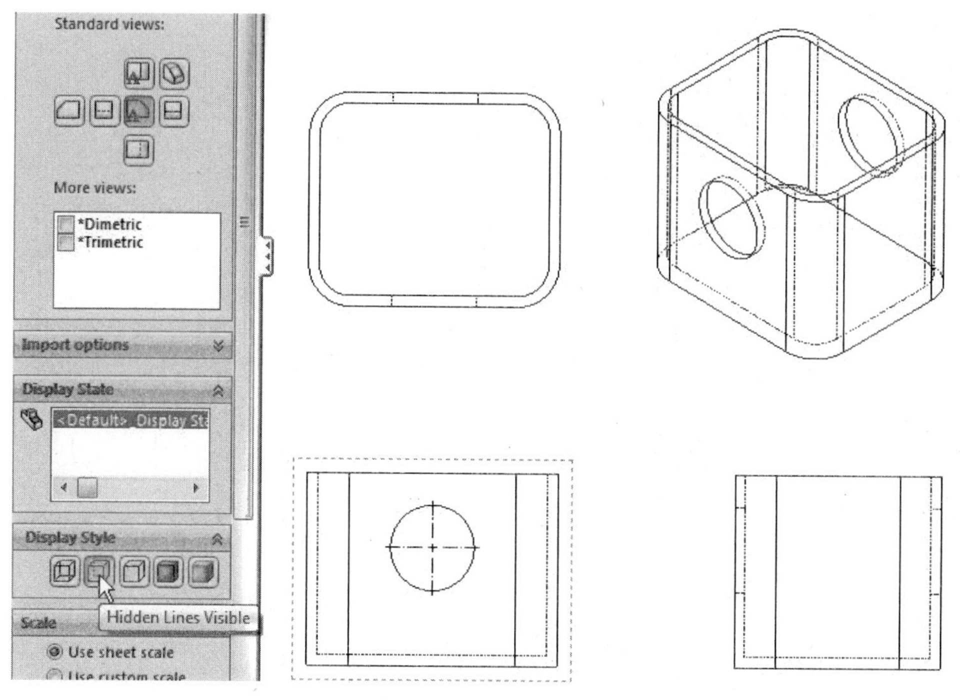

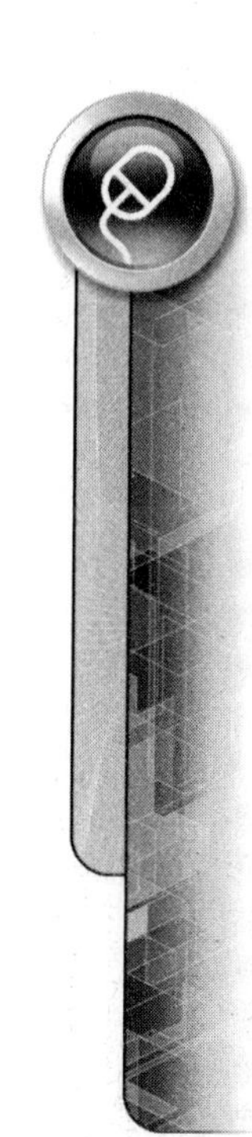

FIGURE 1.55

STEP 16

As hidden lines were placed in all views, this is usually not the case for the isometric view. Click the isometric view and change the Display Style in the Property Manager to Hidden Lines Removed as shown in Figure 1.56 on the left. The results are shown in the following figure on the right with all hidden lines removed from the isometric view; then click the green checkmark to accept.

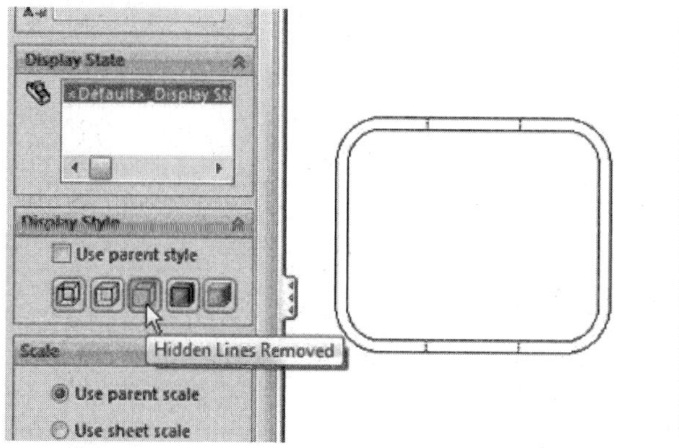

FIGURE 1.56

STEP 17

Tangent edges are placed when a curve such as a fillet begins and changes direction. The tangent edge is normally in the form of horizontal or vertical lines. Tangent edges can also get in the way of interpreting hidden lines especially if the tangent edges overlap the hidden lines. Turn off tangent edges by right-clicking on the dotted front view box. Choose Tangent Edge from the menu as shown in Figure 1.57 in the middle. Then select Tangent Edges Removed from the menu as shown in the following figure on the right. Repeat this procedure if other views need tangent edges removed. Save the file.

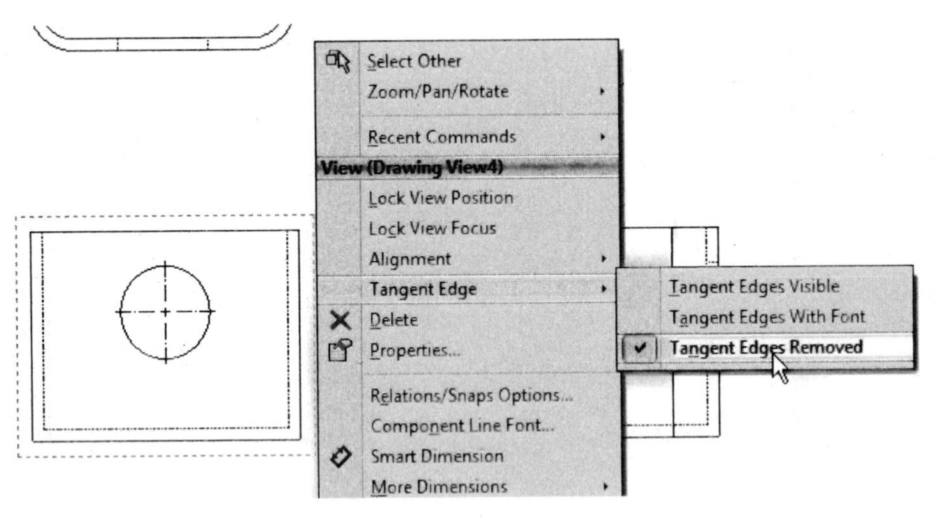

FIGURE 1.57

STEP 18

The completed views are illustrated in Figure 1.58. This completes the tutorial on creating the Container and laying out the orthographic views on a single sheet.

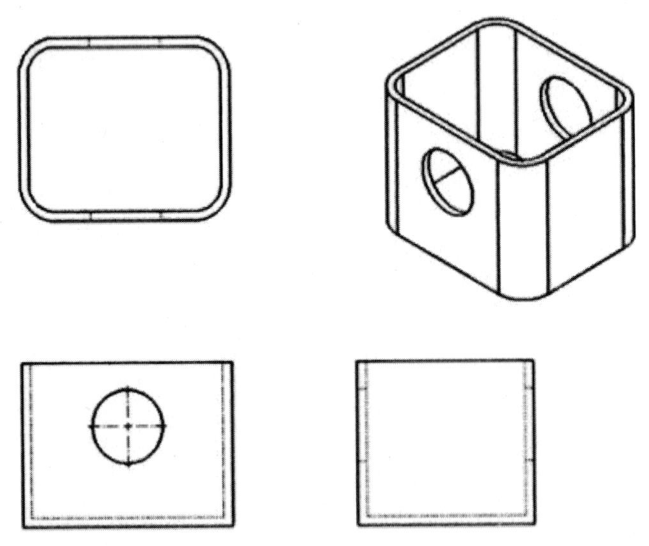

FIGURE 1.58

REVIEW QUESTIONS

1. How can additional toolbars be accessed?
2. What is the function of the Feature Manager?
3. Why is the flyout Feature Manager necessary?
4. What are the three major file types a beginning user would function within?
5. What are the default Units and Drafting Standards settings for a part file?
6. Describe the file-naming structure for this text.
7. What is the advantage of using the rollback function?
8. Describe how mass property calculations are performed.
9. What additional file information is given for the Mass Property function?
10. Whenever a tool is mistakenly selected, can it always be exited by selecting the ESC button?

EXERCISES

1.1

Create a template for English-based units. While in Tools Options, change the Document Properties for an Overall Drafting Standard of ANSI. Also change the Units to IPS (inch, pound, second). Save the template for future use.

1.2

Create a template for Metric-based units. While in Tool Options, change the Document Properties for an Overall Drafting Standard of ISO. Also change the Units to MMGS (millimeter, gram, second). Save the template for future use.

1.3

Open the existing file: SWT-CH01-Base (Figure 1.59).

FIGURE 1.59

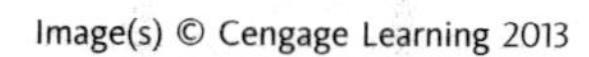

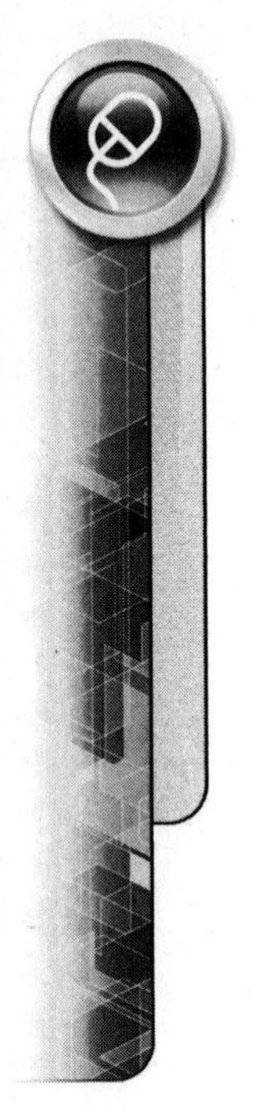

- Using the mouse wheel, zoom in and zoom out. Using the wheel mouse, rotate the part.
- Using the view orientation, view the part from all directions: front, back, top, bottom, left, right, isometric, diametric, ending on trimetric.
- Using the display styles, view the part in the following modes: Shaded with Edges, Shaded, Hidden Lines Removed, Hidden Lines Visible, and Wireframe, ending back on Shaded with Edges.
- Grab the rollback bar and move it just under the Extrude 1. Then roll it down below Extrude 3; what changes were reflected in the model area?
- Continue to move the rollback bar down one feature at a time. Note the changes in the model for every step.

CHAPTER 2

Sketching Techniques

This chapter discusses various sketching techniques used in building parts in SolidWorks and includes the following topics:

- Using the default sketching planes (Front, Top, and Right)
- Sketching entities such as lines, circles, and arcs
- Applying geometric relations to sketch entities
- Applying smart dimensioning techniques to sketch entities
- Creating new sketches from existing model faces
- An introduction to creating extruded boss and cut features

DESIGN INTENT

How dimensions are placed in a sketch is important. The following examples illustrate a rectangular plate with two holes (Figure 2.1). Study each example and explanation regarding how changing the total length dimension affects the location of the holes.

A sketch dimensioned as shown on the right will keep the holes 20mm from each end of the part regardless of how the overall width dimension of 100mm changes.

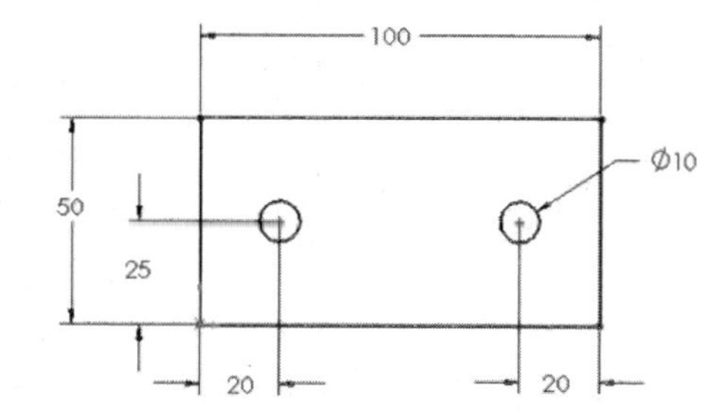

FIGURE 2.1

The example illustrated on the right utilizes baseline dimensions that will keep the holes positioned relative to the left edge of the plate. The position of the right hole to the right edge could be affected by changes to the overall width of the part.

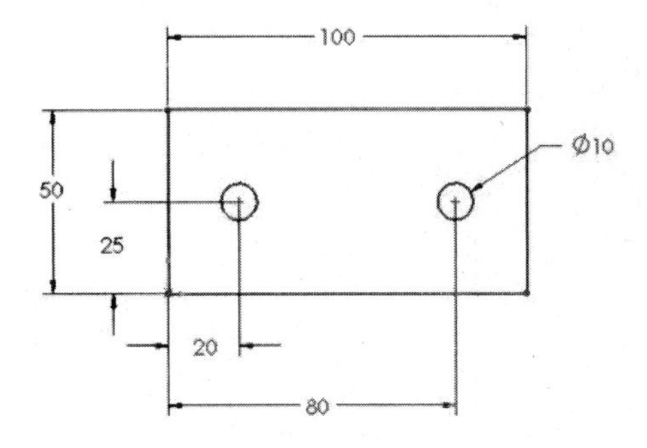

Dimensioning from the edge and from center to center as shown on the right will maintain the distance between the hole centers. If the overall width changes, the holes remain fixed to the left edge based on the edge dimensions.

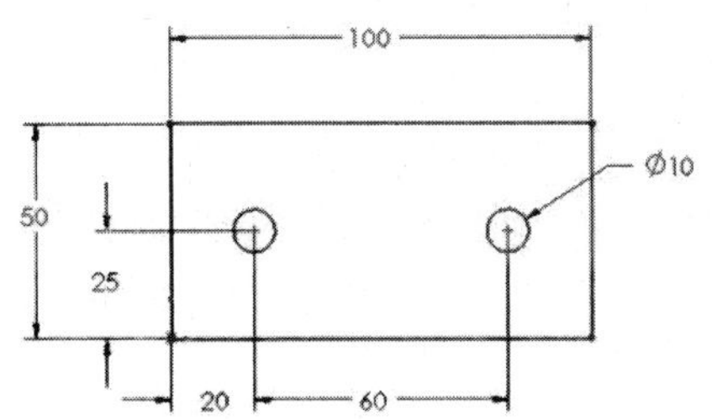

FIGURE 2.1 *Continued*

USING DEFAULT SKETCHING PLANES

The first time you begin a new model and create a new sketch, three standard planes will appear as shown in Figure 2.2. Selecting one of these planes orientates the sketch directly on the surface of this plane. Also, once the plane is selected, your viewing position changes to a normal or perpendicular view.

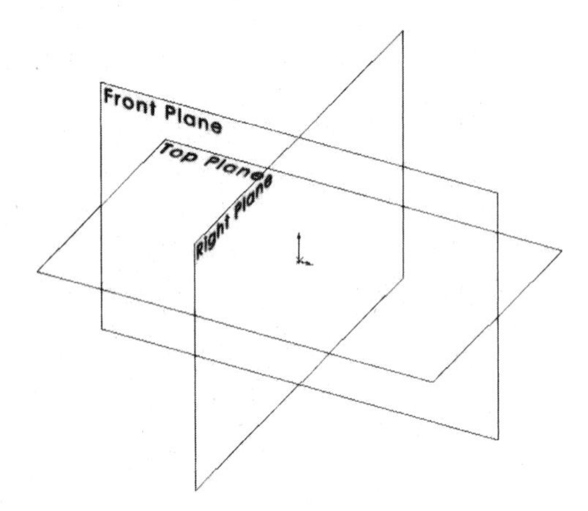

FIGURE 2.2

Figure 2.3 illustrates each plane isolated to get a better idea as to the plane being sketched upon.

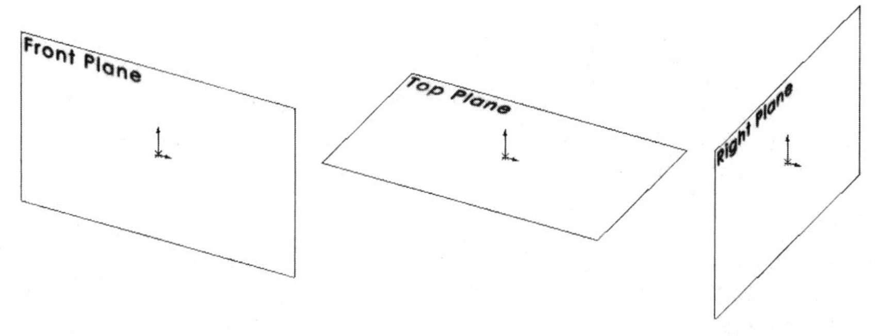

FIGURE 2.3

THE SKETCH ORIGIN

Once you have selected the plane to construct on, you are automatically placed in sketch mode. Components that make up the sketch environment include the red origin and the confirmation corner as shown in Figure 2.4. It is important to construct or reference any sketch geometry to this origin. The confirmation corner is located in the upper right corner of the display screen and is used to indicate the status of the sketch. Clicking on the sketch symbol will save any changes and exits the sketch environment. Clicking on the red X discards any changes and exits the sketch environment.

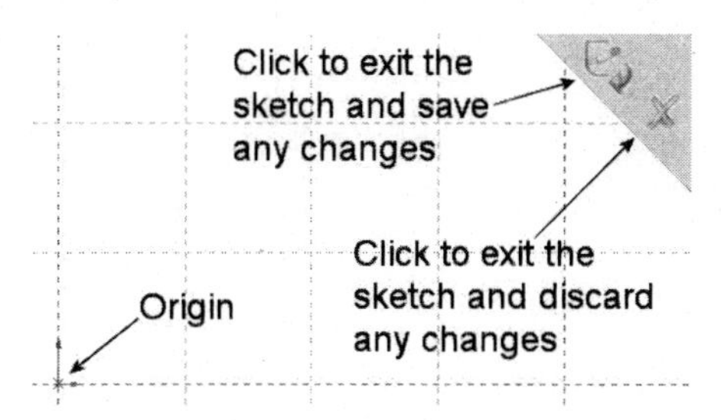

FIGURE 2.4

SKETCH TOOLS

While in sketch mode, you have the full use of all tools related to sketching. These tools are illustrated in Figure 2.5 and are available through the CommandManager.

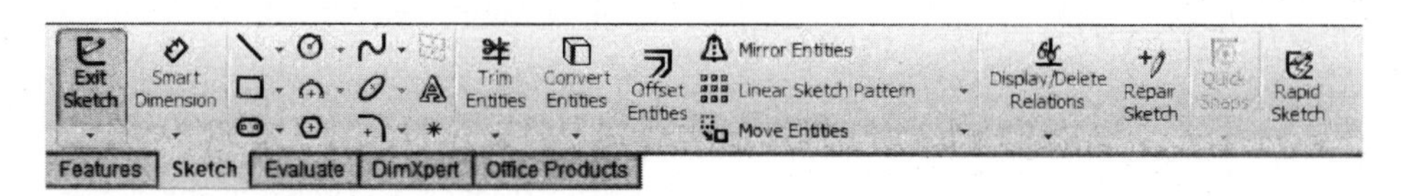

FIGURE 2.5

Some commands have arrows present along size or on the bottom of a command. Clicking these arrows will expose other commands for you to use while working in sketch mode as shown in Figure 2.6.

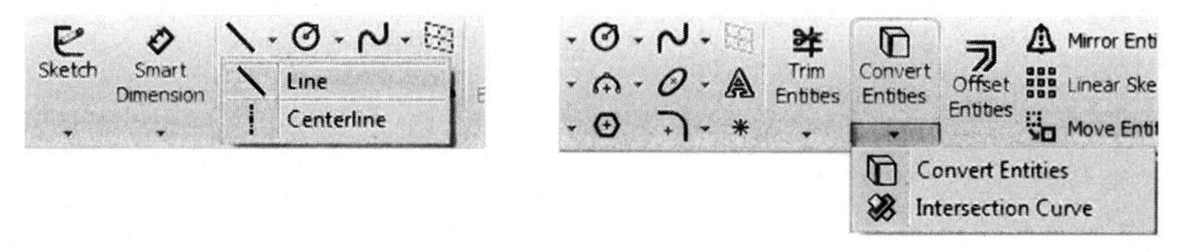

FIGURE 2.6

A more detailed description of each tool is listed as follows.

TABLE 2.1

Sketch Button	Tool	Description
Sketch	Sketch	Sketch commands
Sketch	Sketch	Creates a new sketch or edits an existing sketch.
Smart Dimension	Smart Dimension	Creates a dimension from one or more selected entities.
Line	Line	Sketches a line.
Rectangle	Rectangle	Sketches a rectangle.
Circle	Circle	Sketches a circle. Select the center of the circle, then drag to set its radius.
Centerpoint Arc	Centerpoint Arc	Sketches a center point arc. Set the center point. Drag to place the arc starting point, then to set its length and direction.
Tangent Arc	Tangent Arc	Sketches an arc tangent to a sketch entity. Select the end point of the sketch entity, then drag to create the tangent arc.
3 Point Arc	3 Point Arc	Creates a 3 point arc. Select start and end points, then drag the arc to set the radius or to reverse the arc.
Sketch Fillet	Sketch Fillet	Rounds the corner at the intersection of two sketch entities, creating a tangent arc.
Centerline	Centerline	Creates a centerline. Use centerlines to create symmetrical sketch elements, revolved features, or as construction geometry.
Spline	Spline	Creates a spline. Click to add spline points that shape the curve.

continued

TABLE 2.1 *Continued*

Sketch Button	Tool	Description
Point	Point	Sketches a point.
Add Relation	Add Relation	Controls the size and position of entities with constraints such as concentric or vertical.
Display/... Relations	Display/Delete Relations	Displays and deletes geometric relations.
Mirror Entities	Mirror Entities	Mirrors selected entities about a centerline.
Convert Entities	Convert Entities	Converts selected model edges or sketch entities into sketch segments.
Offset Entities	Offset Entities	Adds sketch entities by offsetting faces, edges, curves, or sketch entities a specified distance.
Trim Entities	Trim Entities	Trims or extends a sketch entity to be coincident to another, or deletes a sketch entity.
Construct... Geometry	Construction Geometry	Toggles sketch entities between construction geometry and normal sketch geometry.
Move Entities	Move Entities	Moves sketch entities and annotations.
3D Sketch	3D Sketch	Adds a new 3D sketch or edits an existing 3D sketch.

THE SKETCH PENCIL

Selecting any sketch tool that draws an entity such as a line, circle, or arc displays the pencil icon as shown in Figure 2.7. Depending on the tool selected, the image of the object being sketched will be attached to this pencil. For example, clicking on the Rectangle tool will display the sketch pencil along with a rectangle shape.

Line Rectangle Circle Arc Spline

FIGURE 2.7

SKETCHING LINES

Clicking on the Line tool allows you to locate points on the screen which will draw line segments between each point location. In Figure 2.8, the first point of the line is located at the origin. As you drag the pencil cursor to a new location, notice the line appearing with the total length being displayed at the end of the line. This gives you an idea as to the approximate length of the line segment.

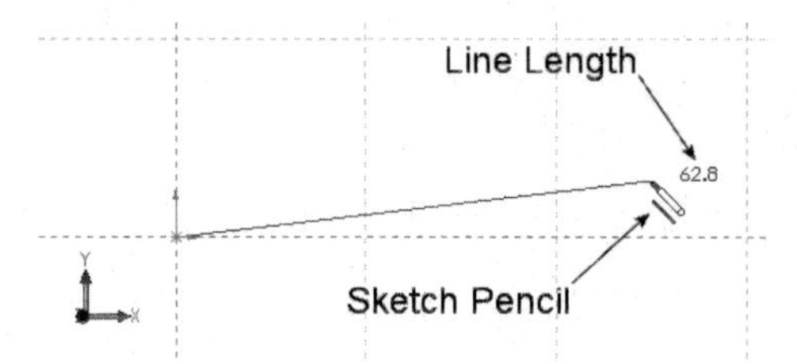

FIGURE 2.8

As you place a second point identifying the length of the line, you can still construct additional line segments. Depending where you move your pencil cursor, inference lines could appear. These apply to certain geometric conditions. For instance, the following example illustrates inference lines that appear to be perpendicular to the previously placed line segment. Also, a symbol or glyph appears alerting you are about to locate this new line perpendicular to the first. This glyph is referred to as a relation and will be discussed in greater detail later in this chapter (Figure 2.9).

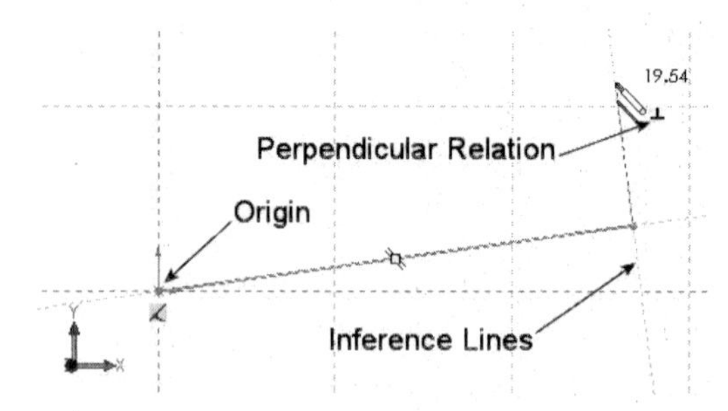

FIGURE 2.9

NOTE

As you sketch objects such as lines and circles, automatic inferencing is enabled. Sometimes this gets in the way of creating the sketch. Press on the CTRL key while sketching these objects to turn off automatic inferencing.

PRECISION INPUT FOR LINES

When the Line tool is activated, the FeatureManager changes to the PropertyManager as shown in Figure 2.10 on the left. Here, the default setting under Orientation is As sketched, meaning you construct lines close to their intended lengths. When you change the Orientation to Horizontal, the PropertyManager activates additional controls for lines. Under Parameters, you can enter a length for the line. When you click the start of the line and move your cursor to the right, a horizontal line of 50mm appears as shown in the following figure on the right. You can also check the box next to Add dimensions. When the second point of the line is picked, the exact dimension will also be placed.

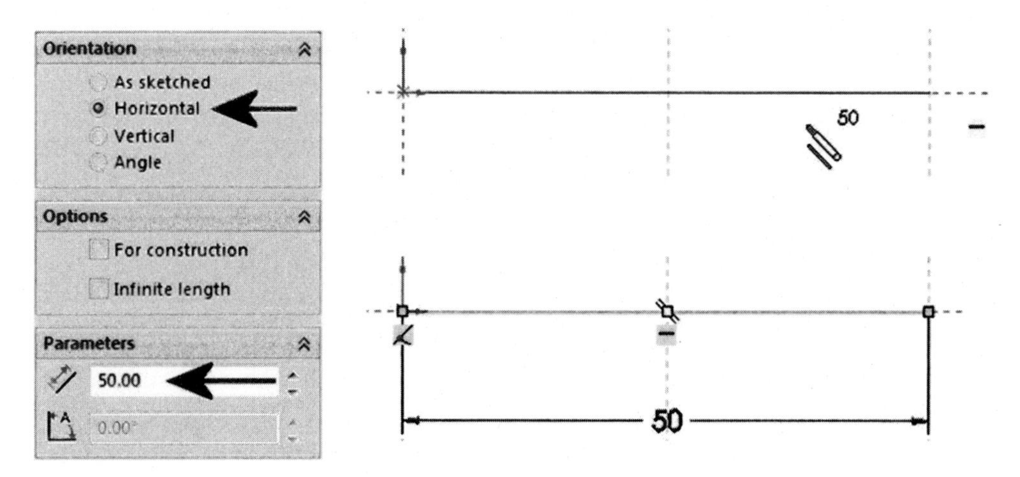

FIGURE 2.10

A similar capability is available for constructing lines at an angle. First, change to Angle under the Orientation heading. Under the Parameters heading, change the length to 50mm and the angle to 30° as shown in Figure 2.11 on the left. Picking a first point and dragging your cursor will display the line at the correct angle. Picking to locate the second point will place the line and add a length and angle dimension as shown in the following figure on the right.

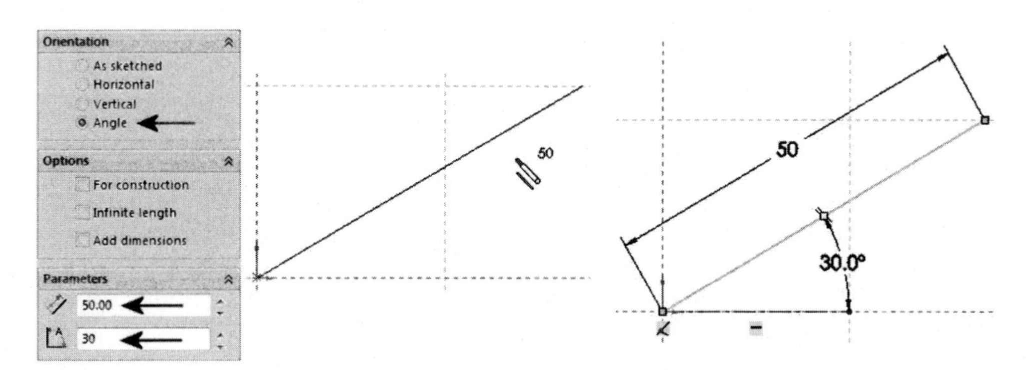

FIGURE 2.11

SKETCH STATUS

As you sketch lines, feedback or sketch status is available at the lower right corner of your display screen. This area displays the current position of the pencil cursor and has status indicators that tell if the sketch is under or fully defined (Figure 2.12).

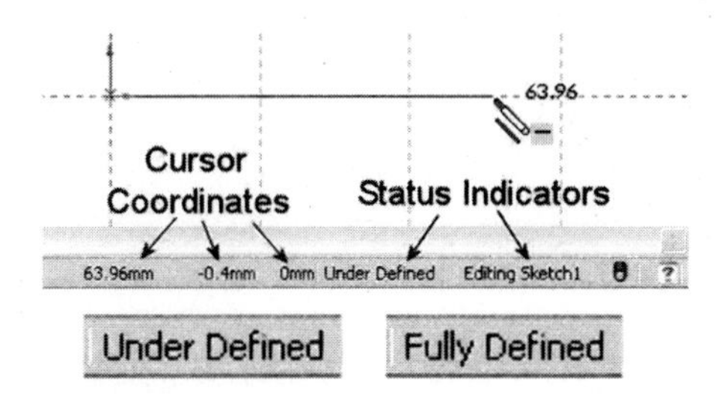

FIGURE 2.12

Sketch Status Colors

Additional sketch status is available through colors. For example, when first sketching entities such as lines, they will appear blue in color. This color signifies that the sketch is underdefined. As you place relations and add dimensions to the sketch, certain entities will change to a black color; this signifies the sketch entity is fully defined. In fact, you will need to turn all entities black in order for the sketch to be considered fully defined. Other colors that affect the quality of the sketch are displayed in the following table.

TABLE 2.2

Color	Meaning	Description
Blue	UnderDefined	The sketch is not fully defined but can still be used to generate a solid.
Black	Fully Defined	The sketch has all complete information.
Red	OverDefined	The sketch may consist of duplicate dimensions or relations that conflict with the design. Offending dimensions and/or relations must be deleted.
Brown	Dangling	Offending dimensions and/or relations creating this condition must be repaired.
Pink	Not Solved	Offending dimensions and/or relations creating this condition must be repaired.
Yellow	Invalid	Offending dimensions and/or relations creating this condition must be repaired.

RELATIONS

Sketch relations are geometric constraints that are applied between sketch entities. Typical relations include but are not limited to Horizontal, Vertical, Tangent, Perpendicular, and others. Relations can also act to control the geometric relationships between entities. For example, Figure 2.13 illustrates two arcs connected by two line segments. You want both arc segments to be equal in radius. You could add two radius dimensions, or you could apply an equal relation to both arcs. Then when a radius dimension is added, both arcs are affected.

To add relations to sketch entities, follow these steps:

1. Click the Add Relation button found in the Sketch CommandManager.
2. Select the entities to apply relations to arc segments such as "A" and "B."
3. When a listing of all valid relations appears in the FeatureManager, choose the proper relation from this list. From this list the equal relation is selected.

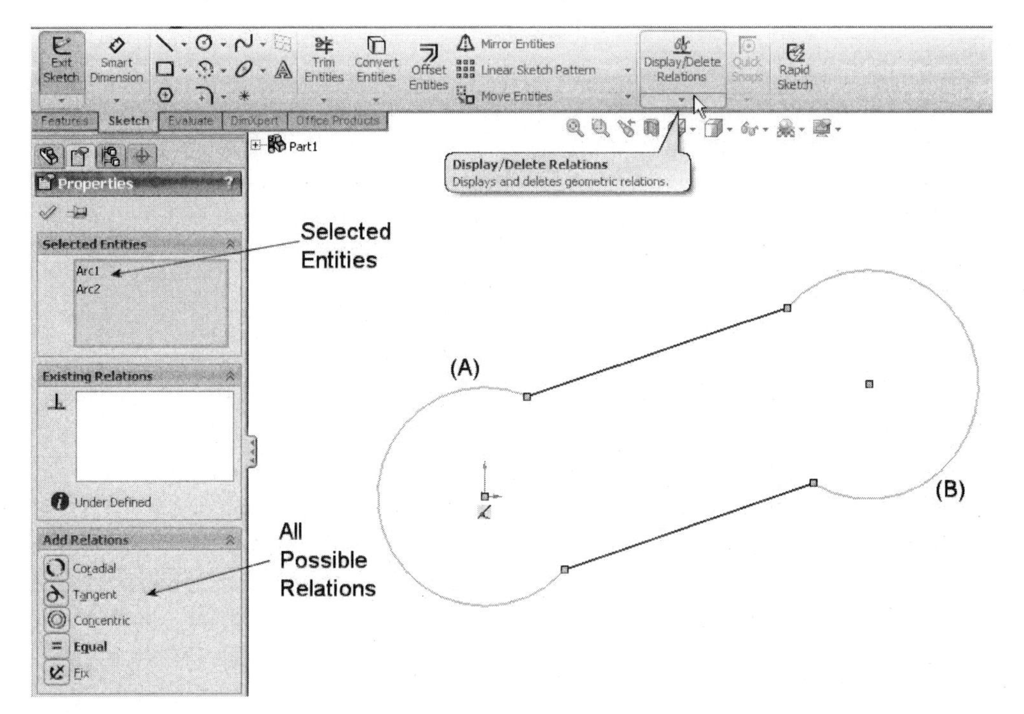

FIGURE 2.13

The results of performing this operation are illustrated in Figure 2.14. A smart dimension has also been added to one of the arcs; the other arc is also affected by this dimension value. When applying a relation, a glyph or symbol will appear on both entities, in this case both arcs. Notice also that the equal relations have a number assigned, namely 5. This allows you to see in complicated sketches which entities are affected by this relation.

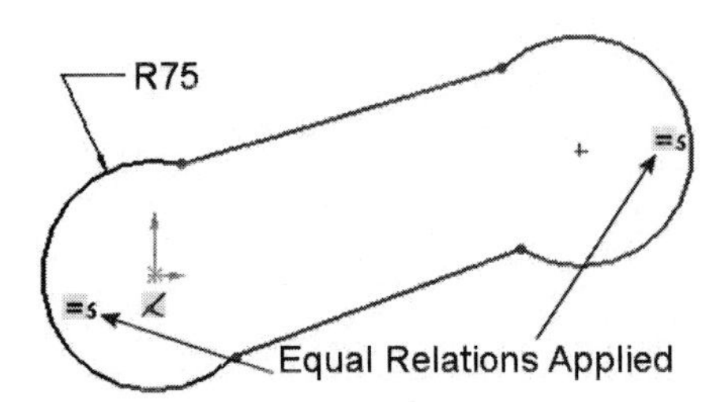

FIGURE 2.14

Press and hold down the CTRL key while picking entities. This will display all possible relations in the FeatureManager. Choose the proper relation from this list.

QUICK VIEW RELATIONS

Whenever selecting objects for the purpose of applying relations, a quick view toolbar appears at your cursor location. This enables you to apply the selected relation to the object in a more convenient manner (Figure 2.15).

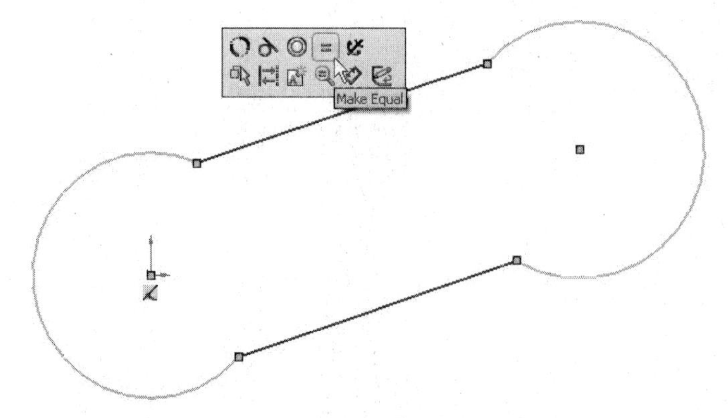

FIGURE 2.15

SMART DIMENSIONS

Along with relations, smart dimensions are used to define the size requirements of the sketch. They are called smart dimensions because the dimension is created based on the entity you select. For instance, selecting the edge of the circle automatically creates a diameter dimension while selecting the edge of an arc creates a radius dimension. Horizontal, vertical, and even angular dimensions can be placed using smart dimensions. With smart dimensions being used in conjunction with relations, these methods both combine to fully solve the sketch as shown in Figure 2.16.

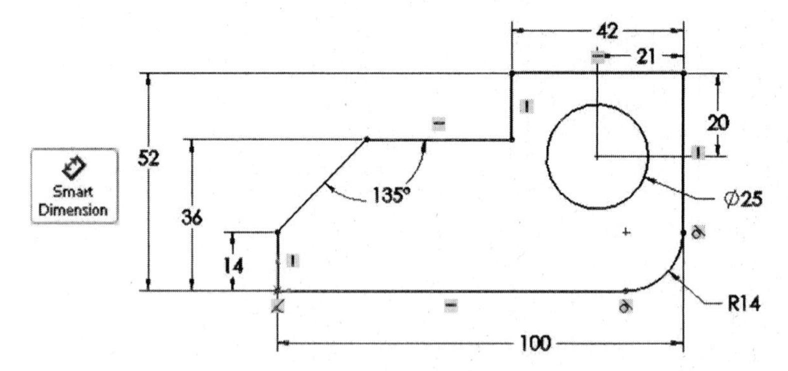

FIGURE 2.16

Whenever placing smart dimensions, care needs to be taken when determining their correct values. For example, the illustration in Figure 2.17 on the left shows an arc that has a smart radius dimension of 2.00 inches. You may be tempted to leave this value alone since it appears correct. However, with double-clicking on the dimension, the actual value is listed in the Modify dialog box shown in the following figure on the right. While this value is close to 2.00, it is not the exact value. Never take placed dimensions for granted; always double-click on the dimension to display its correct value in the Modify dialog box. You can also change the dimension value by dragging on the Virtual Thumb Wheel to the left or to the right as shown in the following figure on the right.

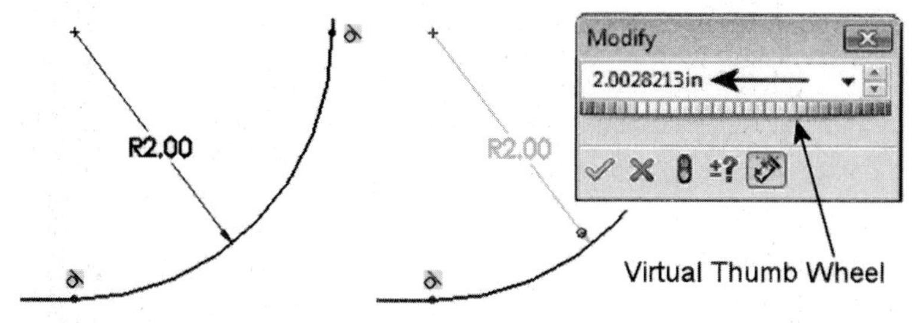

FIGURE 2.17

LEVEL I RELATIONS

The topic of geometric relations has already been introduced in a previous section. Rather than discuss all relations, two groups will be used to apply these relations to short exercises and real-world examples. The first grouping of relations involves Horizontal, Vertical, Collinear, Perpendicular, Parallel, Merge, and Coincident.

TABLE 2.3

Relation Icon	Relation Meaning	Description
	Horizontal	When a line is selected, the line will become horizontal. Selecting individual points will allow points to be aligned horizontally.
	Vertical	When a line is selected, the line will become vertical. Selecting individual points will allow points to be aligned vertically.
	Collinear	When two or more lines are selected, this relation allows the lines to lie along the same plane. This relation does not affect points.
	Perpendicular	When two lines are selected, the two entities are made perpendicular to each other.
	Parallel	When two or more lines are selected, these entities are made parallel to each other. When a line and a plane (or planar face) are selected in a 3D Sketch, the line is made parallel to the selected plane.
	Merge	Selecting two points or the endpoints of lines or arcs will merge the points into a single point.
	Coincident	Selecting a point and a line, arc, or ellipse will locate the point on the line, arc, or ellipse.

The Horizontal and Vertical Relation

Open the drawing file SWT-Relation-01 (Horiz-Vert). This exercise is designed to apply horizontal and vertical relations to line segments.

TRY IT!

Holding down the CTRL key, pick lines "A," "B," and "C." When the list of all possible relations appears as shown in Figure 2.18 on the left, click on the Horizontal relation to force these lines to be horizontal. Continue by holding down the CTRL key as you select lines "D" and "E" and choose the vertical relation to force these two lines vertical.

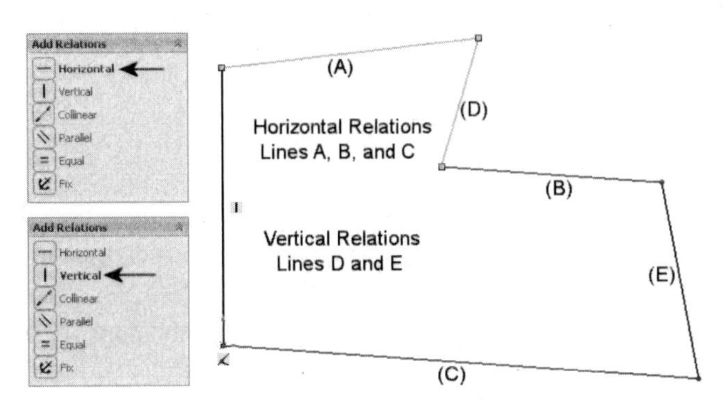

FIGURE 2.18

The results of applying horizontal and vertical relations to the sketch are illustrated in Figure 2.19 on the left. This model is solved by adding the two sets of horizontal and vertical dimensions as shown in the following figure on the right.

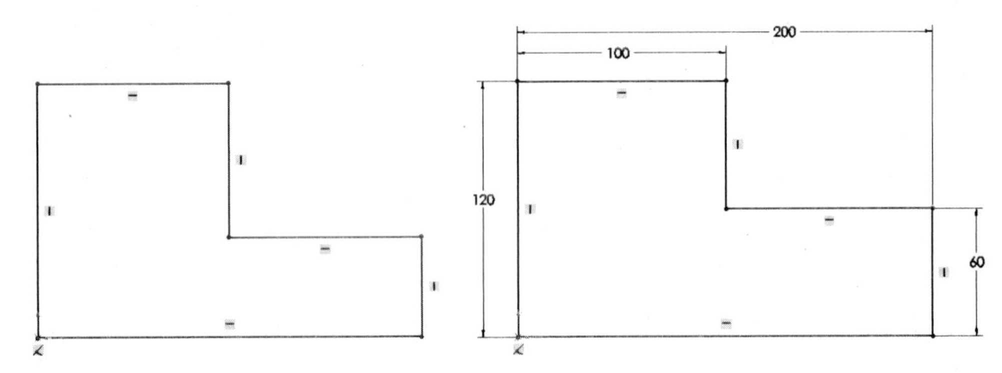

FIGURE 2.19

The Merge Relation

TRY IT!

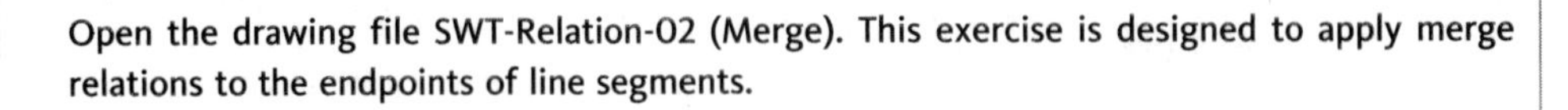
Open the drawing file SWT-Relation-02 (Merge). This exercise is designed to apply merge relations to the endpoints of line segments.

Click on the Add Relations button and select the points at "A" and "B" as shown in Figure 2.20 on the left. Select Merge from the list of relations in the PropertyManager. You could also drag one endpoint on the top of the other endpoint. The results are illustrated in the following figure on the right.

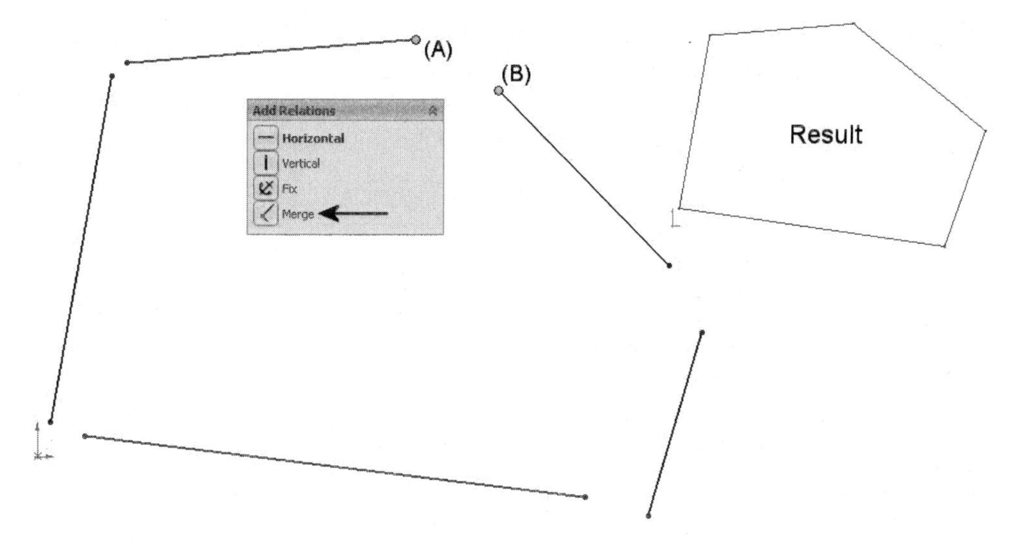

FIGURE 2.20

The Collinear Relation

Open the drawing file SWT-Relation-03 (Collinear). This exercise is designed to apply collinear relations to line segments.

TRY IT!

The collinear relation is defined as two or more lines that lie along the same infinite line. First click the Add Relations button and choose lines "A" and "B." When the list of relations appears in the PropertyManager, select Collinear as shown in Figure 2.21 on the left.

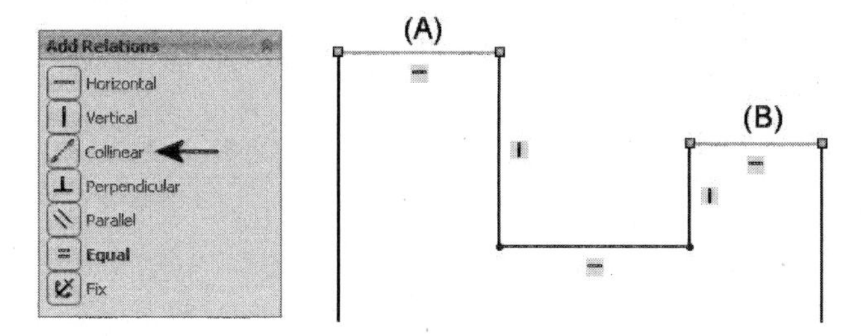

FIGURE 2.21

Both lines "A" and "B" will now lie along the same line. Notice also the appearance of collinear guidelines to show the lines are even with each other (Figure 2.22).

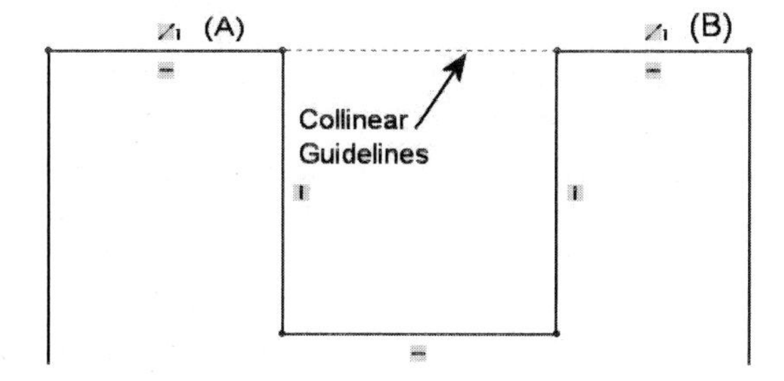

FIGURE 2.22

The Perpendicular and Parallel Relations

TRY IT!

Open the drawing file SWT-Relation-04 (Per-Par). This exercise is designed to apply perpendicular and parallel relations to the following line segments.

Two lines perpendicular to each other will form a 90° angle. Click the Add Relations button and select lines "A" and "B." When the list of possible relations appears in the PropertyManager, select Perpendicular from this list as shown in Figure 2.23.

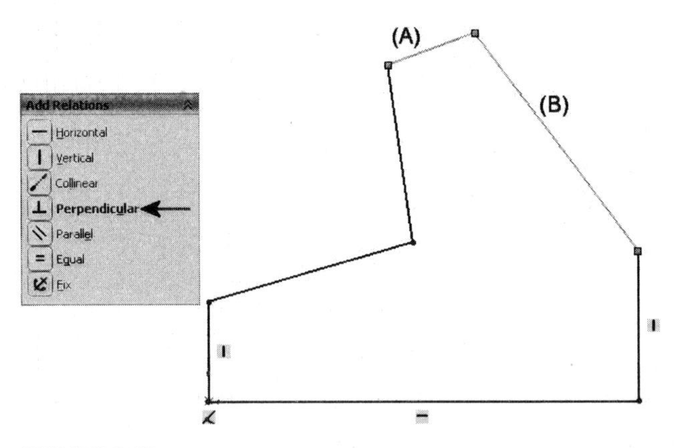

FIGURE 2.23

Next click on lines "A" and "B"; when the list of possible relations appears, click on the Parallel relation. The results should appear similar to Figure 2.24 on the right.

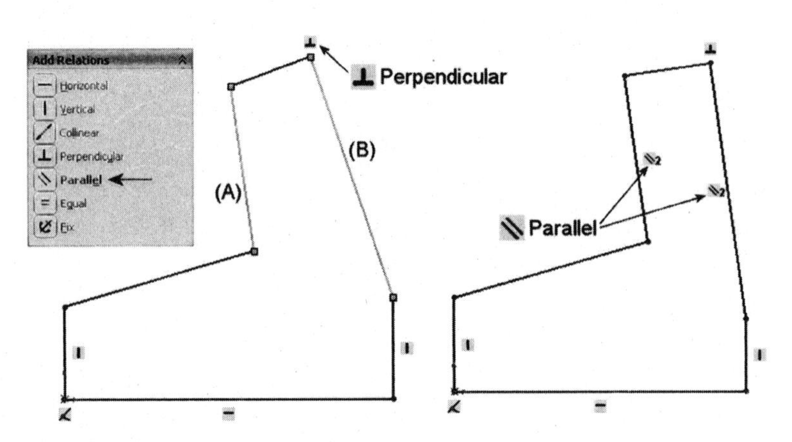

FIGURE 2.24

Applying the Horizontal Relation to Points

TRY IT!

Open the drawing file SWT-Relation-05 (Horiz-Points). This exercise is designed to apply horizontal relations to points.

Points "A" and "C" need to lie along the same horizontal plane as the origin point at "B." These points cannot be aligned using a collinear relation; however, they can be made horizontal. First select the three points as shown in Figure 2.25. When the Add Relations list appears in the PropertyManager, pick the Horizontal relation.

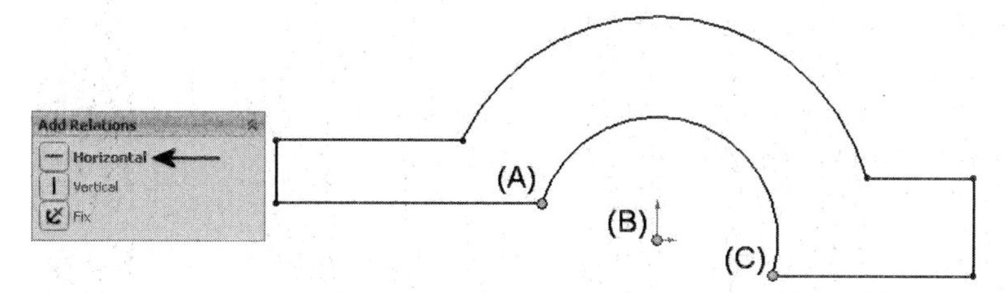

FIGURE 2.25

The results are illustrated in Figure 2.26 with the horizontal relation being applied to the three points making them lie in the same plane. The two base horizontal lines shared by these points cannot be dragged vertically as a result of this operation.

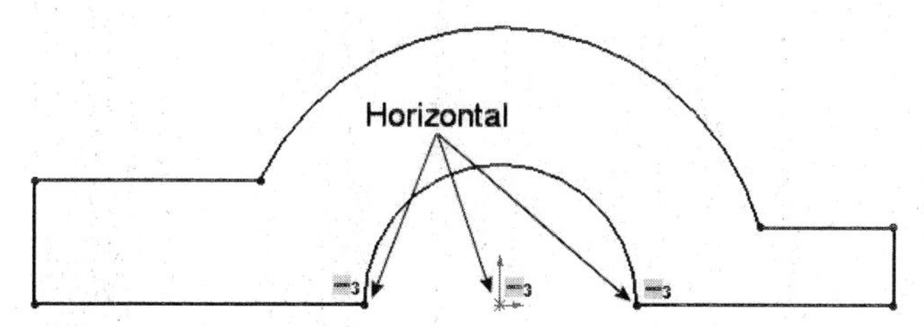

FIGURE 2.26

Dealing with Relations That Overdefine a Sketch

Open the drawing file SWT-Relation-06 (OverDefined). This exercise is designed to fix a sketch that is overdefined.

TRY IT!

When opening up this sketch, an error dialog box appears alerting you that the sketch is overdefined. The message continues on by suggesting that you delete some overdefining dimensions or relations. Delete the overall vertical dimension as shown in Figure 2.27.

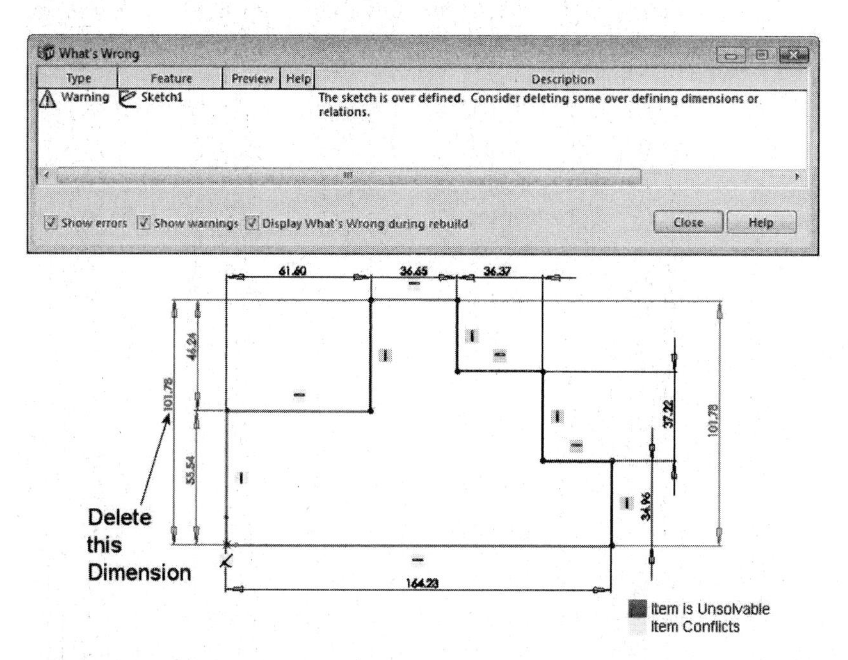

FIGURE 2.27

Deleting the overall vertical dimension displays a dialog box alerting you that the sketch is no longer overdefined. Notice the presence of another overall vertical dimension on the right size of the sketch. This dimension should appear gray in color on your screen and is referred to as a driven dimension or reference dimension. It cannot be changed. However, if other dimensions change, the driven dimension will update its value (Figure 2.28).

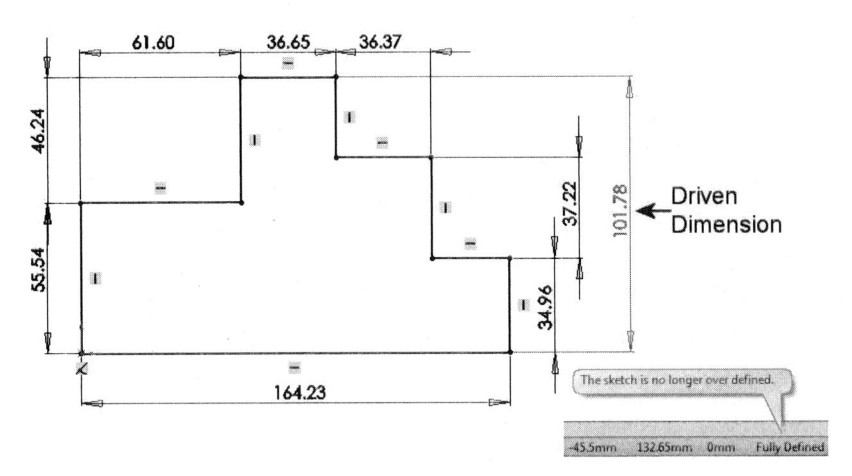

FIGURE 2.28

DELETING RELATIONS THAT OVERDEFINE A SKETCH

TRY IT!

Open the drawing file SWT-Relation-06 (Delete). This exercise is designed to delete a relation due to the sketch being unsolvable.

Due to errors such as overdefining a sketch, it may be necessary to remove or delete relations. Illustrated in Figure 2.29 on the left is a sketch that immediately displays an error message. With a diameter of .50 and an equal relation applied to all circles, the small diameter cannot be solved using the existing relations. Clicking on the Equal relation in the top circle and deleting it will resolve the error. One more diameter dimension needs to be added to one of the bottom circles in order to fully define the sketch as shown in the following figure on the right.

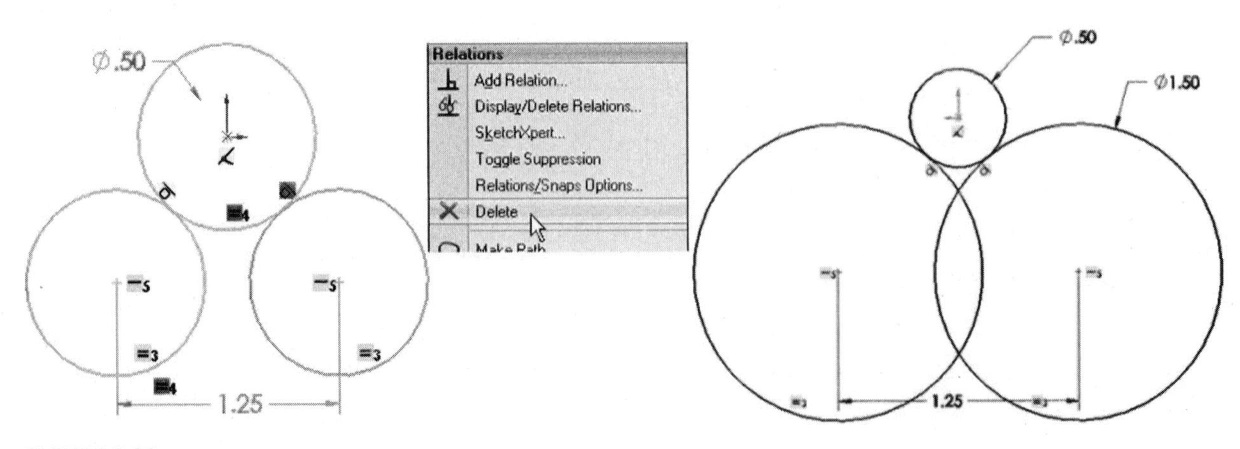

FIGURE 2.29

CREATING AN EXTRUDED BOSS FROM A SKETCH

Once the sketch is fully defined, the next step is to convert it into a solid object or feature. The Extrude tool will be used to accomplish this task. Clicking on the Extruded Boss/Base button activates the PropertyManager that contains information related to this operation as shown in Figure 2.30 on the left. Generally the first time you extrude a fully defined sketch, the area being extruded highlights giving you a preview of the current extruded distance, in this example 10mm.

Open the drawing file Extrude-01 (Blind). Practice performing an extruded feature using the Extruded Boss/Base tool as shown in Figure 2.30.

TRY IT!

When performing the initial extrusion, the PropertyManager will prompt you for information related to the first extrusion direction. You can change the direction of the extrusion and the distance as shown in Figure 2.30 on the left. A preview of the extrusion will be displayed as shown in the following figure on the right.

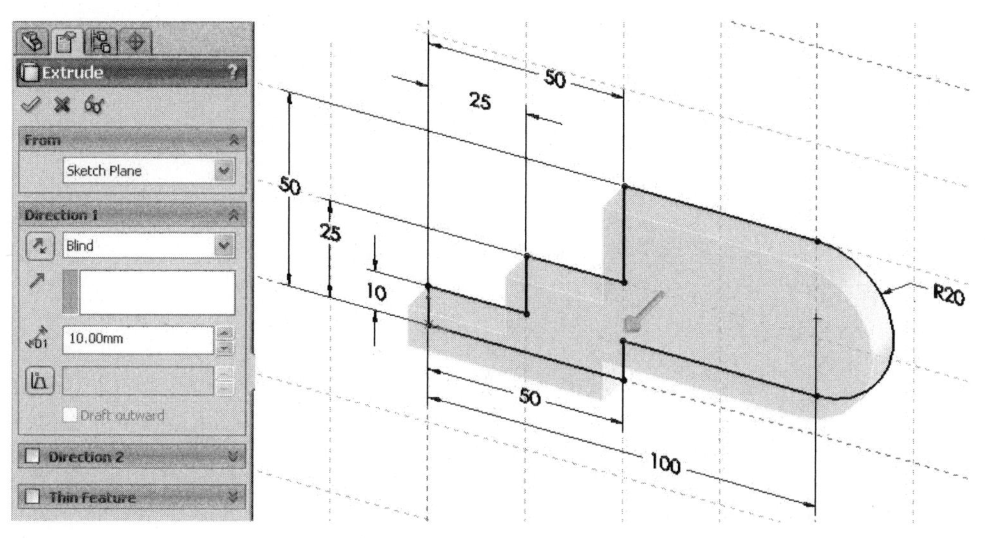

FIGURE 2.30

The results of creating an extruded boss are illustrated in Figure 2.31. There are many more options available when extruding sketches. These will be covered in chapter 3.

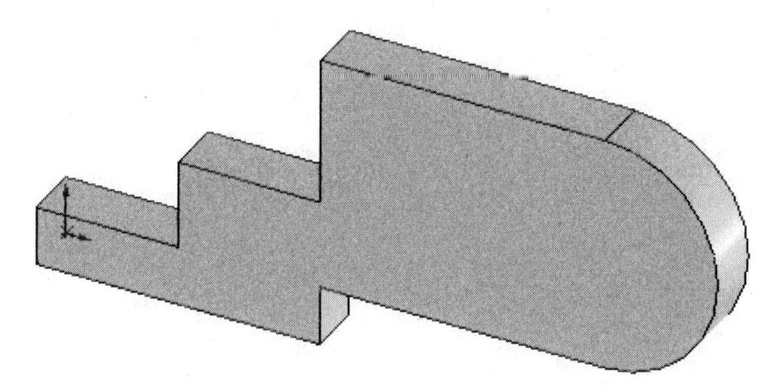

FIGURE 2.31

TUTORIAL EXERCISE: SWT-CH02-01.SLDPRT

Part Number: SWT-CH02-01.SLDPRT

Description: Sketch Phase I

Units: English

This tutorial exercise is designed to create a simple sketch consisting of line entities as shown in Figure 2.32. Once the sketch is completed, an extruded base feature will be created. Finally the mass properties will be calculated to determine the volume of the object.

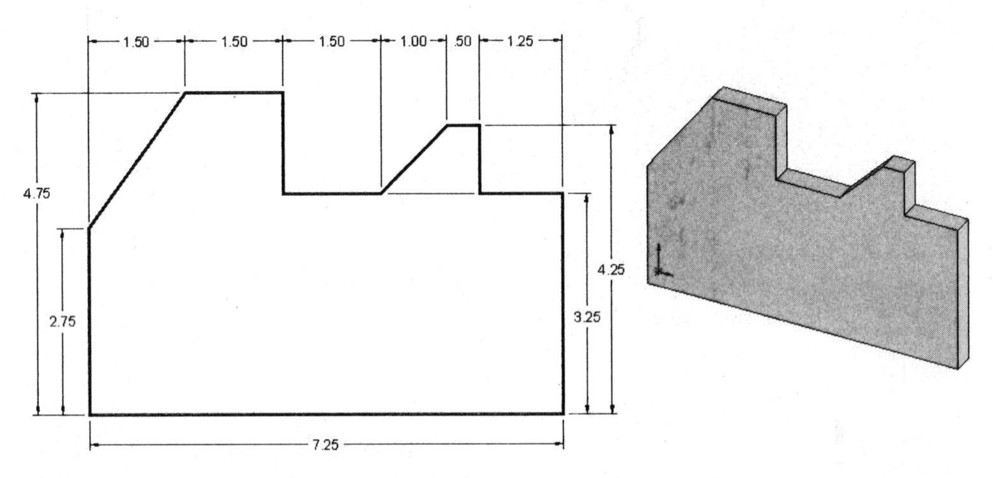

FIGURE 2.32

STEP 1

Begin by creating a *New* SolidWorks Part Document and change to English Units under the *Tools* pull-down menu then *Options*. Change the *Document Properties, Units setting to IPS*. After entering the SolidWorks Design Screen, click the Sketch icon. When the three default sketch planes appear, click the edge of the Front Plane to make this the current sketch plane as shown in Figure 2.33.

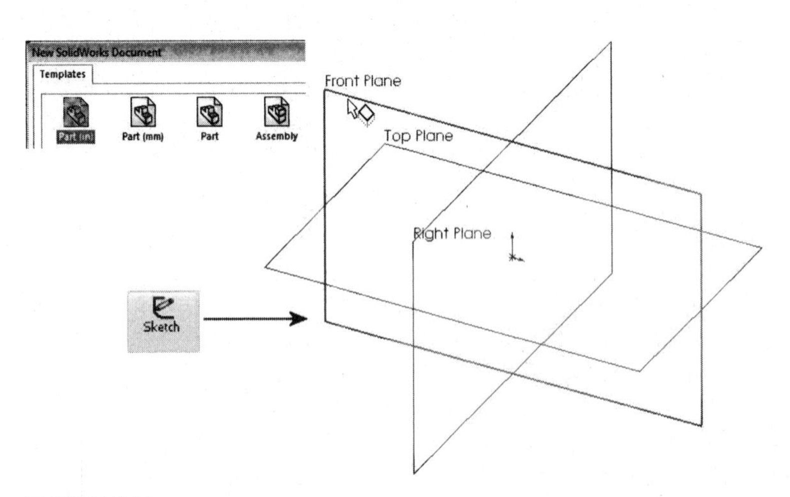

FIGURE 2.33

STEP 2

Select the *Line* tool and construct the entities that make up the sketch as shown in Figure 2.34. Be sure to use the status bar located in the lower right of the display screen to create the sketch as close to the finished dimensions as possible. You can also use the distance readout at your cursor location to determine the closest distance per entity. As you created the sketch, notice the automatic horizontal and vertical relations placed on most lines. You should also apply the Collinear relation to the horizontal lines at "A" and "B" as shown in the following figure. Do this by holding the CTRL key and selecting lines A, B, and then *Collinear*.

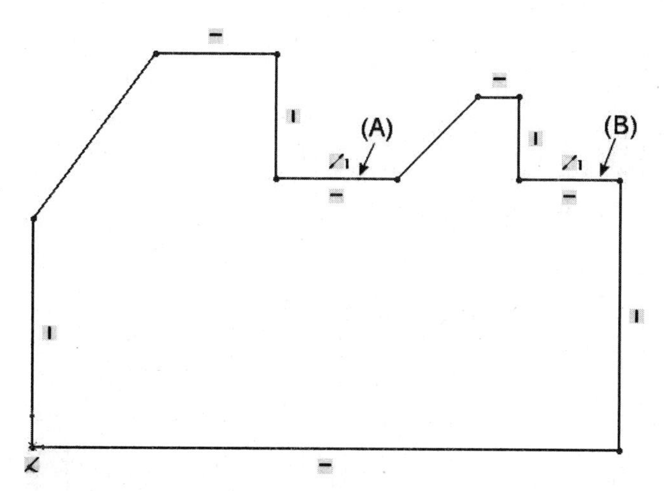

FIGURE 2.34

STEP 3

Add dimensions to the sketch using the Smart Dimension tool. Select the endpoints to place the 4.25 and 4.75 dimensions. The 1.25 dimension was placed last and technically does not need to be added but is driven and placed for reference purposes as shown in Figure 2.35. Select *Make This Dimension Driven*, then select *OK*, and then select the green checkmark. It is identified by the gray color it is created in. You cannot change a driven dimension. However, as you make changes to other sketch dimension, the driven dimension will change to reflect its new value.

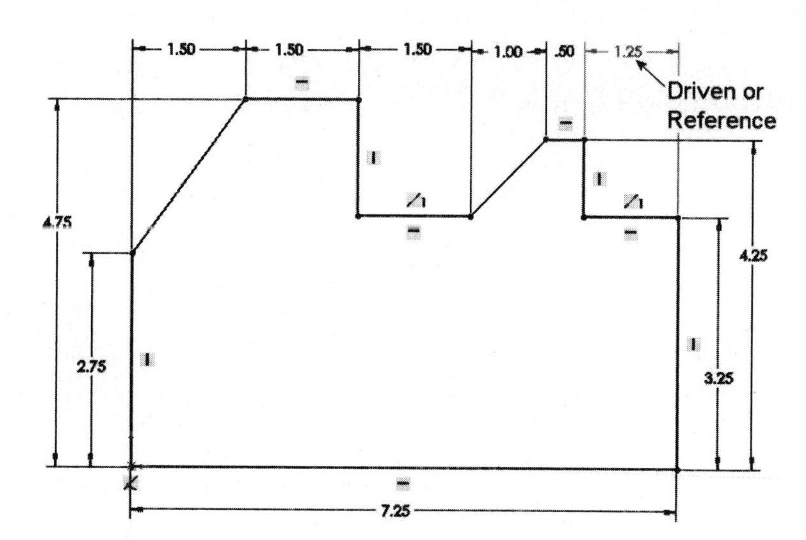

FIGURE 2.35

STEP 4

Click the Features tab located in the CommandManager and pick the Extruded Boss/Base tool. When the extrude settings appear in the PropertyManager, change the extrusion distance to .50 as shown in Figure 2.36 on the left. This distance should preview as shown in the following figure on the right.

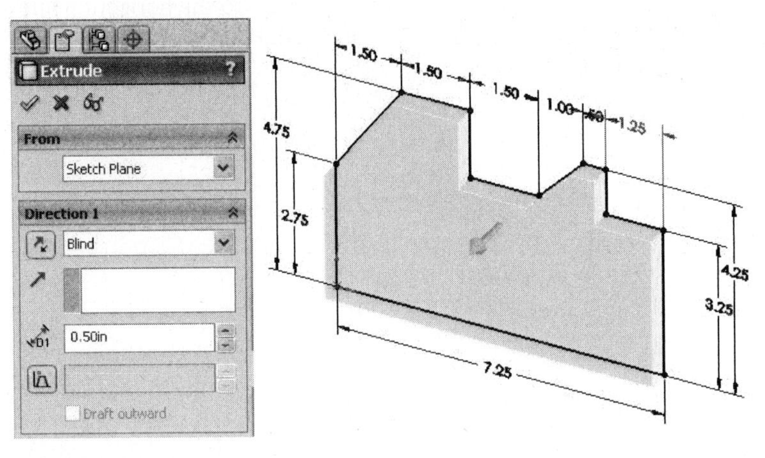

FIGURE 2.36

STEP 5

Clicking the green checkmark at the top of the PropertyManager will create the extruded feature as shown in Figure 2.37.

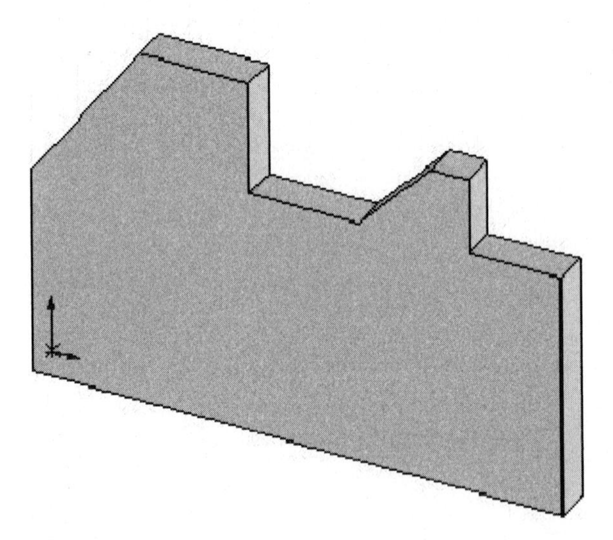

FIGURE 2.37

STEP 6

With the object created, check the model for accuracy by performing a mass property calculation. Start by clicking Mass Properties from the Tools Menu Bar as shown in Figure 2.38 on the left. When the Mass Properties dialog box appears, take note of the Volume value. You will also notice a triglyph appearing on the model that signifies the location of the centroid or location on the model where it is considered balanced as shown in the following figure on the right. Save the file for future use.

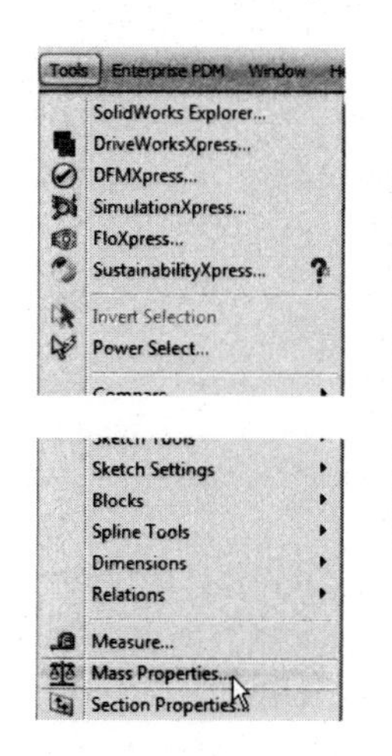

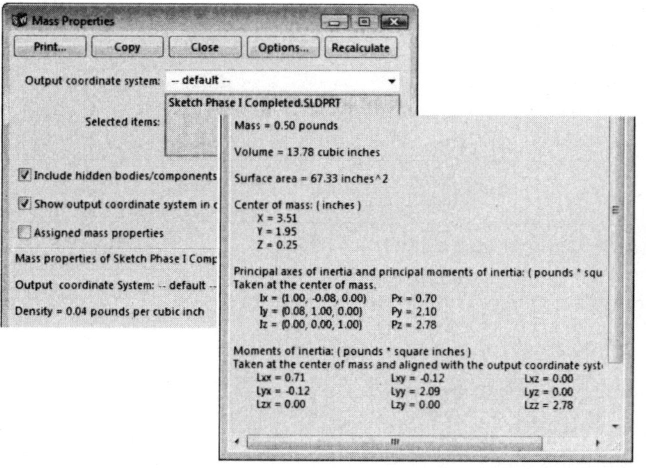

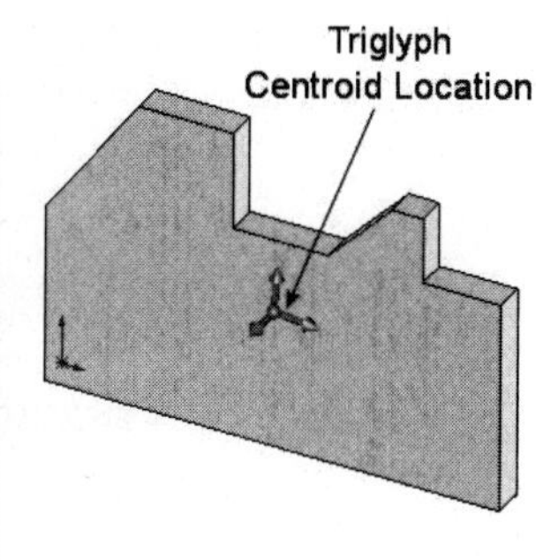

FIGURE 2.38

CREATING AN ADDITIONAL SKETCH PLANE

Part models are not limited to a single sketch. Rather they consist of multiple sketch planes in order to create multiple features of a part. Once an initial feature is created, you create additional sketch planes by clicking on existing part faces. Once the new sketch plane is created, you then construct new geometry, add relations and/or dimensions, and create the new feature. The next exercise will demonstrate the creation of a new sketch plane.

TRY IT!

Open the drawing file SWT_Extrude_Cut01 (Through). This exercise will involve the creation of a new sketch plane on which to create geometry in the form of a rectangle. After the proper relations and dimensions are placed, the rectangle is cut through the part creating a rectangular void in the part.

First, click on an existing face and activate the Sketch tool as shown in Figure 2.39.

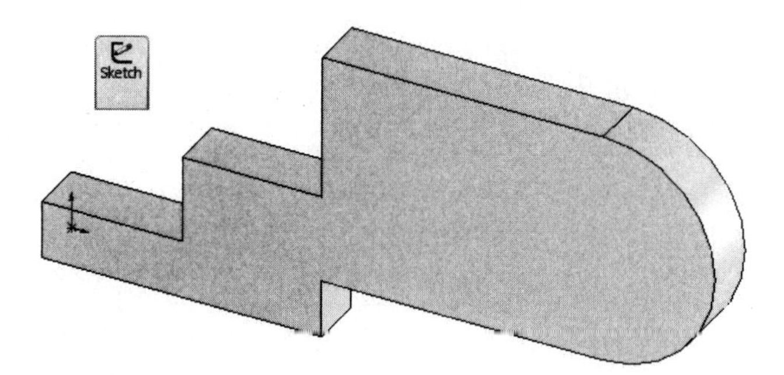

FIGURE 2.39

Notice the creation of the new sketch plane as shown on the face of the object. Also notice the appearance of (-)Sketch2 located in the FeatureManager as shown in Figure 2.40 on the left. The presence of the (-) indicates a sketch that it is underdefined. Once you fully define the sketch, the (-) will disappear from the FeatureManager screen.

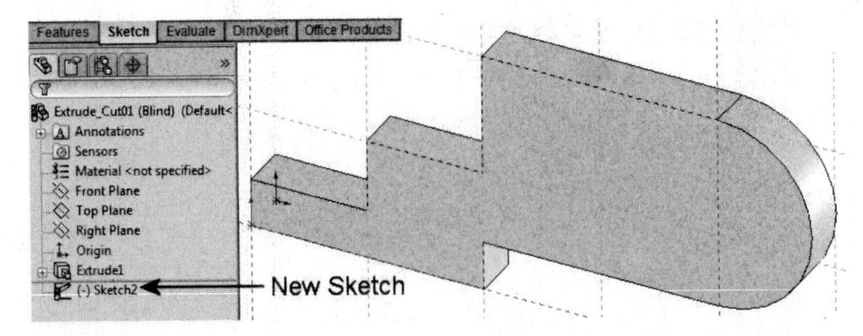

FIGURE 2.40

Another technique to use when defining sketches is to flatten or have the newly created sketch appear in 2D as shown in Figure 2.41 on the left. This will allow you to create geometry shapes in better proportion to the geometry in the sketch. Having the sketch appear in 2D is easily accomplished by right-clicking in the model space away from the part and activating the View Orientation dialog box and double-clicking on *Normal To as shown in the following figure on the right. Also, you may place the Standard Views tool bar with Tools and Customize.

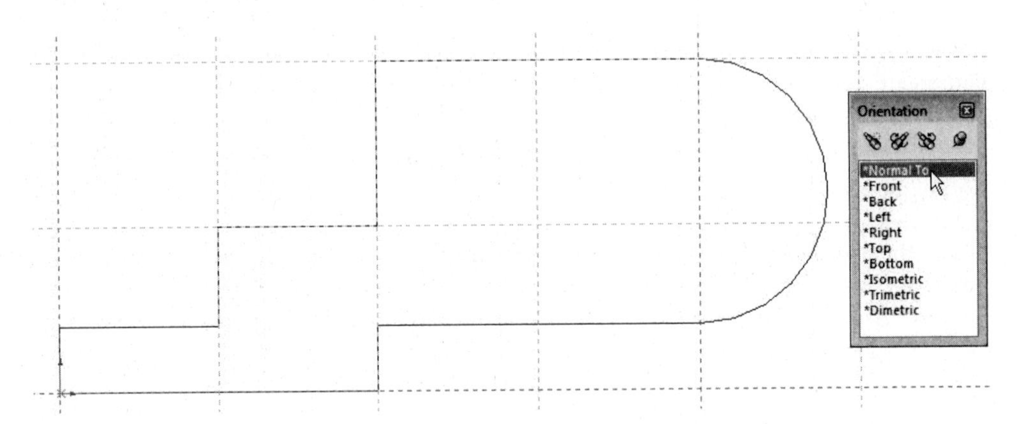

FIGURE 2.41

With the newly created sketch positioned, the next step is to create the geometry used for sketching as shown in Figure 2.42 on the left. Then, add relations or dimensions until the sketch is fully defined as shown in the following figure on the right.

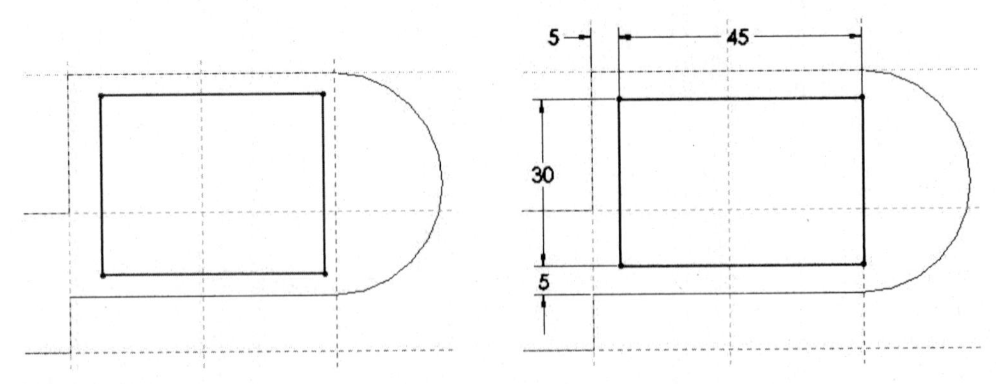

FIGURE 2.42

CREATING AN EXTRUDED CUT FROM A SKETCH

Once a new sketch plane is created, you then will convert this sketch into a feature. The first sketch of any feature is usually an Extruded Boss/Base where you will add material to the feature. The next feature could either add or subtract material from a part; this next example will deal with material removal. After the geometry was located on a new sketch plane, you will create an extruded cut by defining the distance of the extrusion.

TRY IT!

Continue working on the drawing file SWT_Extrude_Cut01 (Through). Practice performing an extruded cut operation using the Extruded Cut tool.

While sketch mode is still active, click on the Extruded Cut tool. Also, change the orientation of your view to *Trimetric as shown in Figure 2.43 on the right. Your model should appear similar to the following figure. Activate the Extruded Cut tool; you will notice a preview of the feature being cut. When the PropertyManager appears, change Direction 1 to Through All as shown in the following figure on the left. This option is designed to cut the rectangular shape completely through the part as shown in the following figure in the middle.

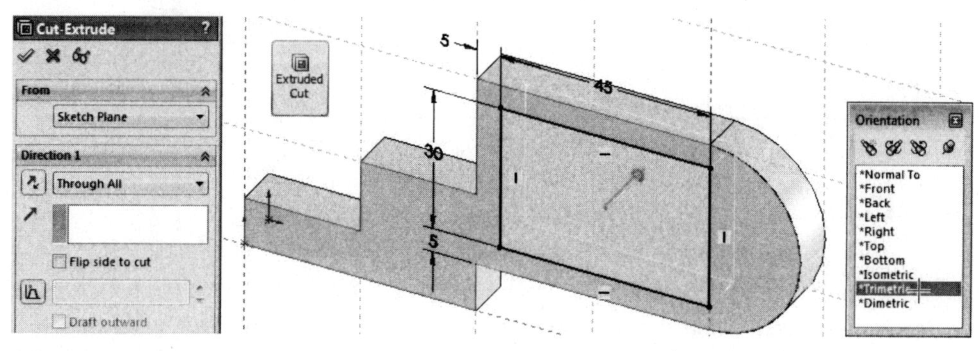

FIGURE 2.43

The results of creating an extruded cut are illustrated in Figure 2.44. As with creating extruded bosses or base features, there are many more options available when creating extruded cuts.

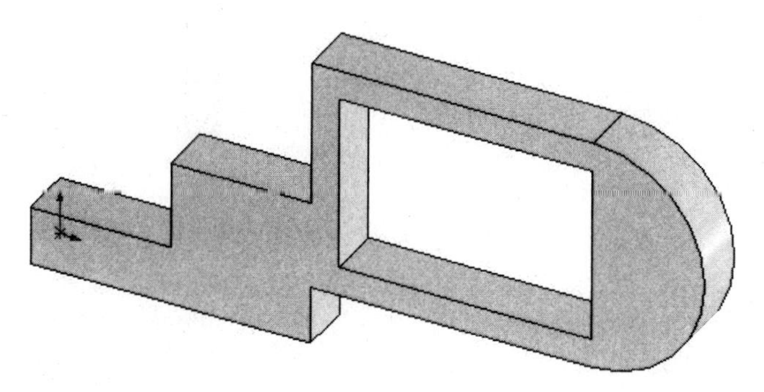

FIGURE 2.44

TUTORIAL EXERCISE: SWT-CH02-02.SLDPRT

Part Number: SWT-CH02-02.SLDPRT

Description: Sketch Phase II

Units: English

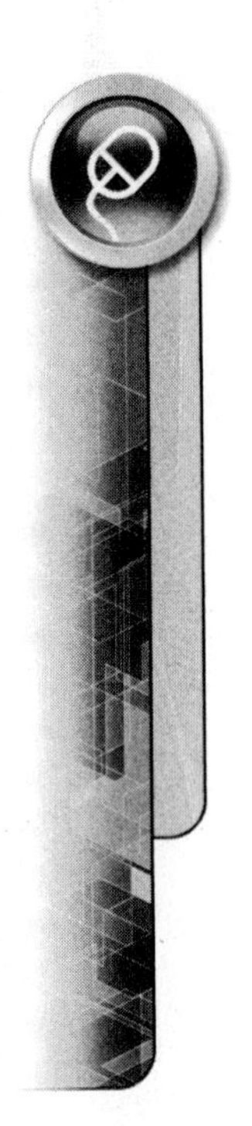

This tutorial is a continuation of the previous exercise and is designed to create a cut-out as shown in Figure 2.45. A new sketch plane will be created. A sketch will define the outline of the shape to be cut. This sketch will then be used to create an extruded cut. Finally the mass properties will be calculated to determine the volume of the object.

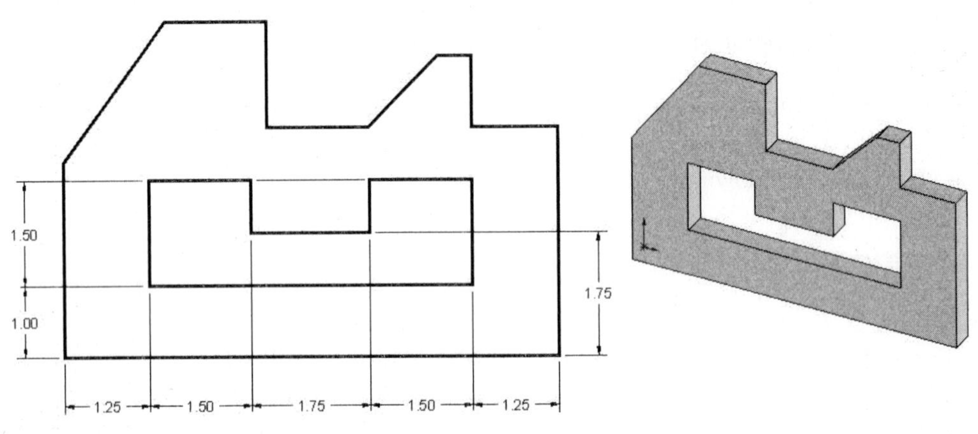

FIGURE 2.45

STEP 1

Select a new sketch plane on the front face of the existing part as shown in Figure 2.46.

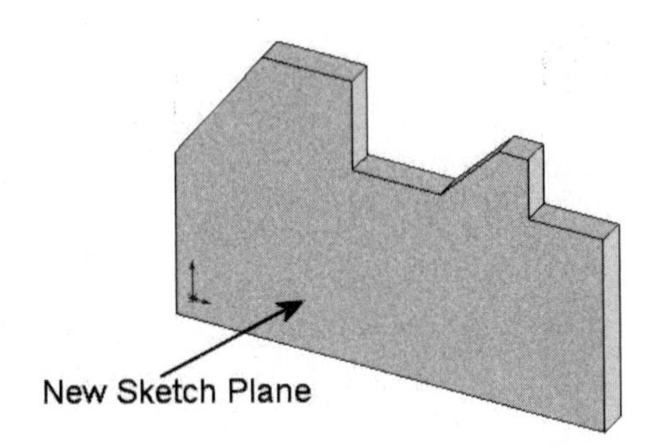

FIGURE 2.46

STEP 2

Right-click on *Normal To in the View Orientation dialog box to view the model in a flat position as shown in Figure 2.47.

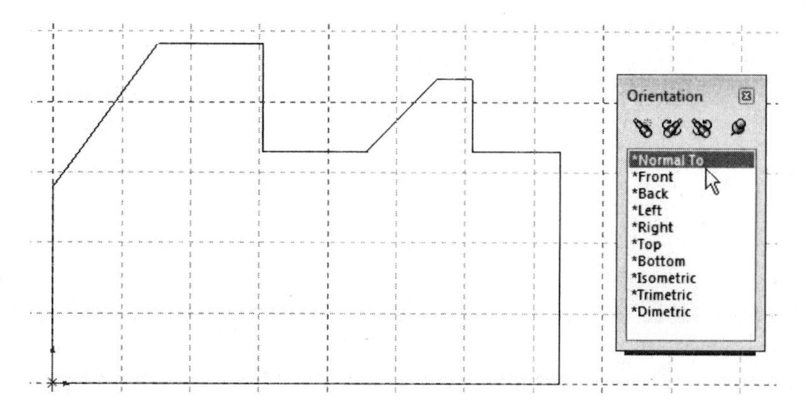

FIGURE 2.47

STEP 3

Sketch the shape on the part as shown in Figure 2.48 on the left. Add Smart Dimensions to the sketch until the sketch turns black, which is considered fully defined.

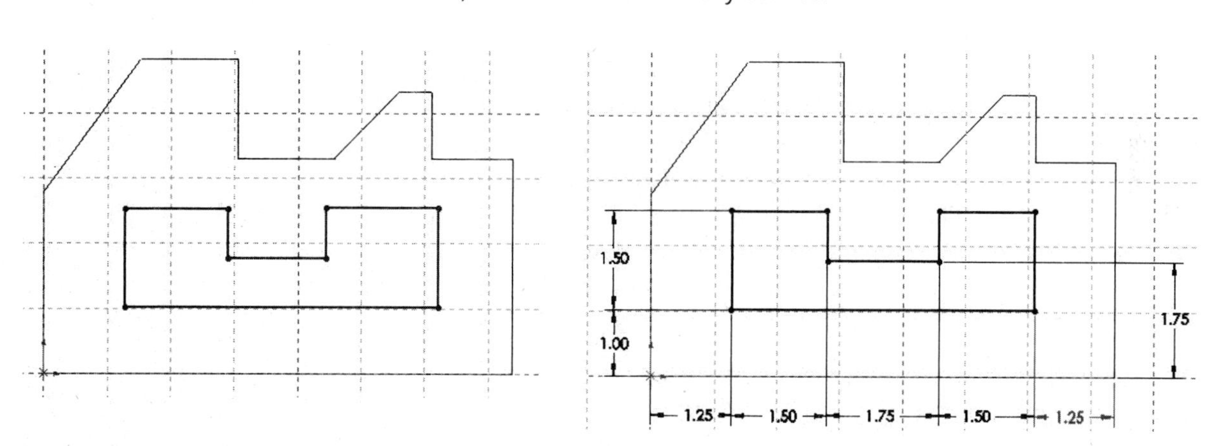

FIGURE 2.48

STEP 4

Switch to the *Trimetric orientation and create an extruded cut (from the Features tab) using the Through All option as shown in Figure 2.49.

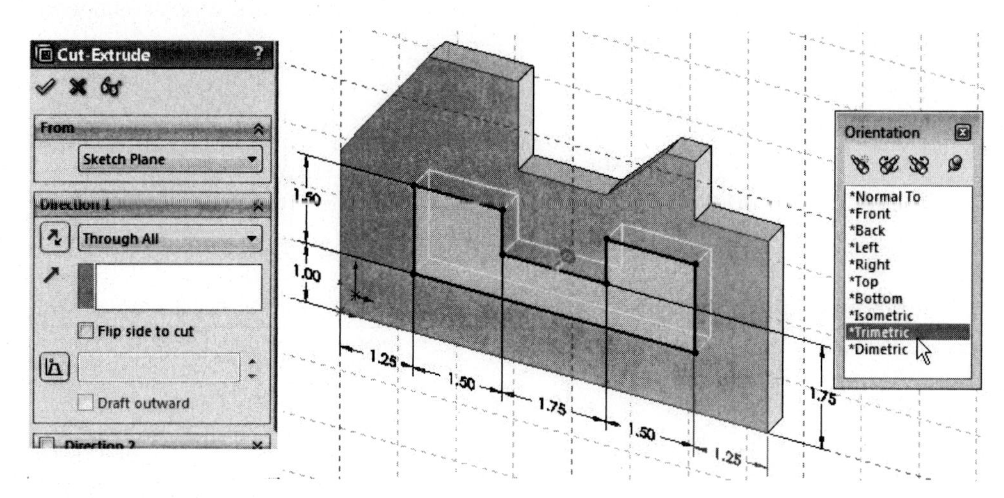

FIGURE 2.49

STEP 5

The cut operation and completed part model is shown in Figure 2.50.

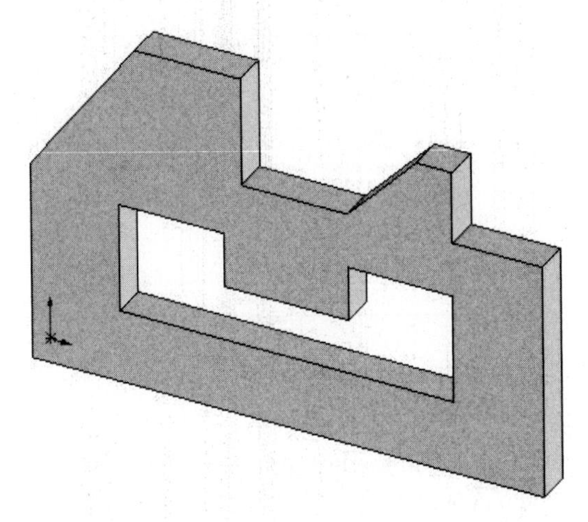

FIGURE 2.50

CREATING CURVED ENTITIES IN SKETCHES

The first series of sketching problems dealt with line segments. Here you learned how to create extruded boss or base shapes from a sketch and even create cuts in a sketch. The next segment of sketches will involve incorporating curved shapes. These will first take the form of transitioning from lines to arcs. A series of arc segments that include creating arcs by center point, tangent arcs, and 3-point arcs will be demonstrated. Circles will also be discussed as a means of creating extruded shapes or cuts.

TANGENT ARC TRANSITION

While sketching line segments, you can transition to drawing arcs without exiting the Line tool as shown in Figure 2.51. Use the following steps for performing this task:

1. Click the Line tool on the Sketch toolbar and sketch a line.
2. Click the endpoint of the line and then move the sketch pencil away.
3. The preview shows another line.
4. As a preview shows another line, move the pointer back to the original endpoint, then move the cursor away again.
5. The preview shows a tangent arc.
6. Click to place the arc.
7. Move the pointer away from the arc endpoint.
8. The preview shows a line. Draw a line or change to an arc using the technique explained in step 4.

NOTE **To toggle between line and arc without returning to the endpoint, press A.**

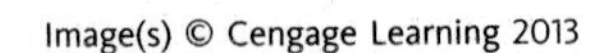

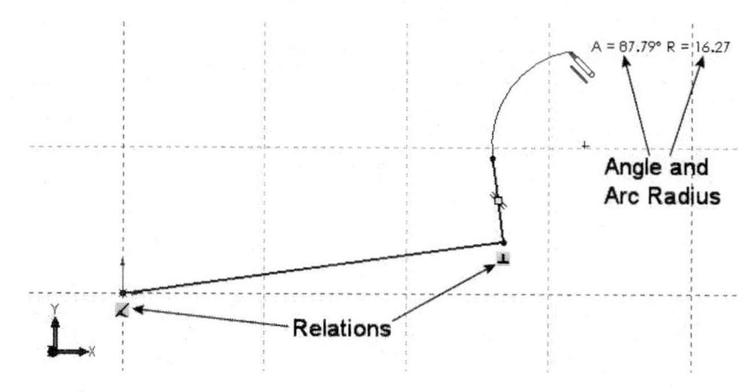

FIGURE 2.51

The transitioned arc is displayed in Figure 2.52. Continue drawing lines or transition back into arc mode.

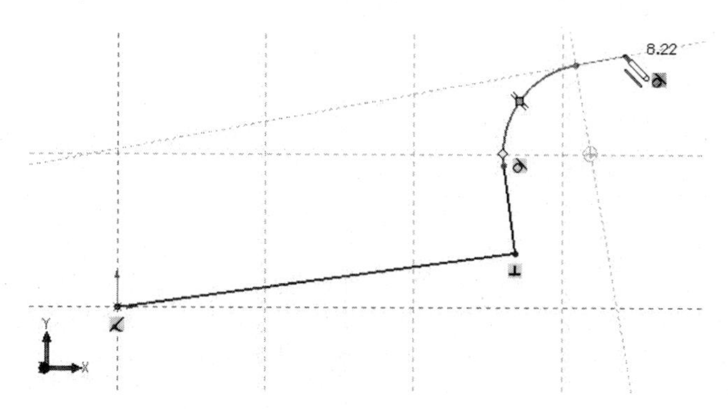

FIGURE 2.52

CONSTRUCTING ARCS

In addition to constructing arcs as part of line segments, various arc modes are available to construct dedicated arc segments. Arcs can be constructed by a center point and two additional points. Choose these commands from either menu as shown in Figure 2.53.

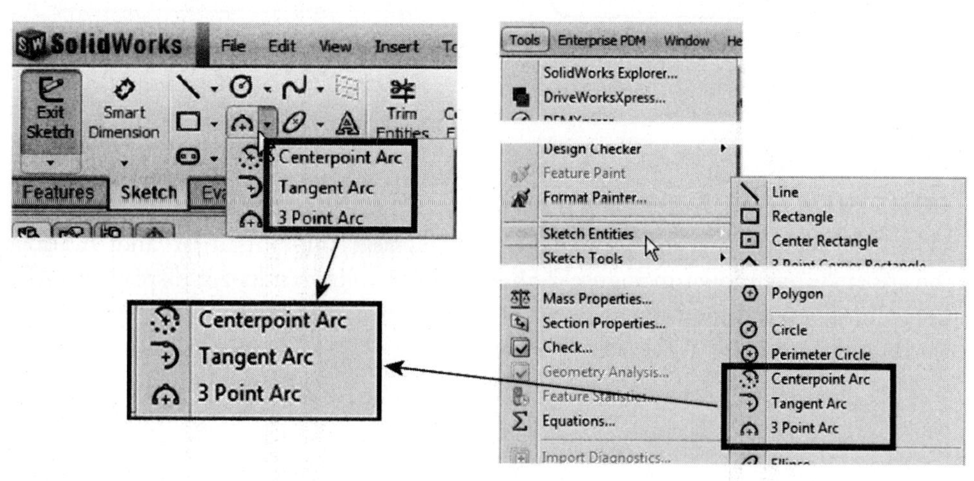

FIGURE 2.53

CONSTRUCTING AN ARC BY CENTER POINT

When you click the Centerpoint Arc tool, begin by picking the center point at "A" followed by endpoints at "B" and "C" as shown in Figure 2.54 on the left. The results are displayed in the following figure on the right. The arc is constructed from the origin and intersects one endpoint. At no time are tangencies formed at these endpoints.

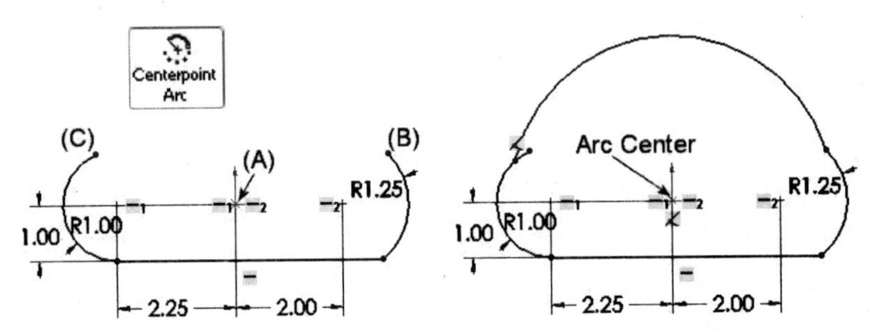

FIGURE 2.54

CONSTRUCTING A TANGENT ARC

Arcs can be constructed tangent to the endpoint of one entity. When you click the Tangent Arc tool, pick the endpoint at "A." As you move your cursor, the next arc will be tangent to this endpoint. Next pick the endpoint at "B" as shown in Figure 2.55 on the left. The results are displayed in the following figure on the right. The arc is constructed from one endpoint to the other. The arc that began from the first endpoint is tangent to that entity.

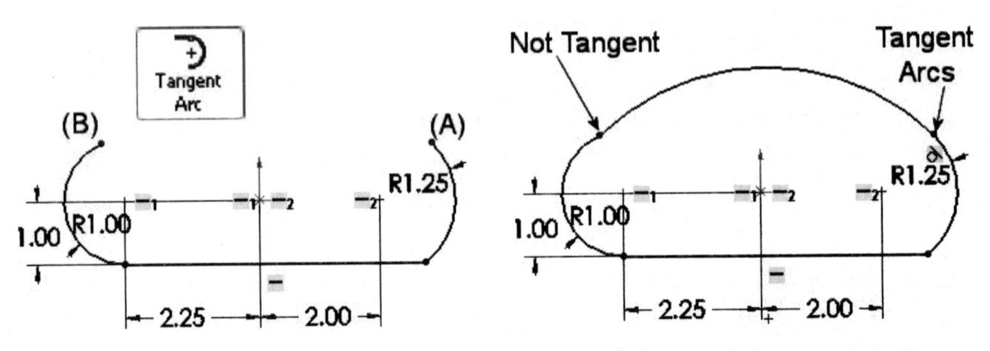

FIGURE 2.55

CONSTRUCTING A 3-POINT ARC

When you click the 3 Point Arc tool, begin by picking endpoints at "A" and "B"; drag the arc at "C" as shown in Figure 2.56 on the left. The results are displayed in the following figure on the right. The arc is constructed from the 3 points. At no time are tangencies formed at these endpoints.

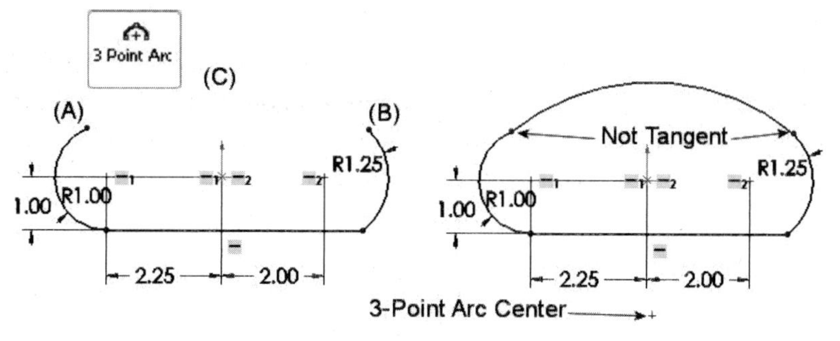

FIGURE 2.56

CONSTRUCTING CIRCLES

Circles are an important entity when constructing solid models. Two types of circles are available, namely, Circle by Radius and Perimeter Circle, as shown in Figure 2.57. Choose these commands from either menu as shown in the following figure.

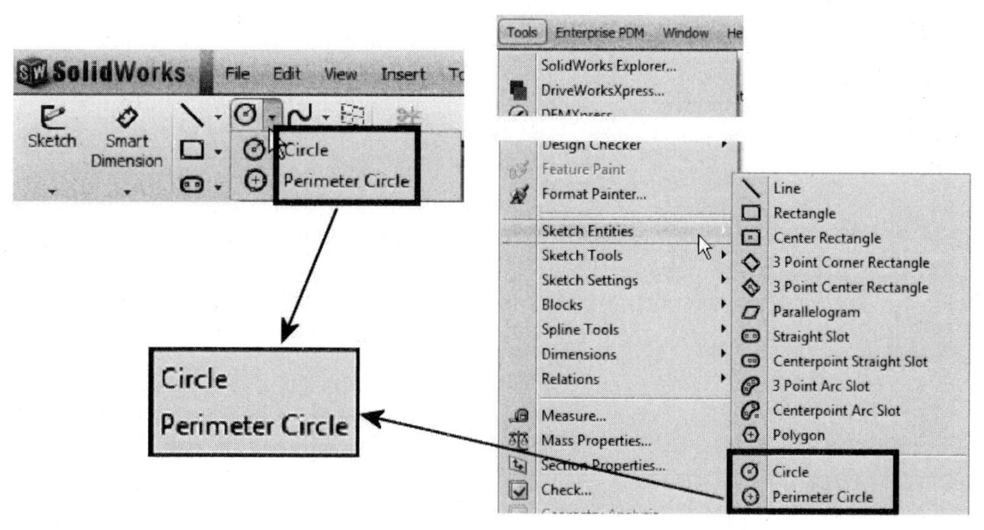

FIGURE 2.57

One popular technique for locating the center point of a circle based on an existing arc is to activate the Circle tool. Then, before picking the center point of the circle, move your cursor over the top of the arc and notice the center point of the arc displayed as shown in Figure 2.58 on the left. Pick this center point to construct the circle as shown in the following figure in the middle. The results are illustrated in the following figure on the right after a dimension is applied to the circle fully defining the circle.

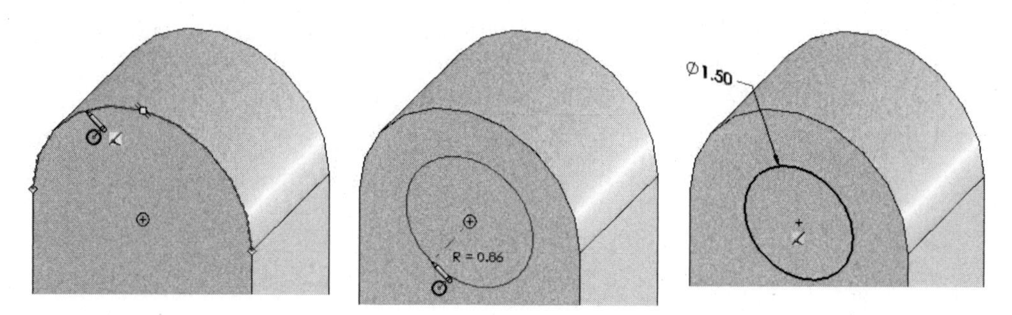

FIGURE 2.58

LEVEL II RELATIONS

This next grouping of relations will also be used to apply short exercises and real-world examples. The second grouping of relations involve Tangent, Equal, Concentric, Coradial, Symmetric, Midpoint, Intersection, and Fix.

TABLE 2.4

Relation Icon	Relation Meaning	Description
	Tangent	Selecting an arc, ellipse, or spline and a line or arc will apply the tangent relation to the two entities.
	Equal	Selecting two or more lines or arcs will make the arc radii or line lengths equal.
	Concentric	Selecting two or more arcs or a point and an arc will make the arcs share the same center point.
	Coradial	Selecting two or more arcs will make the arcs share the same center point and radius.
	Symmetric	Selecting a centerline and two points, lines, arcs, or ellipses will make the selected entities equidistant from the centerline.
	Midpoint	Selecting two points or a point and a line will locate the point along the midpoint of the line.
	Intersection	Selecting one point and two lines will locate the point at the intersection of the two lines.
	Fix	Selecting any entity will fix that entity's size and location. However, the endpoints of fixed lines, arcs, or elliptical segments are free to move.

The Tangent Relation

TRY IT!

Open the drawing file SWT_Relation_07 (Tangent). This exercise is designed to apply tangent relations to the line and arc segments.

Select the *Add Relations* command from the *Display/Delete Relations* pull-down menu; pick arc "A" and line "B." When the list of relations displays in the PropertyManager, click the Tangent relation. Continue adding tangent relations to the other three line and arc locations. The result is illustrated in Figure 2.59 on the right.

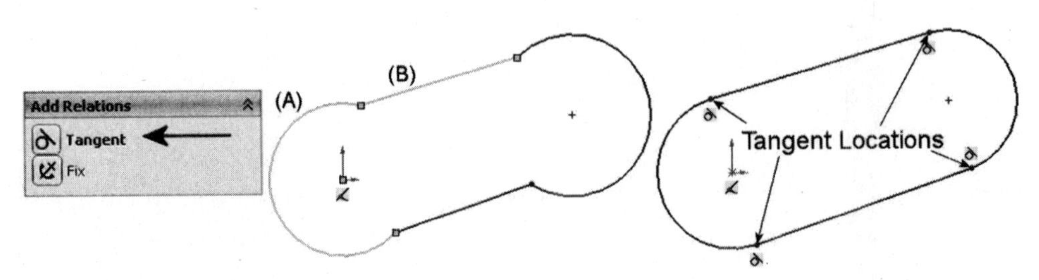

FIGURE 2.59

The Equal Relation

Open the drawing file SWT_Relation_08 (Equal). This exercise is designed to apply equal relations to arcs.

TRY IT!

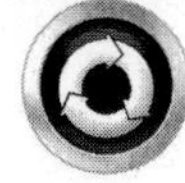

Figure 2.60 illustrates a number of corners that have been filleted. One corner at "A" has a fillet radius of 9mm assigned. You need to assign this radius to all other fillet radius locations. First click the Add Relations tool and pick on all fillet radii identified from "A" to "F." When the list of all possible relations appears in the PropertyManager, click on the Equal relation.

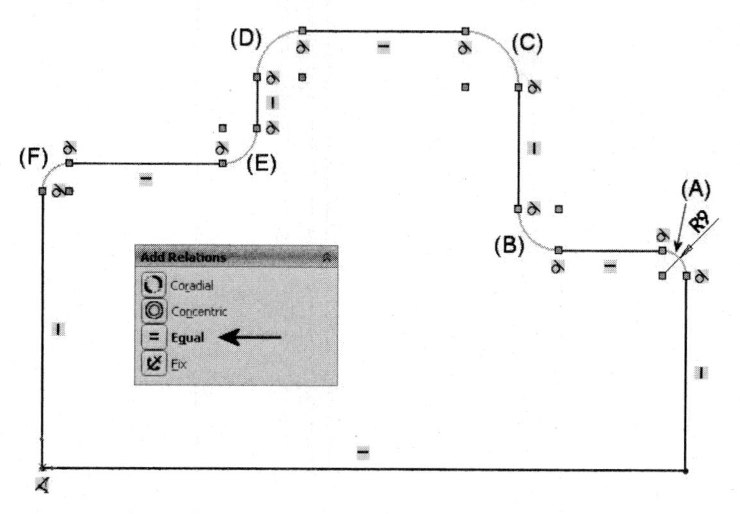

FIGURE 2.60

To test whether all radii values are equal, change the dimensioned fillet radius from 9mm to 35mm. All radii should be updated to the new dimension value as shown in Figure 2.61. Equal relations also work on circle and line entities.

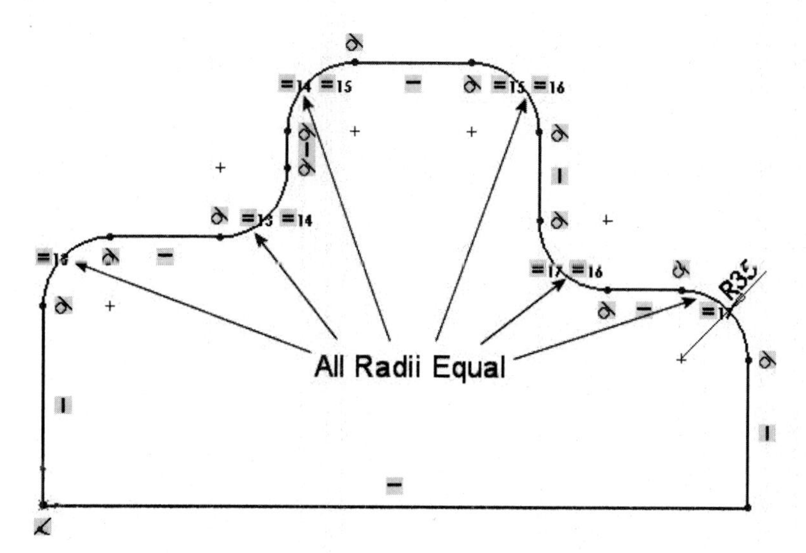

FIGURE 2.61

The Concentric Relation

TRY IT!

Open the drawing file SWT_Relation_09 (Concentric). This exercise is designed to apply concentric relations to arcs and circles.

The concentric relation will allow a circle and arc to share the same center point. To apply this relation, first click the Add Relations tool and select arc "A" and circle "B" as shown in Figure 2.62. When the list of possible relations displays in the PropertyManager, click on the Concentric relation. Make the remaining circles concentric with their corresponding arcs at areas "C," "D," and "E."

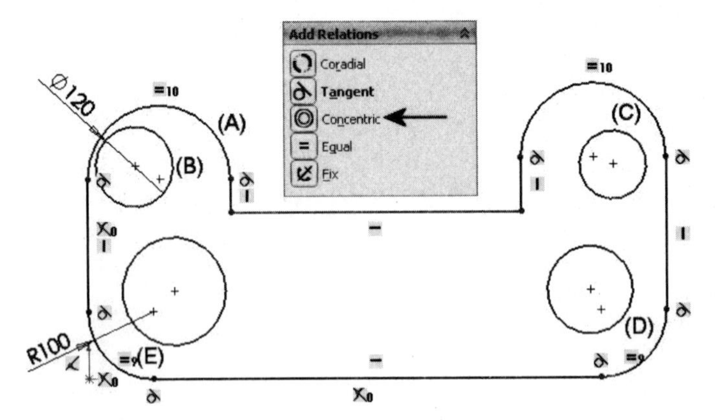

FIGURE 2.62

The results are displayed in Figure 2.63. All circles are now concentric with their respective arcs. Equal relations have also been applied to the top pair of circles and bottom pair.

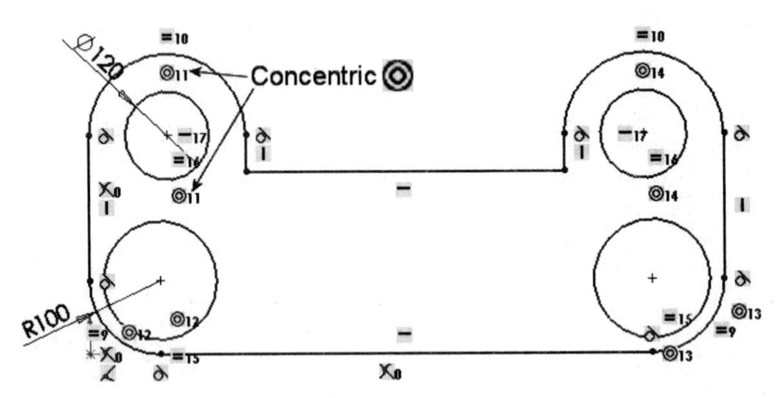

FIGURE 2.63

The Coradial Relation

TRY IT!

Open the drawing file SWT_Relation_10 (Coradial). This exercise is designed to apply horizontal and vertical relations to line segments.

This relation allows two selected arc entities to share the same center point and radius. In Figure 2.64, the top arc entity is already documented with a 50 diameter dimension. Click the Add Relations tool and pick both arc entities "A" and "B." When the list of possible relations appears, click on Coradial as shown in the following figure.

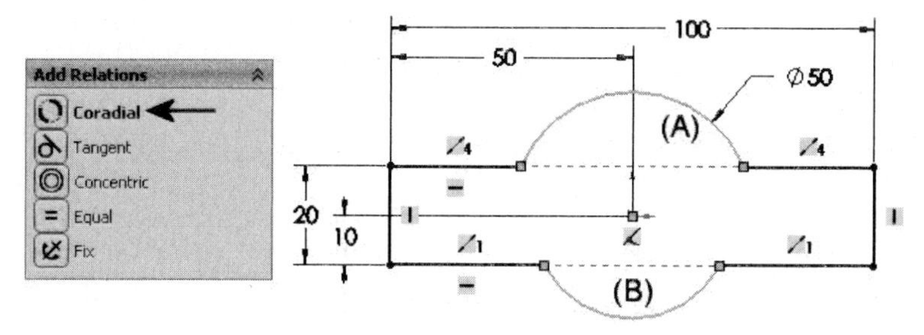

FIGURE 2.64

The results are displayed in Figure 2.65 with both arcs sharing the same radius and center point. Notice also the appearance of the coradial glyph on both arc entities.

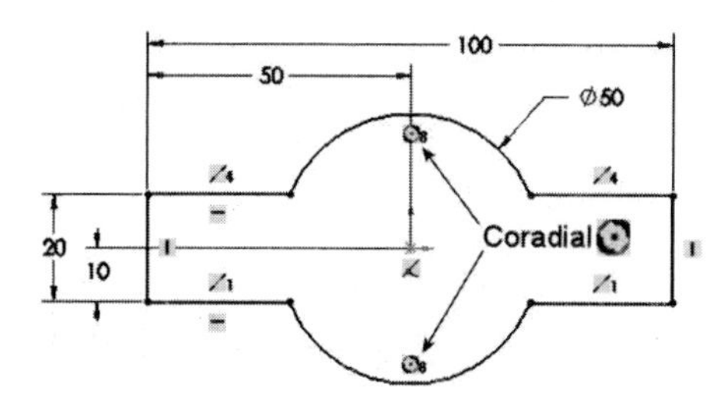

FIGURE 2.65

The Symmetric Relation

Open the drawing file SWT_Relation_11 (Symmetric). This exercise is designed to apply symmetric relations between two circles and a center line segment.

TRY IT!

The Symmetric relation will make two selected entities equidistant from a placed centerline. In Figure 2.66, click the Add Relations tool and pick line segments "A" and "B" in addition to centerline "C" as shown in the following figure. When the list of possible relations appears, click the Symmetric relation as shown in the following figure.

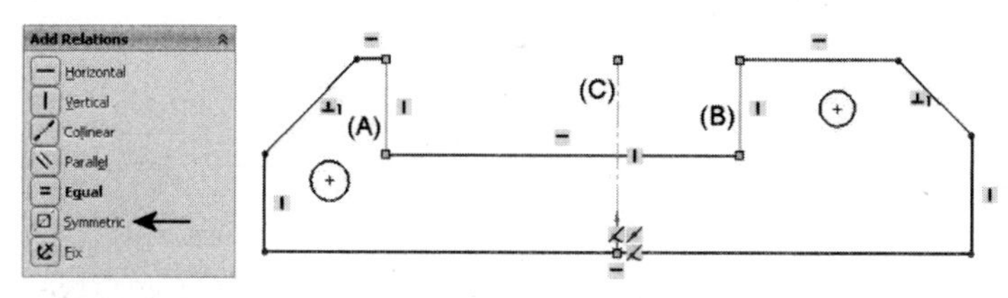

FIGURE 2.66

The results are illustrated in Figure 2.67 with both line segments positioned at equal distances from the common centerline. Experiment by adding this relation to other sets of entities such as both circles. Be sure to pick the centerline in order to display Symmetric in the list of possible relations.

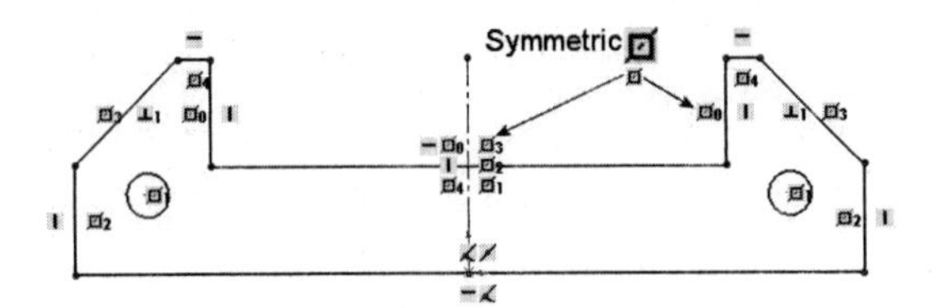

FIGURE 2.67

WORKING WITH CONSTRUCTION GEOMETRY

At times, you want to add geometry to a sketch that is ignored when producing features; this type of entity is called construction geometry. Figure 2.68 illustrates a stair step sketch that is controlled by a construction or centerline. This entity was drawn using the Centerline tool. You could also draw a regular line segment and then select this entity to activate the PropertyManager. Click in the box next to For Construction as shown in the following figure on the left to convert this entity into construction geometry as shown in the following figure on the right.

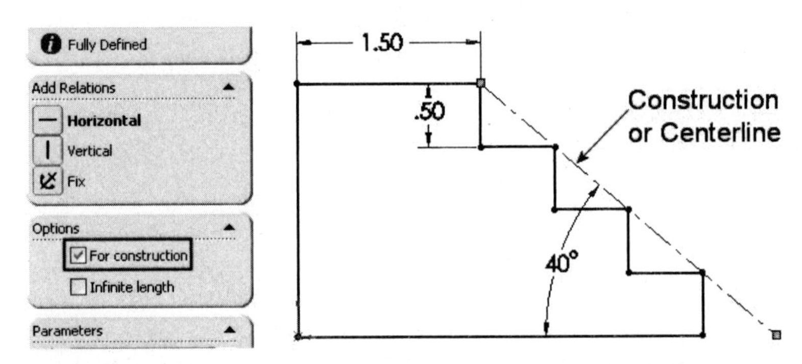

FIGURE 2.68

To better illustrate the power of construction geometry, Figure 2.69 illustrates the same stair step sketch. However, instead of a 40° angle controlling the slope, the angles were changed to 30° (as shown on the left) and 50° (as shown on the right). Notice the stair step sketch changes based on the new angles.

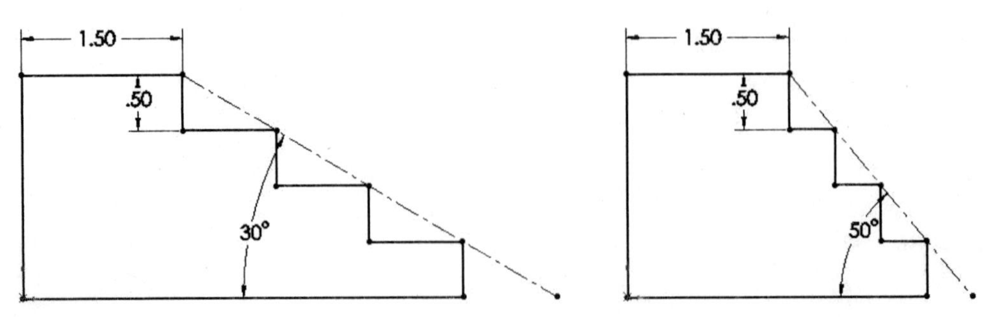

FIGURE 2.69

Figure 2.70 is yet another application of how construction geometry can be used to assist with solving a sketch. A bolt hole needs to be cut through this part. A circle was sketched. Then a line was constructed from the model origin to the center of the circle. Finally, a circle was constructed; the center of the large circle lies at the intersection of the model origin. The large circle also intersects the center of the small circle. Both of these entities were selected and changed to construction geometry. This changes both entities to the center linetype as shown in the following figure on the left. To fully define the sketch, a vertical relation was applied to the construction line and a dimension added to the construction circle as shown in the following figure on the right.

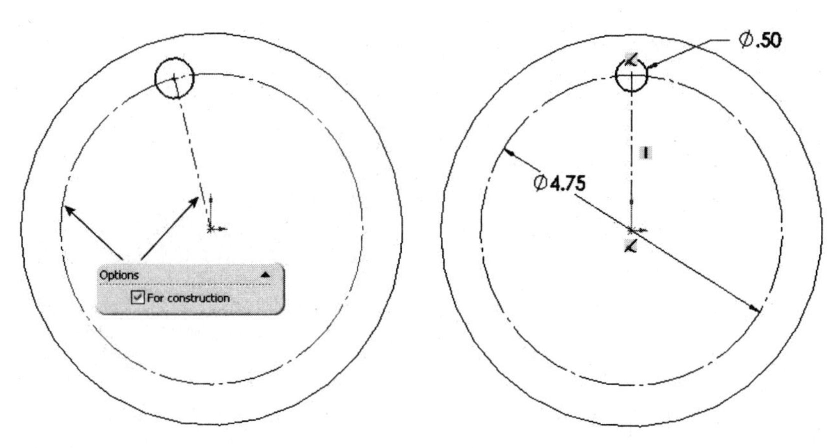

FIGURE 2.70

Open the drawing file SWT_Relation_12 (CenterRec). This exercise is designed to place center lines and apply horizontal and vertical relations to these center line segments for the purpose of centering a rectangle.

TRY IT!

While in sketch mode, construct a centerline from the origin to the midpoint of the horizontal line at "A." Construct a second centerline from the origin to the midpoint of the vertical line at "B" as shown in Figure 2.71 on the left. Next, apply a Vertical relation to line "C" and a Horizontal relation to line "D" as shown in the following figure on the right. Drag edges or corner points and observe that the rectangle remains centered about the origin.

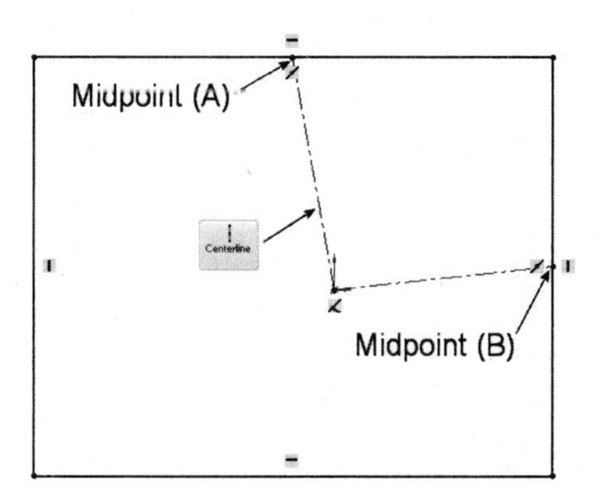

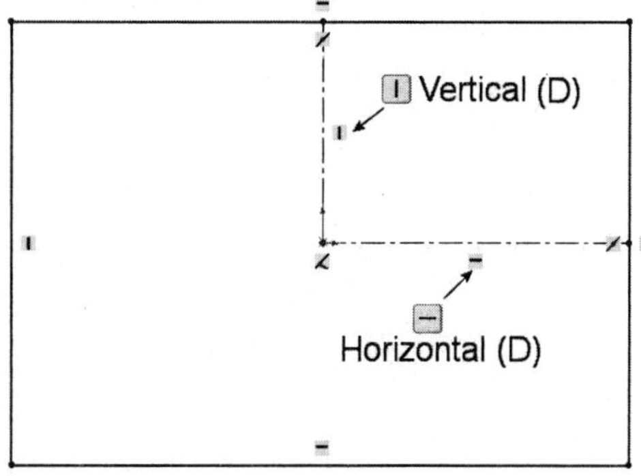

FIGURE 2.71

TUTORIAL EXERCISE: SWT-CH02-03.SLDPRT

Description: Angle Pattern

Units: English

This tutorial exercise is designed to construct a model using angle and aligned dimensions. After constructing the base horizontal line, all other lines are constructed with automatic relations turned off (Figure 2.72). This is accomplished by pressing and holding down the CTRL key as the lines are sketched. Follow the next series of steps for constructing this object.

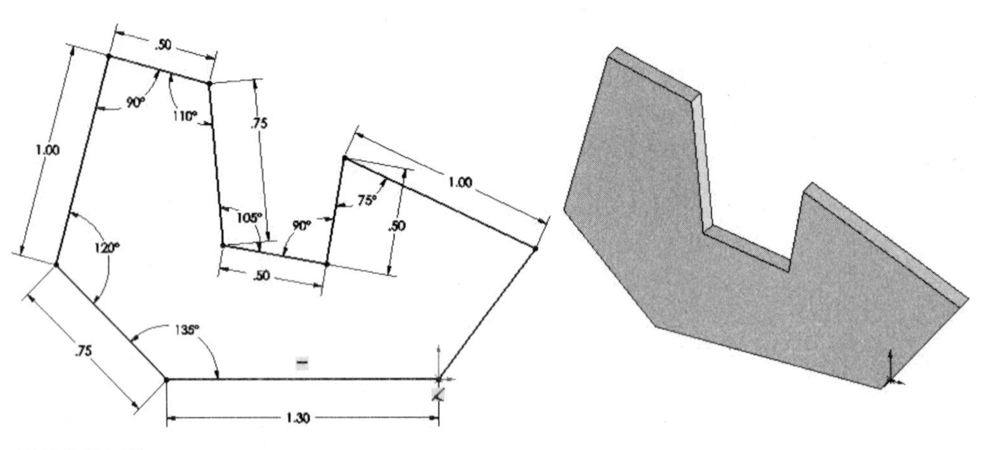

FIGURE 2.72

STEP 1

Create a new sketch plane using the default front plane. Select the *Line* command and create a sketch consisting of the following objects. Due to the nature of all angle entities, be careful not to accidentally apply automatic relations such as parallel or perpendicular. This could cause errors to occur during the placement of dimensions.

NOTE **To disable automatic relations, press and hold down the CTRL key as you sketch.**

Begin this sketch at the origin and draw in a clockwise direction (Figure 2.73).

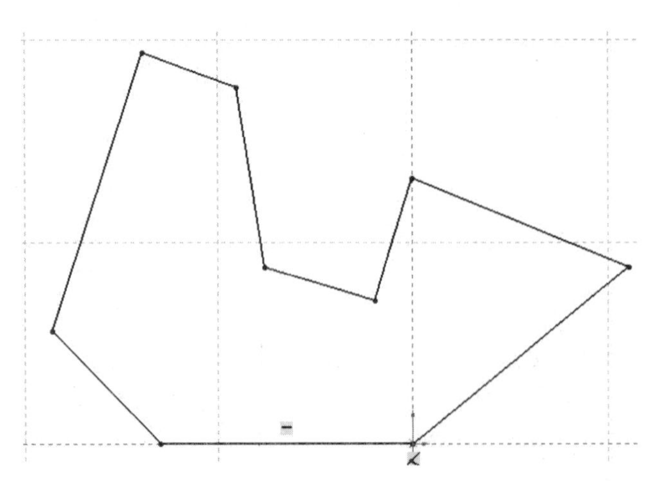

FIGURE 2.73

STEP 2

Create the following angle dimensions as shown in Figure 2.74 on the left. Notice that the majority of these dimensions are incorrect. Initially you are to place the angle dimensions and not worry about their values. Select Smart Dimension, select the two adjoining lines, click the cursor, change to the correct angle, and accept it by selecting the green check-mark. Once all of these angle dimensions are placed, now edit each angle value making sure the exact dimensions are placed as shown in the following figure on the right. If the location of these dimensions does not match exactly with this figure, feel free to locate the dimensions to better positions.

NOTE

If at any time a line is placed incorrectly, press the ESC key on the keyboard to go back one step.

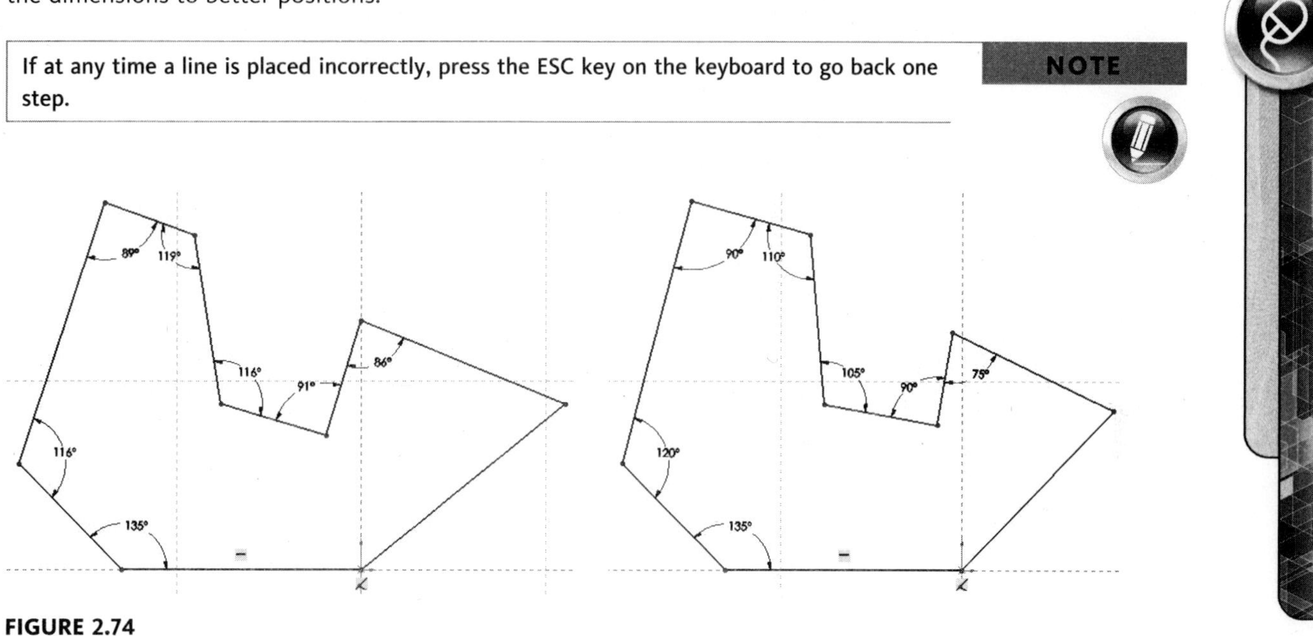

FIGURE 2.74

STEP 3

Use the aligned dimensions for the true size of every line segment. Place the dimensions close to the line segments. To modify the dimension, select it and change the primary value then accept by clicking OK (Figure 2.75).

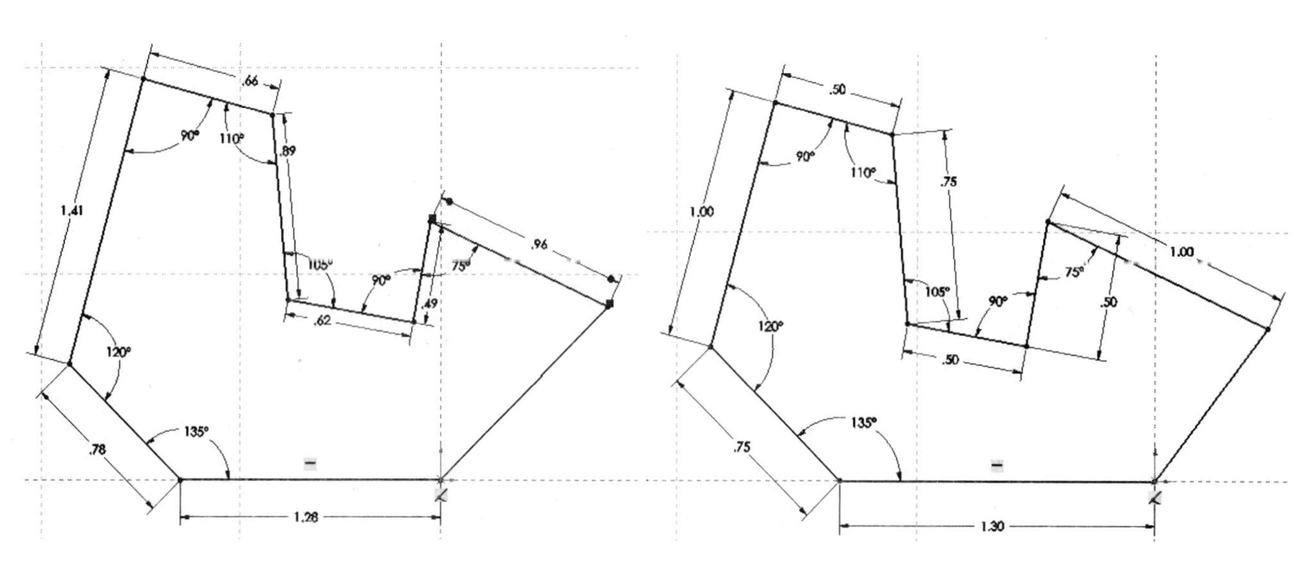

FIGURE 2.75

STEP 4

Create the extruded boss at a distance of .10in and preview the extrusion. The finished object is shown in Figure 2.76 on the right.

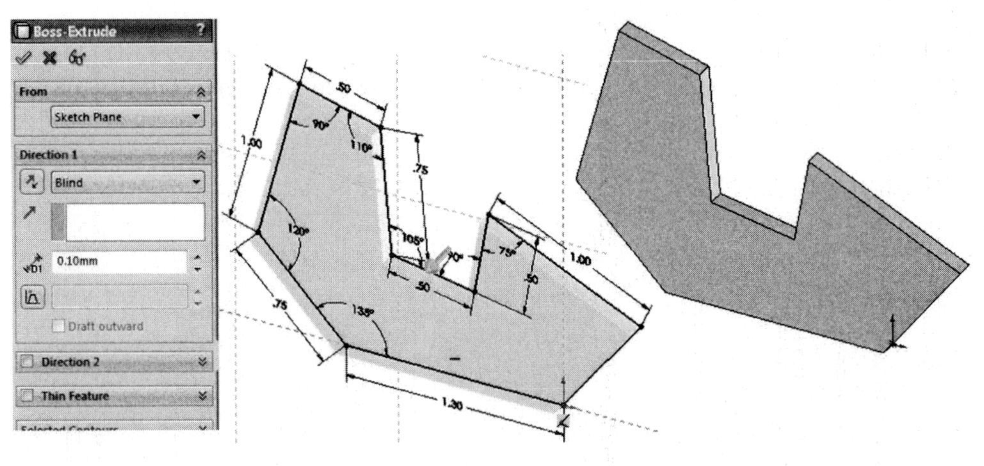

FIGURE 2.76

ADDITIONAL RELATIONS

The following relations may not be used on a regular basis. However, it is good to be aware of these additional relations in the event you need to apply a midpoint, intersection, or a fix relation. Use the following Try It exercises to become better acquainted with these additional relations.

The Midpoint Relation

TRY IT!

Open the drawing file SWT_Relation_13 (Midpoint). This exercise is designed to apply a midpoint relation between a line and a point.

Click the Add Relations button and pick the horizontal baseline at "A" and the origin at "B" as shown in Figure 2.77. When the list of possible relations appears, click Midpoint.

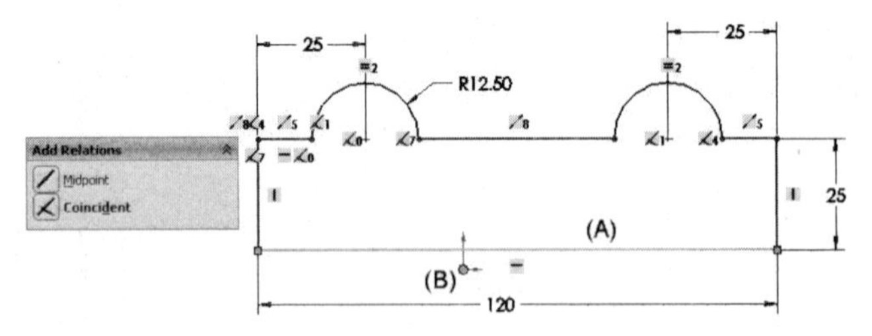

FIGURE 2.77

The results are shown in Figure 2.78. The midpoint of the base horizontal line is now locked on to the origin.

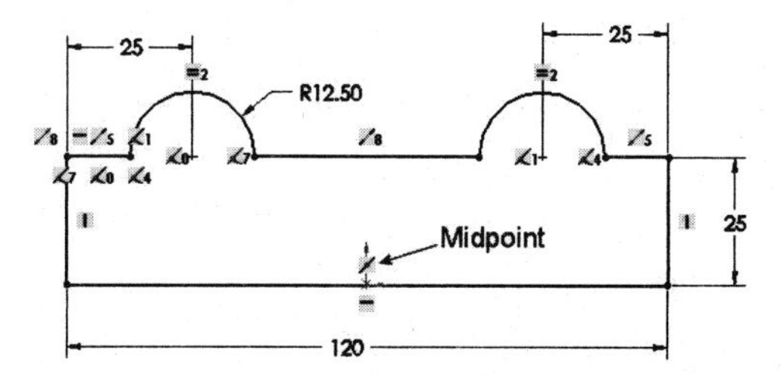

FIGURE 2.78

The Intersection Relation

Open the drawing file SWT_Relation_14 (Intersection). This exercise is designed to apply an intersection relation between three line segments.

TRY IT!

Click the Add Relations button and select the endpoint of the line at "A" followed by lines "B" and "C" as shown in Figure 2.79 on the left. When the list of possible relations appears, click in intersection. The results are illustrated in the following figure on the right. The endpoint of the line at "A" now intersects lines "B" and "C."

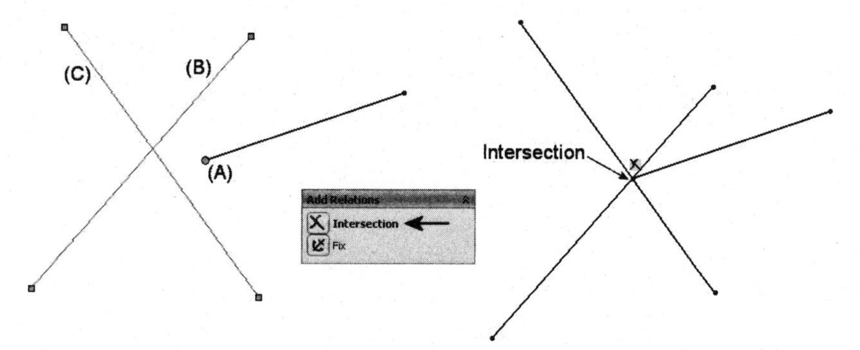

FIGURE 2.79

The Fix Relation

Open the drawing file SWT_Relation_15 (Fix). This exercise is designed to apply a fix relation to all sketch entities.

TRY IT!

The Fix relation locks the size and position of any entity that is selected. In one sense, objects that are fixed do not require dimensions. However, there is no way of making changes to any of the fixed entities. As a result, it is highly recommended to be careful when applying the fixed relation to any entity or entities. To apply this relation, first click the Add Relations button and select all entities. Choose the Fix relation from the PropertyManager as shown in Figure 2.80.

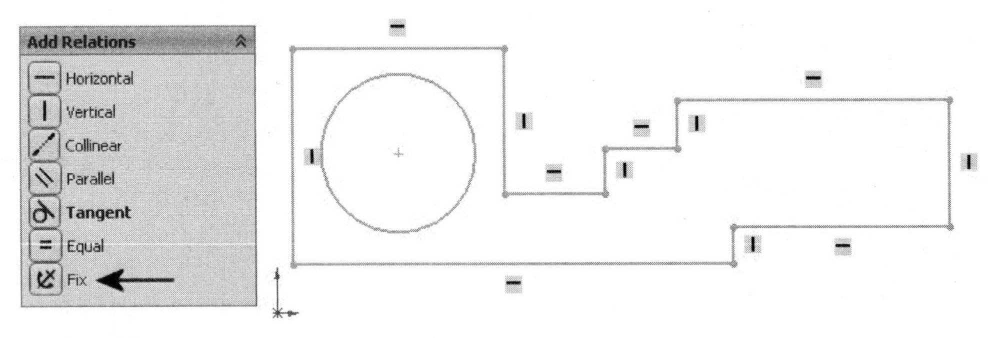

FIGURE 2.80

The results are displayed in Figure 2.81 where all objects display the anchor glyph signifying the fixed relation has been applied. Notice that even though the lower left corner of the sketch does not intersect the origin, all fixed entities have their color changed to black signifying the sketch is fully defined even without dimensions.

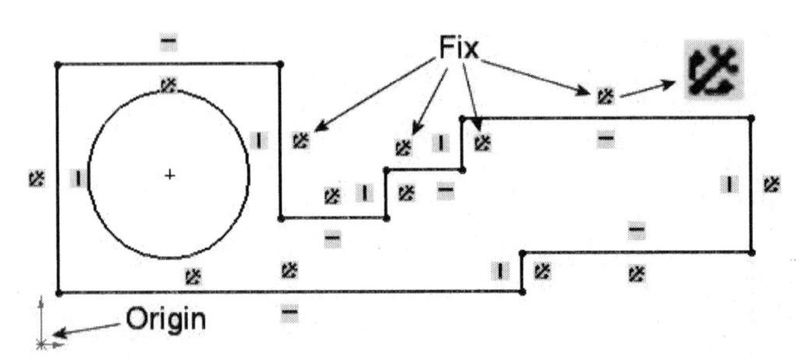

FIGURE 2.81

CREATING EQUATIONS IN SKETCHES

Another powerful feature is the ability to incorporate mathematical formulas in sketches. These formulas are called equations. For instance, you want to control the length of a rectangle such that the height is always half of the length. Use this next exercise to demonstrate how to create equations.

TRY IT!

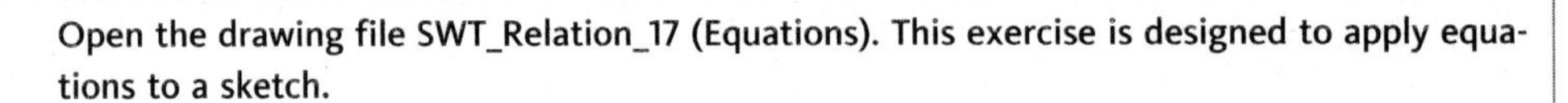

Open the drawing file SWT_Relation_17 (Equations). This exercise is designed to apply equations to a sketch.

The following sketch already exists; also, the 4.00 horizontal dimension is already created. Place another vertical smart dimension as shown in Figure 2.82. However, instead of accepting its value, click the down arrow and select Add Equation from the list.

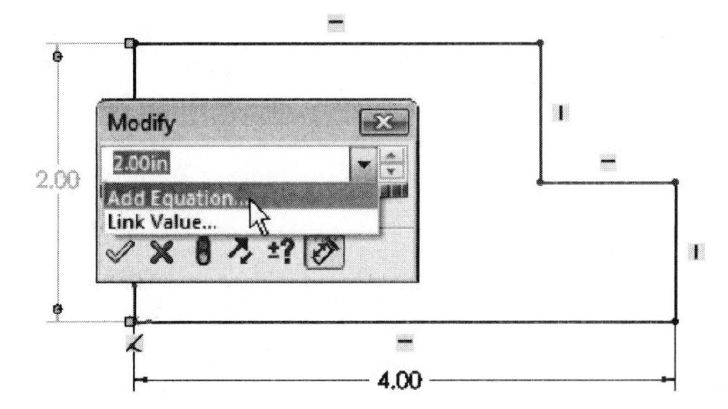

FIGURE 2.82

When the Add Equation dialog box appears, click on the existing 4.00 Horizontal dimension. This will add the parameter on the right side of the equation (D1@Sketch1). Since the height must always remain half of the length, add /2 to the equation as shown in Figure 2.83. The complete equation is shown as the following; "D1@Sketch1"/2.

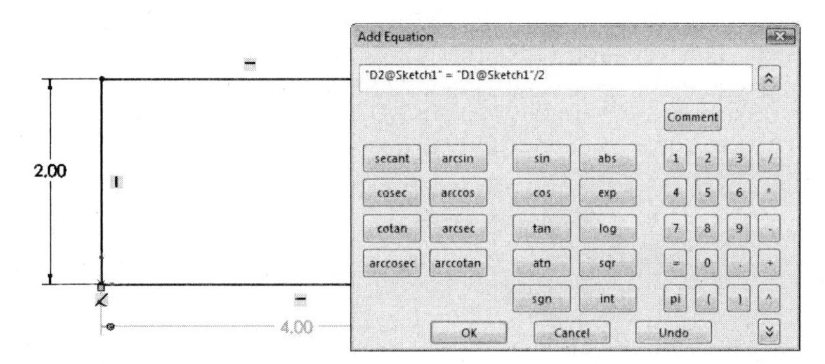

FIGURE 2.83

Clicking the OK button will display the Equations dialog box where you can verify the equation and the value that it evaluates to. Clicking the OK button in the main Equations dialog box will display the sigma symbol next to the 2.00 vertical dimension as shown in Figure 2.84 on the right. This symbol is used to identify a dimension value that is being controlled by an equation.

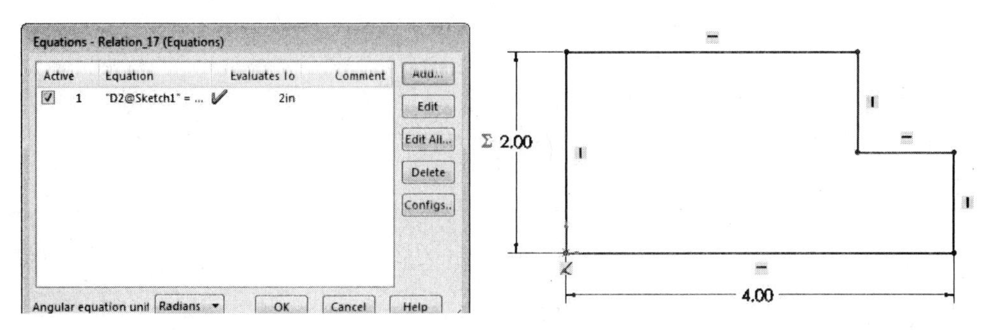

FIGURE 2.84

Place another vertical dimension; this dimension with a value of 1.00 will document the height of the slot. Click Add Equation from the Modify dimension dialog box and click on the 2.00 vertical dimension. In the Add Equation dialog box, add /2 to the end of the parameter to complete the equation. This equation states that the height of the step is to be exactly half of the overall height of the object (Figure 2.85).

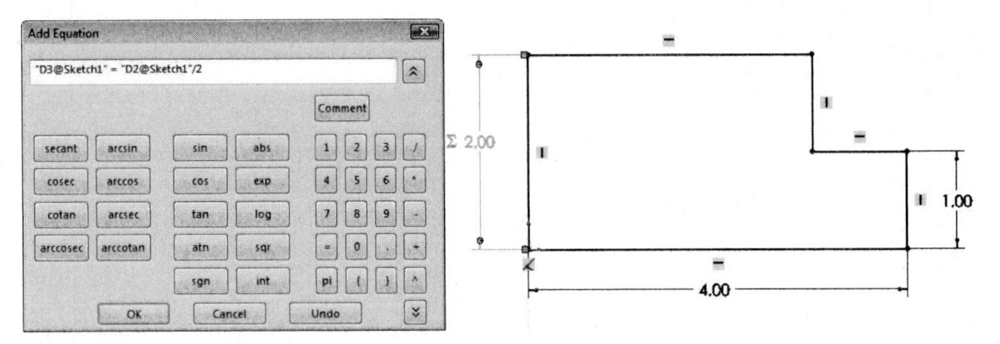

FIGURE 2.85

Add one more dimension that documents the horizontal length of the step. In the Add Equation dialog box, click on the 1.00 vertical dimension that documents the height of the step. This equation states that as the height of the step changes, the length of the step will be the same value (Figure 2.86).

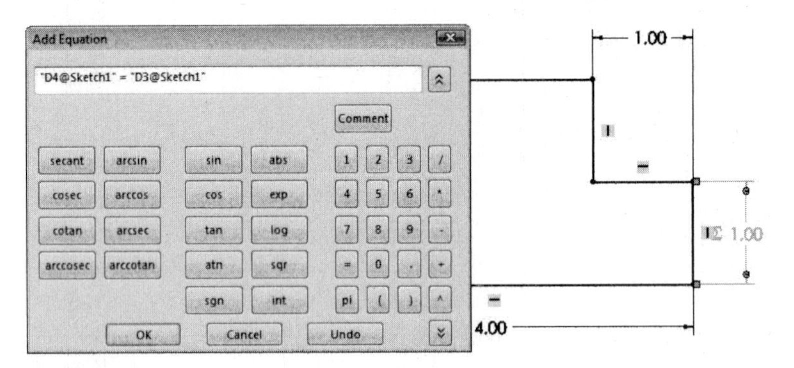

FIGURE 2.86

Verify that three equations are present in the Equations in the dialog box. Also verify the three sigma symbols that are present on the dimensions being controlled by equations (Figure 2.87).

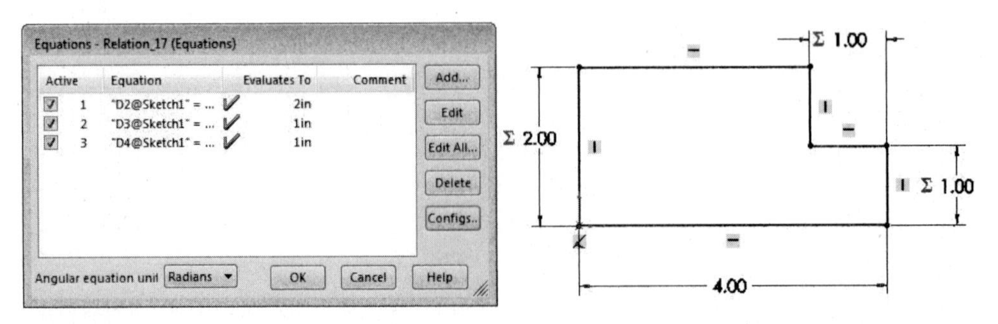

FIGURE 2.87

Check the performance of the equations by changing the overall length of the sketch from 4.00 to 11.00. Notice all values recalculate to new values based on the equations that all reference the length dimensions as shown in Figure 2.88 on the right.

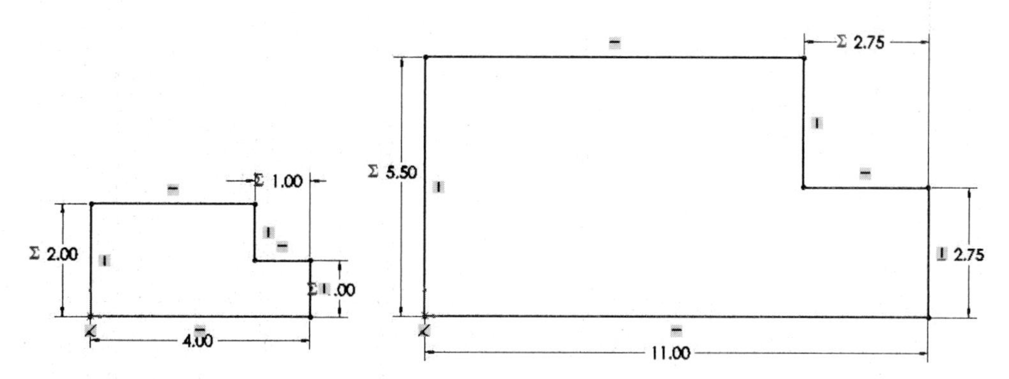

FIGURE 2.88

VISIBILITY OF RELATIONS

Applying numerous relations to a sketch can make the design appear busy and difficult to read especially when smart dimensions are placed. You do have the ability to control the display of relations by clicking View in the pull-down menu followed by Sketch Relations. Clicking this item will turn off all relations in the sketch. Click on Sketch Relations a second time to turn the relations back on as shown in Figure 2.89.

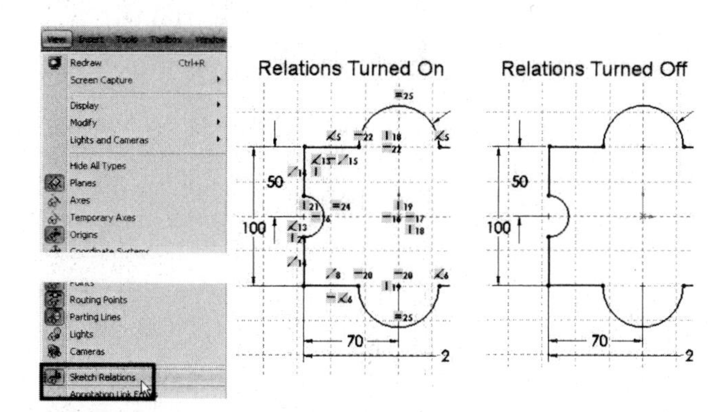

FIGURE 2.89

MODIFYING A SKETCH

After a sketch is converted into a feature, you can easily return to the sketch for making changes by right-clicking on the feature in the FeatureManager and clicking Edit Sketch from the menu as shown in Figure 2.90 on the left. Once changes to the sketch have been made, click the Rebuild tool to return back to the feature.

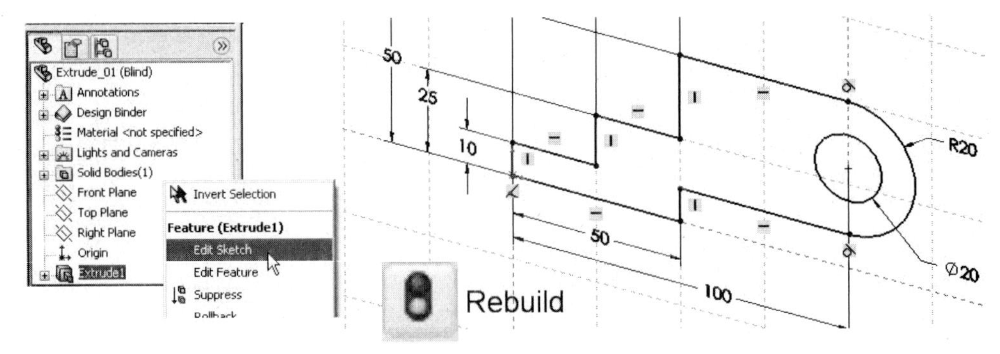

FIGURE 2.90

MODIFYING A FEATURE

You can also make changes to a part feature that was already created by right-clicking on the feature located in the FeatureManager and clicking Edit Feature as shown in Figure 2.91 on the left. This will display a preview of the current feature distance as shown in the following figure on the right. Making changes to such properties as distances and accepting these values will automatically update the feature.

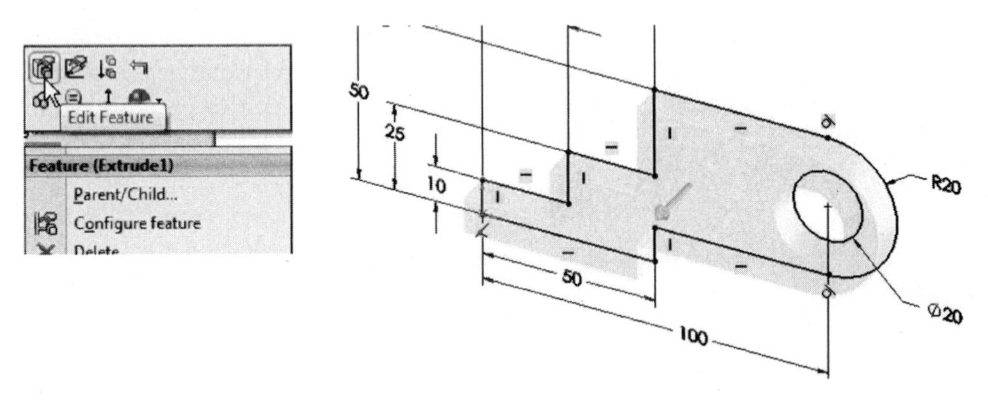

FIGURE 2.91

TRIMMING A SKETCH

Whether you are creating simple or complex sketches, a very operation tool is to partially delete entities by trimming. Choose the Trim tool by clicking this icon from the CommandManager as shown in Figure 2.92 on the left. You could also choose this tool by choosing the Tools pull-down menu followed by Sketch Tools where the Trim tool can be found as shown in the following figure in the middle. Five Try It exercises are available that will allow you to practice the following options, namely, Power trim, Corner, Trim away inside, Trim away outside, and Trim to closest. These will be discussed in greater detail.

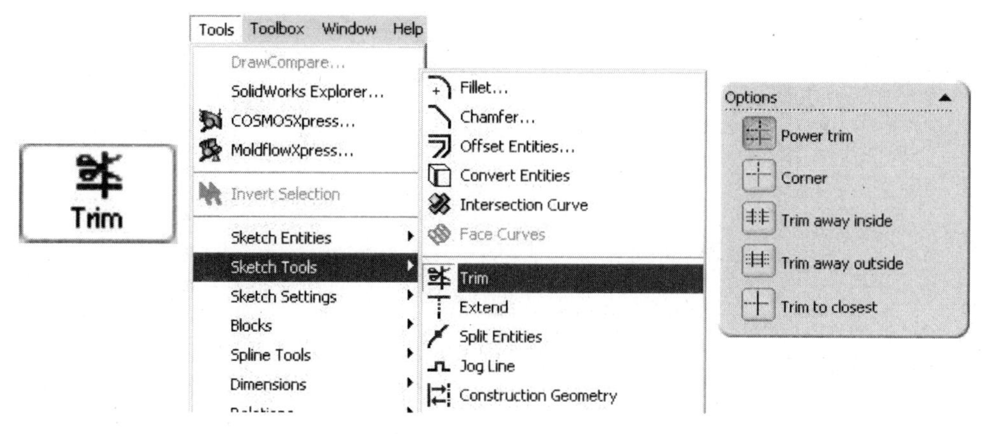

FIGURE 2.92

TRY IT!

Open the drawing file SWT_Trim_Sketch. This exercise is designed to use various options for trimming a sketch.

Power Trim

Activate the Trim tool and choose Power trim from the PropertyManager. Press and hold down the left mouse button as you drag across the area of the cut curve as shown in Figure 2.93. As the cut curve hits an entity, it will disappear from the screen. Drag the cut curve through the four line entities in order to trim these entities. The results of this trimming operation are displayed in the following figure on the right.

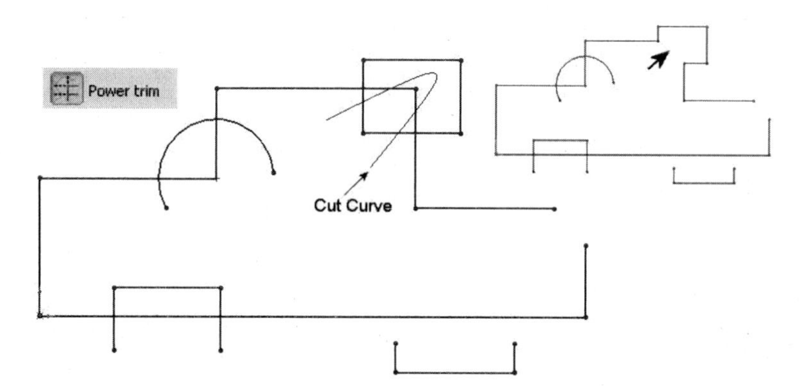

FIGURE 2.93

Corner Trim

This next Trim option is designed to create a corner between two entities. Activate Trim and pick Corner from the list of options from the PropertyManager. Pick the edge of the lines at “A” and “B” as shown in Figure 2.94. This will join both endpoints of the selected lines to form a corner as shown in the following figure on the right.

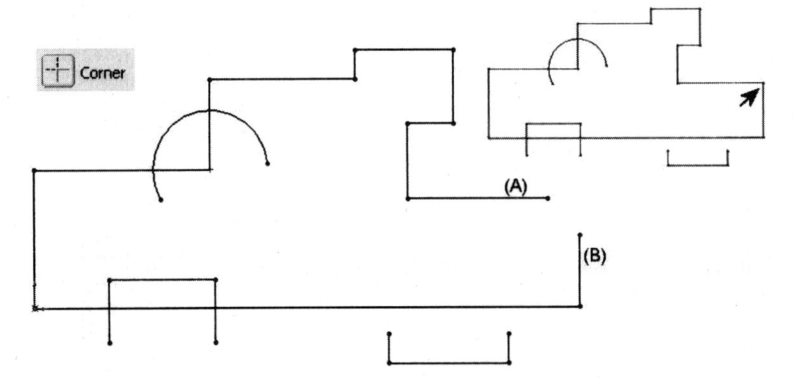

FIGURE 2.94

Trim Away Inside

This next option will demonstrate how to trim away any entities considered inside of two cutting edges. Activate Trim and pick the Trim away inside option. Pick vertical lines "A" and "B" as the cutting edges. Then pick line "C" as the entity to trim away inside as shown in Figure 2.95. The results are illustrated in the following figure on the right.

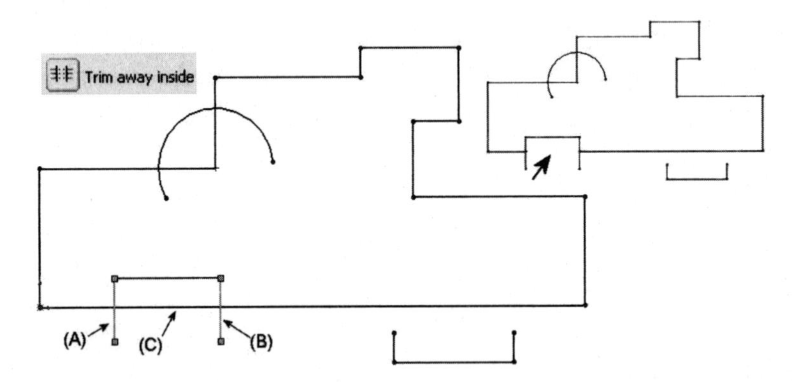

FIGURE 2.95

Trim Away Outside

This next Trim option will illustrate how to trim away outside edges based on two cutting edges. Activate Trim and pick the Trim away outside option. Pick lines "A" and "B" as cutting edges and select the edge of the arc at "C" as shown in Figure 2.96. The results are illustrated in the following figure on the right with the outer part of the arc being removed from the sketch.

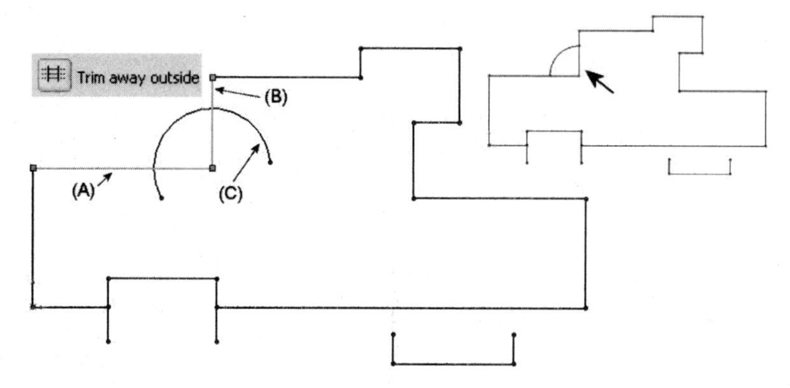

FIGURE 2.96

Trim to Closest

This last trim option will illustrate how to trim to the closest object. This operation will actually demonstrate how to extend objects through this option. Activate Trim and select the Trim to closest option. Instead of picking lines, you will need to press and drag the endpoint of a line to intersect with another. Press and drag on the point at "A" and drag your cursor to intersect line "C"; the line should extend. Perform this same operation on the endpoint at "B." The results should appear similar to Figure 2.97 on the right.

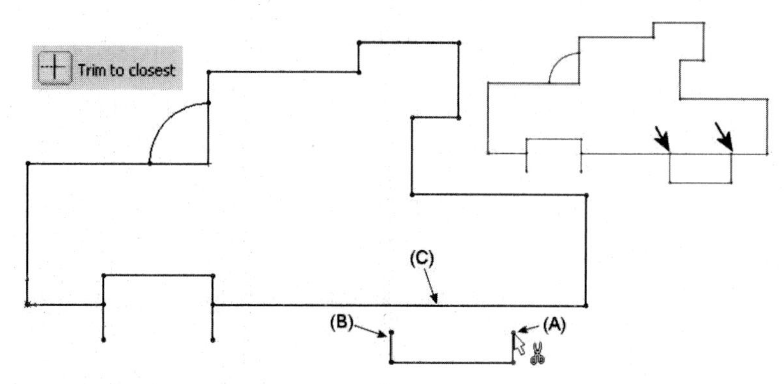

FIGURE 2.97

Finally, use Power Trim on the remaining objects until your sketch appears similar to Figure 2.98.

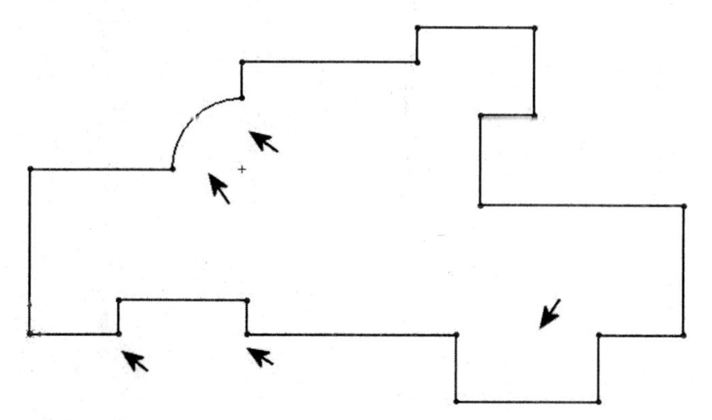

FIGURE 2.98

CREATING ADDITIONAL SKETCH ENTITIES

In addition to sketching, lines, circles, and arcs, additional sketching tools are available to help construct other entity types as shown in Figure 2.99.

Parallelogram allows you to construct a rectangle at an angle. The lines that make up the rectangle remain perpendicular to the angle controlling the rectangle.

Polygon allows you to construct a multiside shape based on a circle reference. Edges of the polygon are automatically made tangent to the edge of the circle

Ellipse allows you to construct an ellipse based on a major (2.00) and minor diameter (1.00) as shown in the following figure.

Text allows you to incorporate words or sentences of text in a model as shown in the following figure. The text entity requires a curve to sketch the text on. This curve can be either in the form of line or arc segments. The placing of text in a part model will be covered in greater detail in a later chapter.

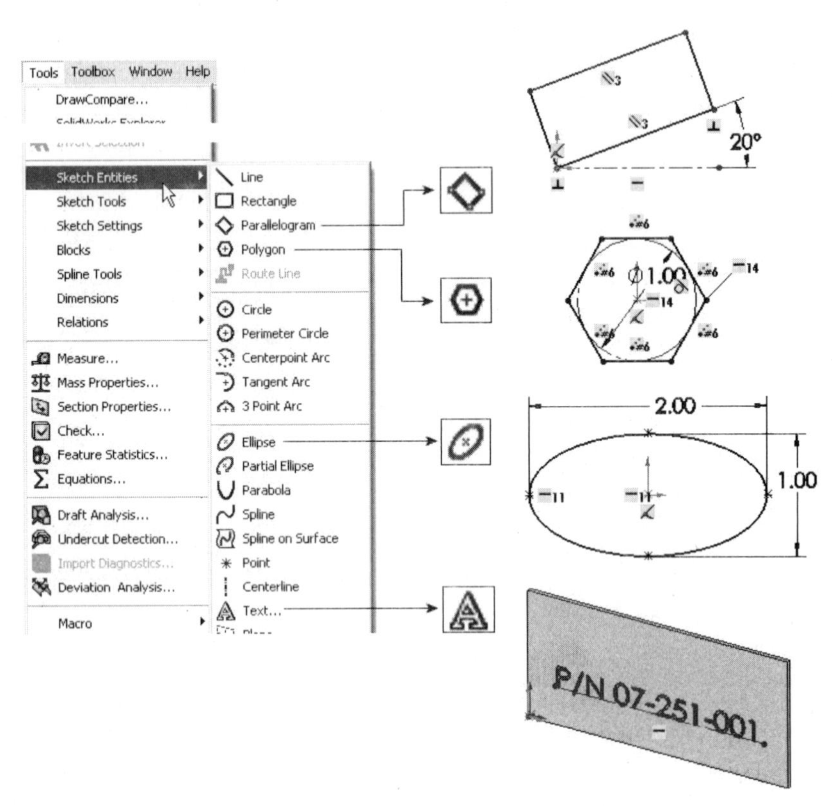

FIGURE 2.99

TUTORIAL EXERCISE: SWT_CH02_04.SLDPRT

Description: Rocker Arm

Units: English

This tutorial exercise is designed to create a part model that consists of numerous geometric shapes (Figure 2.100). Proper relations need to be applied along with dimensions to fully define the shape.

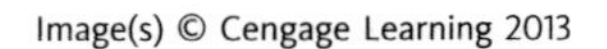

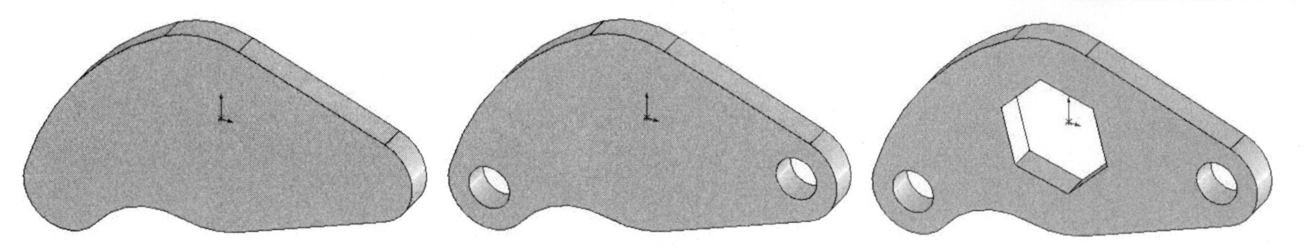

FIGURE 2.100

STEP 1

Start a new sketch on the front plane. Construct three circles as shown in Figure 2.101.

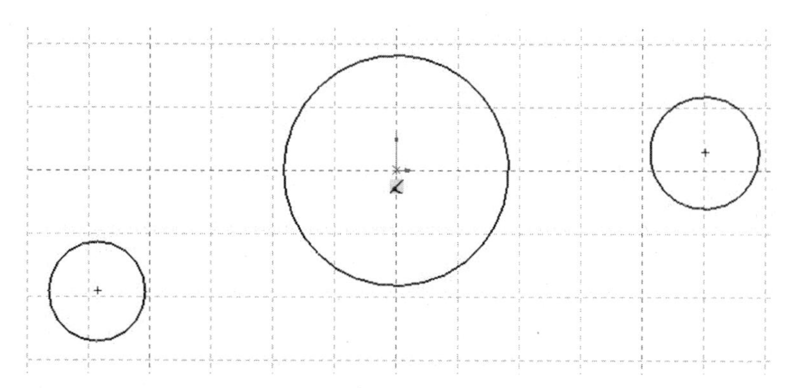

FIGURE 2.101

STEP 2

Construct the two line segments as shown in Figure 2.102. Apply tangent relations between these lines and the circles as shown in the following figure. Select *Add Relations* for each of the desired tangents. Select Tangent and then accept it with the green checkmark.

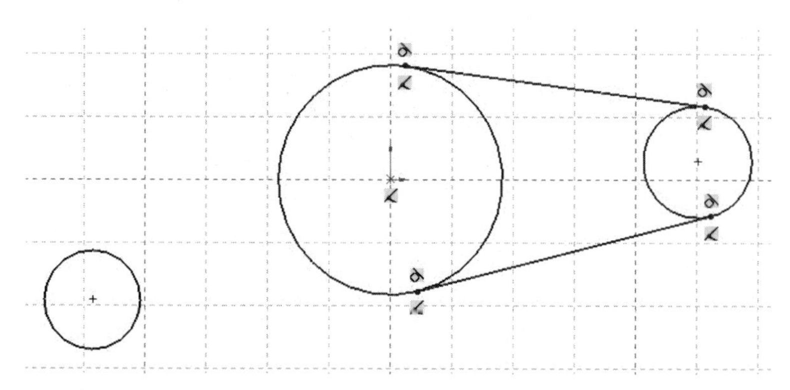

FIGURE 2.102

STEP 3

Construct two arc segments using the 3 Point Arc tool as shown in Figure 2.103. Select *Trim Entities*, then *Trim To Closest*, and trim away arc segments "A," "B," "C," and "D" as shown in the following figure.

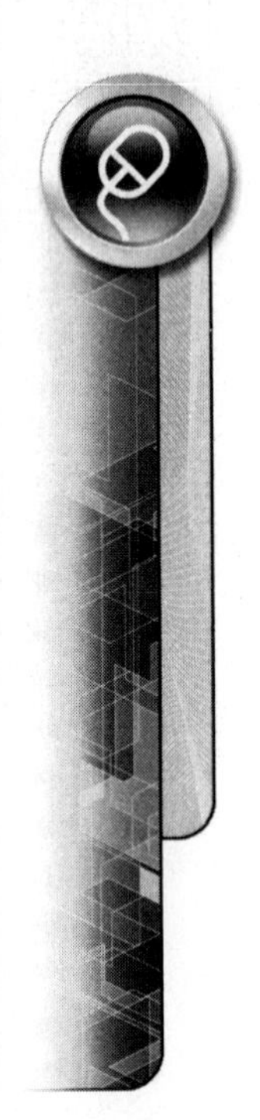

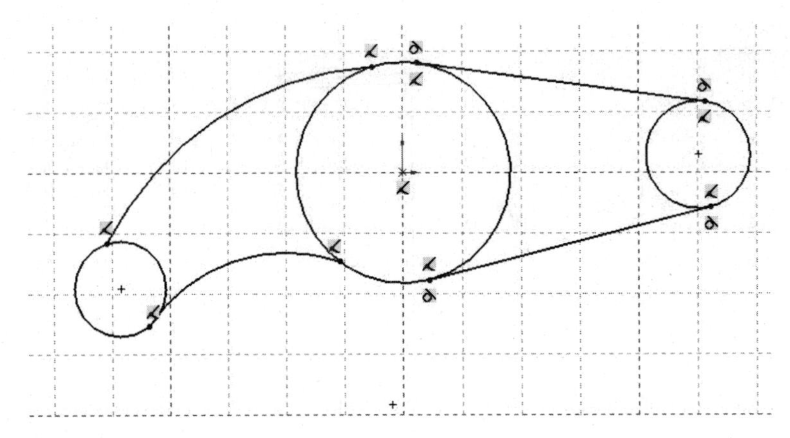

FIGURE 2.103

STEP 4

Your model should appear similar to Figure 2.104.

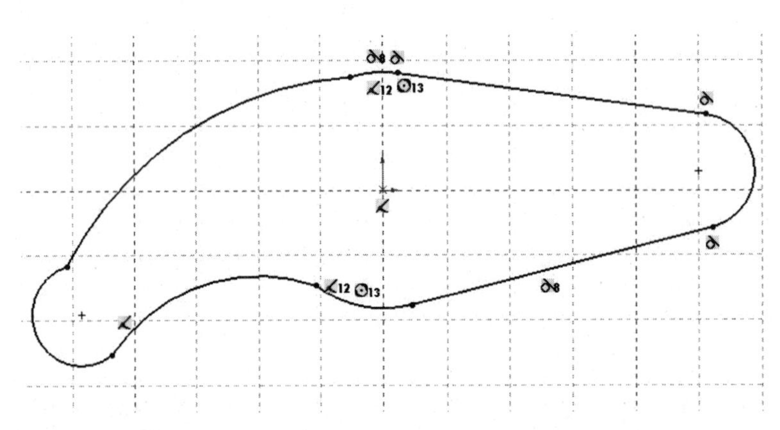

FIGURE 2.104

STEP 5

Place tangent relations (with the Add Relations command) between all arc segments. Also apply a horizontal relation between points 28 (the center of the right arc) and the origin (Figure 2.105).

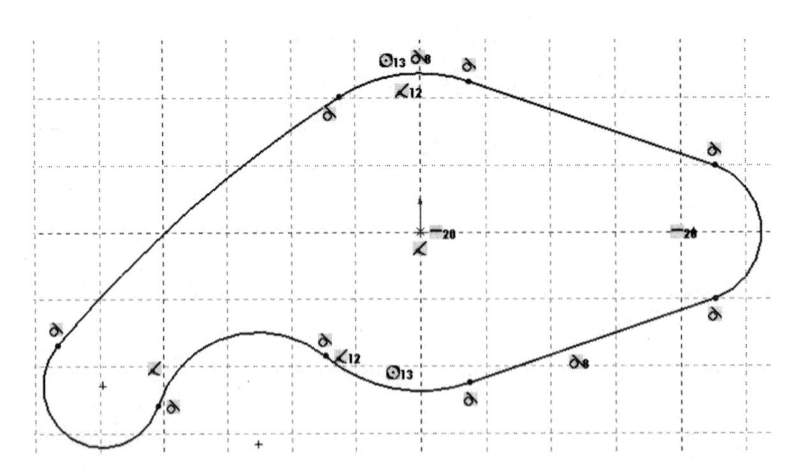

FIGURE 2.105

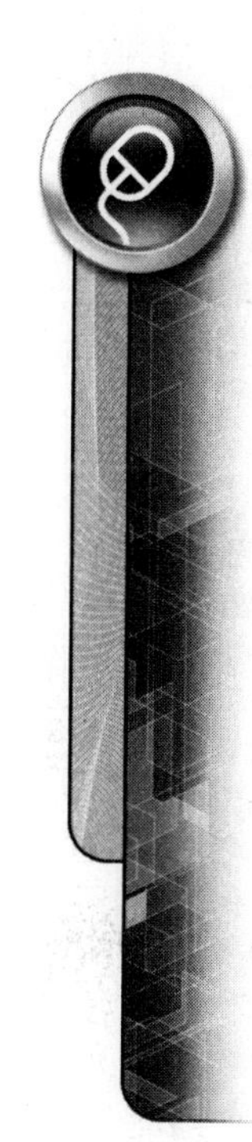

STEP 6

Apply an equal relation between both smaller arc segments (far left and far right) (Figure 2.106).

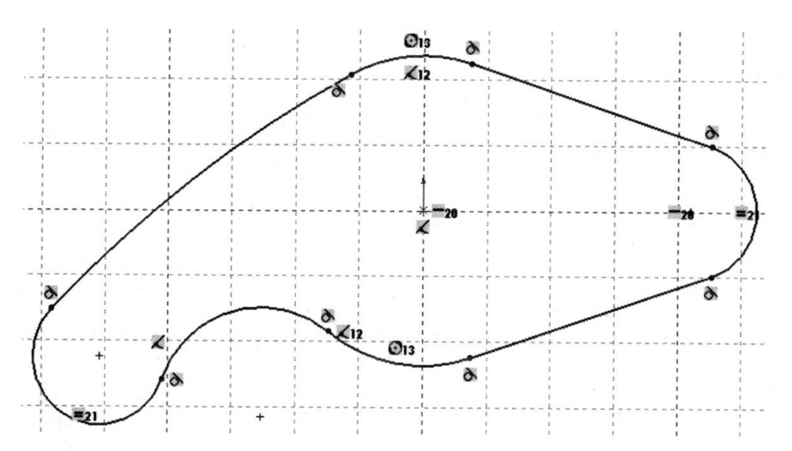

FIGURE 2.106

STEP 7

Add radius dimensions to large and small arc segments as shown in Figure 2.107.

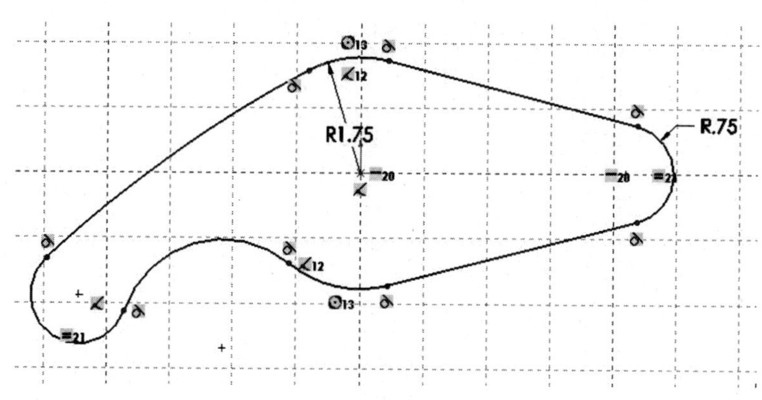

FIGURE 2.107

STEP 8

Add linear dimensions identifying the centers of both small arcs in relation to the center of the larger arcs as shown in Figure 2.108.

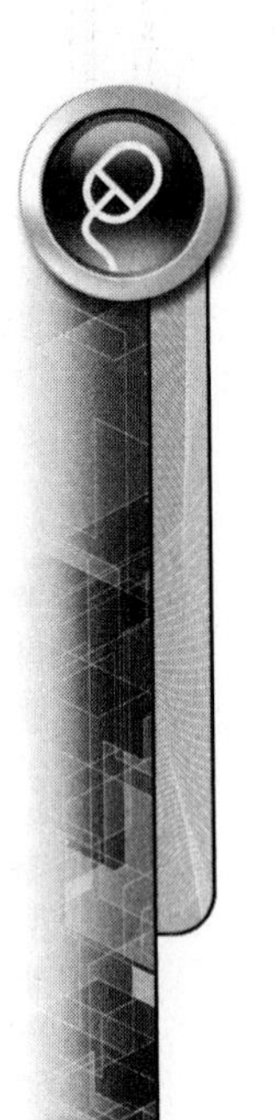

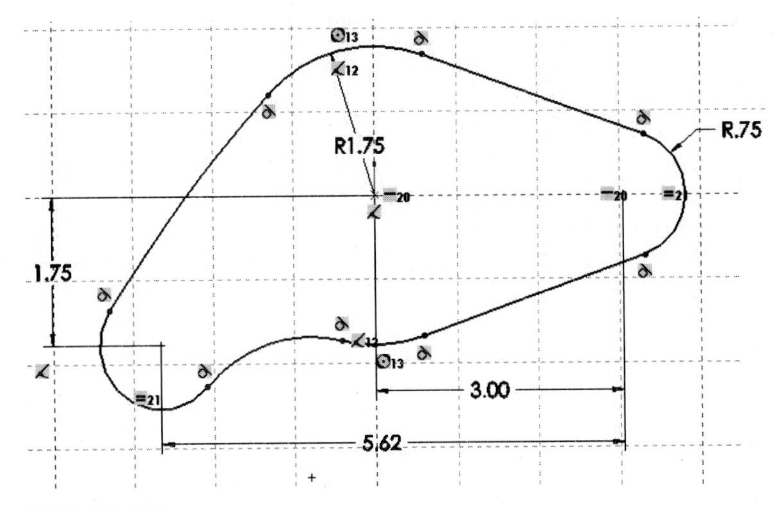

FIGURE 2.108

STEP 9

Add the R5.00 and R1.50 radius dimensions as shown in Figure 2.109. This completes the outline of the rocker arm.

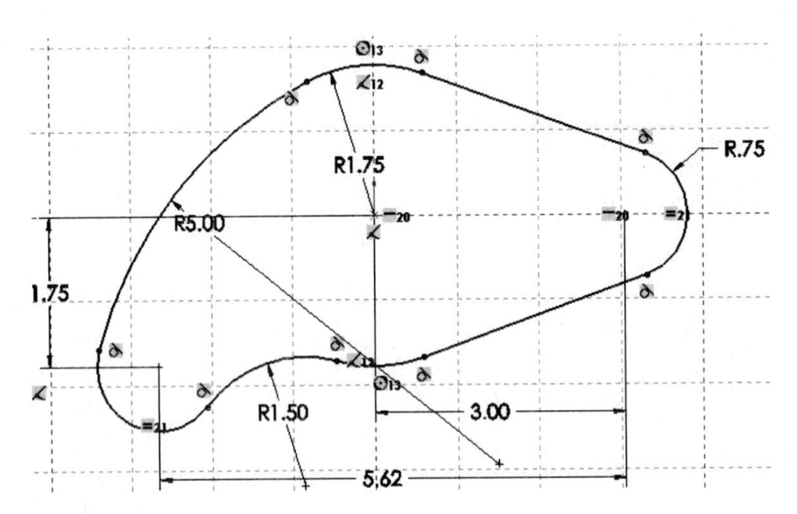

FIGURE 2.109

STEP 10

Create an extruded boss (Extrude command on the Features tab) from the fully defined sketch as shown in Figure 2.110. Set Direction 1 to Blind and the Distance to .50 as shown in the PropertyManager.

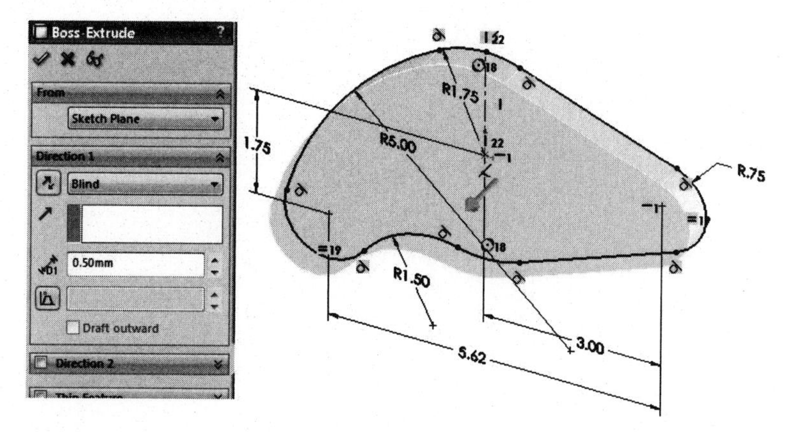

FIGURE 2.110

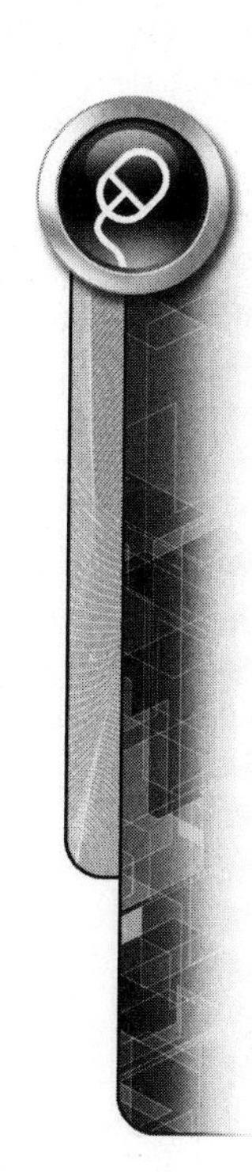

STEP 11

The complete perimeter outline of the rocker arm is shown in Figure 2.111. Then create a new sketch plane on the top face of the rocker arm.

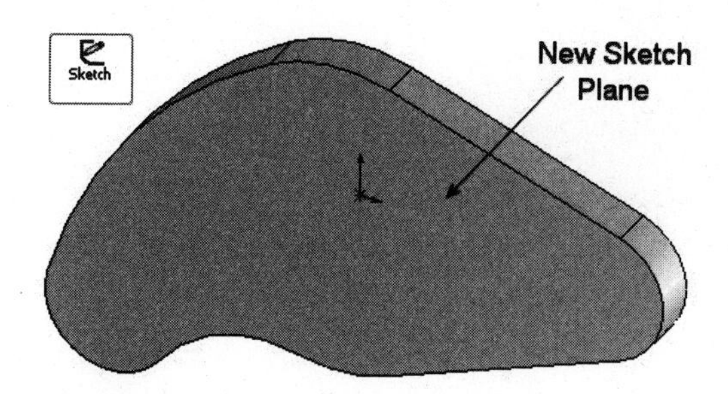

FIGURE 2.111

STEP 12

Use the Orientation dialog box to switch to the **Normal To* view as shown in Figure 2.112.

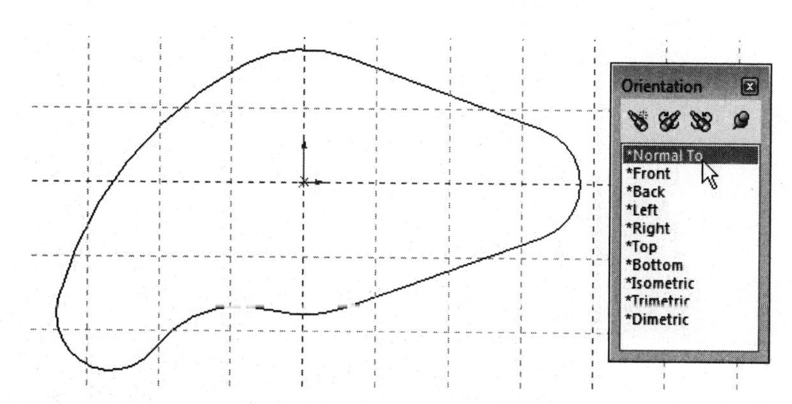

FIGURE 2.112

STEP 13

Add two circles using the center points of the R.75 arcs as shown in Figure 2.113. Add equal relations to both circles. Add the .75 diameter smart dimension as shown in the following figure.

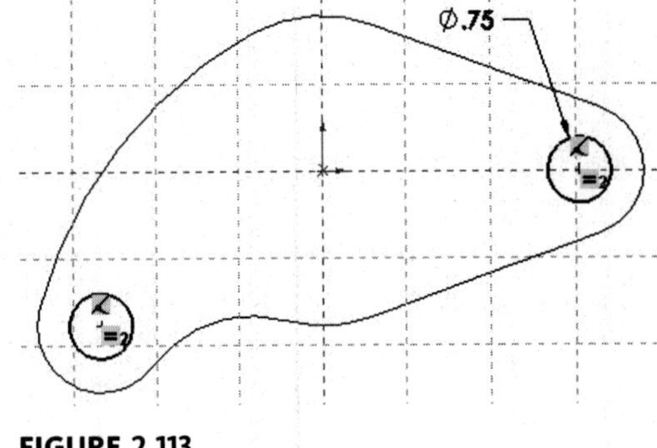

FIGURE 2.113

STEP 14

Switch to the Trimetric orientation view and activate the Extruded Cut tool. Set Direction 1 found under the PropertyManager to Through All. This will cut both holes through the entire part. You will notice a preview of this cutting operation as shown in Figure 2.114.

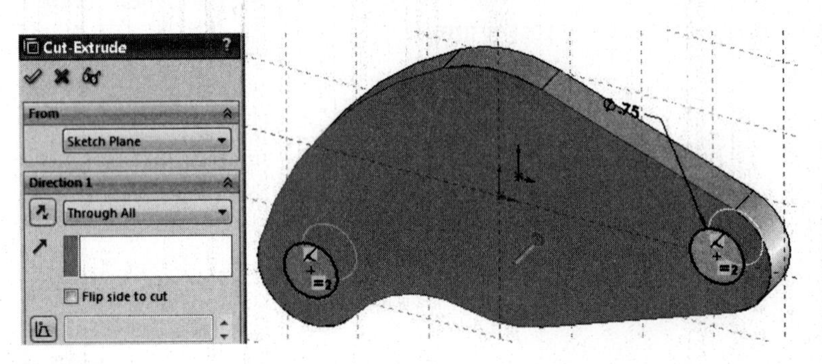

FIGURE 2.114

STEP 15

With both holes cut and extruded, create another sketch plane by selecting the front face of the part as shown in Figure 2.115 on the left. Switch your orientation to **Normal View* as shown in the following figure on the right. This last sketch plane will be used to create a hexagon in the middle of the part.

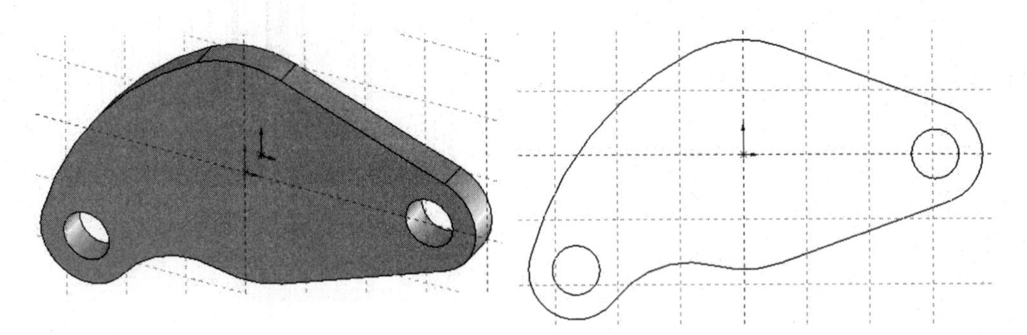

FIGURE 2.115

STEP 16

Add a six-sided hexagon using the Polygon tool as shown in Figure 2.116. Polygon can be selected from the Tool pull-down menu in the Sketch Entities or can be selected from the sketch menu.

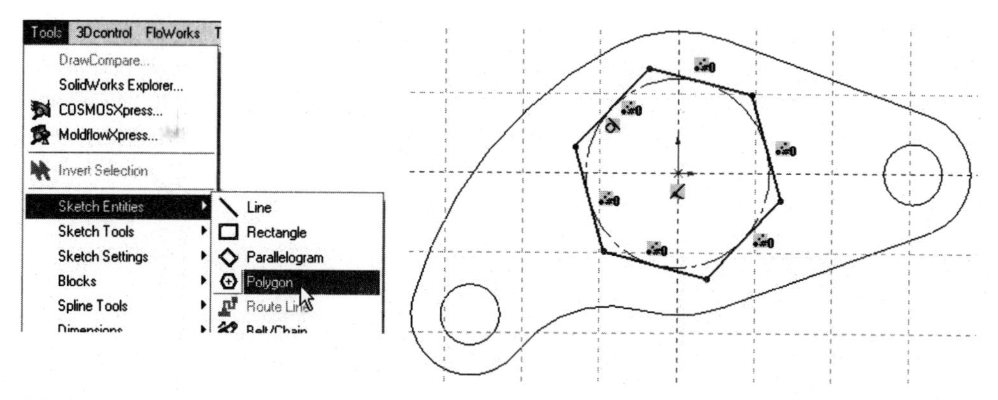

FIGURE 2.116

STEP 17

Add two construction lines (Line pull-down menu, Centerline) from the center of the polygon, one to the top of the arc, the other to the intersection of the polygon as shown in Figure 2.117 on the left. Add an angle dimension to the centerlines and a linear dimension that represents the distance across flats of the hexagon as shown in the following figure on the right. This should fully define the hexagon sketch.

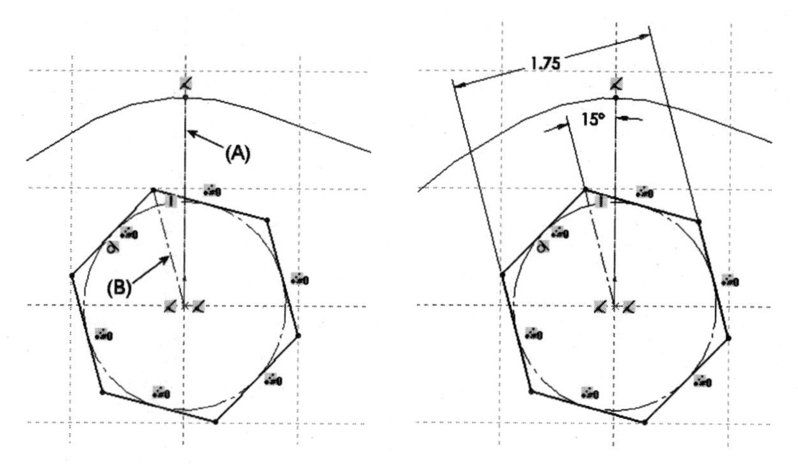

FIGURE 2.117

STEP 18

Switch back to the Trimetric orientation view and use the Extruded Cut tool to cut the hexagon shape *Through All* as shown in Figure 2.118.

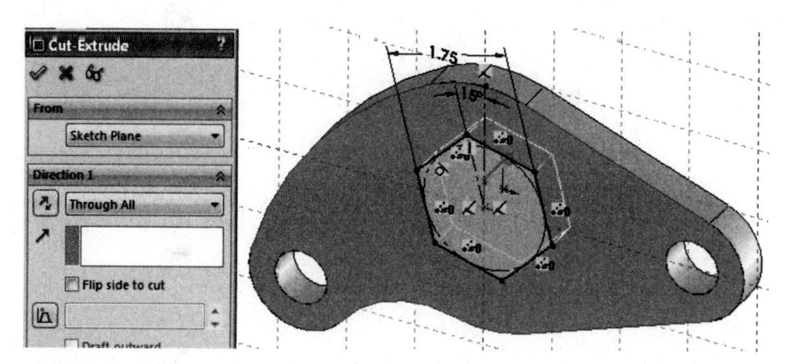

FIGURE 2.118

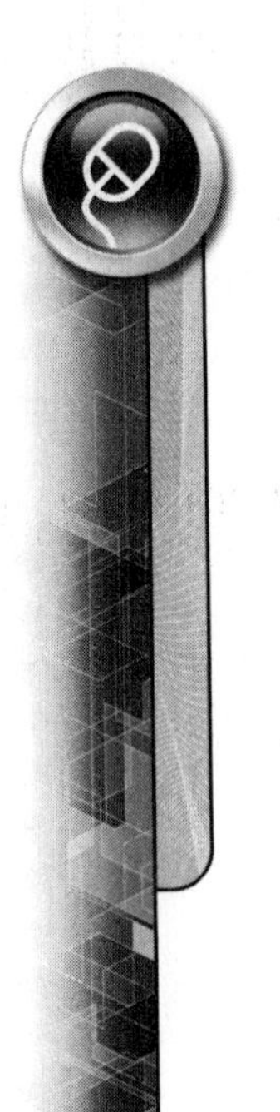

STEP 19

The completed rocker arm part is shown in Figure 2.119. It is considered good practice to rename features as they are being created to better names. In this way, you will be able to better distinguish the features from each other. Names are changed by double-clicking the feature name, keying in the new name, and pressing the Enter key.

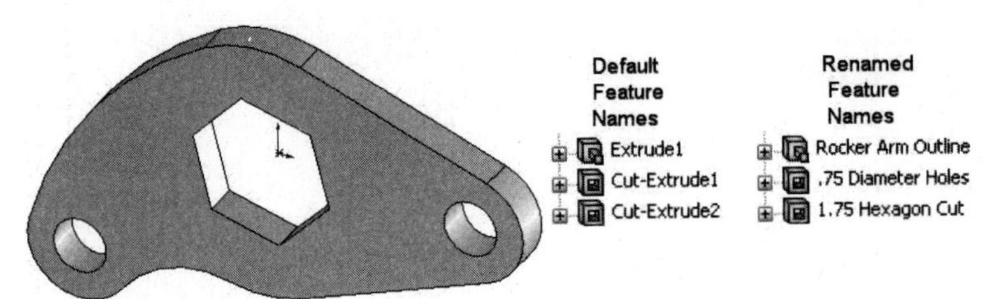

FIGURE 2.119

TUTORIAL EXERCISE: SWT_CH02_05.SLDPRT

Description: Cradle

Units: English

This tutorial exercise is designed to create a part model that consists of numerous geometric shapes. Proper relations need to be applied along with dimensions to fully define the shape as shown in Figure 2.120.

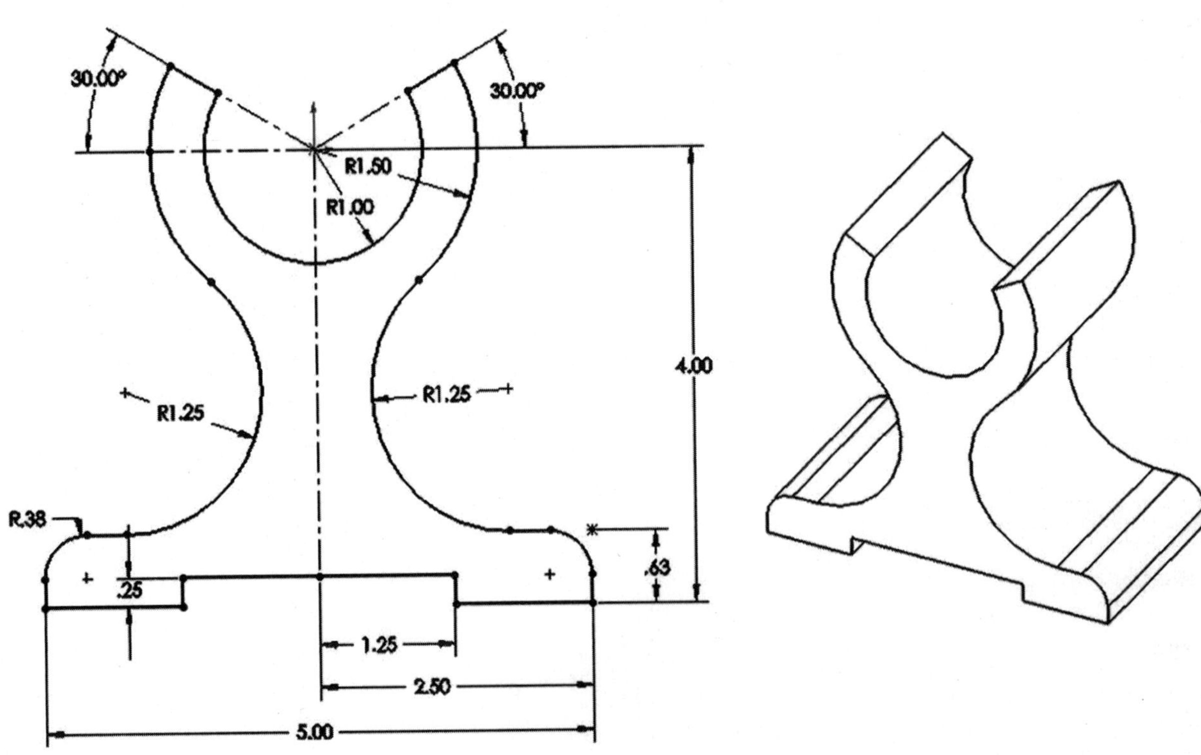

FIGURE 2.120

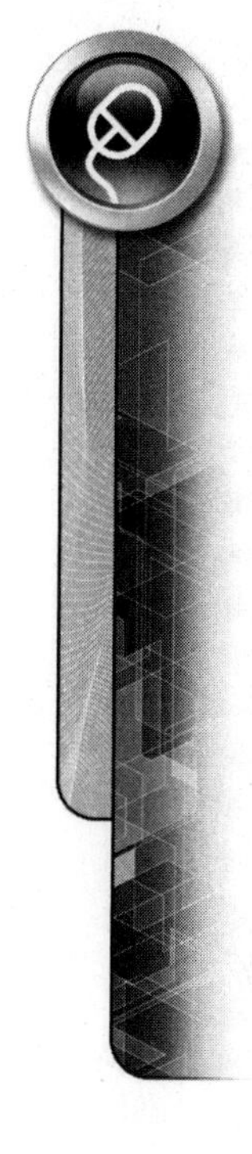

STEP 1

Create a new file setting the units to English (IPS). Then, in the front plane construct both arcs (create circles) with radii of 1.00 and 1.50 with the centers at the red origin as shown in Figure 2.121. Add dimensions to both circles. To change to radius dimensions, select a dimension, right-click, and choose *Display As Radius* from the *Display Options* menu.

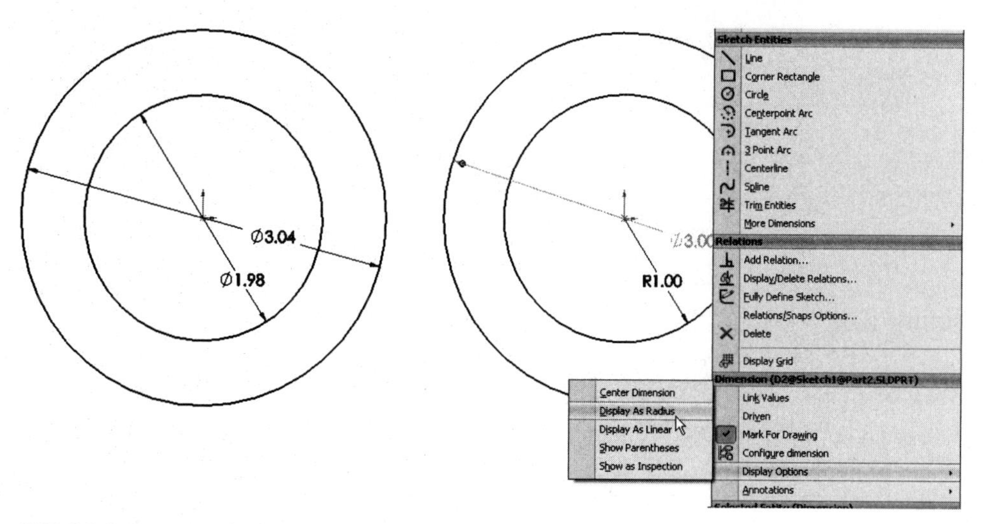

FIGURE 2.121

STEP 2

Construct the bottom shape of the cradle using lines as shown in Figure 2.122. See this image for the approximate position of this shape in relation to the upper circles. Then construct a center line with one endpoint at the red origin and the other at the midpoint of the lower line as shown in the following figure. Apply a Vertical Relation to this center line.

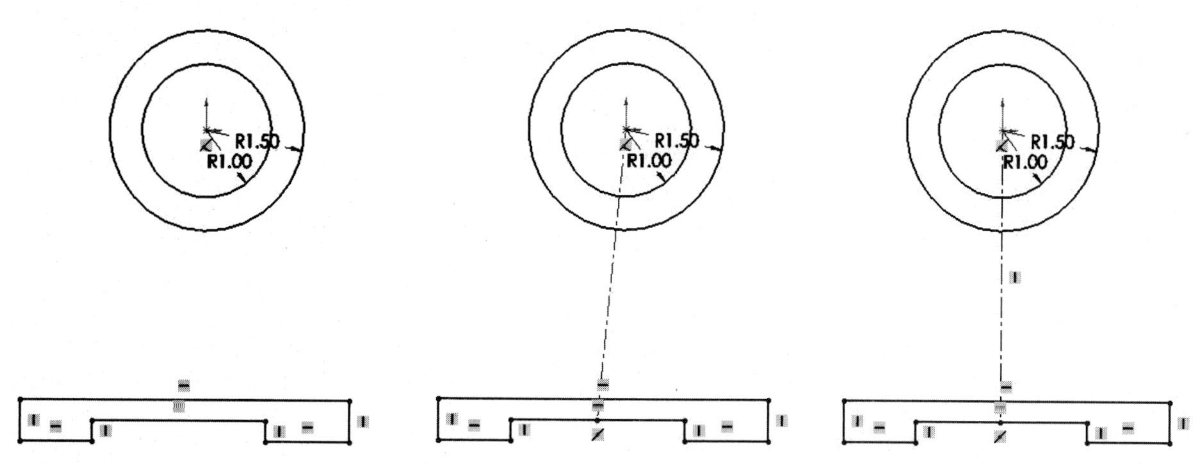

FIGURE 2.122

STEP 3

As relations are added to an object, at times, it can become difficult to work on adding geometry with the vast number of relations displaying. You can turn off the display of relations through the Hide/Show Items button (Glasses icon). Clicking on this button will activate a menu for turning off or on different display items. The following *Perpendicular* icon will turn off all relations (Figure 2.123).

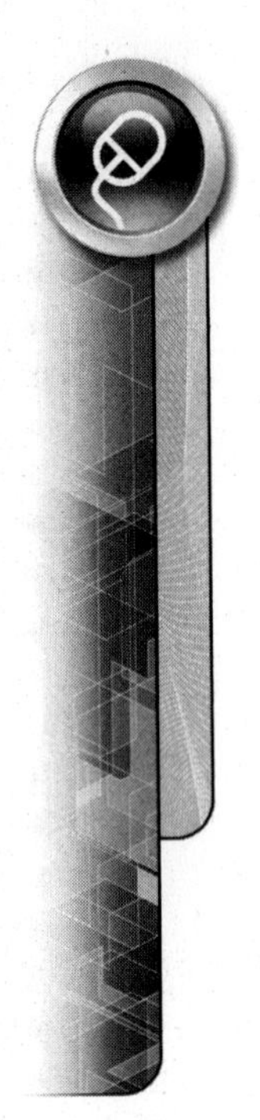

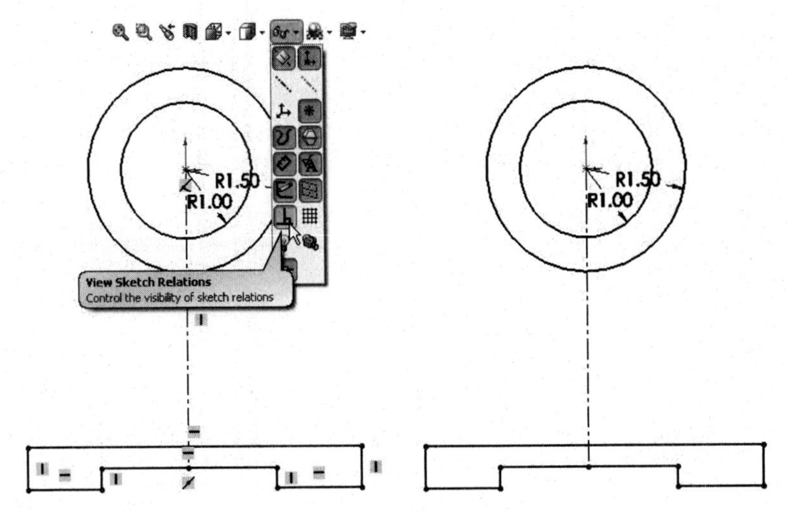

FIGURE 2.123

STEP 4

Add the 4.00 dimension to the sketch. *Zoom to Fit* the sketch if necessary in order that it appears similar as shown in Figure 2.124. Using *Add Relations* makes both bottom lines collinear with each other as shown in the following figure.

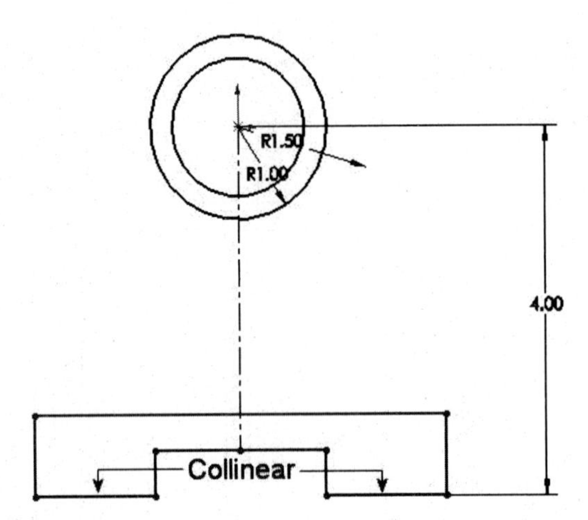

FIGURE 2.124

STEP 5

Add the remaining dimensions to the bottom segment of the part as shown in Figure 2.125.

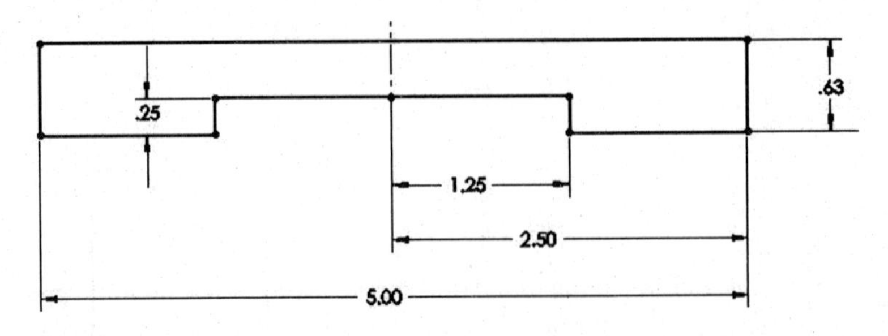

FIGURE 2.125

STEP 6

Add two R.38 sketch fillets to the two corners of the object as shown in Figure 2.126.

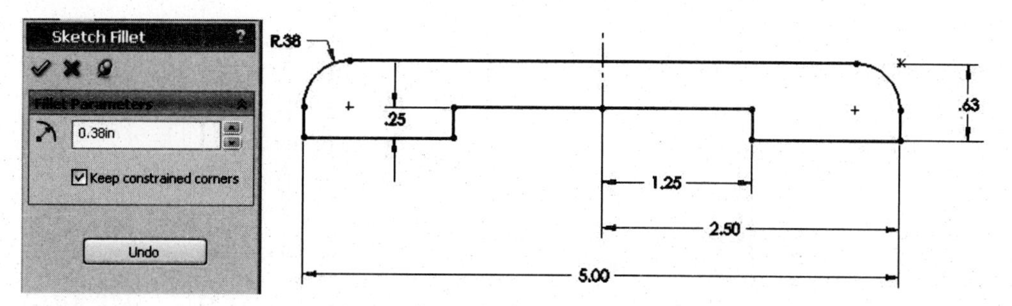

FIGURE 2.126

STEP 7

Add both arcs using the 3 Point Arc tool as shown in Figure 2.127 on the left. Add two 1.25 radius dimensions to both arcs as shown in the following figure on the right. You could also dimension one arc and make the other equal using an Equal Relation.

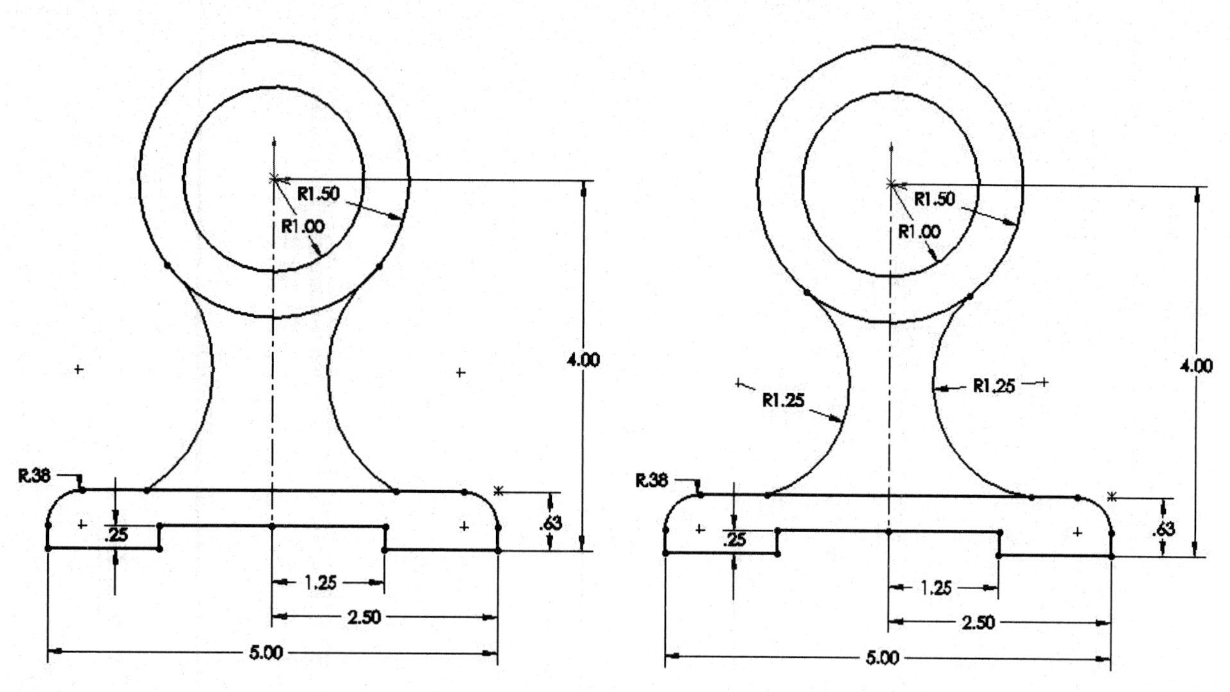

FIGURE 2.127

STEP 8

Select the *Add Relations* command and make both arcs tangent to the 1.50 radius circle at the top and the horizontal line segment at the bottom as shown in Figure 2.128.

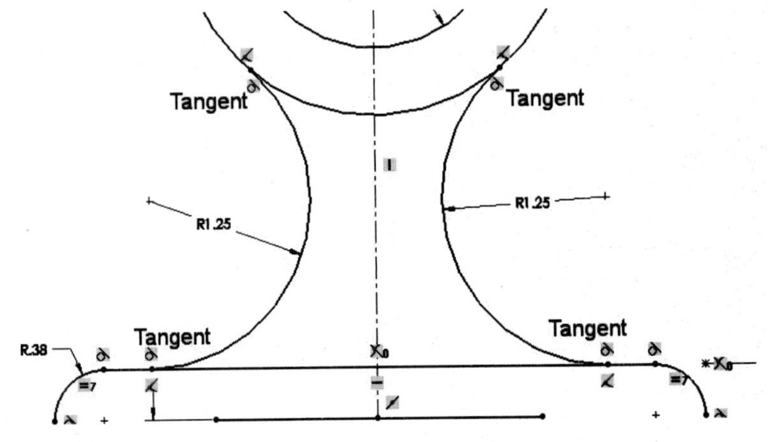

FIGURE 2.128

STEP 9

Select the *Trim Entities* tool to trim the arc at the top and line at the bottom as shown in Figure 2.129 on the left. Your display should appear similar to the illustration in the following figure on the right.

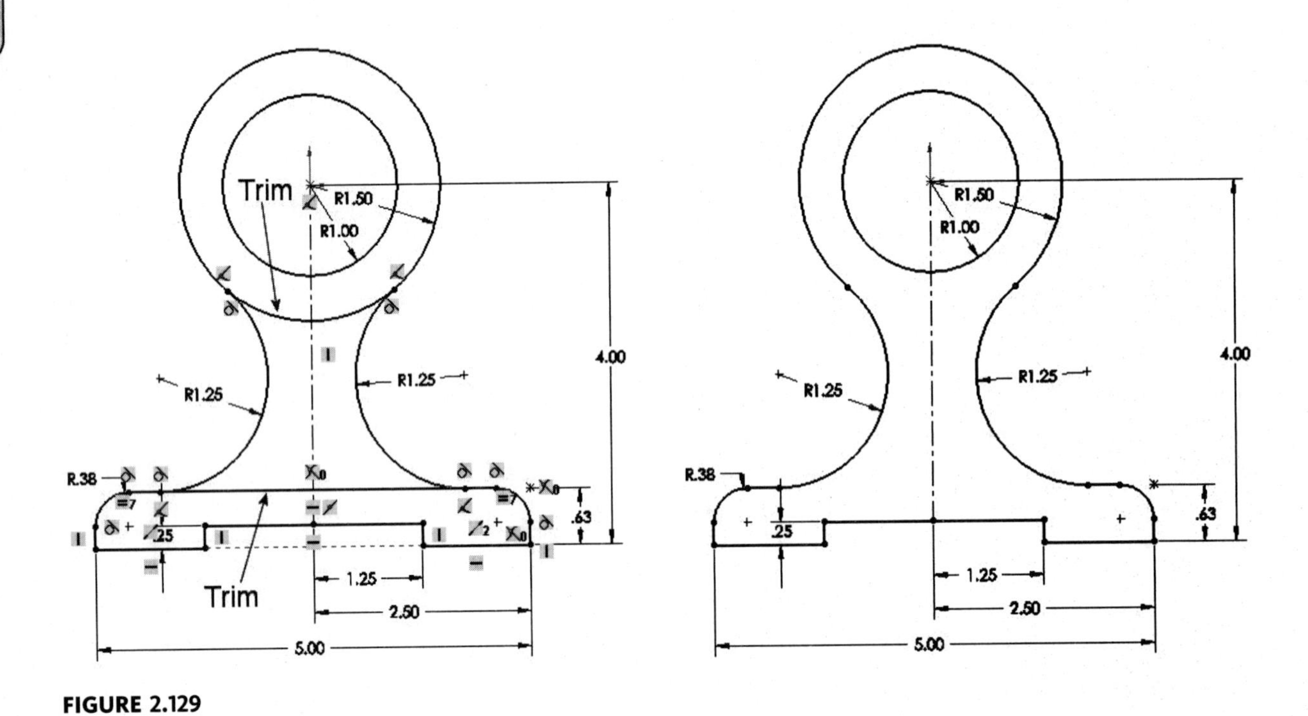

FIGURE 2.129

STEP 10

Sketch the two lines in Figure 2.130 on the left. Then trim these lines as shown in the following figure on the right.

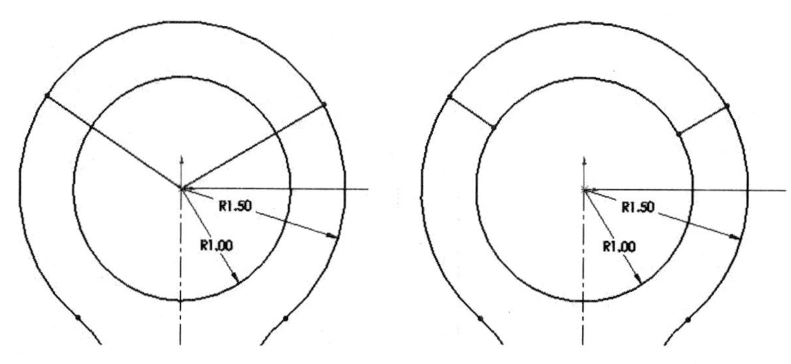

FIGURE 2.130

STEP 11

Construct the three center lines in the upper area of the part as shown in Figure 2.131 on the left. Make each of the center lines collinear with the object lines as shown in the following figure on the right. The collinear relation controls the angle of the short object line segments when the center lines are dimensioned with an angle. (If a relationship is added by mistake, right-click on it and select Delete.)

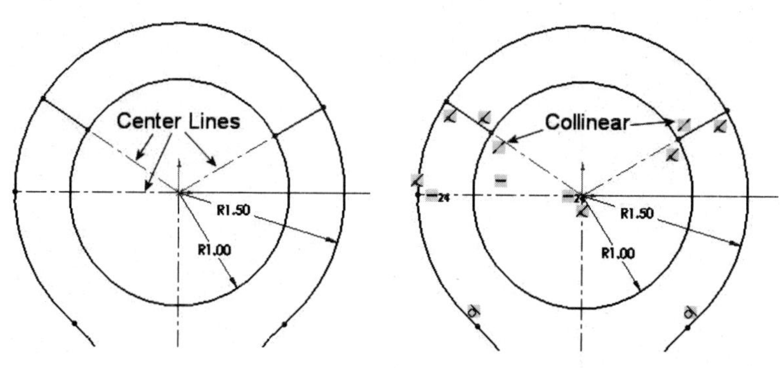

FIGURE 2.131

STEP 12

Add angle dimensions to the two inclined line segments of 30 degrees each (or 30 and 120 degrees). Then trim away the upper portion of the two circles to form the arc segments as shown in Figure 2.132.

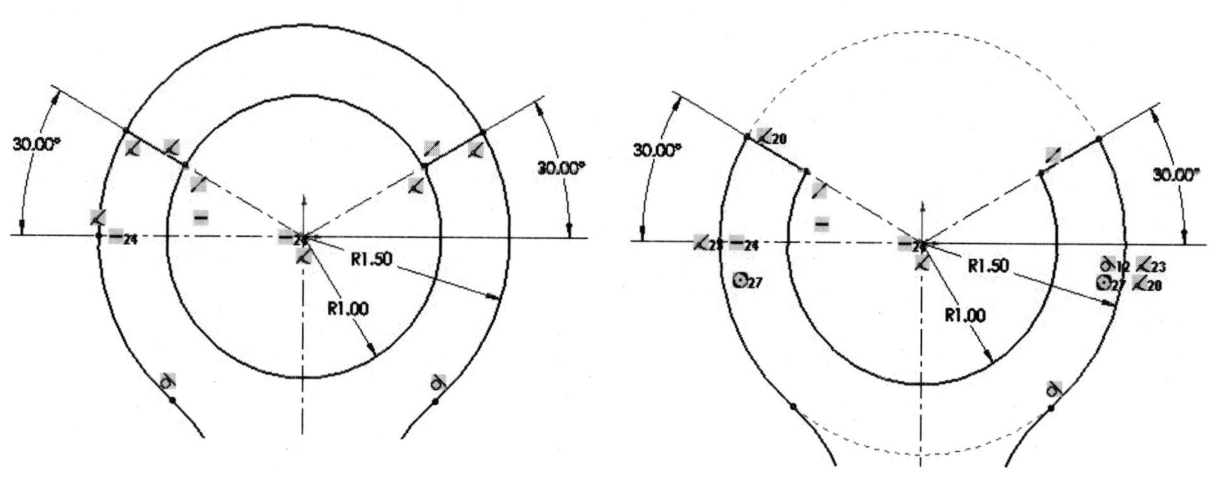

FIGURE 2.132

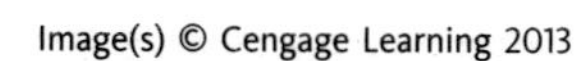

STEP 13

Your object should appear similar to Figure 2.133 on the left. Create an *Extruded Boss/Base* (from the Features tab) using a distance of 2.75 as shown in the following figure on the right to complete this object.

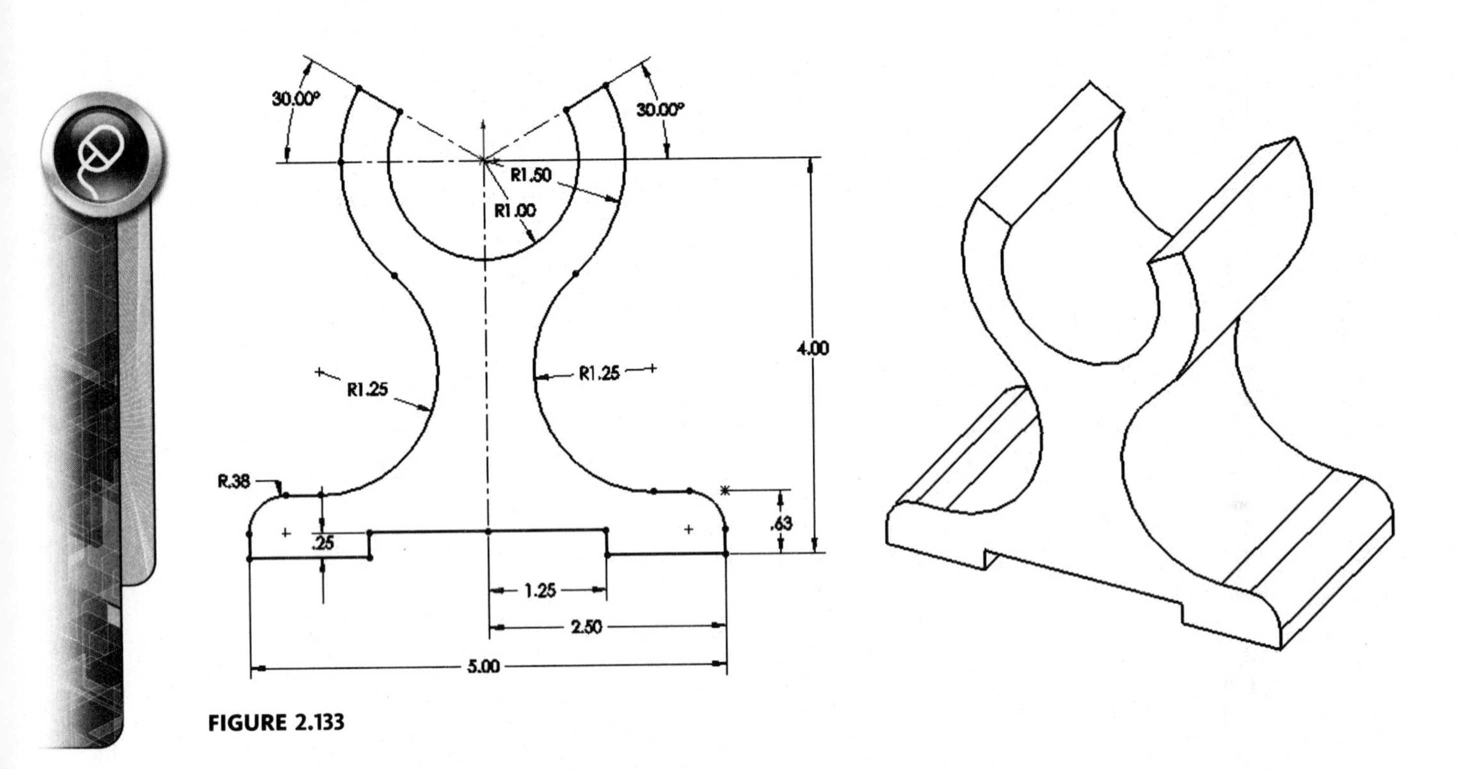

FIGURE 2.133

TUTORIAL EXERCISE: SWT_CH02_06.SLDPRT

Description: Lever

Units: English

This tutorial exercise is designed to create a part model that consists of numerous geometric shapes (Figure 2.134). Proper relations need to be applied along with dimensions to fully define the shape.

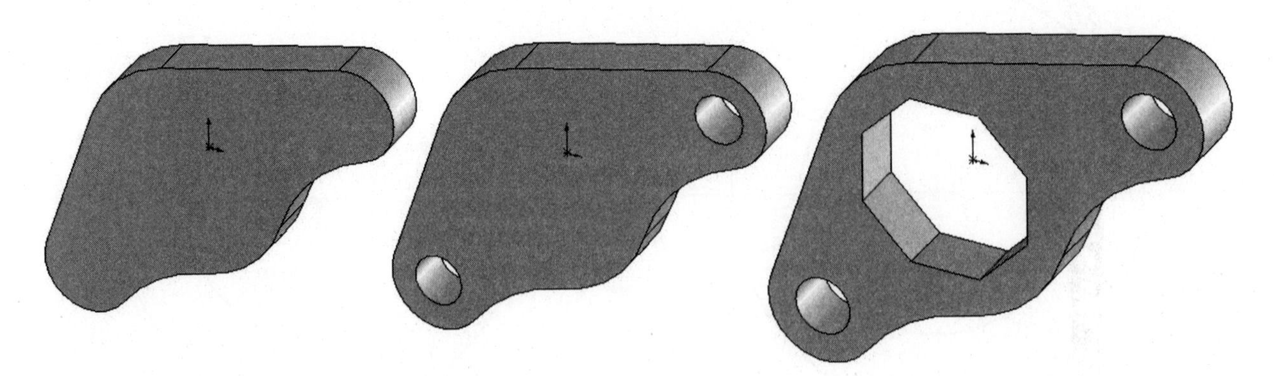

FIGURE 2.134

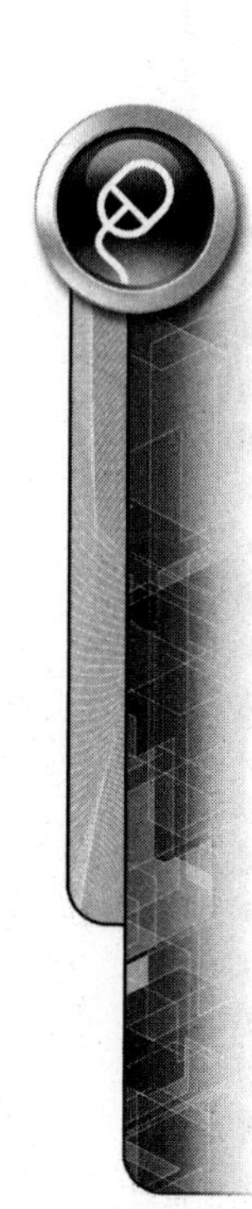

STEP 1

Construct the large 2.25 radius circle at the red origin location. Sketch the remaining two circles. Add diameter dimensions but change them to radius dimensions. Add all linear dimensions identifying the centers of the R1.00 circles. Also, apply the Equal Relation to both small circles to make their radii equal. Your display should be similar to Figure 2.135.

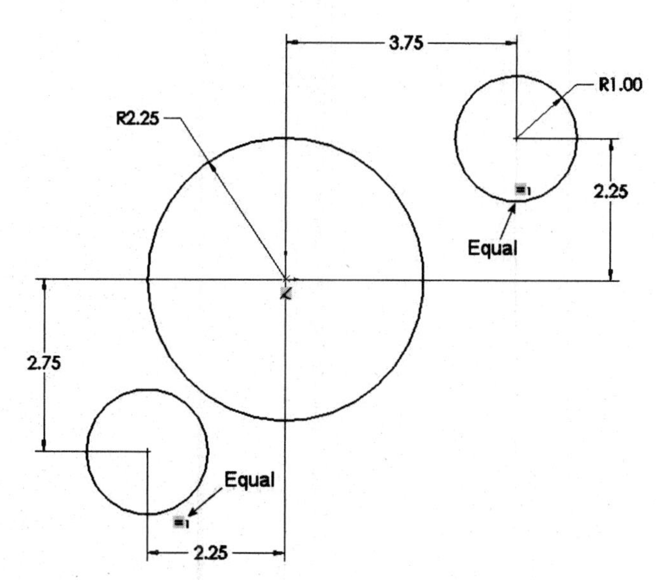

FIGURE 2.135

STEP 2

Add the two lines to the sketch. Apply Tangent Relations if necessary between the lines and arcs (Figure 2.136).

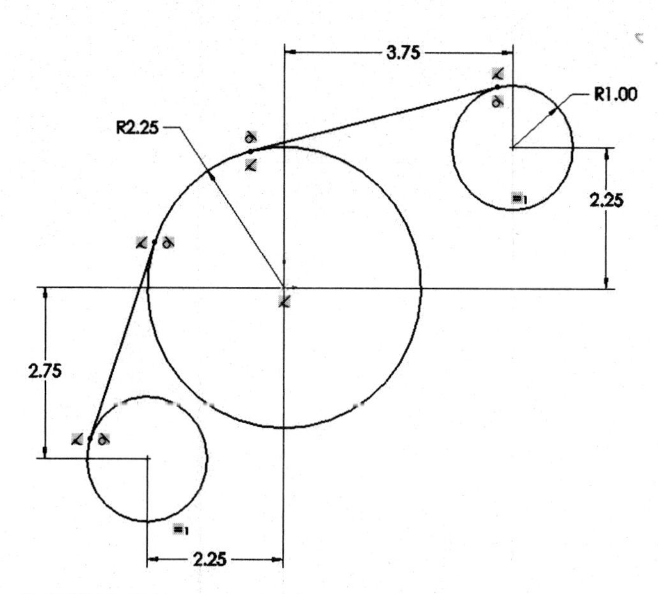

FIGURE 2.136

STEP 3

Add two arcs to the sketch as shown in Figure 2.137 on the left using the 3 Point Arc tool. Then make both arcs tangent to the circles and add one 3.00 radius dimension. Make the other arc equal to the dimensioned arc as shown in the following figure on the right.

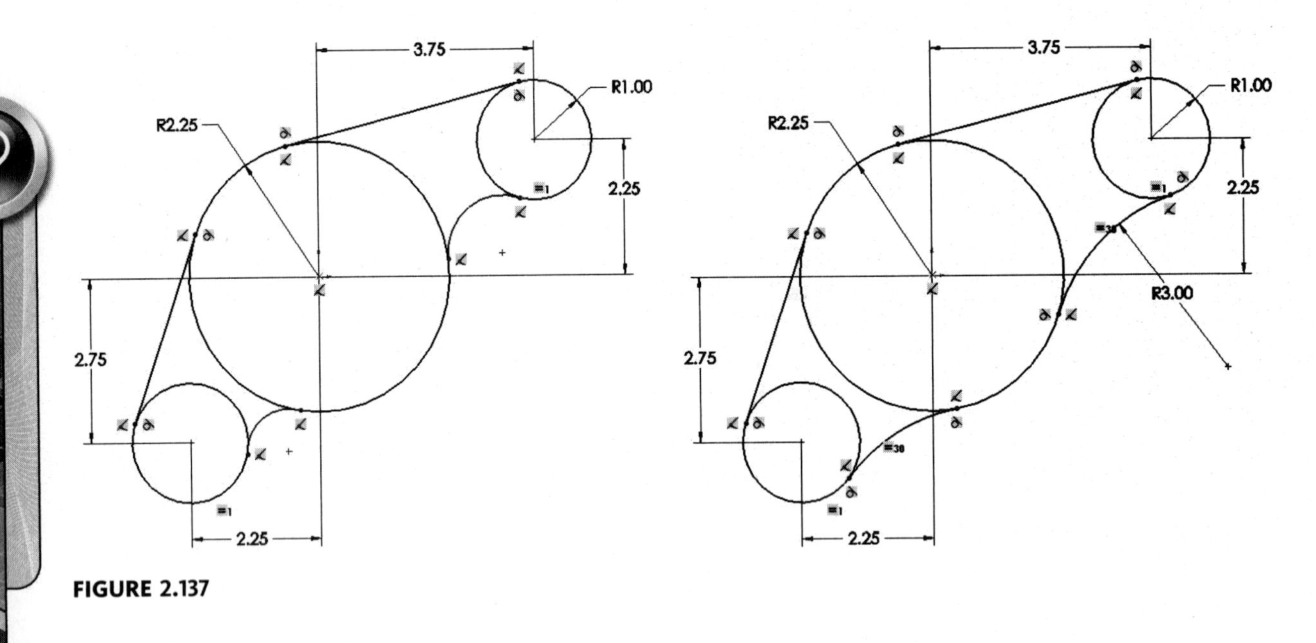

FIGURE 2.137

STEP 4

Trim the circles to form the outer profile of the lever as shown in Figure 2.138.

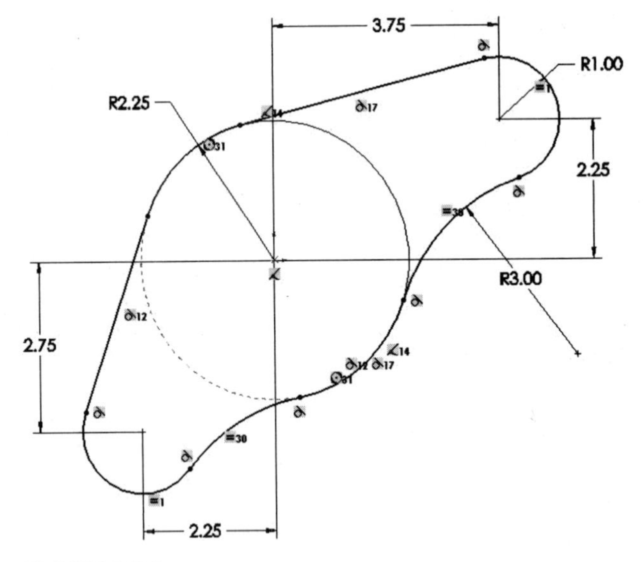

FIGURE 2.138

STEP 5

As the display of relations can get busy, use the Heads Up display menu (Hide/Show Items) and click on the View Sketch Relations button to turn the display of relations off as shown in Figure 2.139. Clicking on this button a second time redisplays all relations.

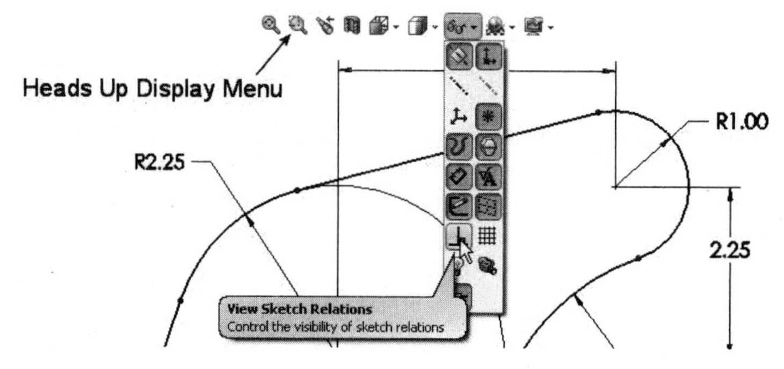

FIGURE 2.139

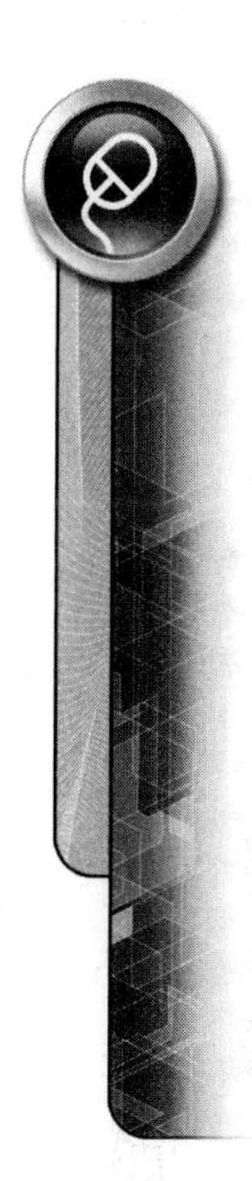

STEP 6

When all entities that make up the outer profile turned black signifying the sketch is fully defined, click on the Features tab and select Extruded Boss/Base from the menu. Extrude the shape to a distance of 1.00 as shown in Figure 2.140.

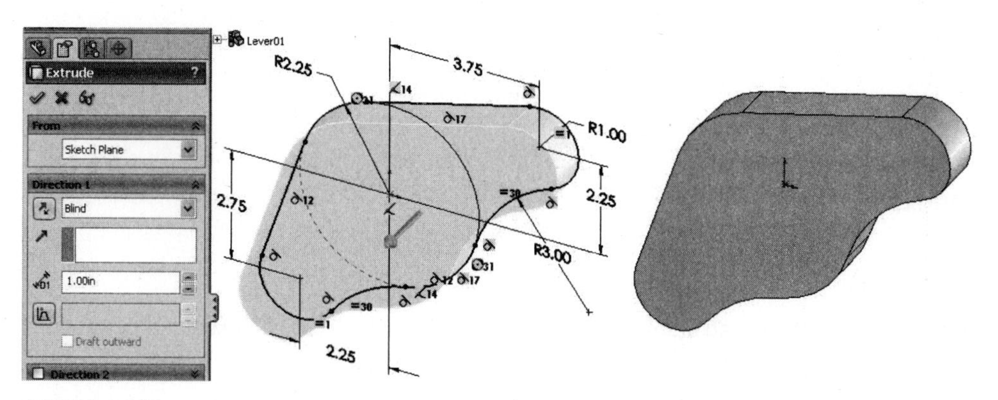

FIGURE 2.140

STEP 7

Create a new sketch plane on the front of the part as shown in Figure 2.141 on the left. Right-click on *View Orientation* and select **Normal To*; this will change to the plane view as shown in the following figure on the right.

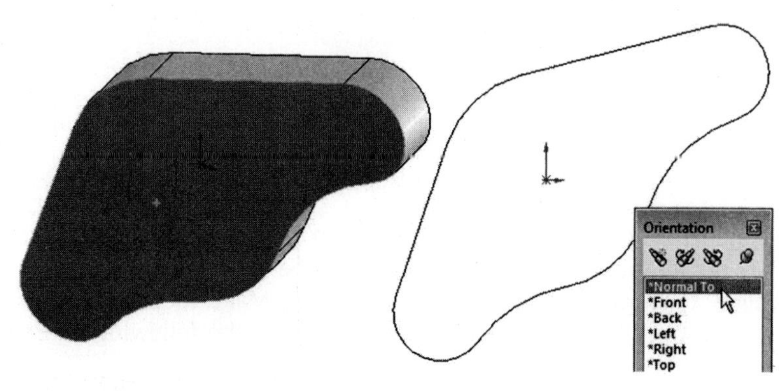

FIGURE 2.141

STEP 8

Add the two circles that represent holes to be cut from the lever. Then apply the Concentric Relation to both circles and the outer arcs as shown in Figure 2.142 on the left. While in the Circle tool, you could also move your cursor over the arc and pause; this will expose the center of the arc where you can draw the circle from as shown in the following figure on the right. Dimension one circle using a value of Ø1.00 and apply the Equal Relation to the other circle to make it equal to the first.

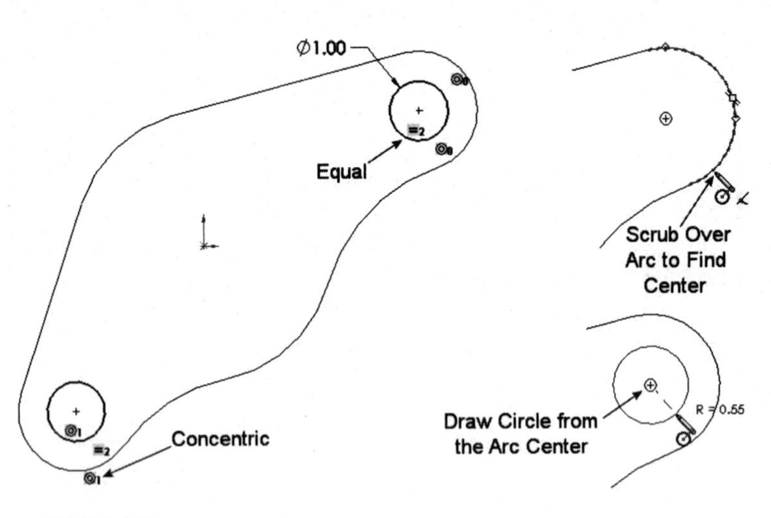

FIGURE 2.142

STEP 9

Click on the Features tab and click the Extrude Cut tool to create an extruded cut of both circles using the Through All option as shown in Figure 2.143.

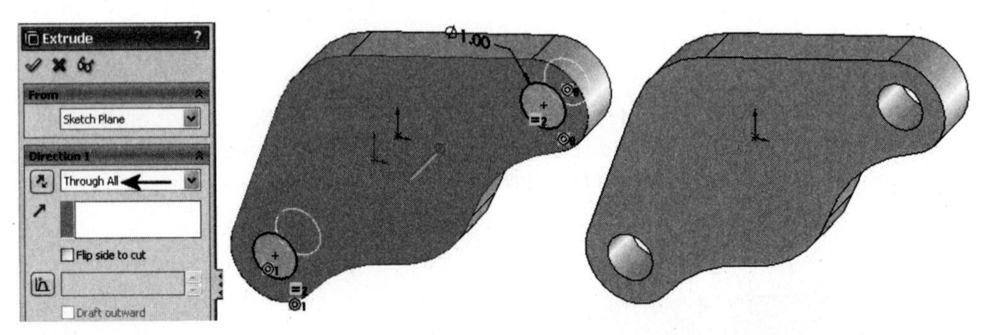

FIGURE 2.143

STEP 10

Create a new sketch plane on the front of the part. Click on **Normal To*. Place a polygon (Octagon) with its center located at the red origin. Add the 3.00 dimension across the flats of the octagon as shown in Figure 2.144 on the left. Make the bottom line of the octagon horizontal to fully define the sketch as shown in the following figure on the right.

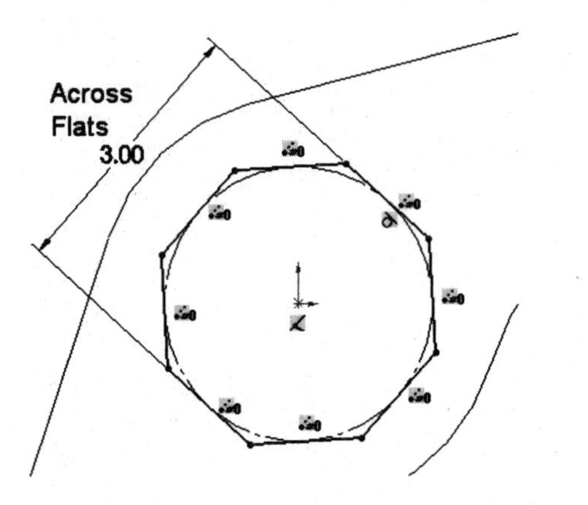

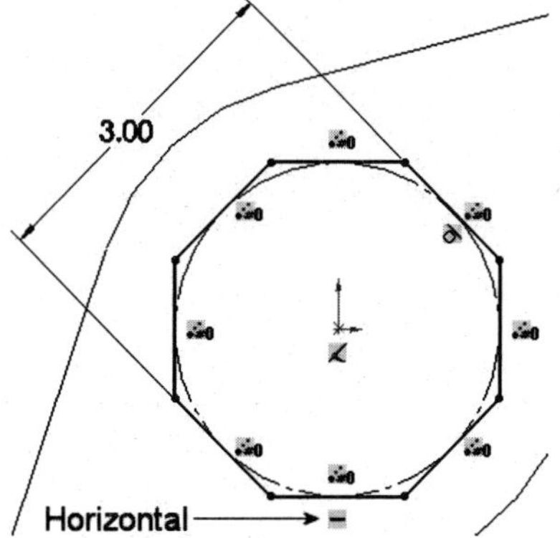

FIGURE 2.144

STEP 11

Create an *Extruded Cut* (from the Features tab) of the octagon using the Through All option (Figure 2.145).

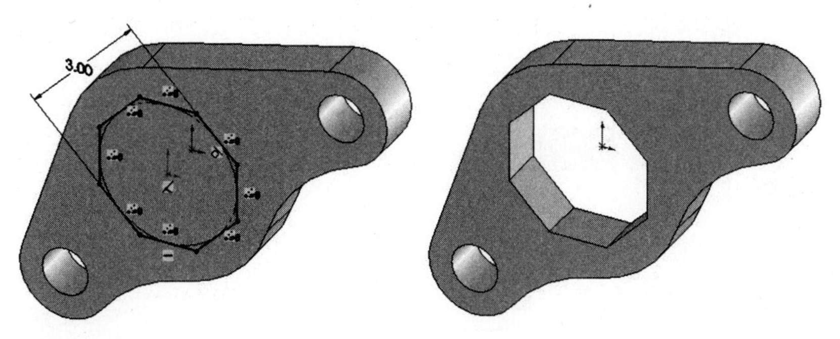

FIGURE 2.145

STEP 12

Notice how all three extrude features are present in the FeatureManager as shown in Figure 2.146 on the left. It is considered good practice to rename the extrusions to better names such as the three shown in the following figure on the right.

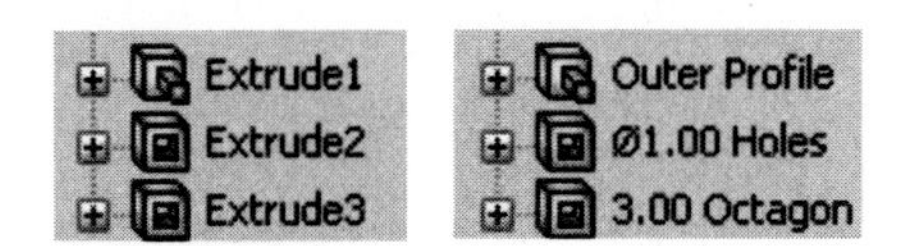

FIGURE 2.146

USING ADDITIONAL SKETCH TOOLS

In addition to sketching, lines, circles, and arcs, additional sketching tools are available to help construct other entity types. A few of these additional tools control how items are copied in a linear or circular pattern. Linear sketch patterns are created in a rectangular shape while circular patterns are created from a common pivot point. The menu used to activate is shown in Figure 2.147.

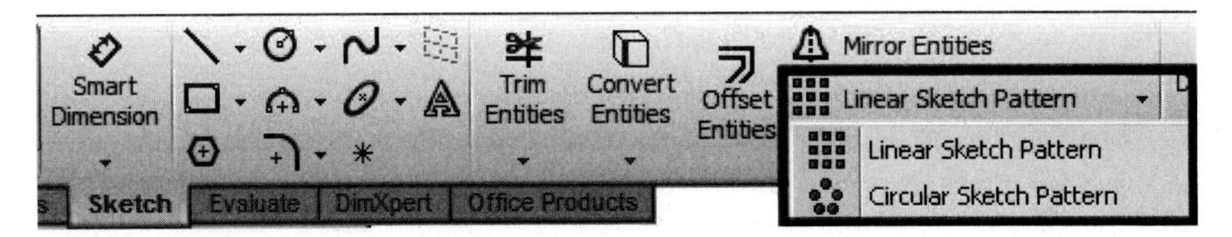

FIGURE 2.147

Creating Linear Patterns

Linear patterns are used when the pattern resembles a rectangular shape. You first select the object to pattern. Then, you select edges, which creates the directions of the pattern. In Figure 2.148, the circle is selected as the entity to pattern, and the bottom horizontal and left vertical lines act to direct the pattern. Buttons are available in the PropertiesManager to change the direction of the pattern. You also enter the number of occurrences or instances of the circle and the distance between the circles. A preview will display showing the pattern that you can either accept or make additional changes to.

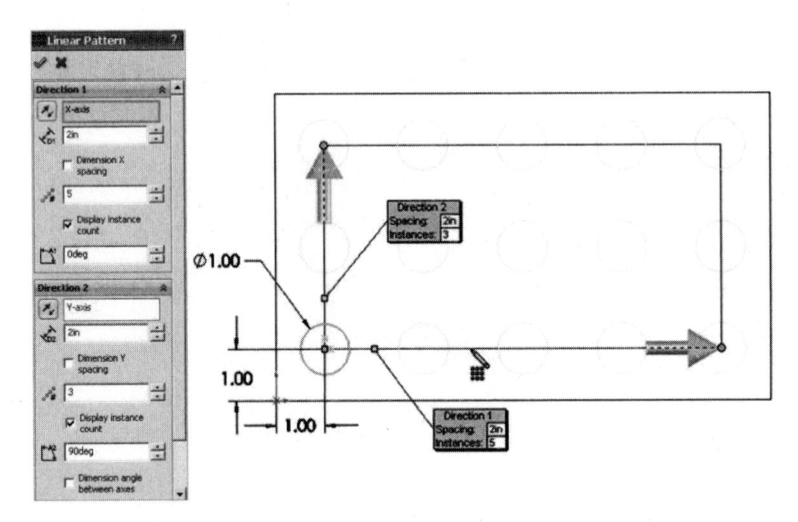

FIGURE 2.148

The results of creating the linear pattern are illustrated in Figure 2.149. Notice how dimensions appear to document the distances of the pattern.

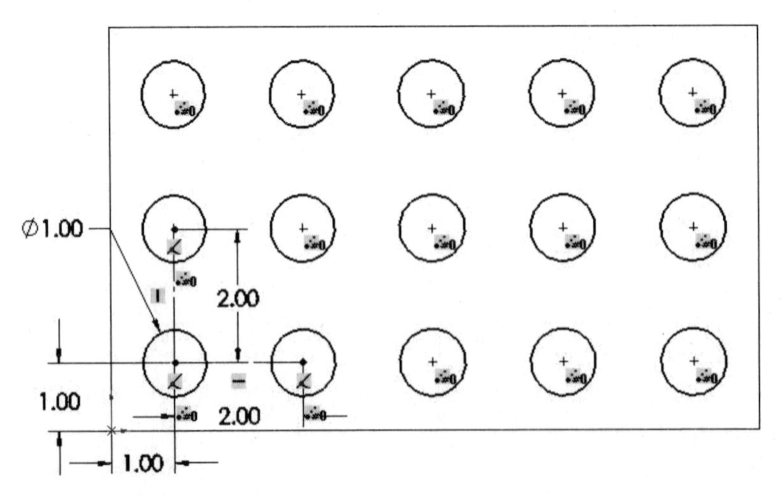

FIGURE 2.149

The resulting linear pattern is illustrated in Figure 2.150.

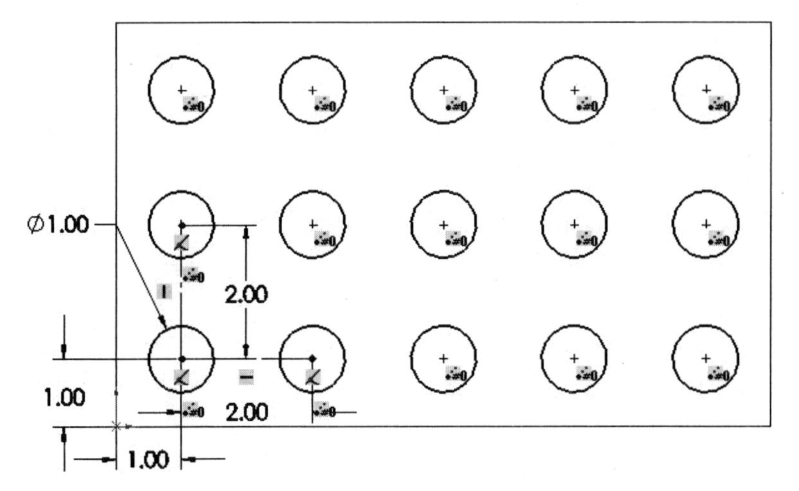

FIGURE 2.150

Additional tools for manipulating sketch objects are displayed in Figure 2.151. Clicking on Tools followed by Sketch Entities displays this menu.

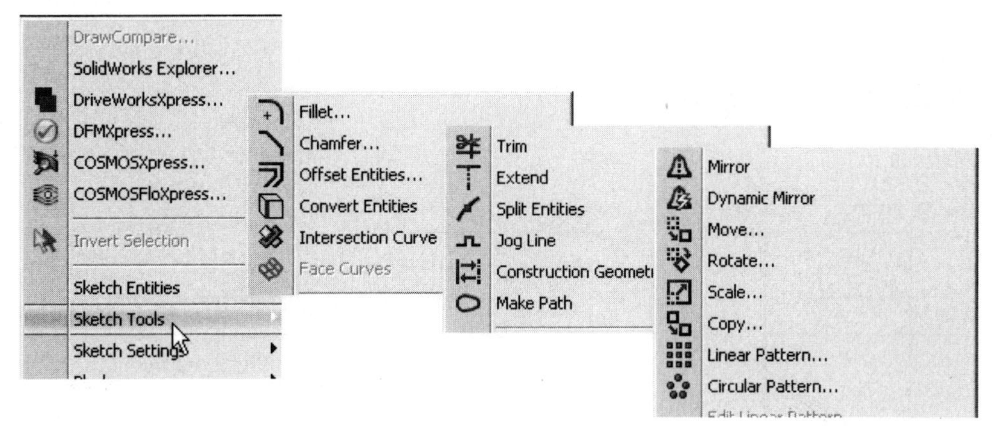

FIGURE 2.151

REVIEW – SKETCHING GUIDELINES

- Select an outline that best represents the part. It is usually easier to work from a flat face.
- Construct the geometry close to the finished size. For example, if you want a 10mm square, do not construct a 1000mm square.
- Create the sketch proportionately in size to the finished shape.
- Draw the sketch so the lines do not overlap. All of the connecting endpoints should be coincident with each other.
- Do not allow sketches to have any gaps.
- Keep the sketch simple. Leave out any fillets or chamfers. These can be placed more easily after the initial feature is created. The simpler the sketch, the fewer the number of relations and dimensions will be required to solve the sketch.

REVIEW QUESTIONS

1. After drawing a first line, what types of relations might be available for a second line?
2. Before exiting a sketch, what sketch status should be satisfied?
3. What sketch color defines unsolved geometry?
4. What sketch color defines solved geometry?
5. What sketch colors arise if a sketch becomes overdefined?
6. What are the two types of constraints that are put on a sketch to define it?
7. If a sketch is overdefined, what must be done?
8. What is the advantage of construction geometry?
9. What is the advantage of using equations in a sketch?
10. What two types of sketch patterns can be created?

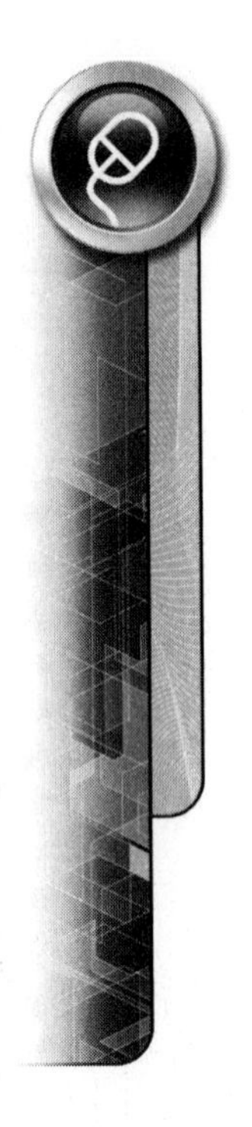

EXERCISES

2.1 a thru d

Construct these geometric construction figures using SolidWorks sketch tools. Create an extruded base feature using the extrusion distance provided. Perform a mass property calculation on each object.

2.1A (ENGLISH UNITS)

Extrude to 1.75

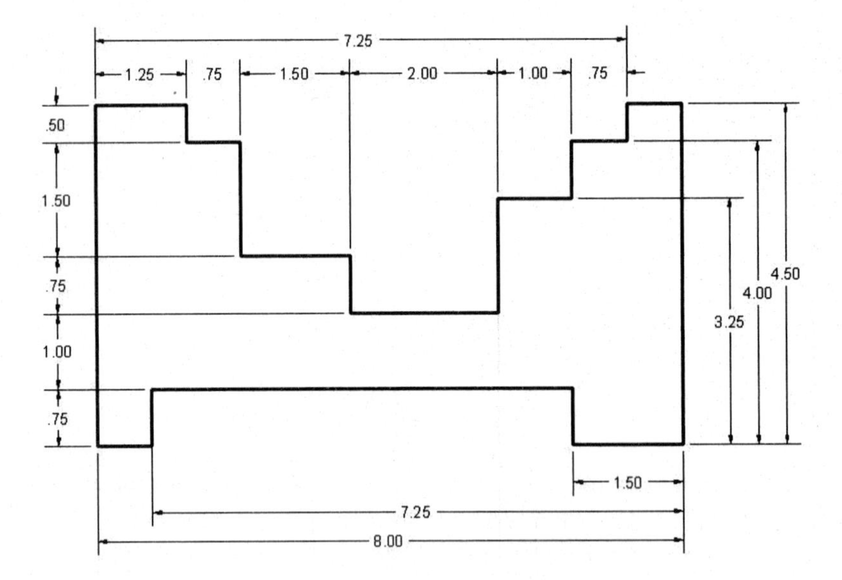

2.1B (ENGLISH UNITS)

Extrude to 3.25

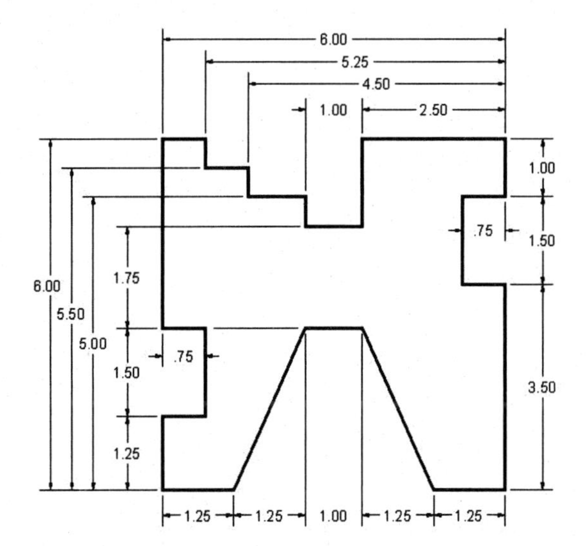

2.1C (ENGLISH UNITS)

Extrude to 2.25

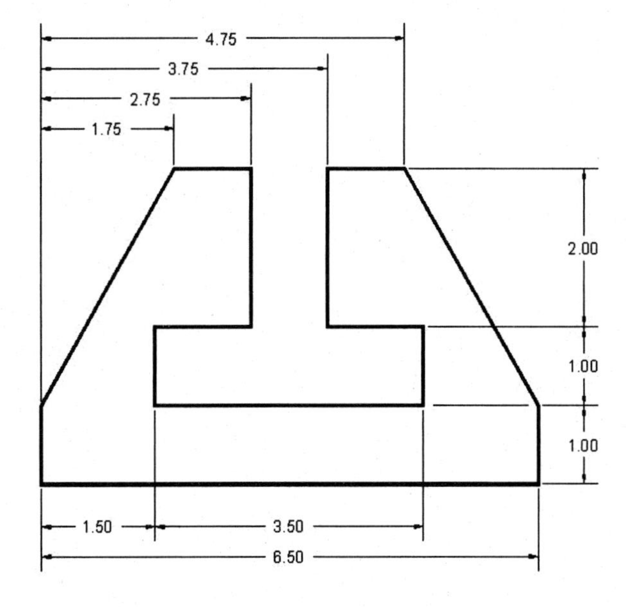

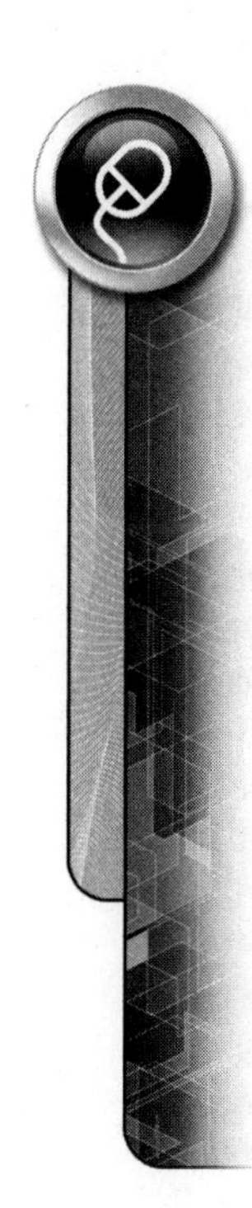

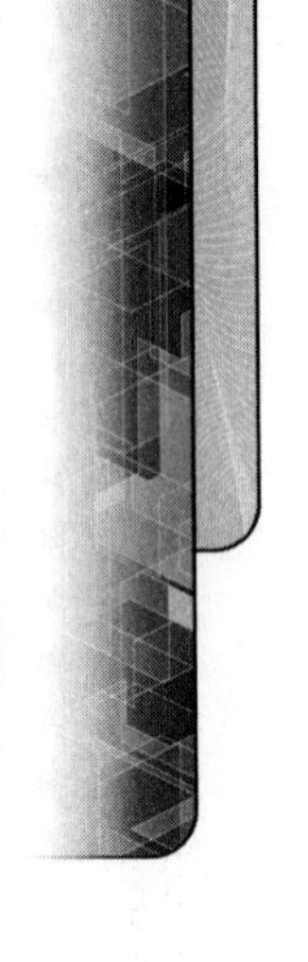

2.1D (ENGLISH UNITS)

Extrude to 2.75

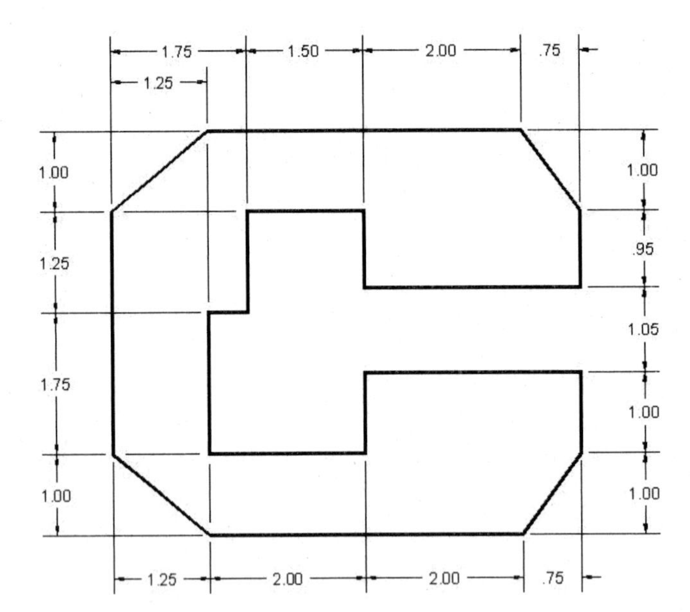

2.2 a thru e

Construct these geometric construction figures using SolidWorks sketch tools. Create an extruded base feature using the extrusion distance provided. Perform a mass property calculation on each object.

PROBLEM 2-2A (ENGLISH UNITS)

Extrude to .75 in.

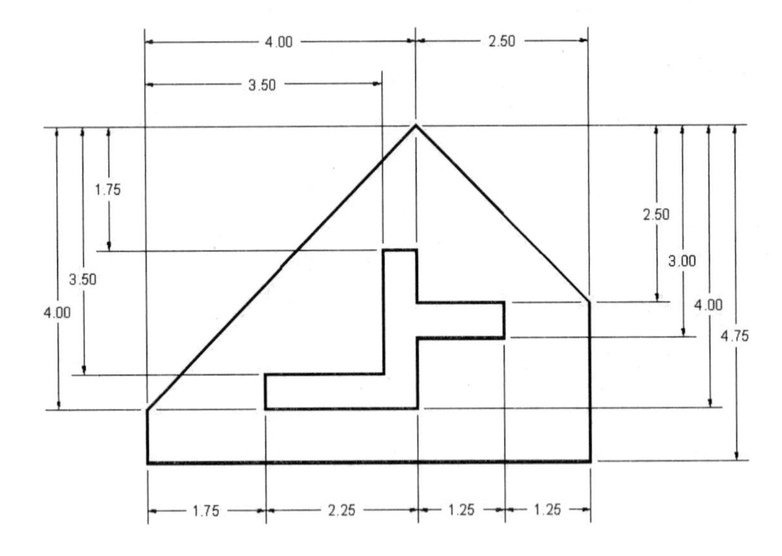

PROBLEM 2-2B (ENGLISH UNITS)

Extrude to 1.50 in.

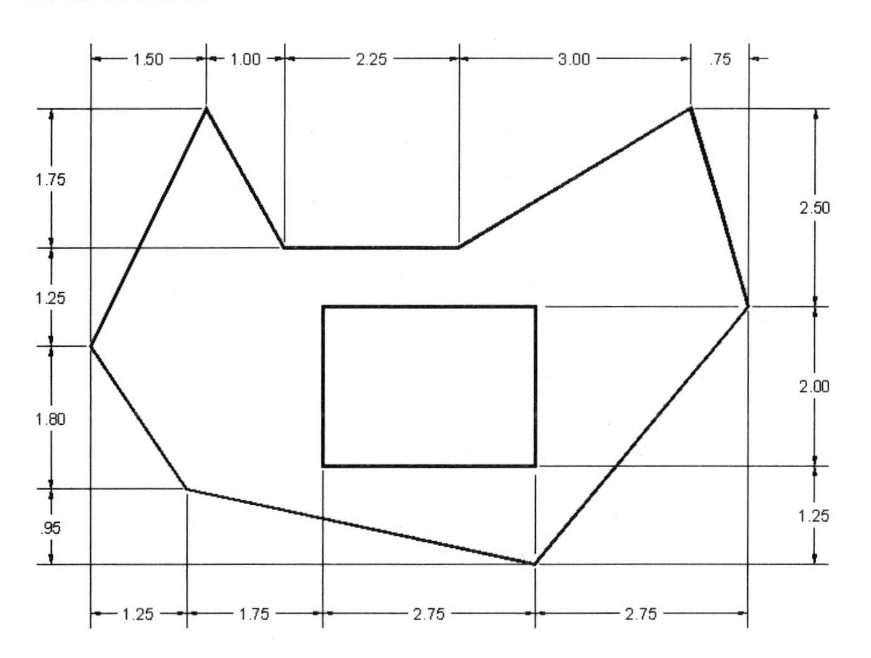

PROBLEM 2-2C (ENGLISH UNITS)

Extrude to 1.25 in.

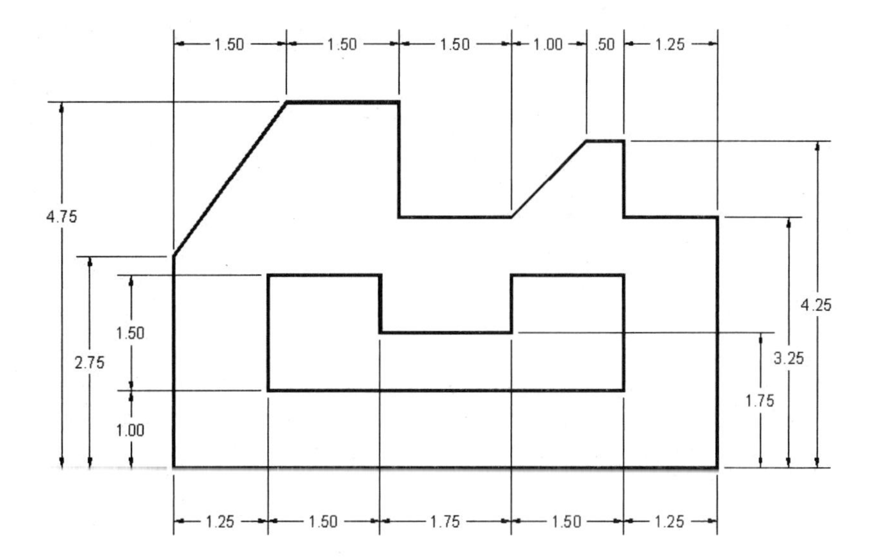

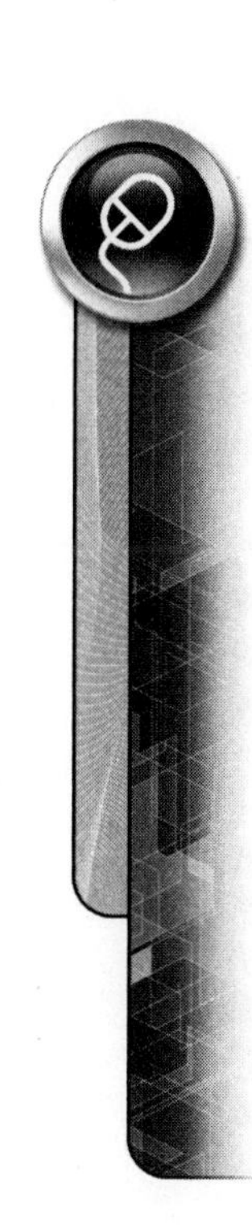

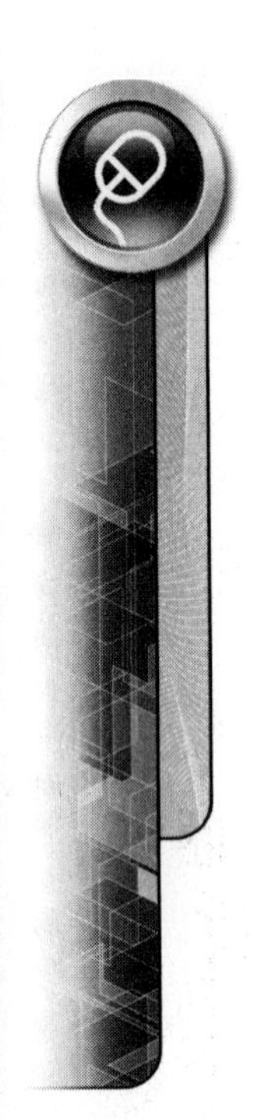

PROBLEM 2-2D (ENGLISH UNITS)

Extrude to .50 in.

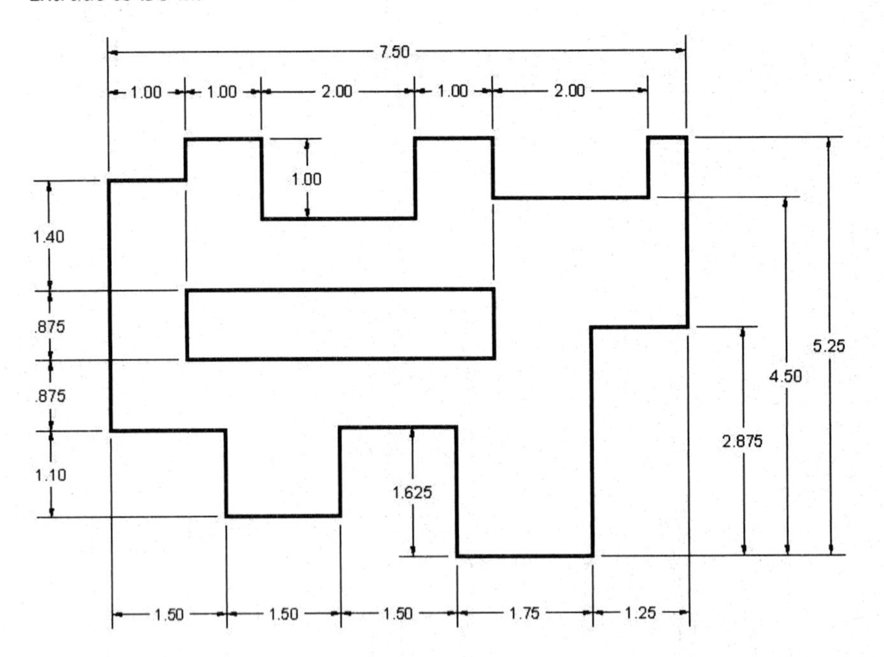

PROBLEM 2-2E (METRIC UNITS)

Extrude to 20mm

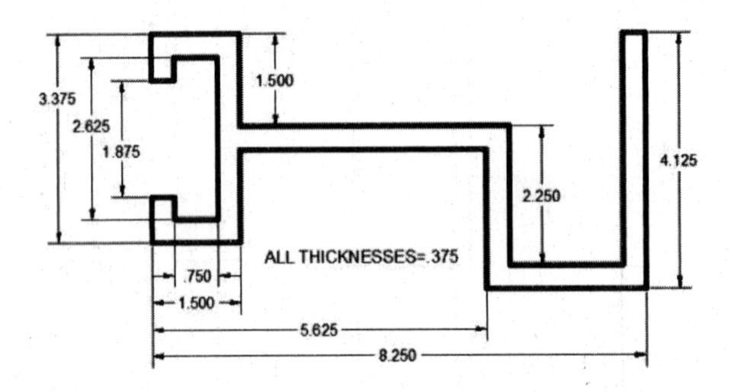

2.3 a thru e

Construct these geometric construction figures using SolidWorks sketch tools. Create an extruded base feature using the extrusion distance provided. Perform a mass property calculation on each object.

PROBLEM 2-3A (ENGLISH UNITS)

Extrude Initial Sketch to .125
Cut Second Sketch Through All

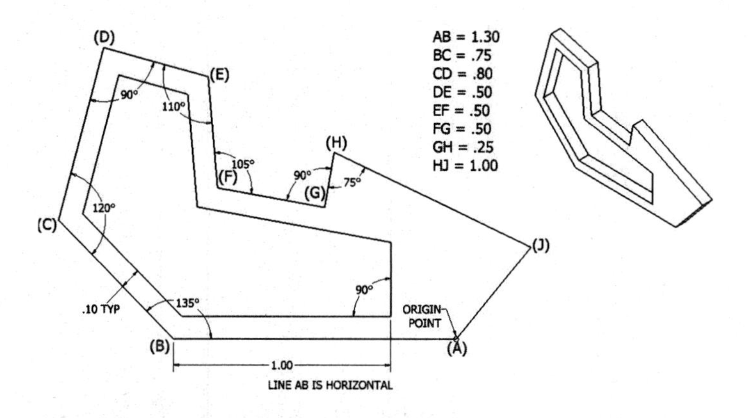

PROBLEM 2-3B (ENGLISH UNITS)

Extrude Initial Sketch to .25
Cut Second Sketch Through All

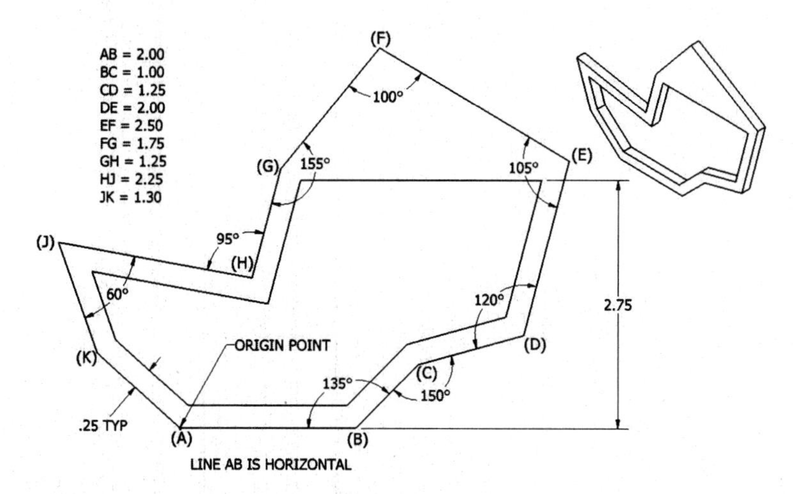

PROBLEM 2-3C (METRIC UNITS)

Extrude to a distance of 5mm

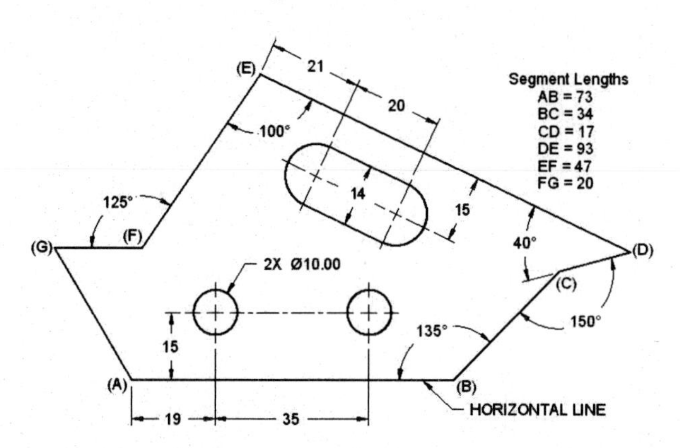

PROBLEM 2-3D (ENGLISH UNITS)

Extrude to a distance of .50

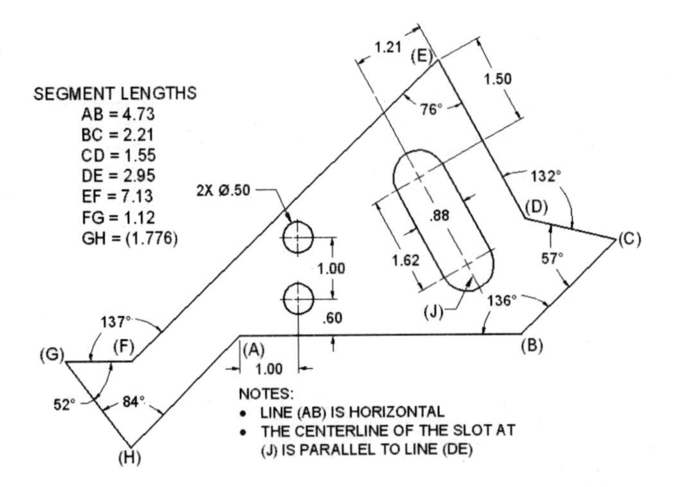

PROBLEM 2-3E (ENGLISH UNITS)

Extrude to a distance of 5mm

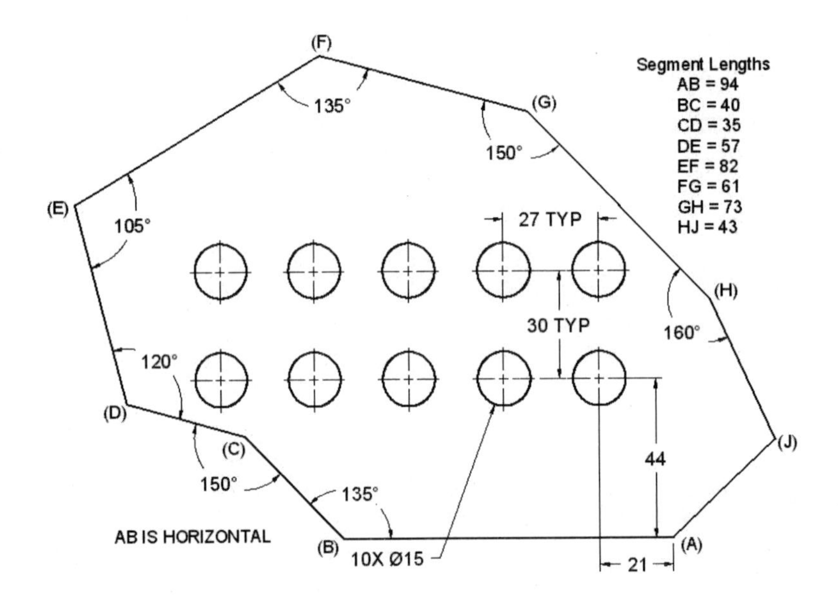

2.4 a thru m

Construct these geometric construction figures using SolidWorks sketch tools. Create an extruded base feature using the extrusion distance provided. Perform a mass property calculation on each object.

PROBLEM 2-4A (ENGLISH UNITS)

Extrude to a distance of .65 inches

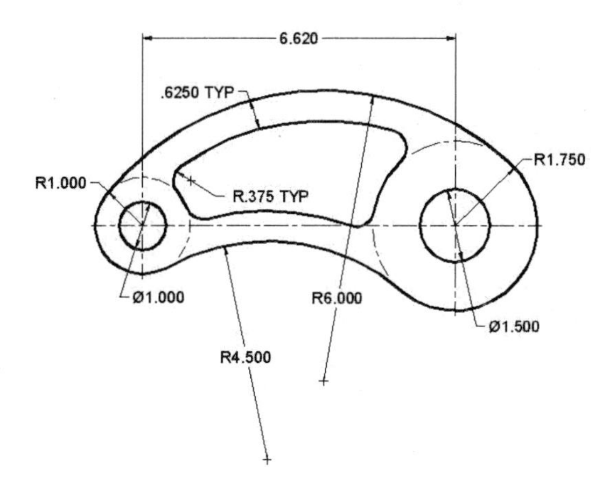

PROBLEM 2-4B (ENGLISH UNITS)

Extrude to a distance of .55 inches

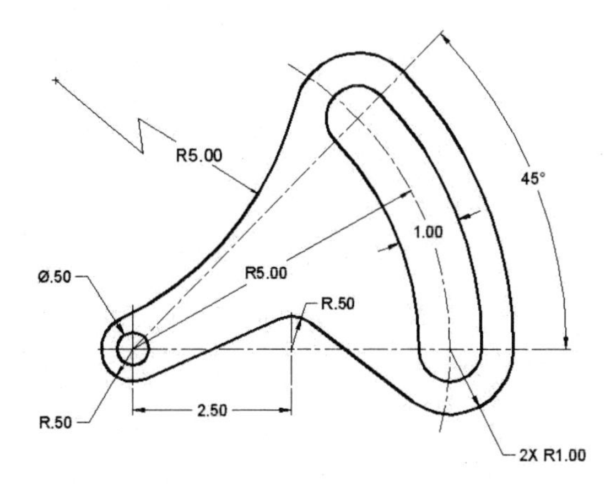

PROBLEM 2-4C (ENGLISH UNITS)

Extrude to a distance of .75 inches

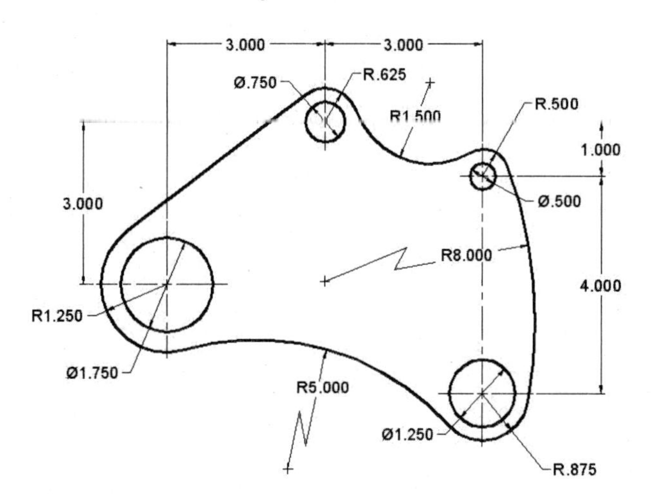

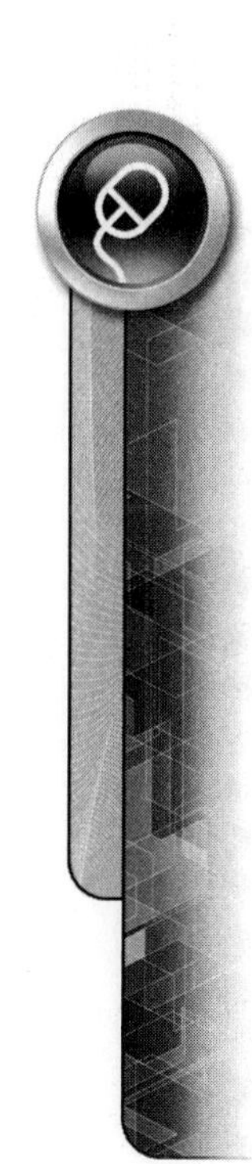

PROBLEM 2-4D (ENGLISH UNITS)

Extrude to a distance of .75 inches

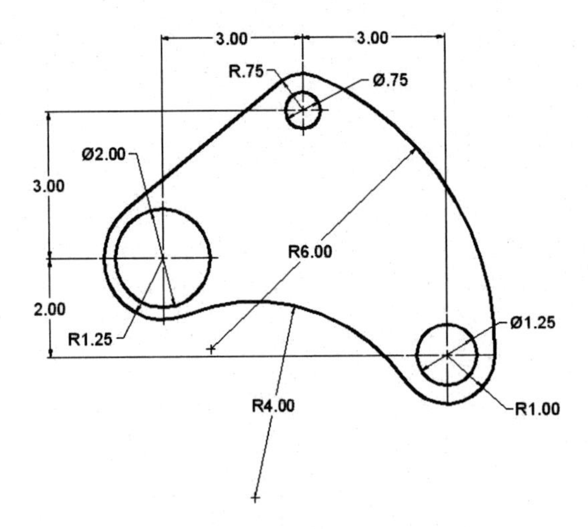

PROBLEM 2-4E (METRIC UNITS)

Extrude to a distance of 15mm

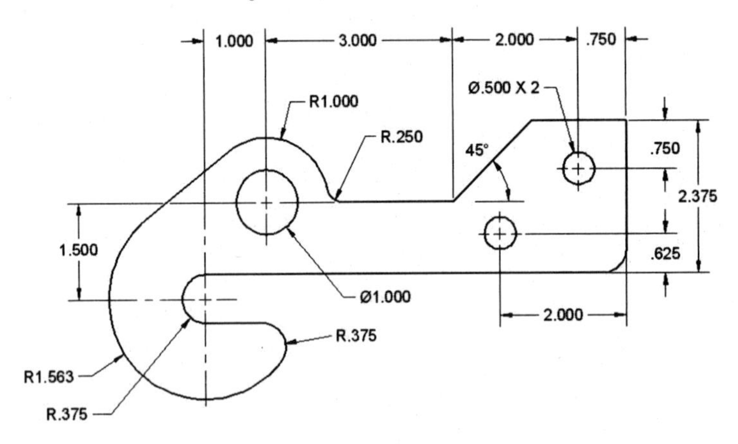

PROBLEM 2-4F (ENGLISH UNITS)

Extrude to a distance of .75 inches

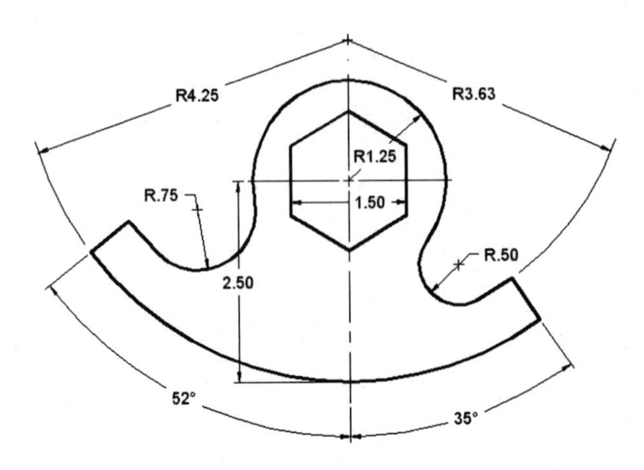

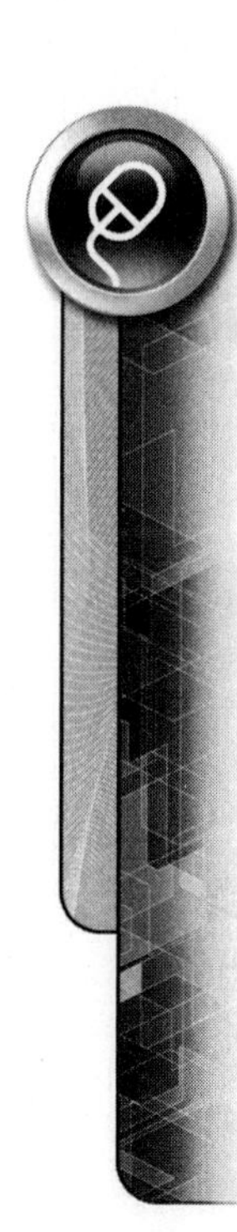

PROBLEM 2-4G (METRIC UNITS)

Extrude to a distance of 15mm

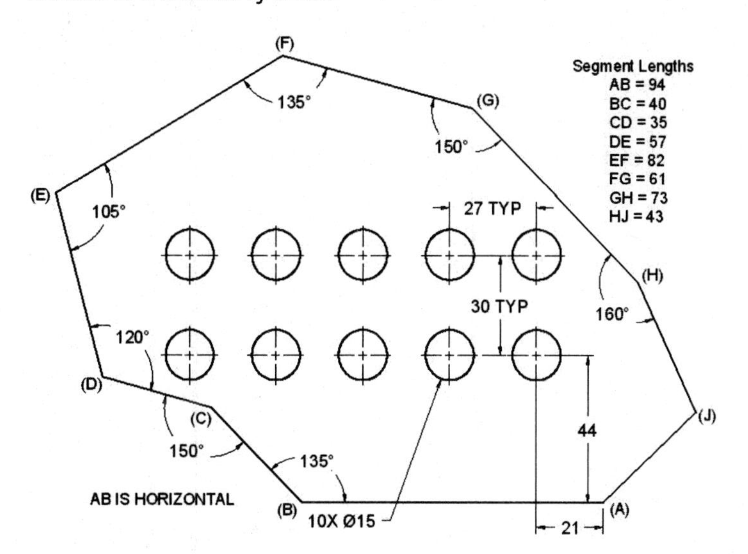

PROBLEM 2-4H (METRIC UNITS)

Extrude to a distance of 17mm

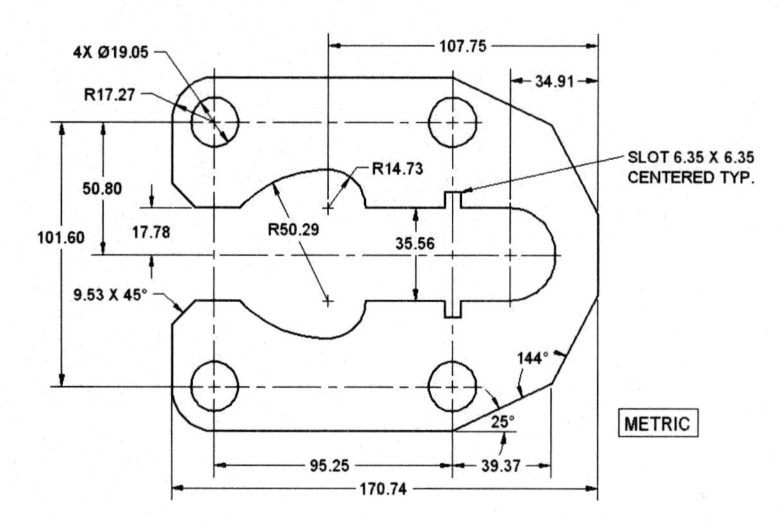

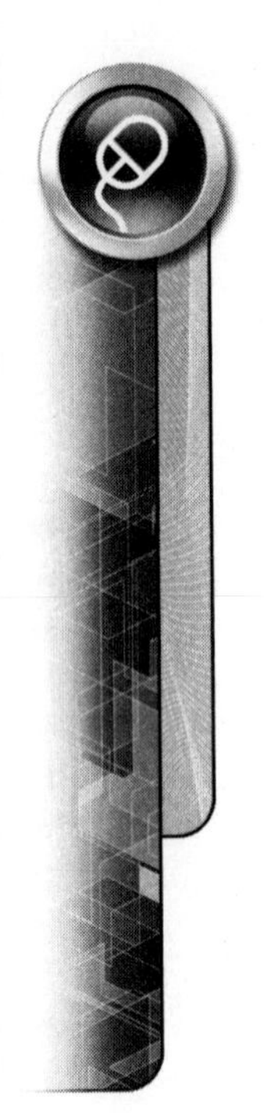

PROBLEM 2-4I (METRIC UNITS)

Extrude to a distance of 15mm

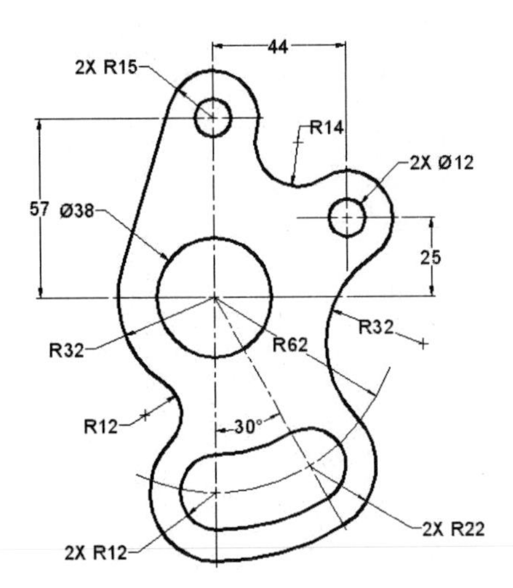

PROBLEM 2-4J (ENGLISH UNITS)

Extrude to a distance of .45 inches

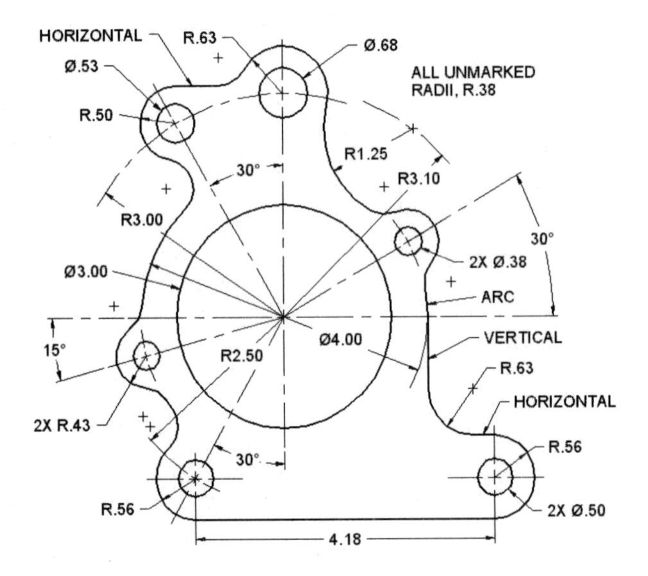

PROBLEM 2-4K (ENGLISH UNITS)

Extrude to a distance of 1.25 inches

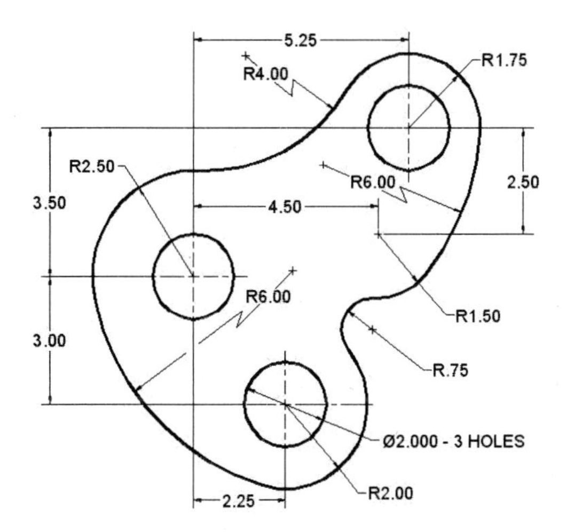

PROBLEM 2-4L (METRIC UNITS)

Extrude to a distance of 12mm

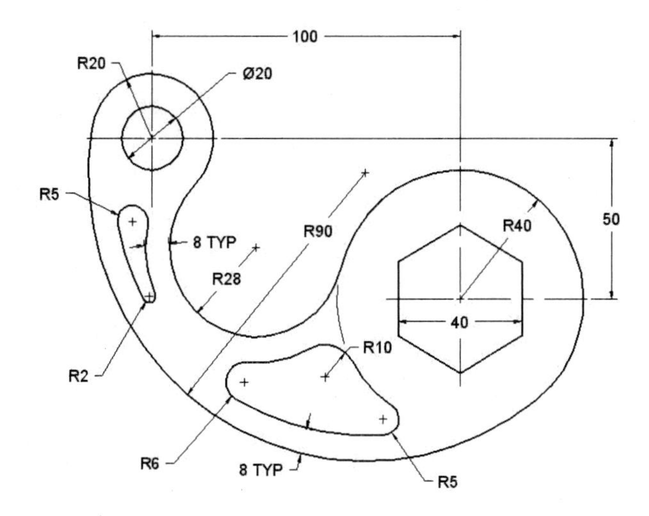

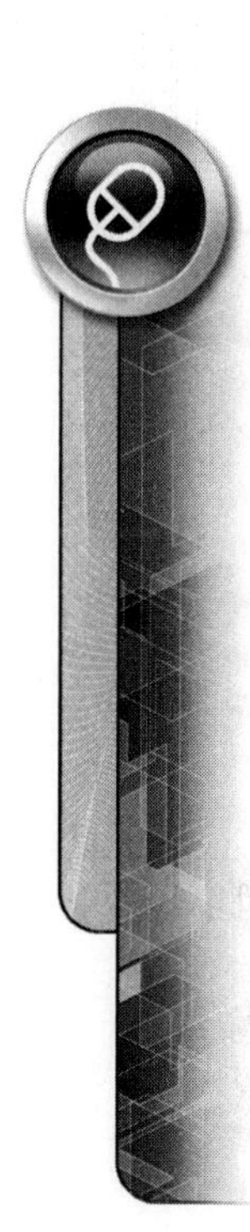

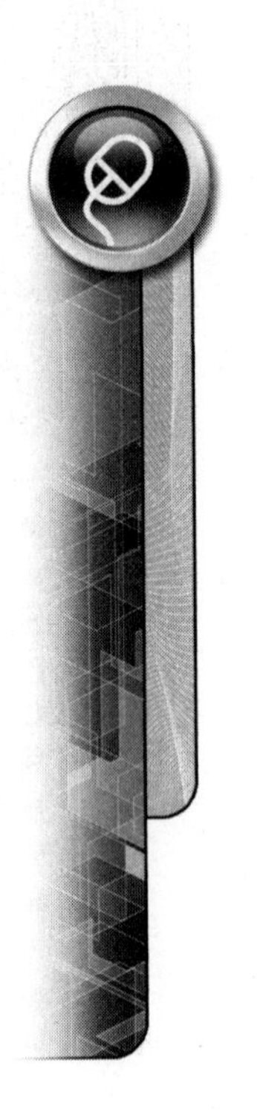

PROBLEM 2-4M (ENGLISH UNITS)

Extrude to a distance of .50 inches

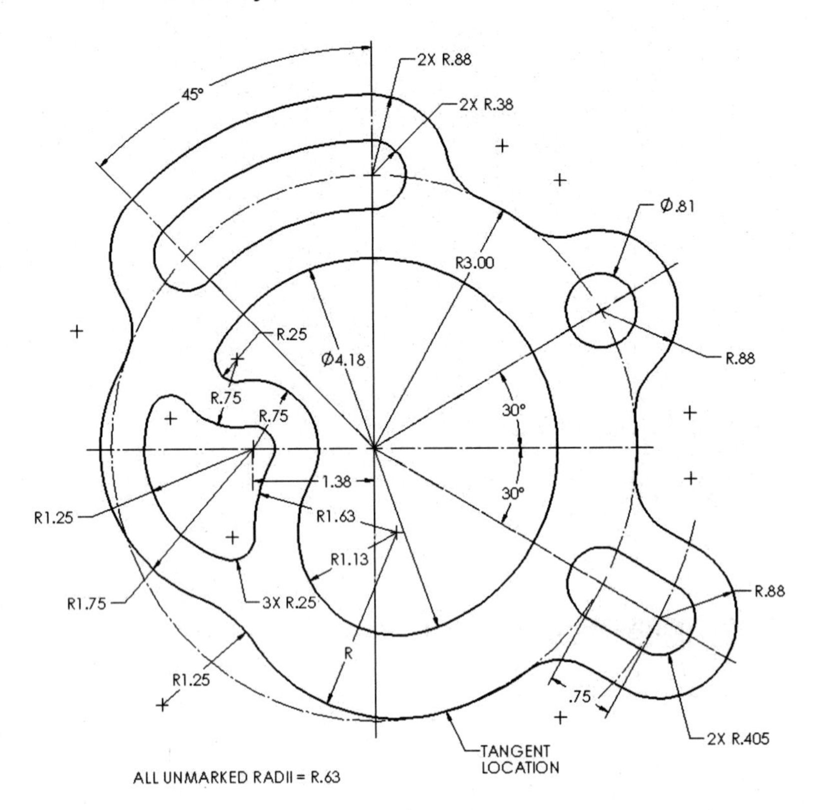

CHAPTER 3

Creating Part Models

This chapter discusses the basic techniques and tools used for creating part models in SolidWorks and includes the following topics:

- The Part Modeling Process
- An overview of all feature tools
- Extrusion options and end conditions
- Creating Extruded Cuts
- Creating Simple Holes
- Creating Fillet and Chamfer features
- Creating Revolved features
- Creating Linear and Circular pattern features

THE PART MODELING PROCESS

No matter how simple or complex a part, the part modeling process is the same. You will need to break down a part into its most basic features. The part in Figure 3.1, for example, consists of a semi-circular base with a solid block on top. This solid block has a square slot cut straight through it. There is also a small hole cut through the whole part in addition to a larger hole cut through the semi-circular base. Each one of the items that make up the solid objects is called a feature. The next series of steps explains how each feature is constructed.

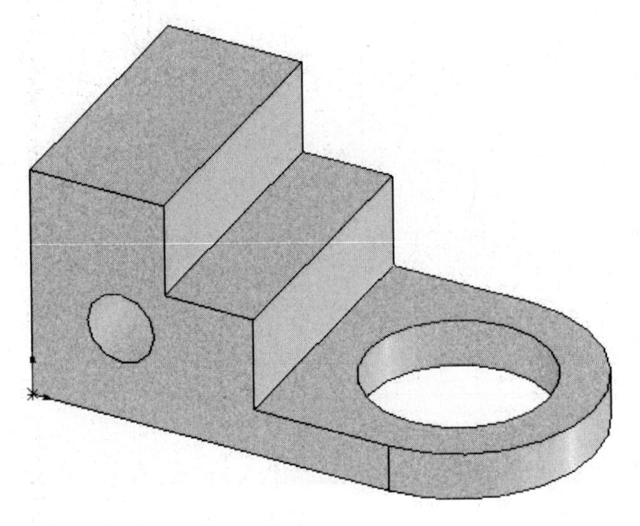

FIGURE 3.1

Begin a *New Part* by first identifying a plane on which to sketch on. Clicking on the sketch button for the first time will display three default planes, namely, Front, Top, and Right. In Figure 3.2, the Top plane will be used to begin the part. After the edge of the Top plane is picked, you are immediately switched to plan or normal view as shown in the following figure on the right.

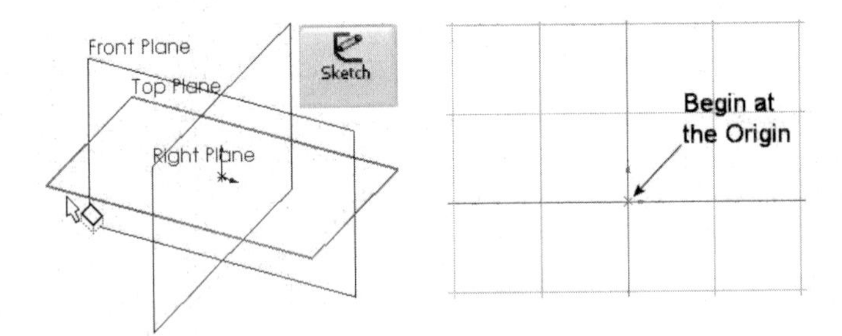

FIGURE 3.2

Begin drawing entities that make up the first sketch. Add relations and smart dimensions as needed as shown in Figure 3.3 on the left. When finished with the sketch, create the first feature by clicking on the Extruded Boss/Base tool. Entering in a height distance of .50 in this example will display the solid base as shown in the following figure on the right.

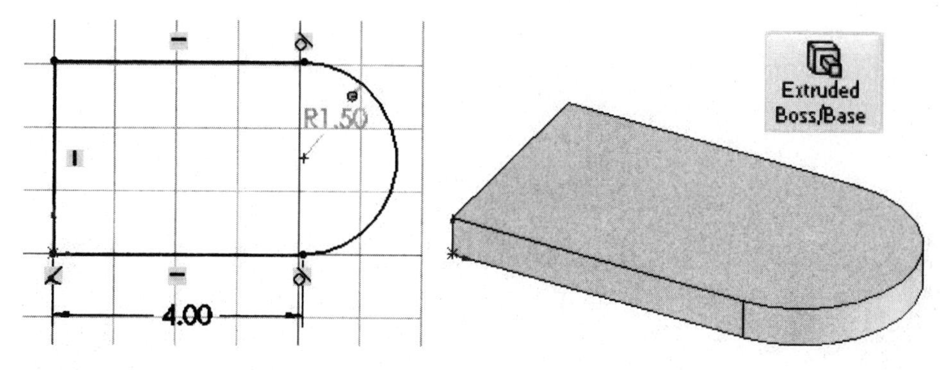

FIGURE 3.3

With this first feature created, begin creating the next feature: a solid box that lies directly on top of the base. Before creating this feature, you must first identify a new sketch plane on which to draw. First click on the top of the solid base. When the top surface highlights, click the Sketch button as shown in Figure 3.4 on the left. Sketch the necessary Lines or a Rectangle, and add relations and dimensions. When finished, switch to the Trimetric view and create an extruded feature (2.00 height) of the box through the Extruded Boss/Base button as shown in the following figure on the right.

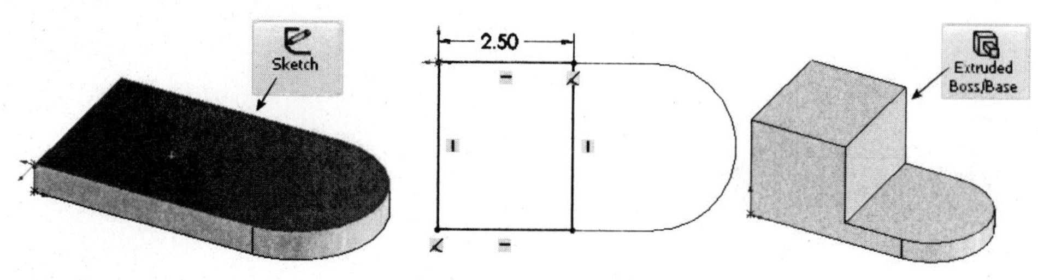

FIGURE 3.4

The next feature to create will be a square cut through the part. Pick the front face of the part and click the Sketch tool as shown in Figure 3.5 on the left. Switch to Normal view, sketch the necessary lines, and add relations and dimensions. When finished, switch to the Trimetric view and cut this sketch through the box using the Extruded Cut button as shown in the following figure on the right.

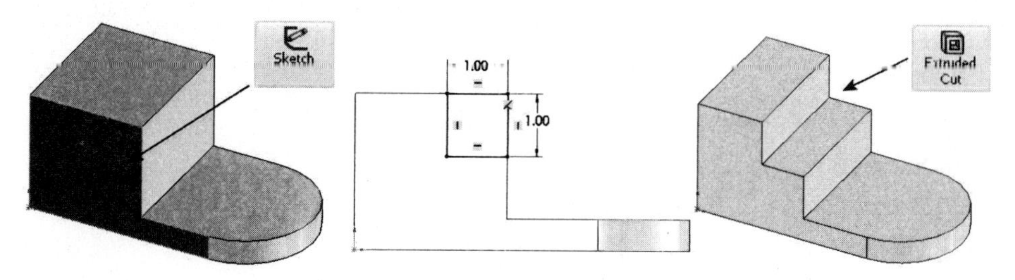

FIGURE 3.5

Create another sketch plane on the front face and switch to Normal view. Create a circle and add the necessary dimensions. Switch to Trimetric View. Cut the circle through the part using the Extruded Cut tool as shown in Figure 3.6 on the right.

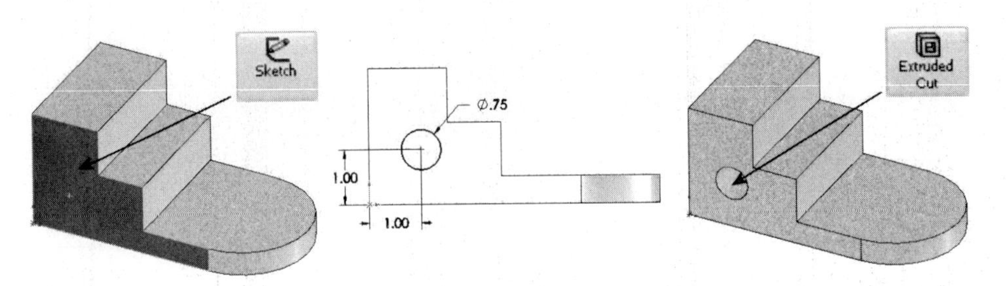

FIGURE 3.6

Add the last feature by creating another sketch plane, this time on the top face of the base feature as shown in Figure 3.7 on the left. Switch to Normal view and create a circle; add the necessary relations and dimensions. Cut this circle through the base feature of the part using the Extruded Cut tool as shown in the following figure on the right. Switch to Trimetric View.

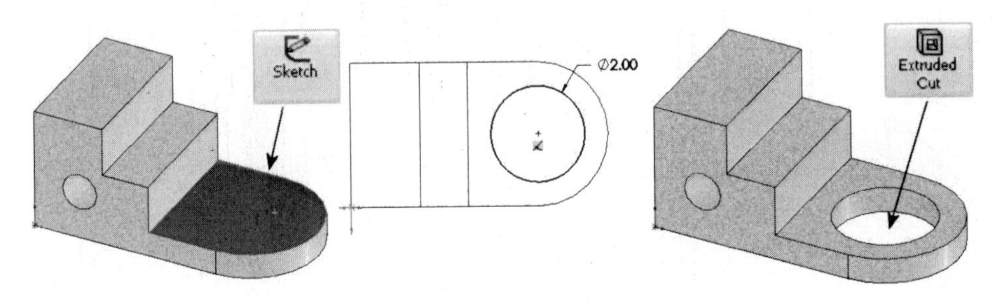

FIGURE 3.7

As features are created, they are listed in the FeatureManager as shown in Figure 3.8 on the left. This figure illustrates the default names that SolidWorks assigns. Each feature can be better identified through renaming as shown in the following figure on the right. In a complicated part file with numerous features, it is considered good practice to rename the features and keep better track of the features as the part model is being built.

Default Feature Names	Renamed Feature Names
Extrude1	Base
Extrude2	2.50x3.00 Block
Cut-Extrude1	1.00x1.00 Cut
Cut-Extrude2	.75 Dia Thru Hole
Cut-Extrude3	2.00 Dia Thru Hole

FIGURE 3.8

FEATURE TOOLS

SolidWorks provides a full compliment of tools for creating features as shown in the CommandManager in Figure 3.9. While the most popular tools are present, additional tools will be explained later on in this chapter.

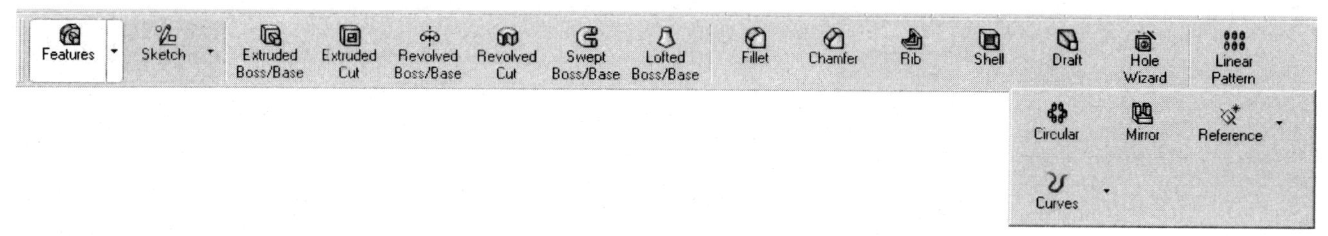

FIGURE 3.9

A more detailed description of each feature tool is listed as follows.

TABLE 3.1

Sketch Button	Tool	Description
Features	Features	Displays a menu for choosing the most popular feature commands.
Extruded Boss/Base	Extruded Boss/Base	Extrudes a sketch in one or two directions to create a solid feature.
Extruded Cut	Extruded Cut	Creates an extruded cut in a solid model by extruding a sketch profile.
Revolved Boss/Base	Revolved Boss/Base	Revolves a sketch or sketch contours around an axis to create a solid feature.
Revolved Cut	Revolved Cut	Cuts a solid model by revolving a sketched profile around an axis.
Swept Boss/Base	Swept Boss/Base	Sweeps a closed profile along an open or closed path to create a solid feature.
Lofted Boss/Base	Lofted Boss/Base	Adds material between two or more profiles to create a solid feature.
Fillet	Fillet	Creates a rounded internal or external face along one or more edges in a solid or surface feature.

continued

TABLE 3.1 *Continued*

Sketch Button	Tool	Description
Chamfer	Chamfer	Creates a bevel feature along an edge, a chain of tangent edges, or a vertex.
Rib	Rib	Adds thin walled support to a solid body.
Shell	Shell	Removes material from a solid body to create a thin wall feature.
Draft	Draft	Tapers model faces by a specified angle, using a neutral plane or a parting line.
Hole Wizard	Hole Wizard	Inserts a hole using a predefined cross section.
Linear Pattern	Linear Pattern	Patterns features, faces, and bodies in one or two linear directions.
Circular Pattern	Circular Pattern	Patterns features, faces, and bodies around an axis.
Mirror	Mirror	Mirrors features, faces, and bodies about a face or a plane.
Reference Geometry	Reference Geometry	Contains plane, axis, coordinate system, and point reference geometry commands.
Curves	Curves	Contains curve commands such as Project Curve, Split Line, and Helix, to name a few.

CREATING EXTRUDED CUTS

Performing an extruded cut on a part removes material. Choose this command from the CommandManager and the Features tab as shown in Figure 3.10 on the left. You can also access the extruded cut tool from the Insert pull-down menu under the Cut heading as shown in the following figure on the right.

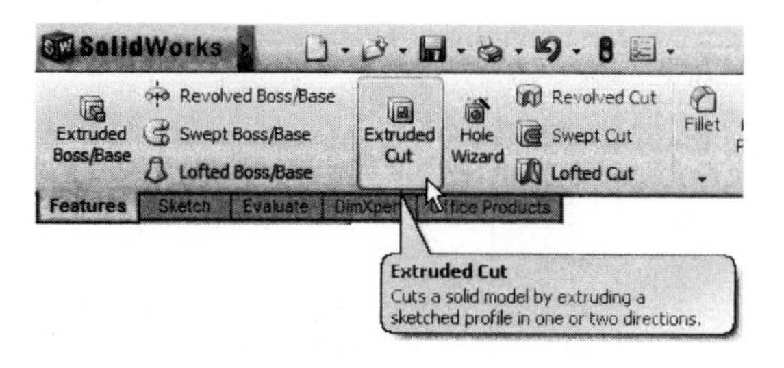

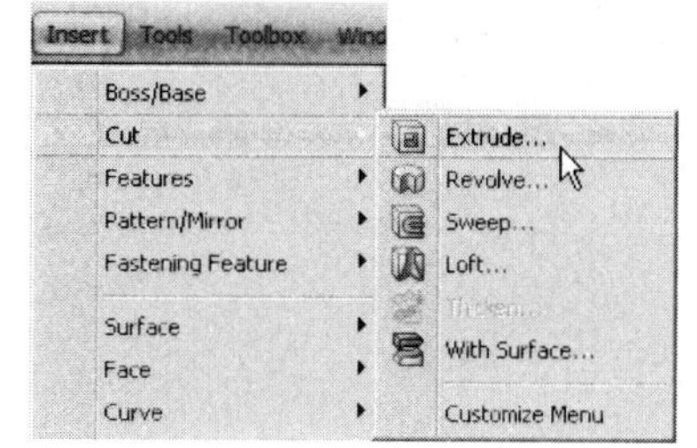

FIGURE 3.10

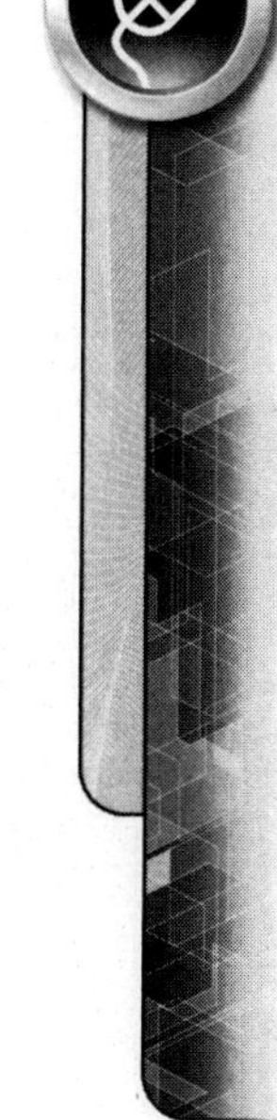

TUTORIAL EXERCISE: 03_EXTRUDED_CUTS.SLDPRT

This tutorial exercise is designed to create a part model that consists of numerous geometric shapes. Proper relations need to be applied along with dimensions to fully define the shape (Figure 3.11).

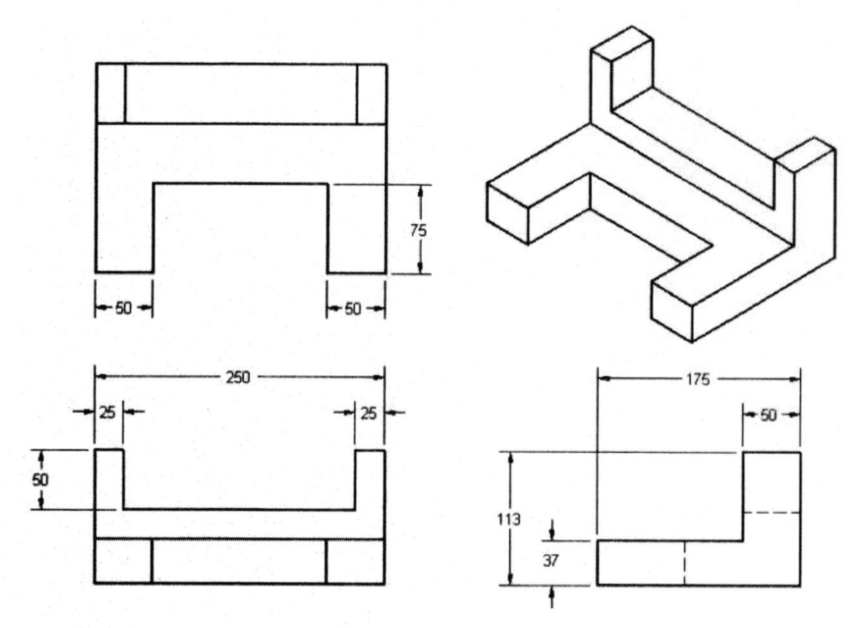

FIGURE 3.11

STEP 1

Begin a new part file and click on the Front Plane as the new sketch plane as shown in Figure 3.12 on the left. Then sketch the shape as shown in the following figure on the right. All lines should either be horizontal or be vertical. Also, add all dimensions to fully define the sketch. Extrude this sketch to a distance of 250.

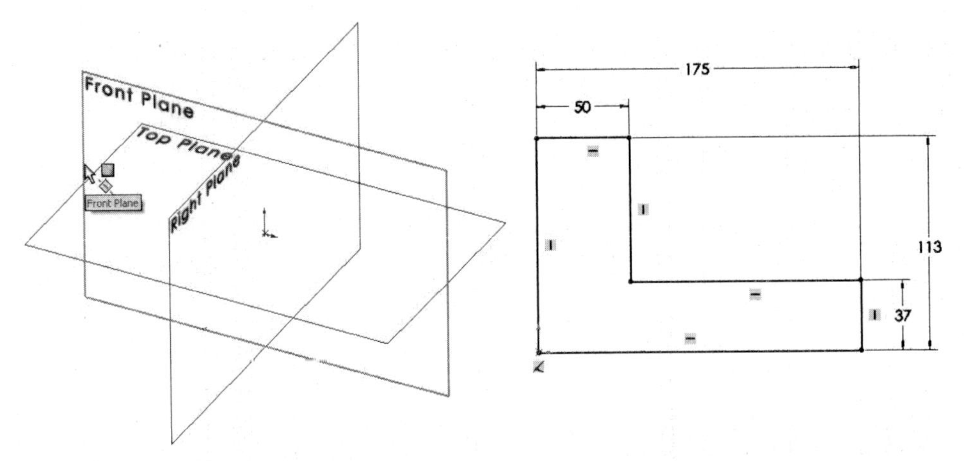

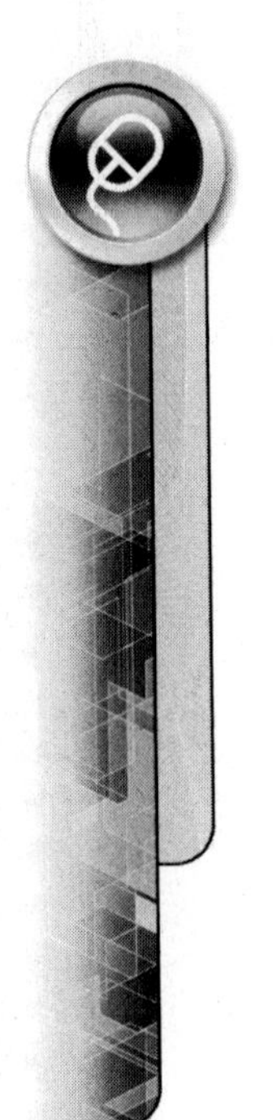

FIGURE 3.12

STEP 2

Once the base feature is created, click the Sketch tool and pick on the face of the part model as shown in Figure 3.13.

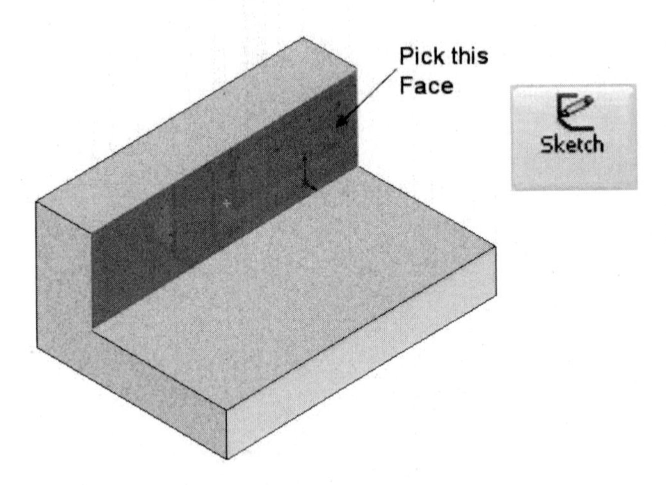

FIGURE 3.13

STEP 3

In the Orientation dialog box, double-click on *Normal To as shown in Figure 3.14 on the right. This should rotate the view allowing you to view the sketch in a perpendicular or normal direction as shown in the following figure on the left.

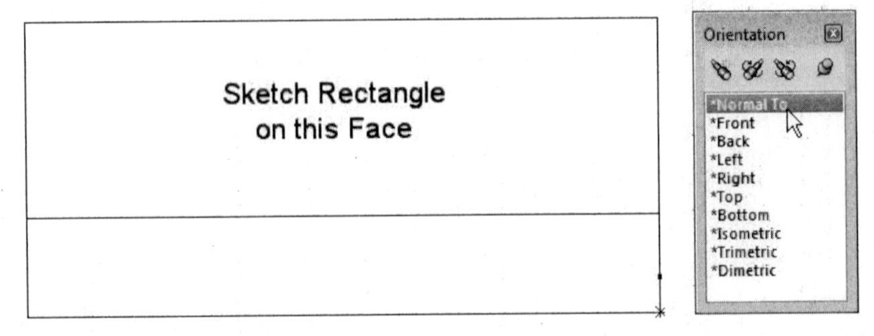

FIGURE 3.14

STEP 4

Construct a Rectangle representing the vertical slot. Add dimensions as shown in Figure 3.15 on the left. If necessary, make the top of the horizontal line collinear with the top edge of the object. The results are shown in the following figure on the right.

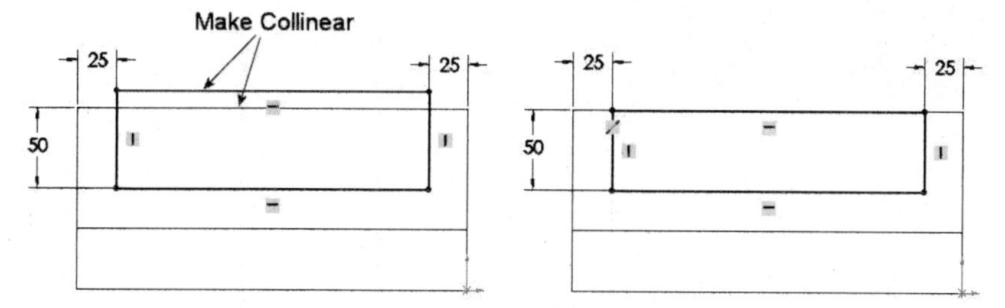

FIGURE 3.15

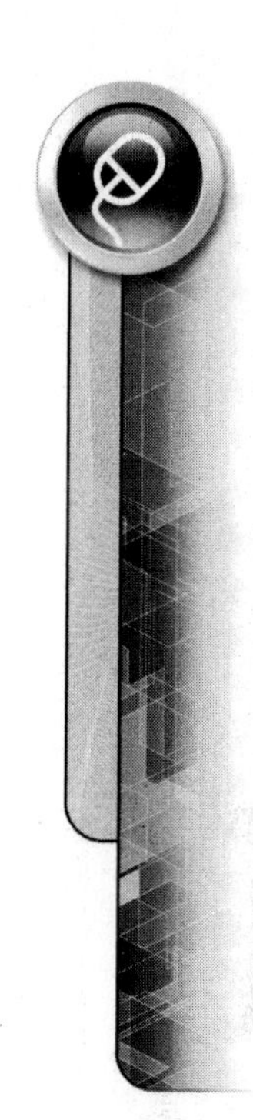

STEP 5

Switch to Trimetric View. Click the Extruded Boss/Base tool. When the Cut-Extrude Properties dialog box appears as shown in Figure 3.16 on the left, set Direction 1 to Through All. A preview of the extruded cut feature will be displayed as shown in the following figure on the right. Click the checkmark to create the extruded cut feature.

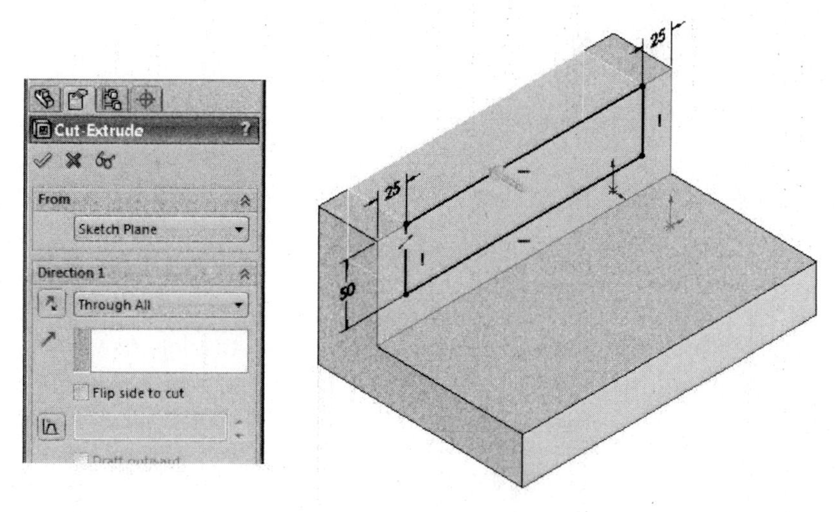

FIGURE 3.16

STEP 6

Create another extruded cut feature by clicking the Sketch tool and picking on the top face as shown in Figure 3.17.

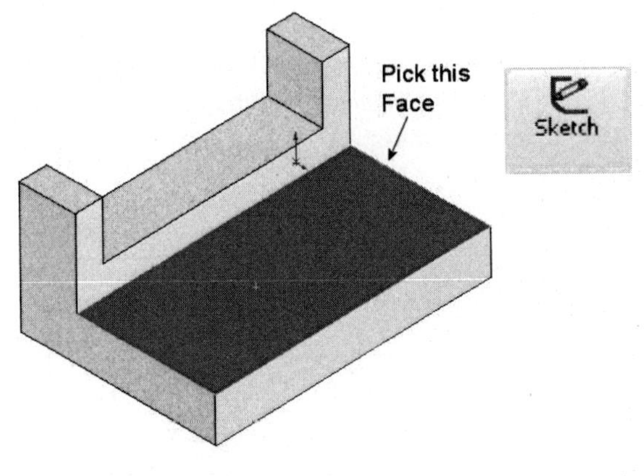

FIGURE 3.17

STEP 7

In the Orientation dialog box, double-click on *Normal To as shown in Figure 3.18 on the right. This should rotate the view allowing you to view the sketch in a perpendicular or normal direction as shown in the following figure on the left.

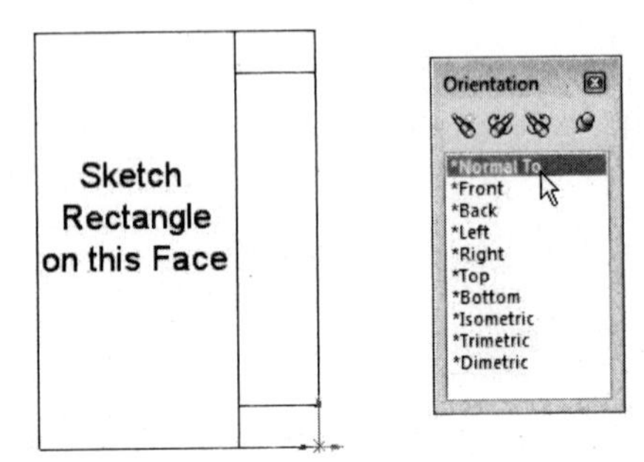

FIGURE 3.18

STEP 8

Create another rectangle; add dimensions and relations as shown in Figure 3.19 on the left. The extruded cut feature will preview as shown in the following figure on the right. Click the checkmark (OK) to create the extruded cut feature. Switch to Trimetric View.

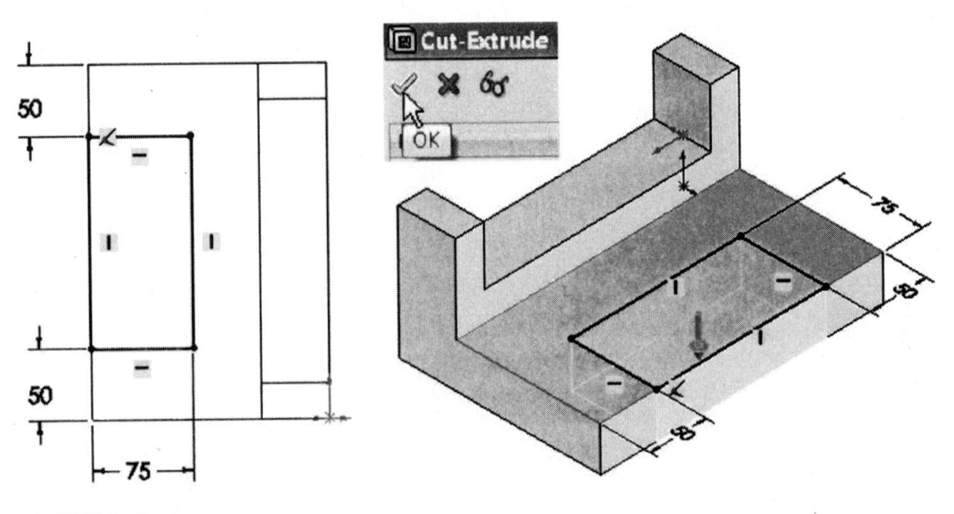

FIGURE 3.19

STEP 9

The finished object consisting of two extruded cuts is illustrated as shown in Figure 3.20 on the left. Notice that the FeatureManager keeps track of all features and assigns default names. You should rename these to names that have more meaning such as the size of the extruded cut features.

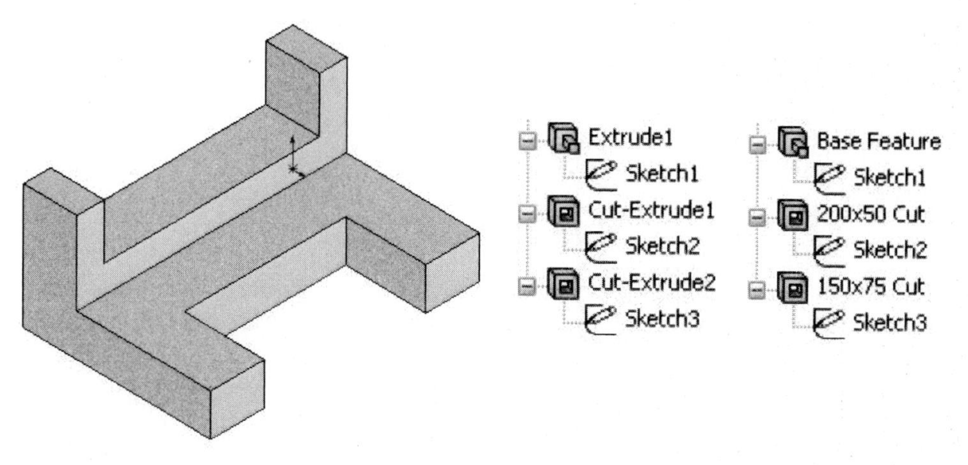

FIGURE 3.20

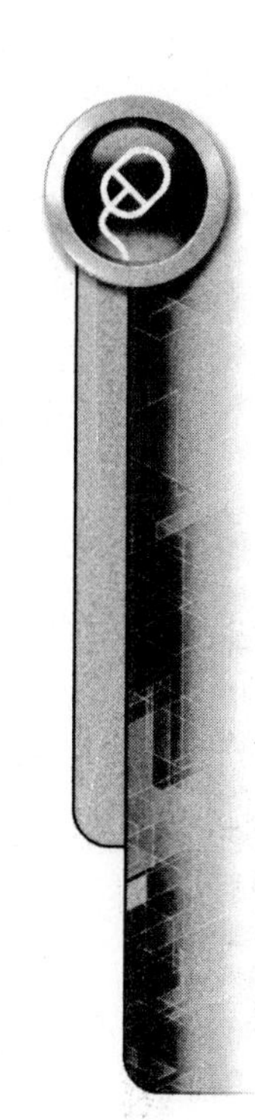

CREATING SIMPLE HOLE FEATURES

A simple hole feature creates a hole based on the diameter and hole depth on a planar face. The center of the hole is placed at an approximate location on the planar face. A sketch plane will automatically be created. Dimensions for accurately locating this hole will be added after the hole is created by editing its sketch plane. Select this tool by clicking Insert > Features > Hole > Simple as shown in Figure 3.21.

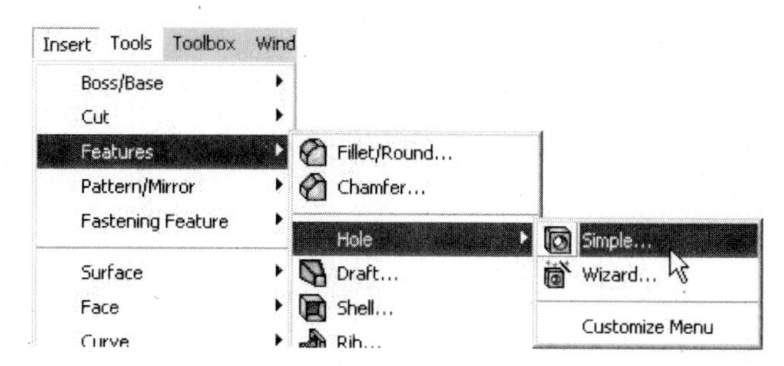

FIGURE 3.21

To practice placing this feature, open the file Simple_Hole01. Clicking this tool located in the Insert pull-down menu will display a message in the PropertyManager stating to select a location for the hole. Move your cursor on the planar face in Figure 3.22 and click this location.

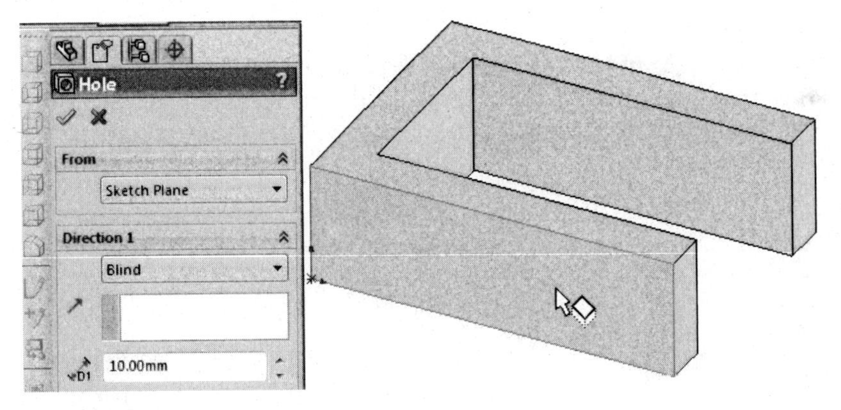

FIGURE 3.22

While in the PropertyManager, set the direction for the hole along with the diameter. In Figure 3.23, the hole direction has been set to Through All with a diameter of 1.00. Notice the hole previewing as shown in the following figure on the right.

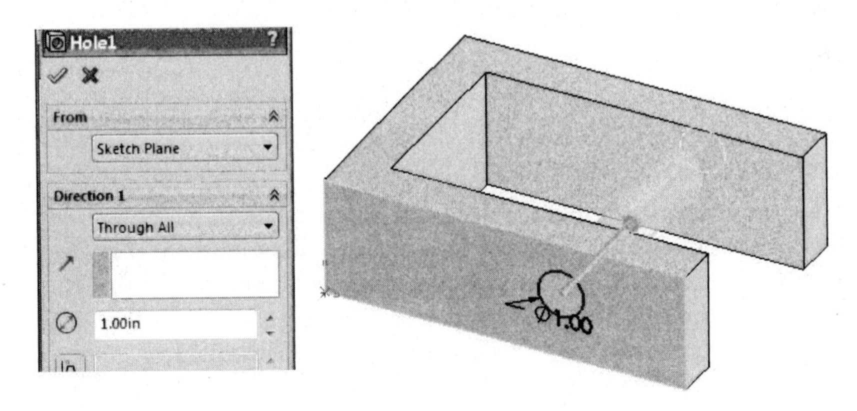

FIGURE 3.23

The hole will be placed in the model based on the settings made in the PropertyManager. This hole, however, is not accurately placed. To better position the hole, right-click on the hole feature sketch in the FeatureManager and click the Edit Sketch option from the menu as shown in Figure 3.24 on the left.

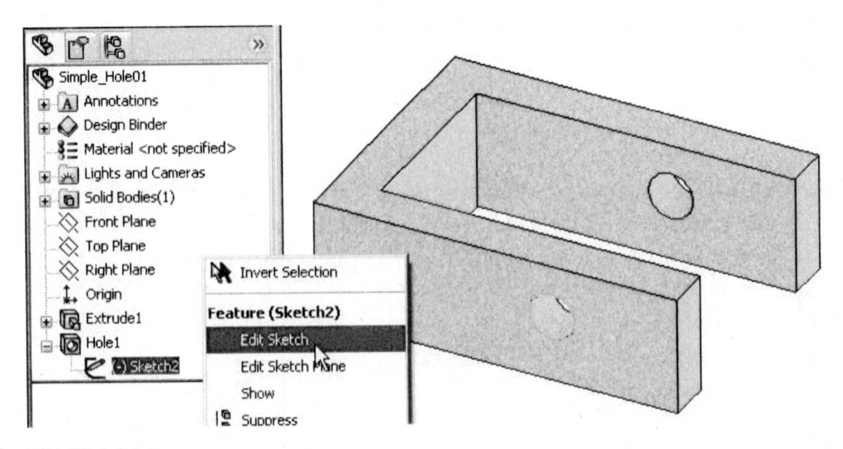

FIGURE 3.24

Add dimensions and/or relations to position the hole. You can also edit the hole diameter as shown in Figure 3.25 on the left. When finished, click the Rebuild button located in the Standard toolbar. This will update the new position of the hole as shown in the following figure.

Note: If you need to change the hole diameter or depth, right-click on the hole feature located in the FeatureManager and select Edit Feature icon from the menu as shown in the following figure on the right.

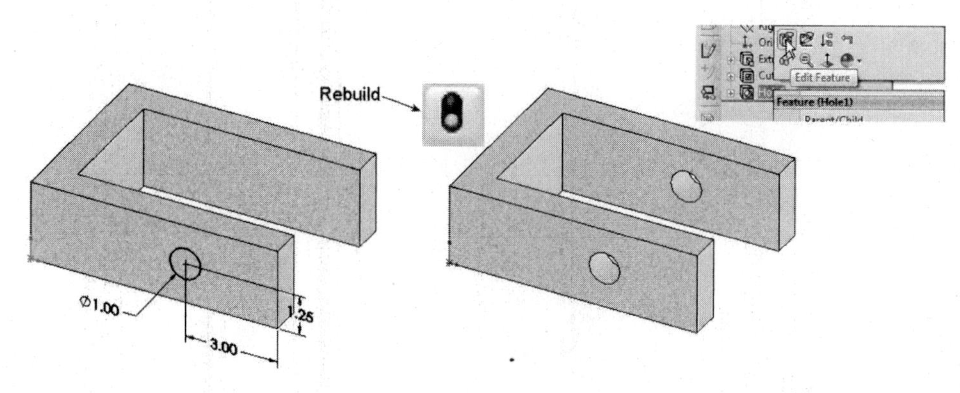

FIGURE 3.25

To prevent the accidental adding of materials to the inside of holes, build holes at the end of the design process.

When placing specialty holes such as counterbore, countersink, or threaded, use the Hole Wizard. This topic will be discussed in chapter 4.

When deciding whether to use the Simple Hole tool or Hole Wizard, choose Simple Hole since this tool provides better model performance.

CREATING FILLET FEATURES

Many objects require highly finished and polished surfaces consisting of extremely sharp corners. Fillets and rounds represent the opposite case where corners are rounded off either for ornamental purposes or as required by the design. Figure 3.26 shows three examples of fillets being applied to a 3D model.

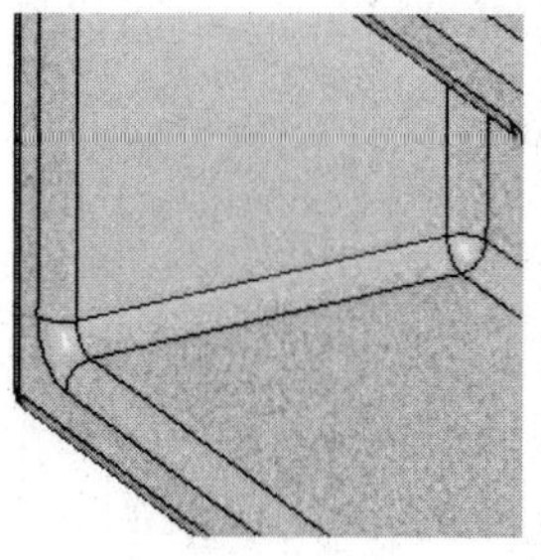
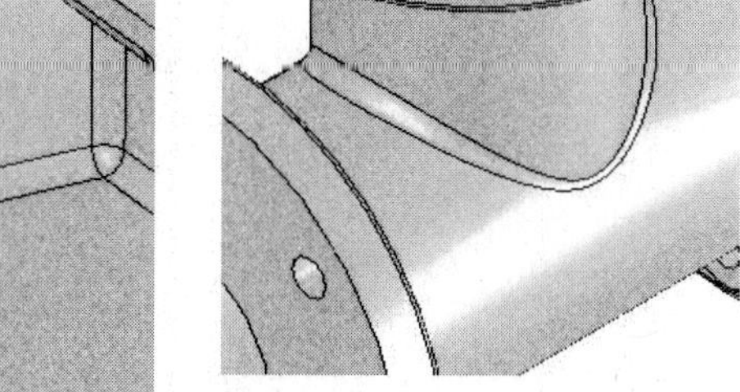
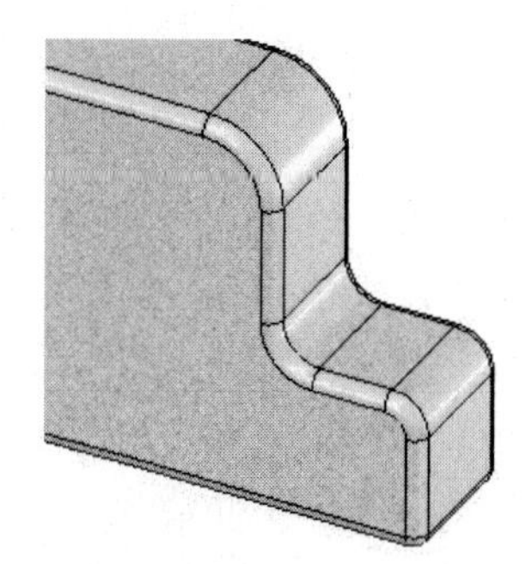

FIGURE 3.26

Activate the Fillet tool in the Feature menu and select the two edges of the model as shown in Figure 3.27 on the left. Set the fillet radius to 30.00mm. The results are illustrated in the following figure on the right. Notice also on the right that a second fillet has been applied to the base of the part; this time a radius of 60mm has been applied.

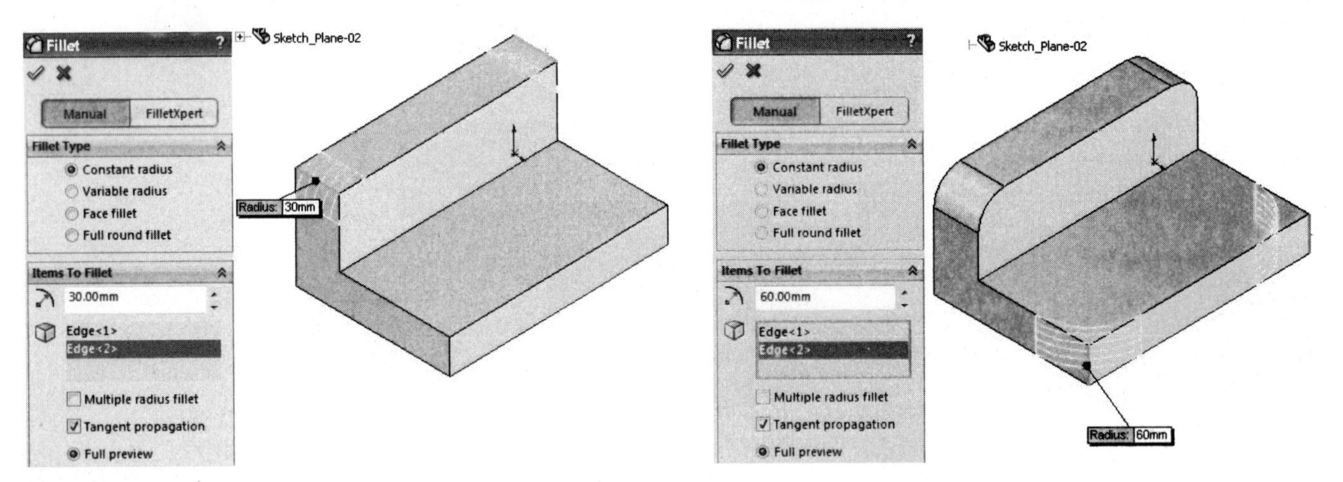

FIGURE 3.27

Constant Radius

To fillet the entire part, use a fillet with a constant radius. Checking the box next to Multiple radius fillet will display all selected fillets as shown in Figure 3.28 on the right.

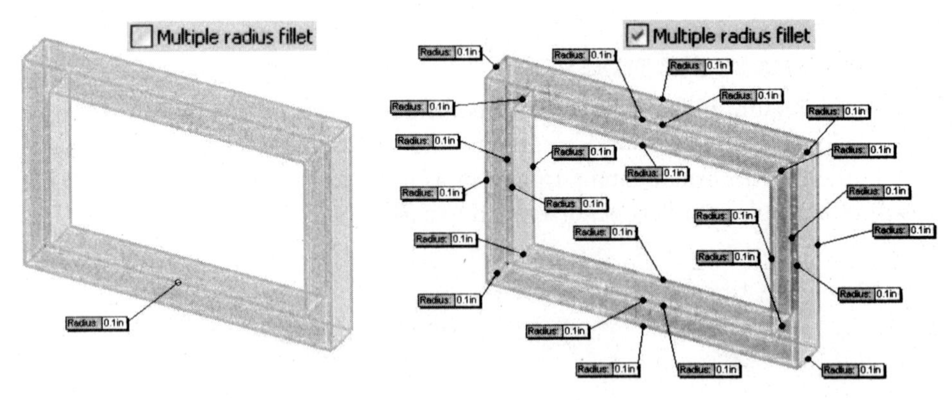

FIGURE 3.28

Variable Radius

Figure 3.29 on the left is an example of applying a variable radius fillet. Notice how the fillet previews a fillet starting with a radius of .40, then applying a fillet radius of .30 to a second location, and so on until the ending fillet measures R.10. The resulting variable radius fillet is illustrated in the following figure on the right.

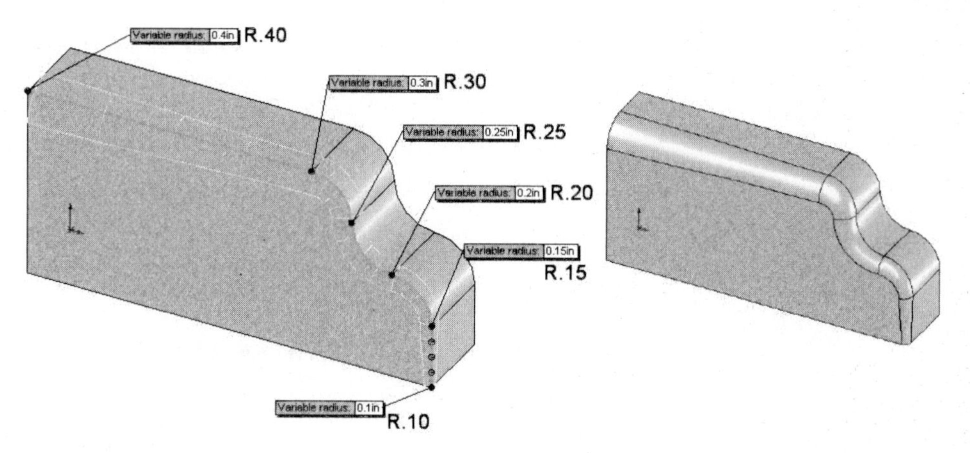

FIGURE 3.29

Fillet Order Is Important

When applying fillets to a model, the order you select edges or faces to be filleted can be important (Figure 3.30).

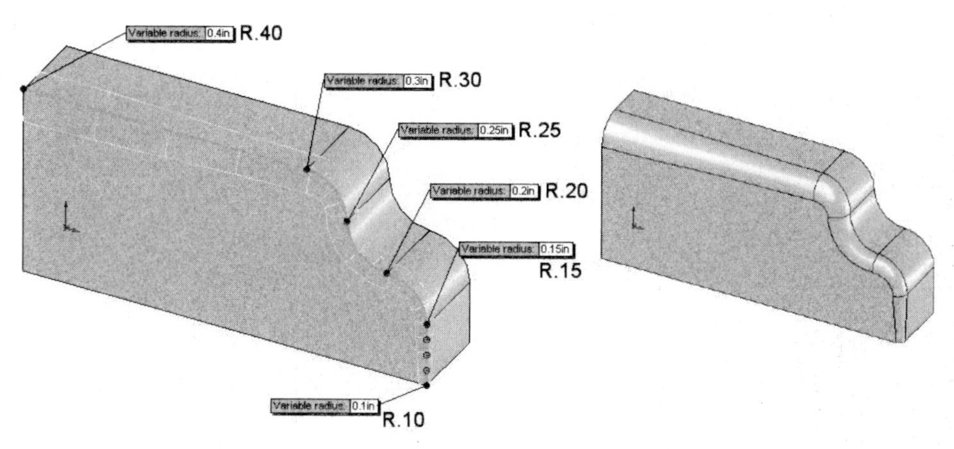

FIGURE 3.30

CREATING CHAMFER FEATURES

Various types of chamfered edges can be created in order to cut the edge of a model at a user-specified angle. Chamfers can also be created by two distances or by clicking on the corner of a vertex. The vertex method prompts you for three distances. Each of these chamfer methods is illustrated in Figure 3.31.

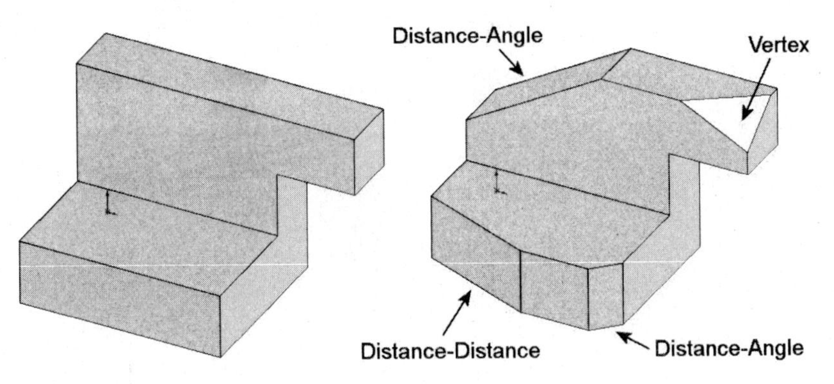

FIGURE 3.31

Distance-Angle (45°)

Clicking on the Chamfer tool button will display a number of chamfer settings in the PropertyManager. The first chamfer mode is Angle distance where you pick an edge on the part model and enter a distance and an angle. In the following example, a distance of 10mm and a 45° angle have been specified. Select the edge of the part model as shown in Figure 3.32. A preview of the chamfer will appear allowing you to determine if this size and angle of the chamfer are correct based on the model. Click the green checkmark to accept these values and create the chamfer.

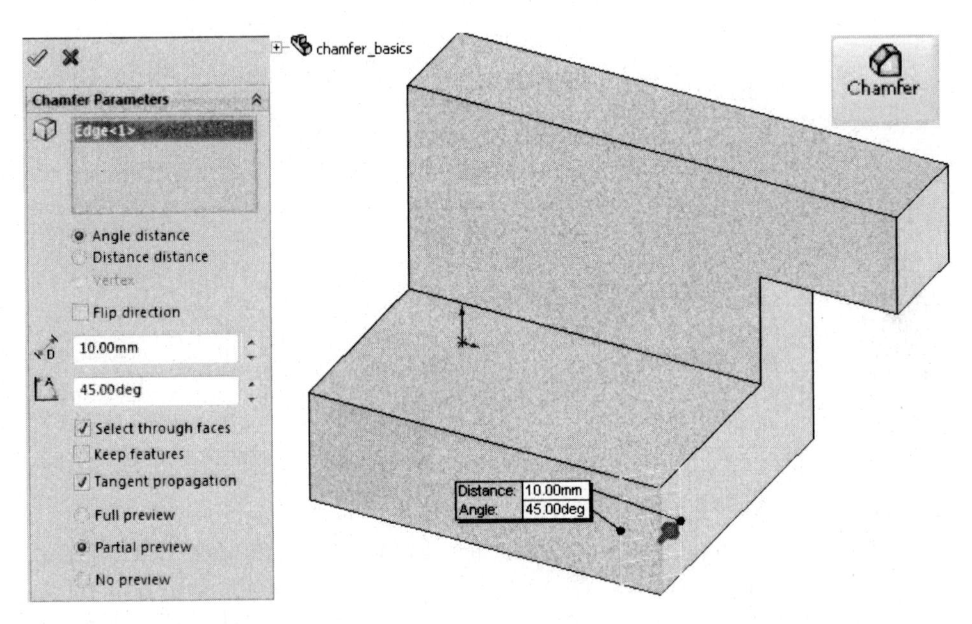

FIGURE 3.32

Distance-Angle (30°)

The following example shows another example of using the Angle distance mode of the Chamfer tool. By definition, the angle of a true chamfer measures 45°. The following example illustrates a chamfer that uses an angle of 30°. This is sometimes referred to as a bevel cut. First select the edge as shown in Figure 3.33 and then change the distance to 60mm and the angle to 30° all through the PropertyManager. You may notice the presence of an arrow which allows you to flip the direction of the chamfer by checking the appropriate box in the PropertyManager. Click the green checkmark to accept these values and create the chamfer.

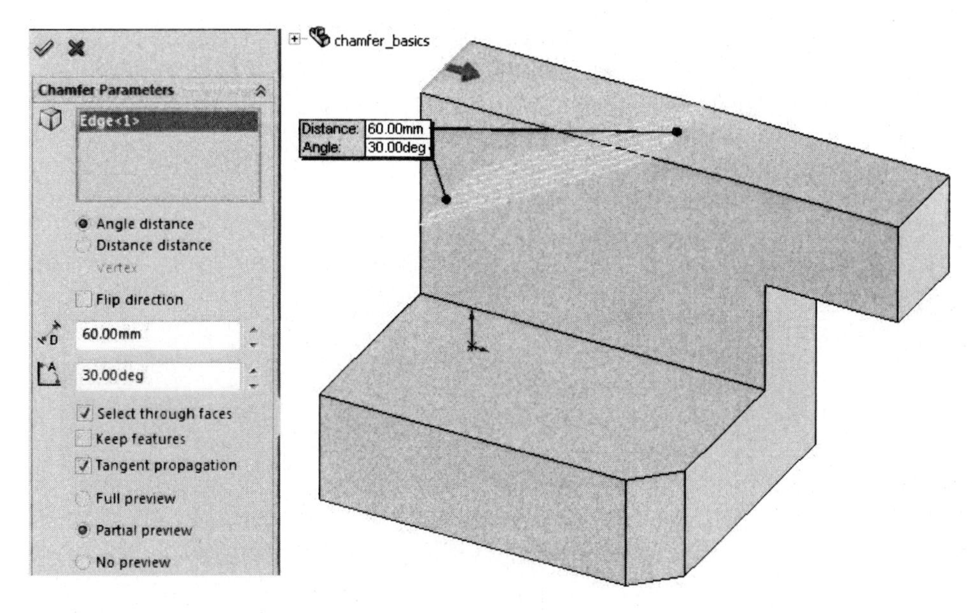

FIGURE 3.33

Distance-Distance

This next example demonstrates the use of the Distance distance mode of the Chamfer tool. Clicking on this mode in the PropertyManager will activate two boxes that will accept distance values. These distances can either be equal or be different. The following example illustrates two separate distances where the first distance is 20mm and the second distance is 50mm. Click the appropriate edge on the part model and click the green checkmark to accept these values and create the chamfer (Figure 3.34).

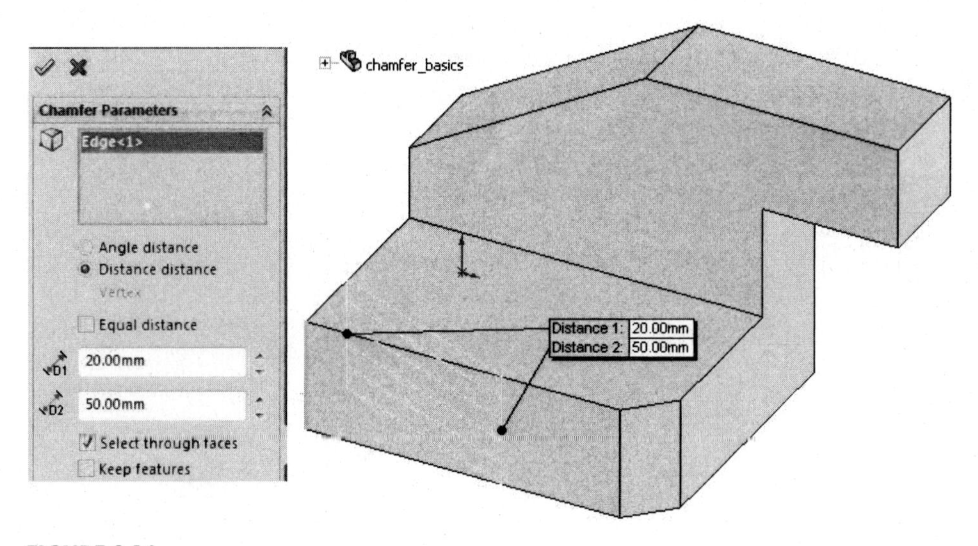

FIGURE 3.34

Vertex (Distance-Distance-Distance)

Another method of creating a chamfer is through the Vertex option. Clicking on this mode in the PropertyManager will prompt you for three distances. First select the corner vertex on the part model as shown in Figure 3.35. Then change distance 1 to 15mm, distance 2 to 20mm, and distance 3 to 30mm. The previewed results are shown in the following figure. All three distances are measured from the common vertex to give a compound or oblique chamfer. Click the green checkmark to accept these values and create this type of chamfer.

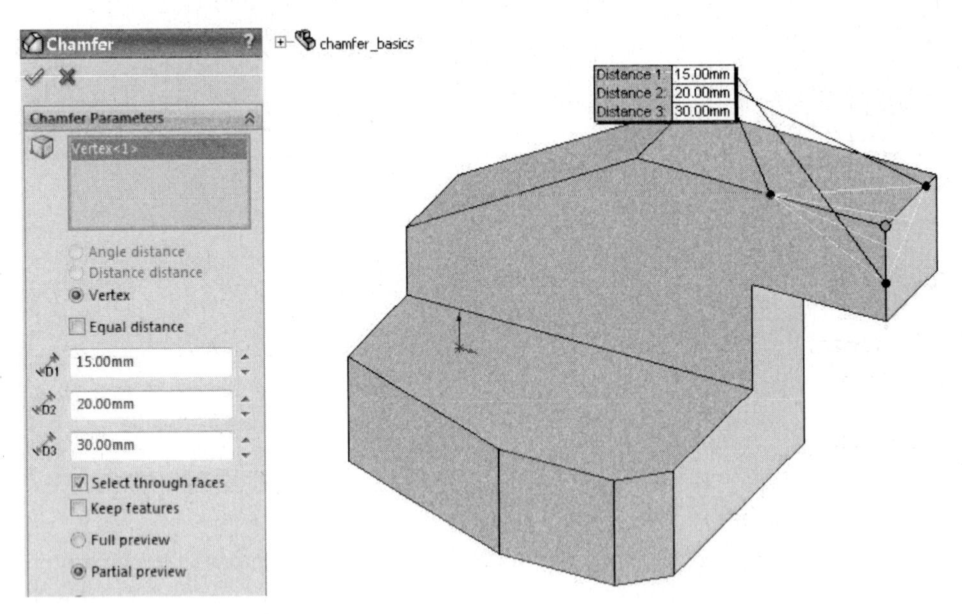

FIGURE 3.35

Keep Features

When chamfers are placed in conjunction to other placed features, the results can be surprising. In the following example, a number of cylinders intersect the top face of a part model. A chamfer with distance of 100mm and an angle of 10° will be created from the top edge of the part model. The results are illustrated in Figure 3.36 on the right. Notice that a number of cylinders were removed as a result of placing this chamfer.

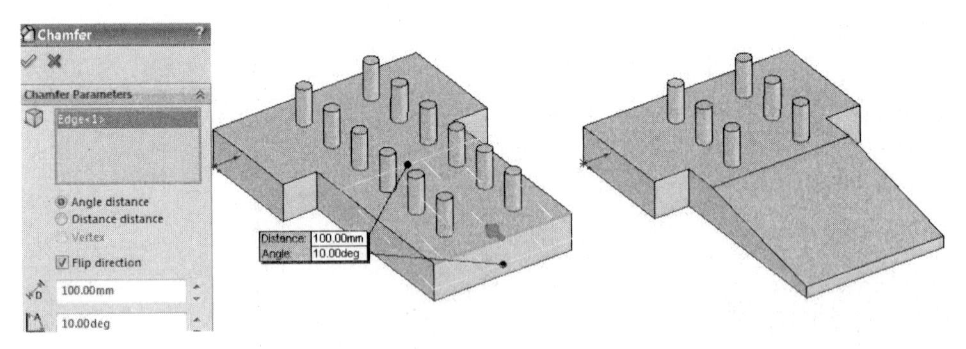

FIGURE 3.36

To create the chamfer and retain all cylinder extrusions, edit this chamfer feature and place a check in the box next to Flip direction in the Property Manager. The results are illustrated in Figure 3.37 on the right with all cylinder features being retained even though the surface was chamfered.

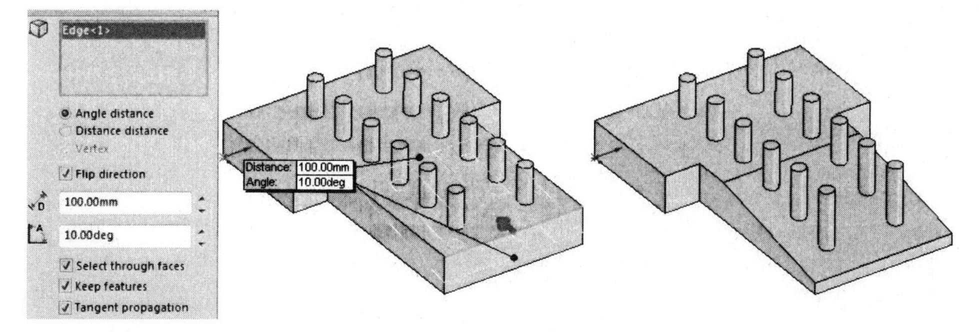

FIGURE 3.37

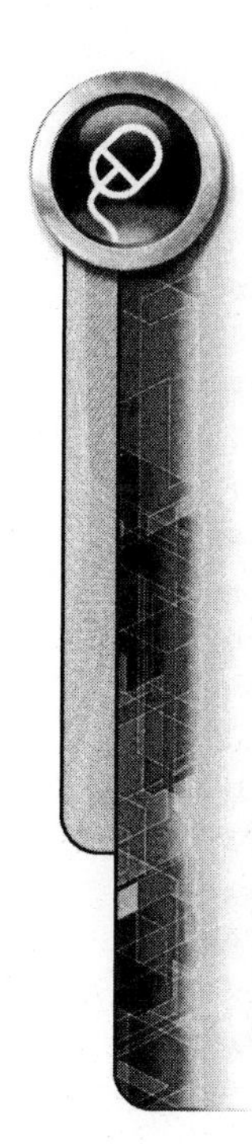

TUTORIAL EXERCISE: 03_TUTORIAL_KEEP.SLDPRT

To assist with this tutorial exercise the following steps are provided to guide you along to the completion of this part model. Begin by making the right plane the current sketch plane and construct the sketch as shown on the left in Figure 3.38. Then extrude this shape a distance of 8 inches as shown in the following figure in the middle. Create another sketch plane on the top face of the object, sketch a rectangle, and add dimensions as shown in the following figure on the right.

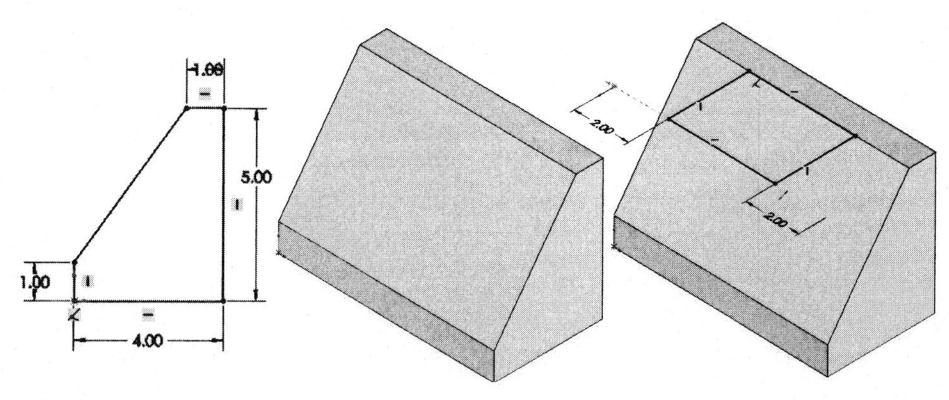

FIGURE 3.38

Activating the *Normal To orientation as shown on the right in Figure 3.39 will display the part model as shown in the following figure on the left.

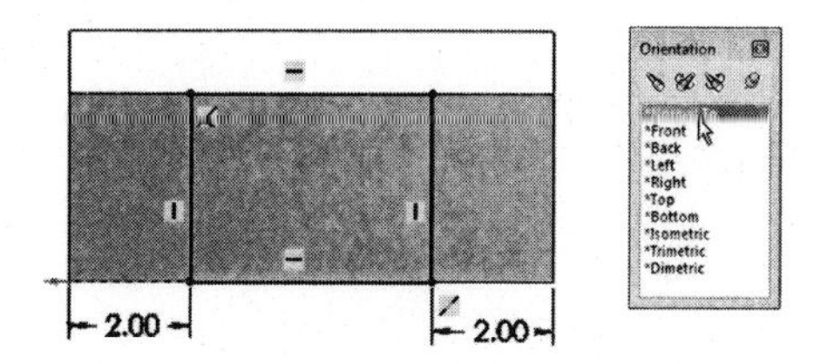

FIGURE 3.39

Use a Blind Extruded Cut, 4.00" deep, to cut the rectangular sketch down into the part to the next face as shown in Figure 3.40.

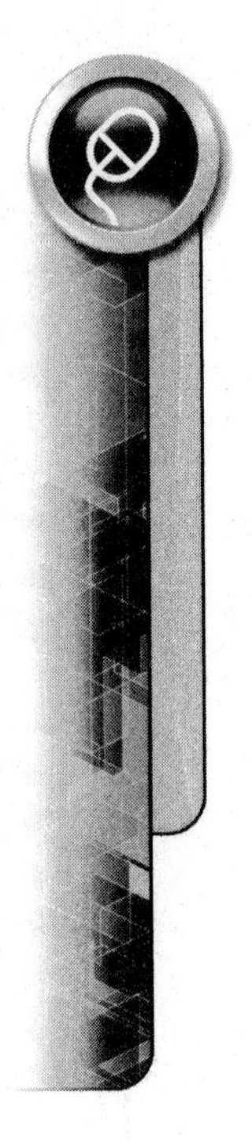

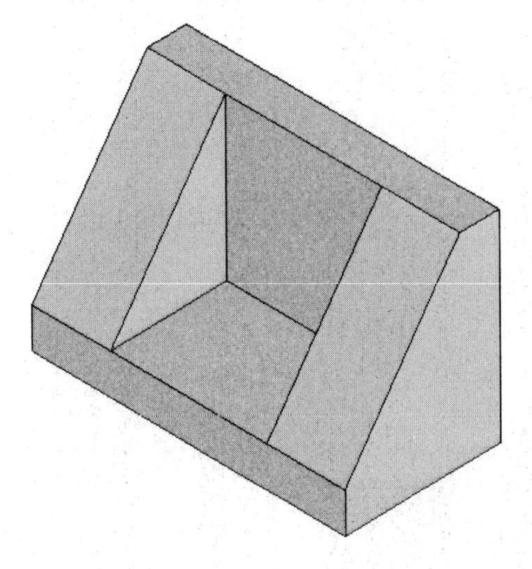

FIGURE 3.40

Create another sketch plane on the part as shown in Figure 3.41 on the left. Add the necessary dimensions to fully define the sketch. Next, cut the triangular shape through the entire part; the results are shown in the following figure on the right. Flip Side may need to be selected to get the desired result.

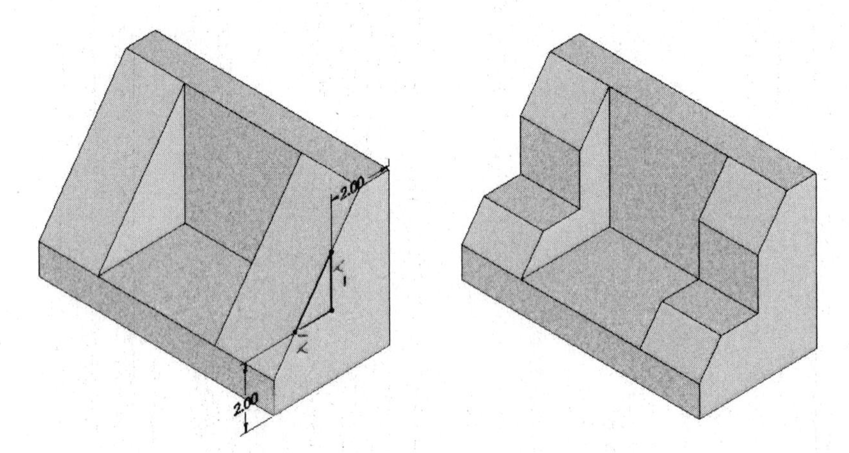

FIGURE 3.41

Create the 2.00" diameter hole in the base area of the part and cut it through all as shown in Figure 3.42.

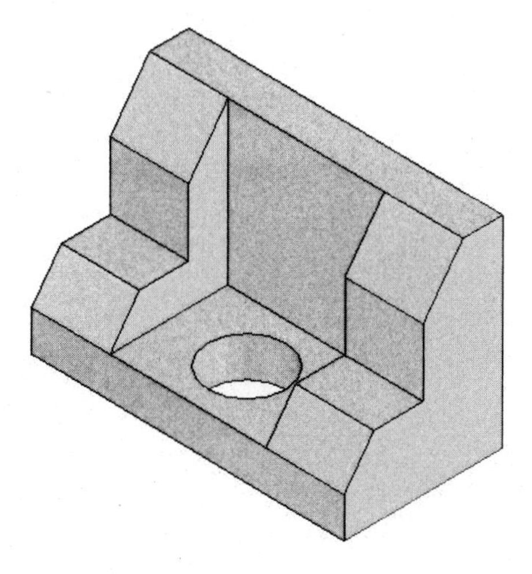

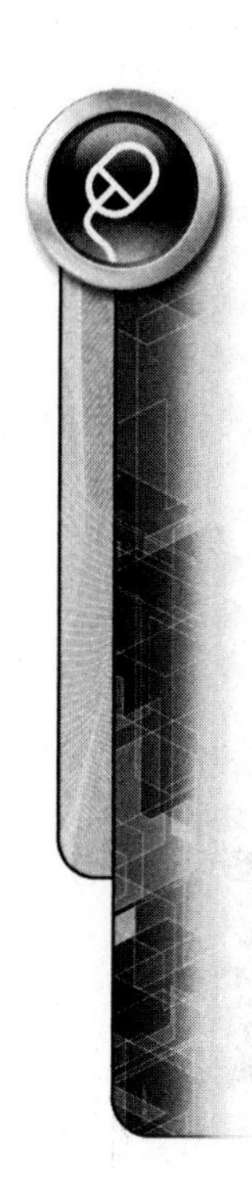

FIGURE 3.42

To construct the next through hole, you will need some construction geometry to center the circle in the space provided. Make this face the current sketch plane and create a construction (centerline) line using the two corner endpoints of the rectangular face. Then, create a circle at the midpoint of the construction line as shown in Figure 3.43 on the left. Both small holes follow similar procedures in their layout. Cut both of these circles through the part as shown in the following figure on the right.

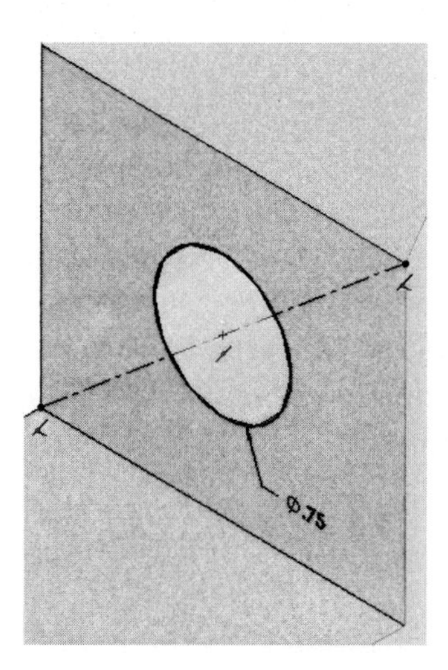

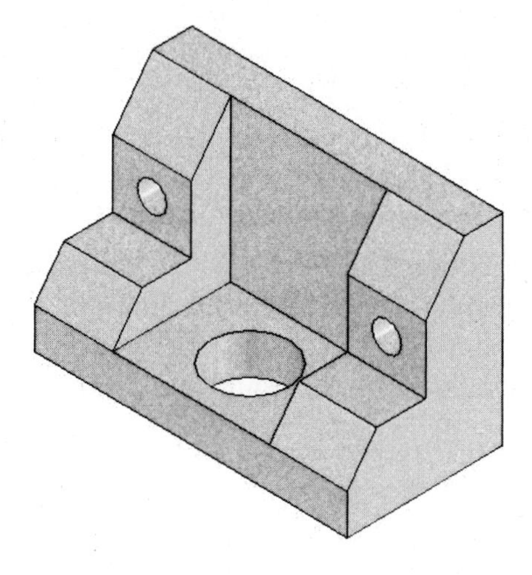

FIGURE 3.43

The last feature to create is the vertical slot in the part. Make the designated back face the current sketch plane, sketch the slot, and add relations and dimensions as shown in Figure 3.44 on the left. Cutting the slot through the part will display the results as shown in the following figure on the right.

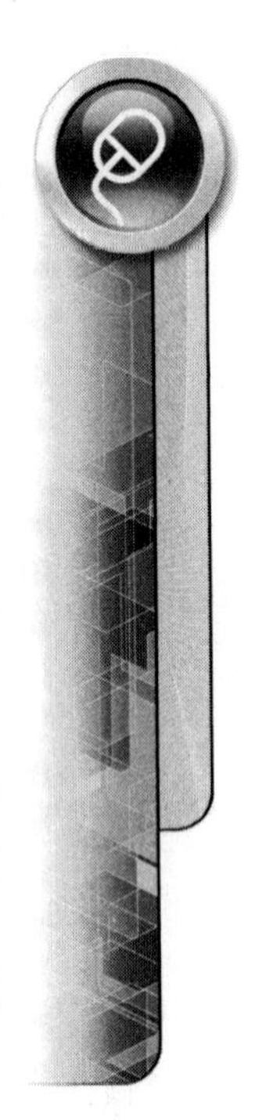

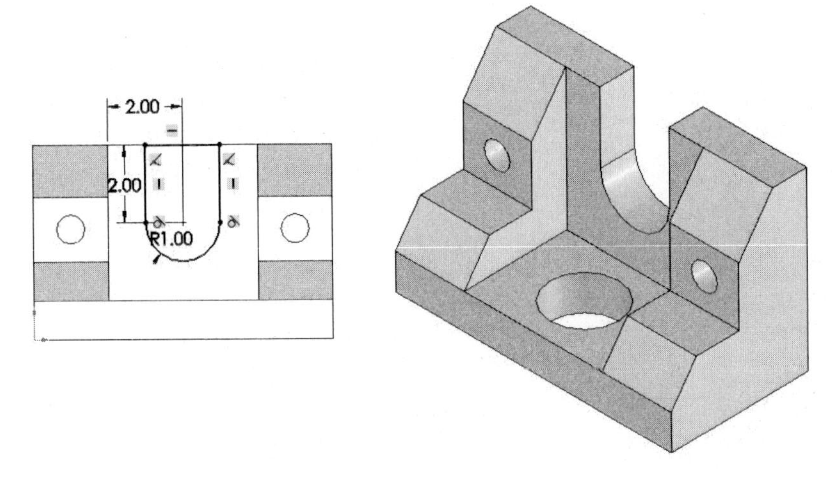

FIGURE 3.44

EXTRUSION OPTIONS AND END CONDITIONS

You have already seen how extruded features are created using the Blind option. Numerous additional options are also available (Figure 3.45).

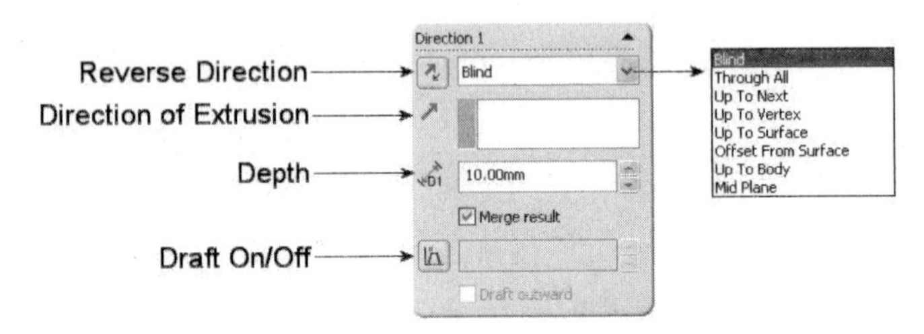

FIGURE 3.45

Extrude – Midplane

Open the file Extrude 02 (Midplane). Click the Extruded Boss/Base tool. When the base area highlights, change the Blind direction to Midplane and enter a depth of 45mm. This will create an extrusion where half of the total depth will be distributed in both directions as shown in Figure 3.46 on the right.

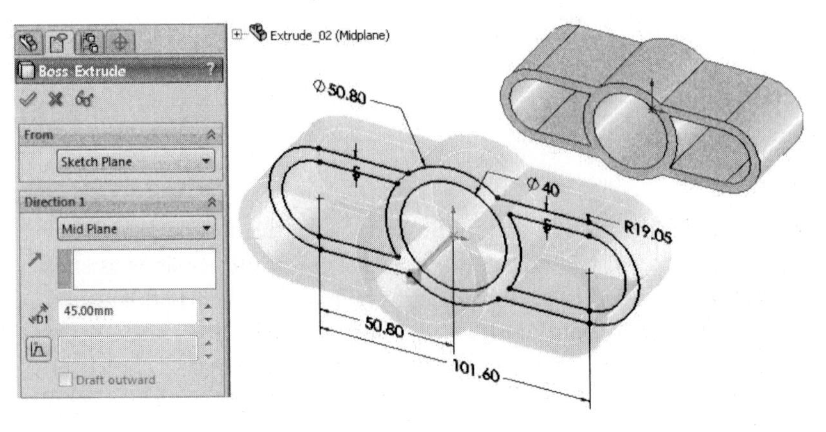

FIGURE 3.46

Extrude – To Vertex

Open the file Extrude 03 (Vertex). Click the Extruded Boss/Base tool. When the base area highlights, change the Blind direction to Up to Vertex and pick Edge<1> as shown in Figure 3.47. This will create an extrusion where the extruded shape will be controlled by this edge as shown in the following figure on the right.

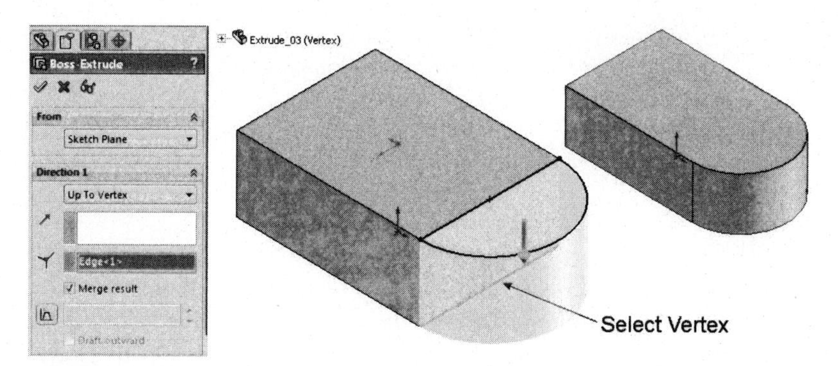

FIGURE 3.47

Extrude – Offset from Surface

Open the file Extrude 04 (Offset). This option will create an extrusion based on a distance that is offset from a face. Click the Extruded Boss/Base tool. When the base area highlights, change the Blind direction to Offset from Surface and pick the inclined face in the model. Then enter a value of 10mm as the distance to create the offset. The results are illustrated in Figure 3.48 on the right with the extruded distance being created at a distance of 10mm parallel to the inclined surface.

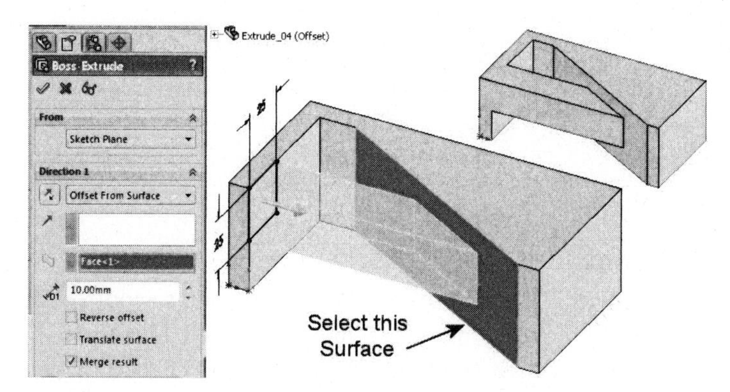

FIGURE 3.48

Extrude – To Surface

Open the file Extrude 05 (Surface). This option allows for a surface to be extruded from a selected face. Click the Extruded Boss/Base tool. When the base area highlights, change the Blind direction to Up to Surface. Then select the surface. This will create an extrusion based on the incline of the face as shown in Figure 3.49 on the right.

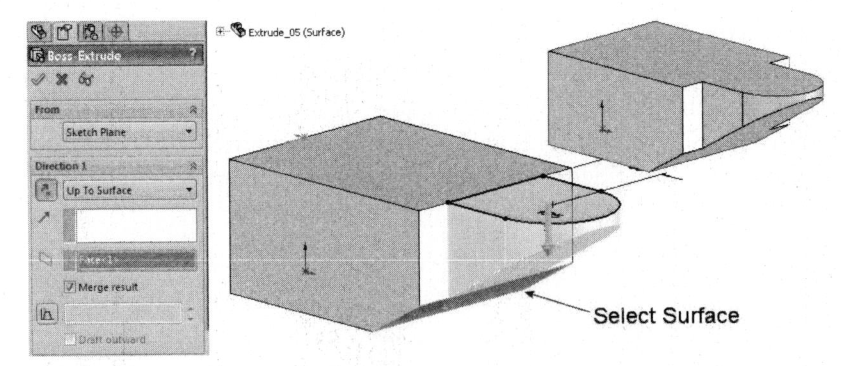

FIGURE 3.49

Extrude – To Body

Open the file Extrude 06 (Body). This option allows for a sketch up to a body. The extruded feature is completely consumed by this body. Click the Extruded Boss/Base tool. When the base area highlights, change the Blind direction to Up to Body. Then select the body. This will merge the sketch to the selected body as shown in Figure 3.50 on the right.

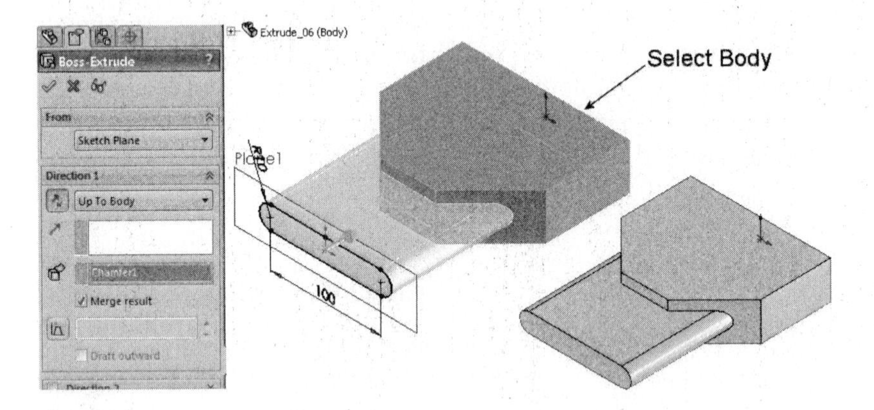

FIGURE 3.50

Extrude – Unequal Directions

Open the file Extrude 07 (Directions). This versatile feature allows you to control an extrusion based on two separate directions, namely, Direction 1 and Direction 2. Click the Extruded Boss/Base tool. When the base area highlights, change to the Blind option and set the value for Direction 1 to 50mm. Next, change to the Blind option and set the value for Direction 2 to 10mm. This will create an extrusion based on two different directions as shown in Figure 3.51 on the right.

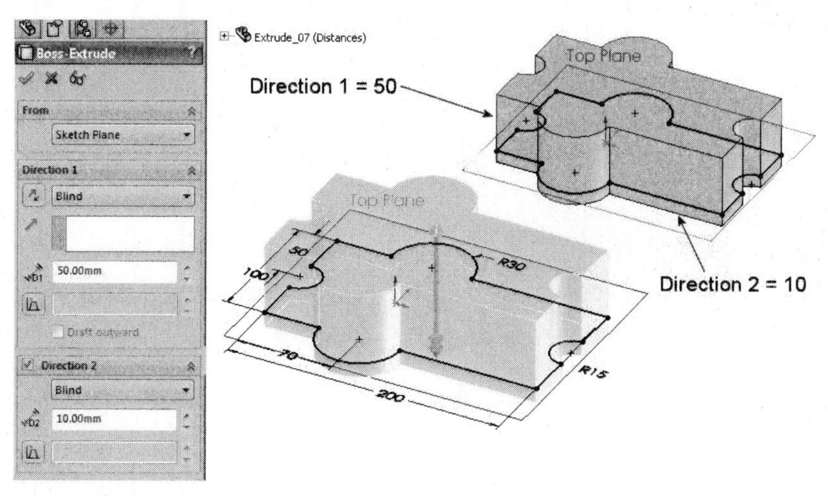

FIGURE 3.51

Extrude – To Next

Open the file Extrude 08 (Next). This option will create an extrusion up to the next surface. This option can also be merged using two different directions. To perform this operation, change the Blind direction to Up to Next under Direction 1. Verify that the Up to Next option is also set for Direction 2. This will create an extrusion based on two different up to next directions as shown in Figure 3.52 on the right.

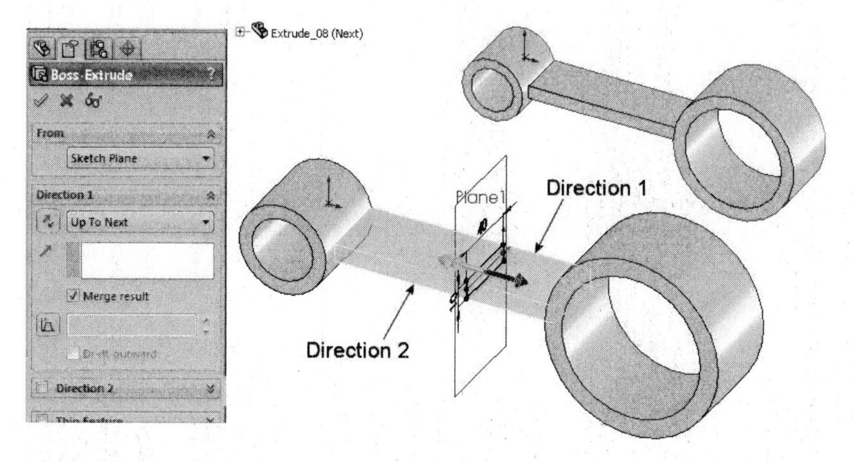

FIGURE 3.52

Extrude – Close Sketch

Open the file Extrude 09 (Close). This next operation is not necessarily an option of an extruded base. Rather it deals with an uncompleted sketch as shown in Figure 3.53. When the extruded tool detects that the sketch is open, an alert box allows you to close the sketch; sometimes you may need to even reverse the direction of the closed sketch. The results of closing an open sketch are shown in the following figure on the right.

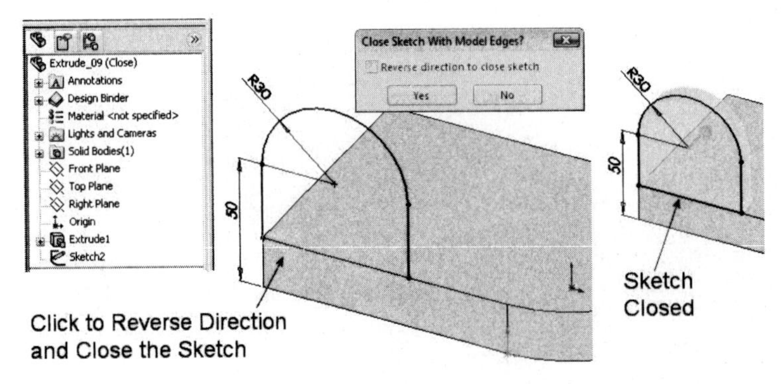

FIGURE 3.53

Extrude – Thin Feature

Open the file Extrude 10 (Thin). This specialized option allows for an extrusion to be created from an open sketch. You first enter information under Direction 1 along with the Blind option and set a distance of 40mm. Expanding the PropertyManager will display more information dealing with the creation of a thin feature. In this example, the Mid Plane option is being used along with a distance of 15mm. The resulting thin feature extrusion is shown in Figure 3.54 on the right.

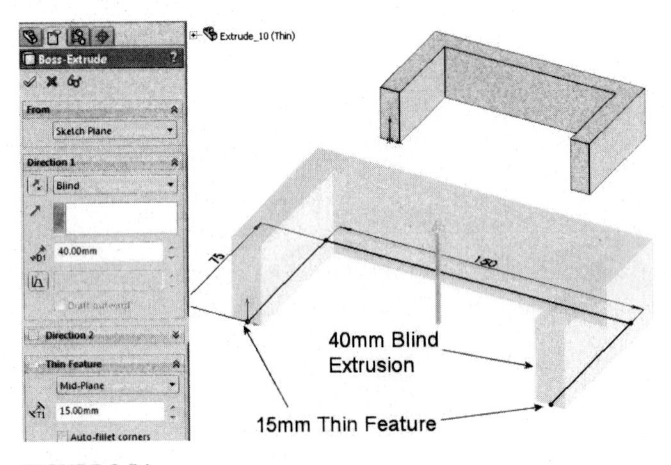

FIGURE 3.54

Extrude – Through All

Open the file Extrude 11 (Through).This option will extend the feature from the current sketch plane through all existing geometry. In Figure 3.55, the square sketch is selected and the Through All option of the Extrude tool is selected. Notice how this sketch extends through all existing features until it ends at the last feature as shown in the following figure.

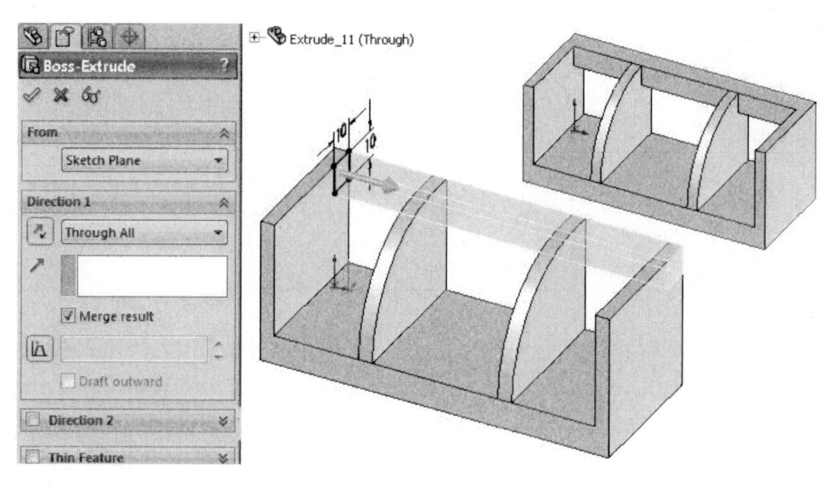

FIGURE 3.55

Extrude – Draft Angle

Open the file Extrude 12 (Draft). This extrude option adds draft to the extruded feature. Draft is associated with an angle where the draft angle is applied either inwardly or outwardly as shown in Figure 3.56. In the FeatureManager, set the draft angle followed by the direction of the draft (inward or outward).

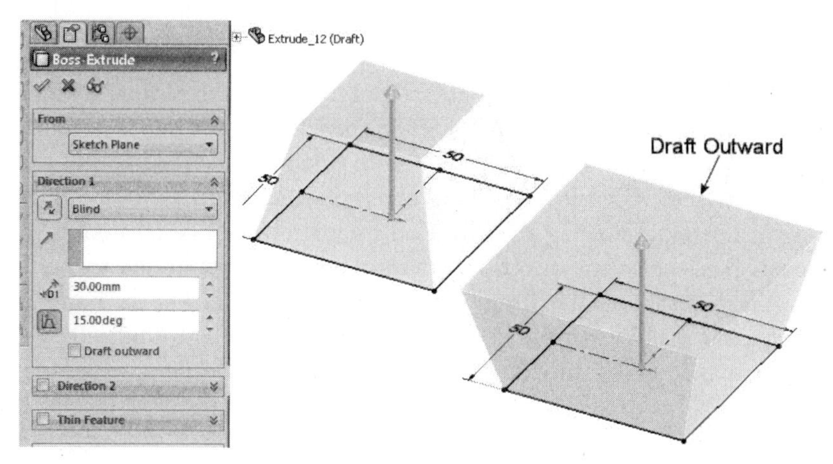

FIGURE 3.56

CREATING LINEAR PATTERNS

Linear patterns allow you to create multiple copies of a feature. You first supply the directions for the patterns. Then you enter the number of copies or instances and the spacing in between each copy. Typical linear patterns are illustrated in Figure 3.57. This provides a very quick way of duplicating features.

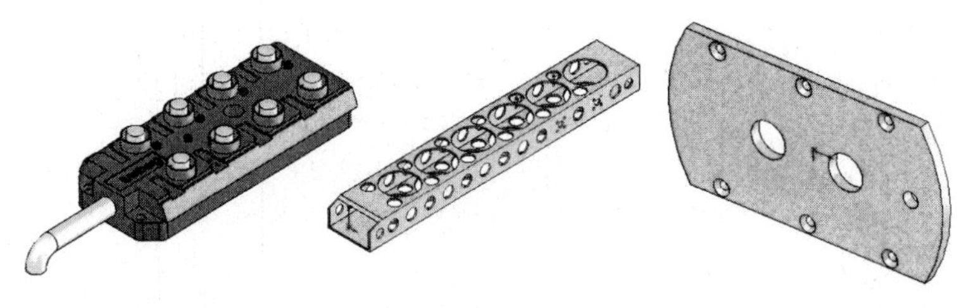

FIGURE 3.57

You can also skip copies or instances of features for special design considerations. Figure 3.58 illustrates a hole that is duplicated along the edges and middle of the plate. Unfortunately the holes need to be removed from the middle of the plate. This too is easily accomplished as shown in the following figure on the right.

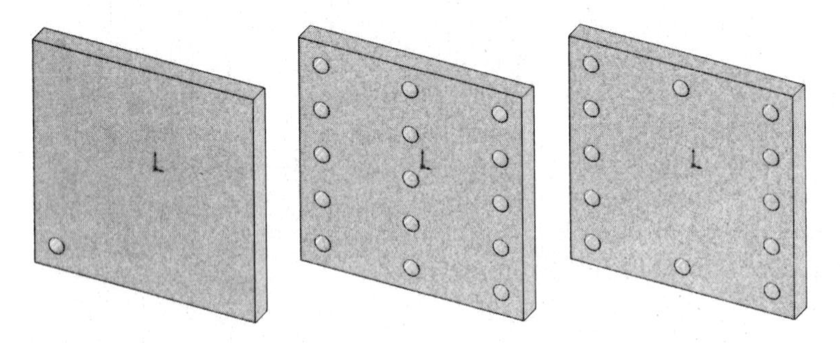

FIGURE 3.58

To create a linear pattern, click the Linear Pattern tool to display the various settings in the Property Manager as shown in Figure 3.59 on the left. While there is no exact order for supplying the information, it is always best to first pick the directions of the linear pattern. This is accomplished by picking an edge which dictates these directions. An arrow will display showing the direction the pattern will be created in. If the direction arrow is in the wrong direction, a button will allow you to flip to a different direction. You then supply the distance between the instances and the number of instances to repeat. Finally select the hole as the Feature to Pattern; this will preview the hole pattern as shown in the following figure on the right.

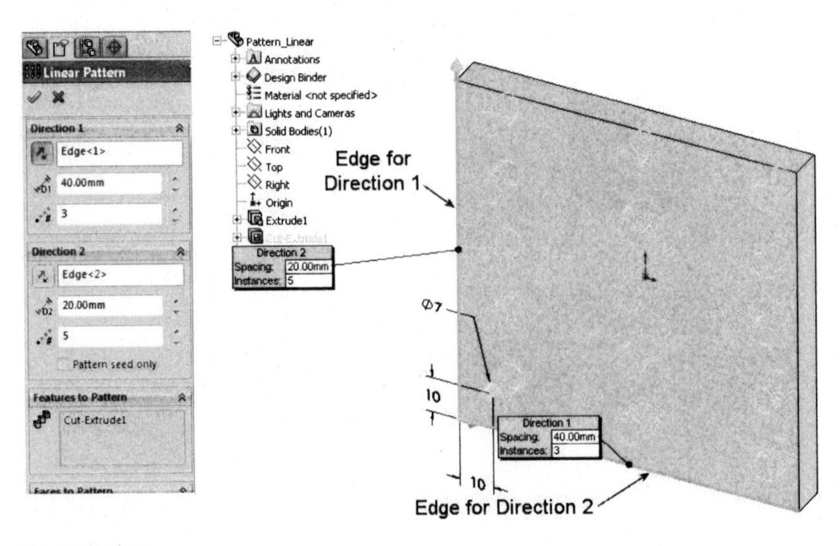

FIGURE 3.59

In the event you need to remove items from the pattern, click in the edit area under Instances to Skip as shown in Figure 3.60 on the left. Instance markers will appear at the location of all features that make up the linear pattern. Clicking on one of these markers will remove the instance from the pattern as shown in the following figure on the right. Click on the instance marker a second time to make the copy reappear.

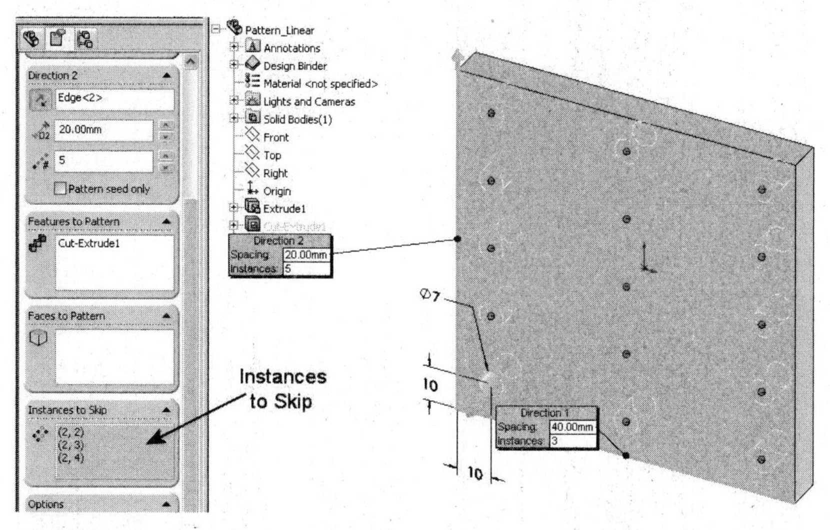

FIGURE 3.60

TUTORIAL EXERCISE: 03_PLASTIC_BEAM.SLDPRT

Create a new part file using Metric units. (The Part (mm).dotprt file could be used.) Use the dimensions in Figure 3.61 for creating the 1×15 Beam.

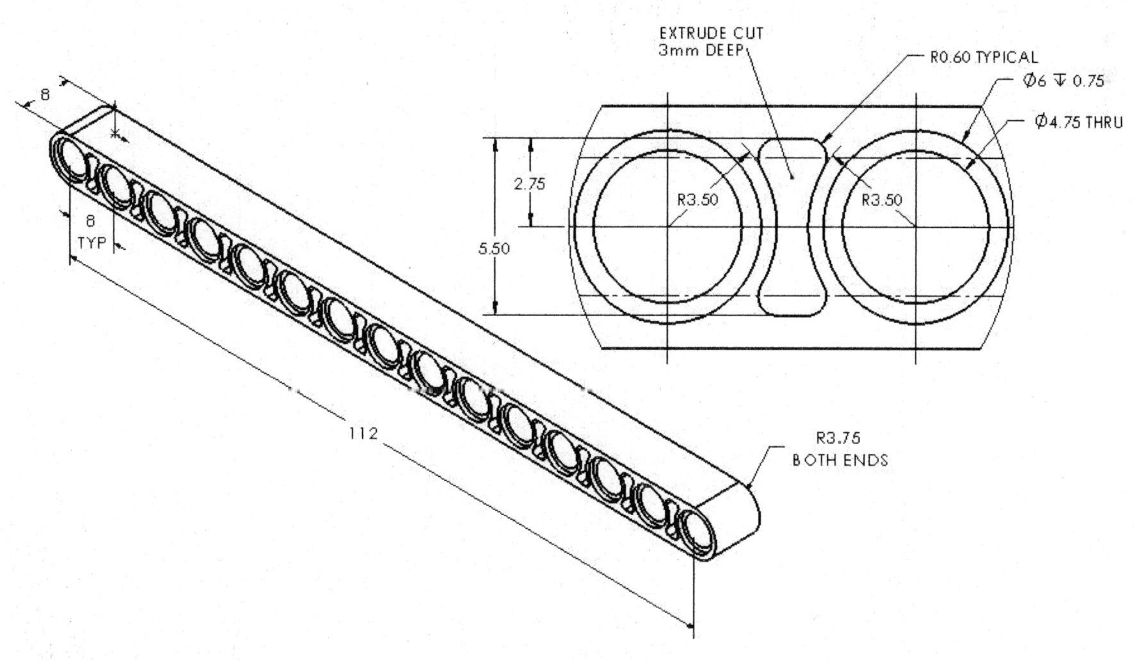

FIGURE 3.61

STEP 1

Create a new sketch plane using the Front plane and sketch the oblong shape. Extrude this shape to a distance of 8mm as shown in Figure 3.62 on the left. Zoom to Area and place the zoom box completely around all of the objects. Create a through hole of 4.75mm in diameter as shown in the following figure on the right.

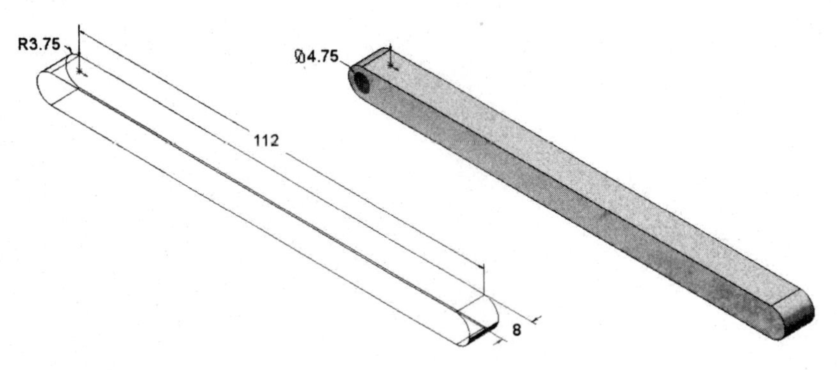

FIGURE 3.62

STEP 2

Create a linear sketch pattern of the 4.75 diameter hole. Click on the edge for Direction 1 as shown in Figure 3.63 on the left. The spacing in between each instance is 8mm and the total number of instances is 15. Extrude Cut all of the holes through the part. Create a new extruded cut feature based on the 6 diameter circle. Make this circle concentric with the through hole. Cut this circle 0.75mm into the part as shown in the following figure on the right.

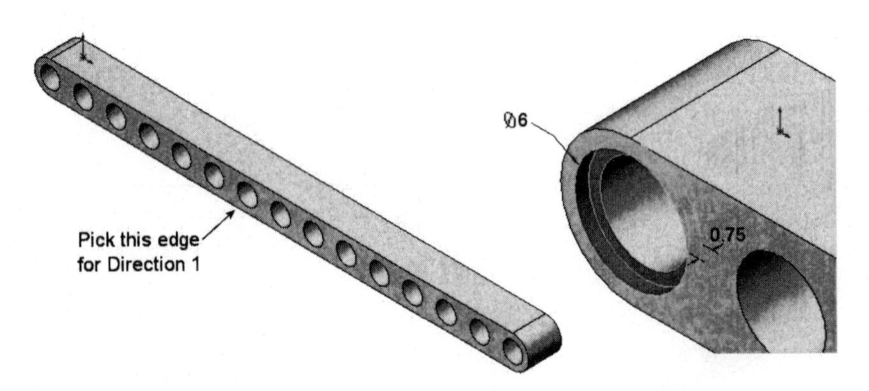

FIGURE 3.63

STEP 3

Create a linear pattern of this extruded cut circle as shown in Figure 3.64. This extruded circle needs to be copied along both the front and rear faces of this part. To accomplish this, two directions need to be identified for creating this type of linear pattern. Click on the long sloping edge for Direction 1 and the short depth right edge for Direction 2 as shown in the following figure. For Direction 1, the spacing in between each instance is 8mm and the total number of instances is 15. For Direction 2, the spacing in between is 7.25 (the total depth of 8mm minus the extruded cut depth of 0.75mm). The completed pattern is illustrated in the following figure.

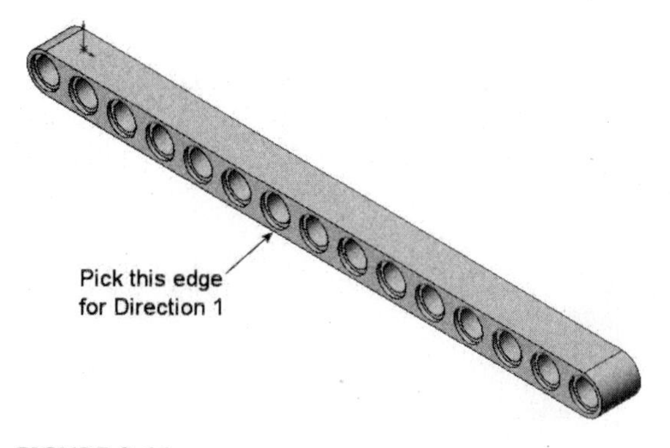

FIGURE 3.64

STEP 4

Create the irregular shape and cut this shape 3mm into the part as shown in Figure 3.65 on the left. Use the dimensions shown on the right for creating the initial sketch.

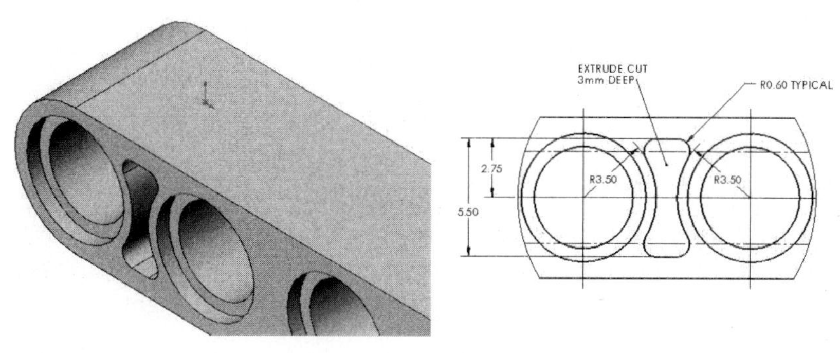

FIGURE 3.65

STEP 5

Create a linear pattern of this cut shape as shown in Figure 3.66. Like the extruded cut circle in a previous step, this extruded shape needs to be copied along both the front and rear faces of this part. Two directions will again need to be identified for creating this type of linear pattern. Click on the long sloping edge for Direction 1 and the short depth edge for Direction 2 as shown in the following figure. For Direction 1, the spacing in between each instance is 8mm and the total number of instances is 14. For Direction 2, the spacing in between is 5 (the total depth of 8mm minus the extruded cut depth of 3mm). The completed pattern is illustrated in the following figure.

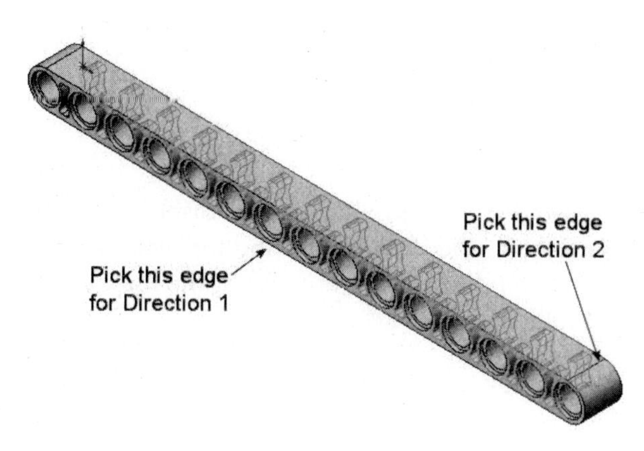

FIGURE 3.66

STEP 6

The completed beam is illustrated in Figure 3.67 that displays the required features on the front and rear faces of the part.

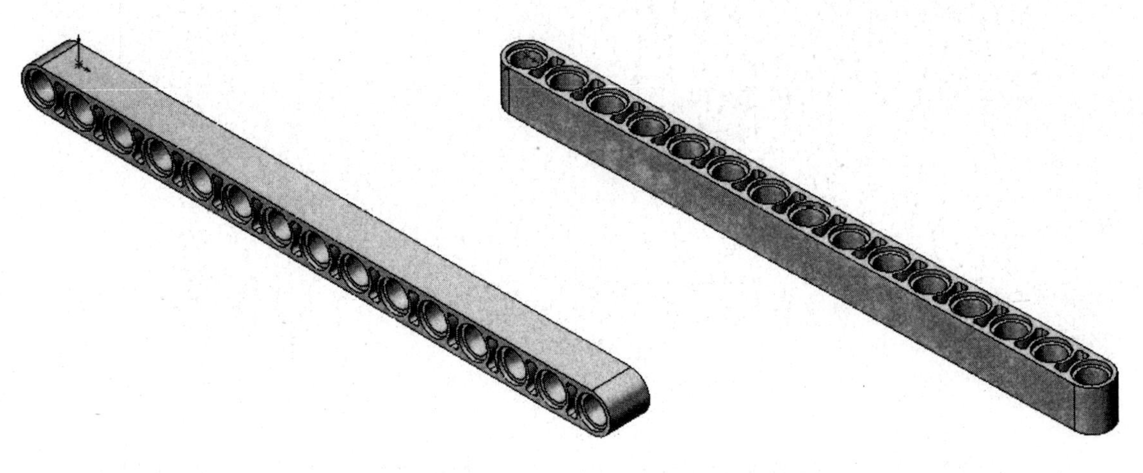

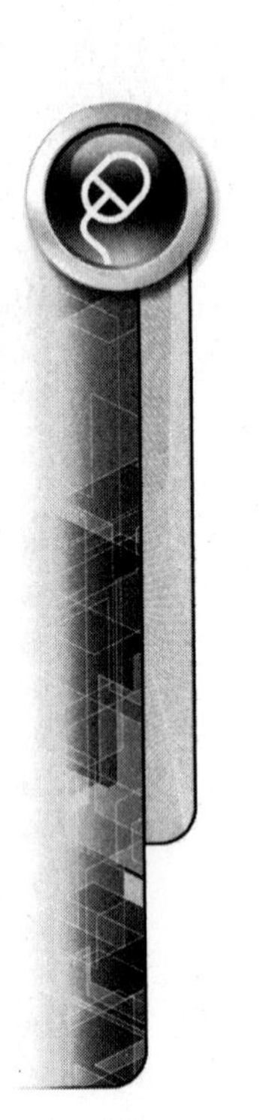

FIGURE 3.67

MEASURING FEATURES

Measuring an Arc

Clicking on the edge of an arc while using the Measure tool displays the arc length, chord length, radius, and angle as shown in Figure 3.68 on the left. The center of the arc is also listed in the quick properties box as shown in the following figure on the right.

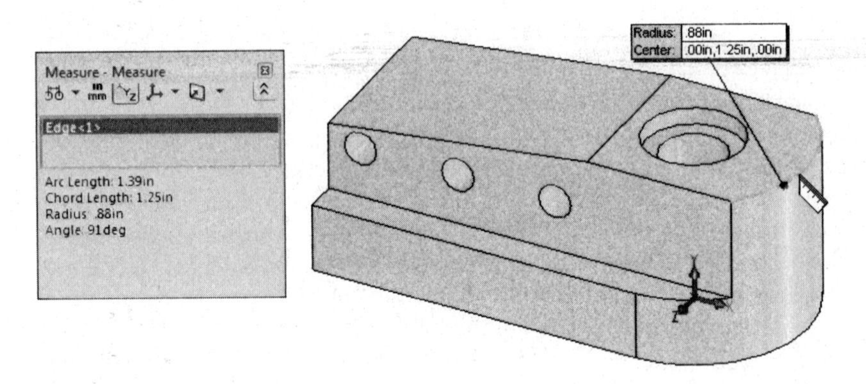

FIGURE 3.68

Measuring Between Circle Centers

Clicking on the edges of two circles while using the Measure tool displays the distance between circle centers, the Delta X, Y, and Z distances, and the *total length* as shown in Figure 3.69 on the left. The center distance of the circles is also listed in the quick properties box as shown in the following figure on the right.

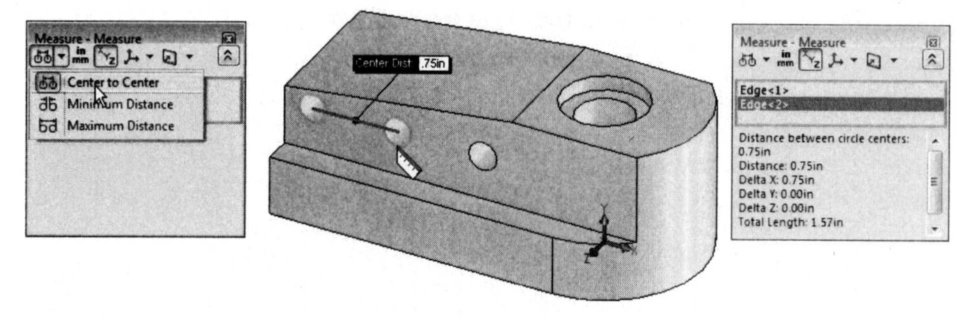

FIGURE 3.69

Measuring a Line

Clicking on the edge of line while using the Measure tool displays the length of the line as shown in Figure 3.70.

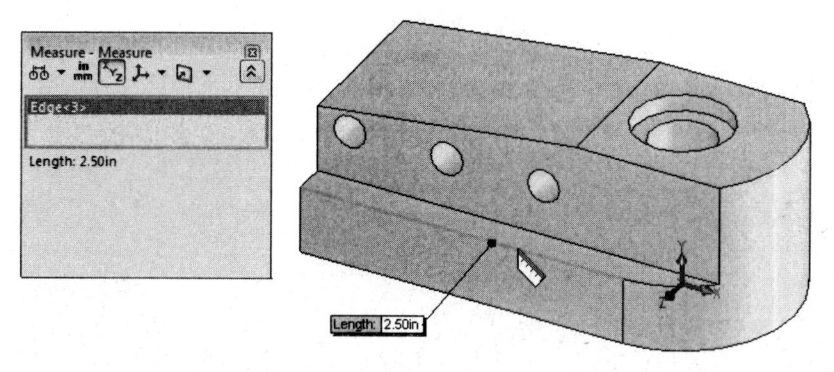

FIGURE 3.70

Measuring the Vertical Length between Two Circles

Clicking on the upper and lower edges of a counterbore hole while using the Measure tool displays the distance, Delta X, Y, and Z distances, and *total length* as shown in Figure 3.71.

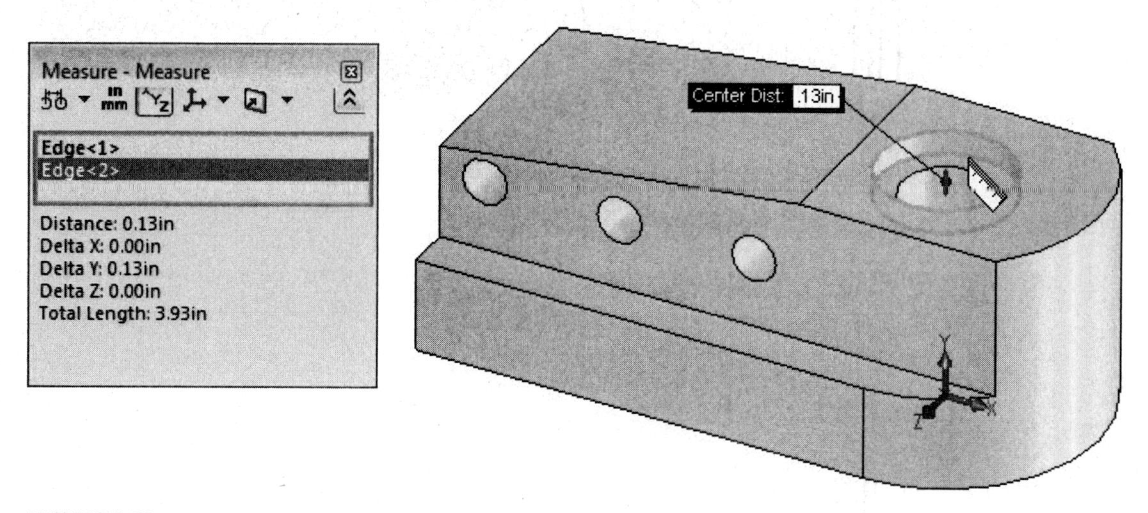

FIGURE 3.71

Measuring the Minimum Distance between Two Circles

Two other modes are available when performing measurements on two circles. By default, the Center to Center option is active. You could also measure the Minimum circle distance (the distance inside two circle edges), or the Maximum circle distance (the outer distance of two circles). The Minimum distance is illustrated in Figure 3.72.

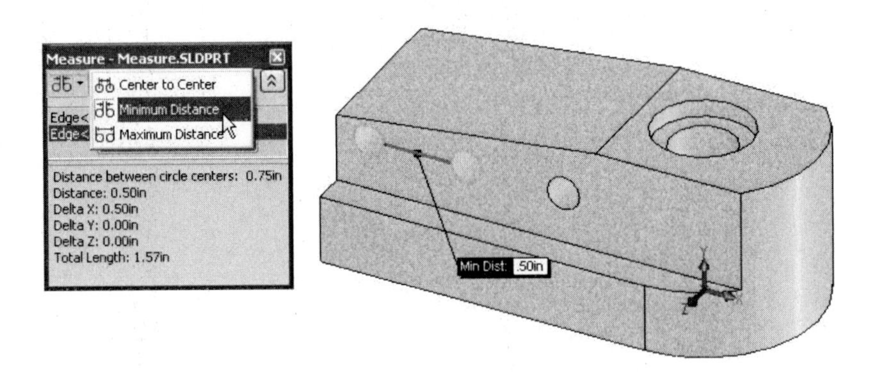

FIGURE 3.72

Measure the distance between two points.

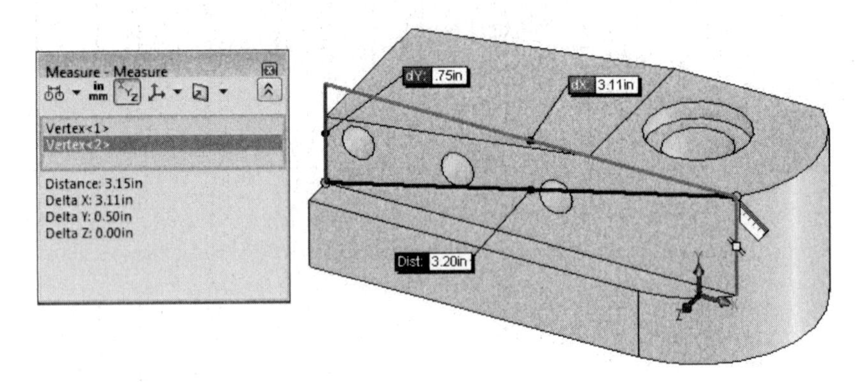

FIGURE 3.73

CREATING SHELL FEATURES

The Shell tool is used to create a thin wall inside of all features that make up a part file. When the shell parameters appear in the PropertyManager, change the thickness of the wall; the default thickness of 0.100 as shown in Figure 3.74 on the left should be changed to 1.0. Notice that a number of faces were removed as shown in the following figure on the right; these faces will not have the shell thickness applied to them. Select the *Show Preview* option.

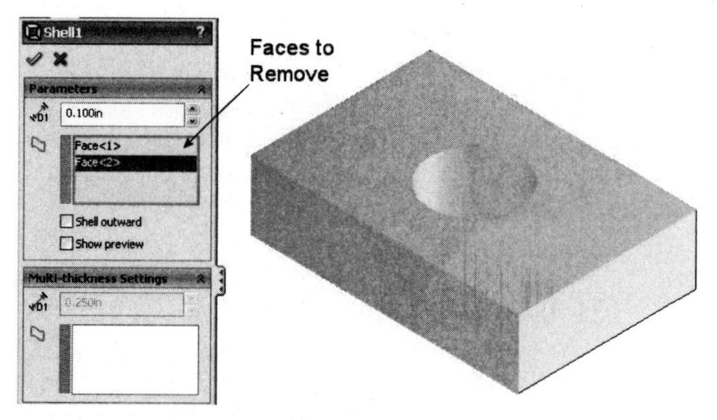

FIGURE 3.74

The results of the shell operation are illustrated in Figure 3.75. Notice that the entire rectangular feature was hollowed out leaving a wall thickness of 1.0 units. The shell also affected the hole that was drilled through the part by adding the same wall thickness to the hole creating a cylinder.

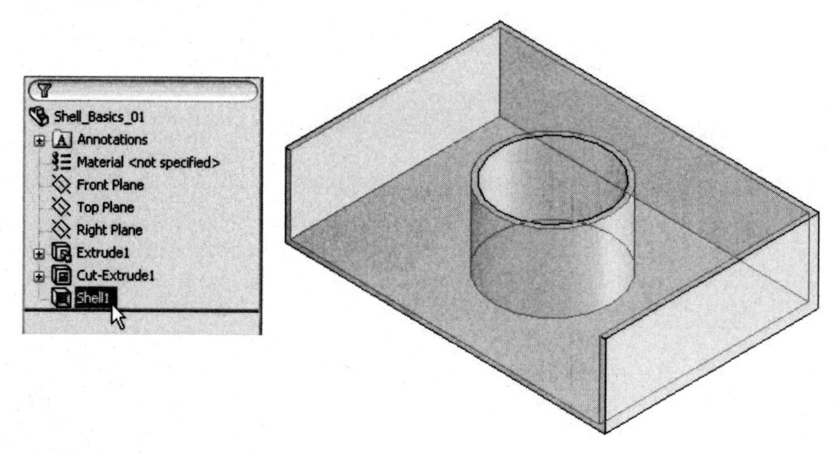

FIGURE 3.75

Suppose you wanted to drill a hole through only the base of the rectangle. Figure 3.76 illustrates a very powerful technique called reordering. In this example, the hole should have been drilled after the shell operation. Rather than delete the Cut-Extrude1 feature, the shell operation is dragged underneath the Extrude1 feature and dropped, causing the reordering and the conversion of the cylinder into a hole.

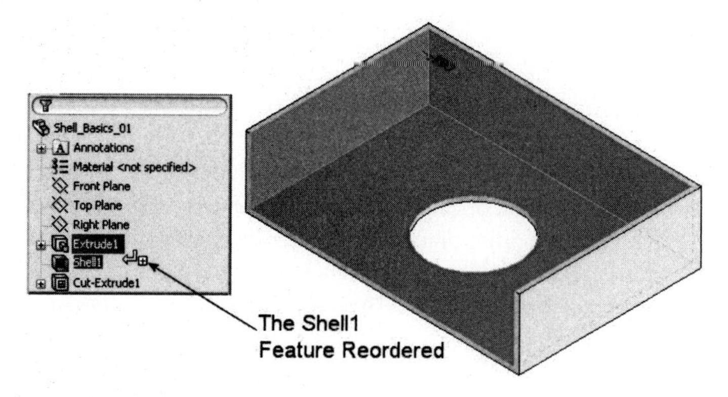

FIGURE 3.76

Another powerful function of the shell tool is to apply more wall thickness to certain parts of the model. In Figure 3.77, a multi-thickness wall distance of 3.0 is applied to the three sides of the rectangular box. Prior to selecting faces, select the appropriate display box in the Feature Manager. Accepting this change will display the box shown in the following figure on the right where the original shell distance of 1.0 is applied to the base surface while the new multi-thickness distance of 3.0 is applied to the sides.

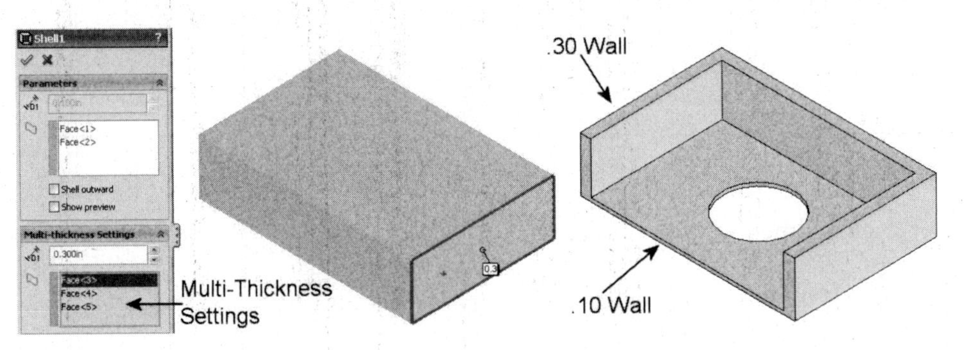

FIGURE 3.77

TUTORIAL EXERCISE: 03_PLASTIC_PLATE.SLDPRT

This exercise will allow you to practice creating linear patterns and shell features. Create a new part file using metric units. You will first create a base extrusion followed by a pattern of the short cylinders. The bottom of the part will be shelled out and another pattern consisting of hollow cylinders will be created to provide strength (Figure 3.78).

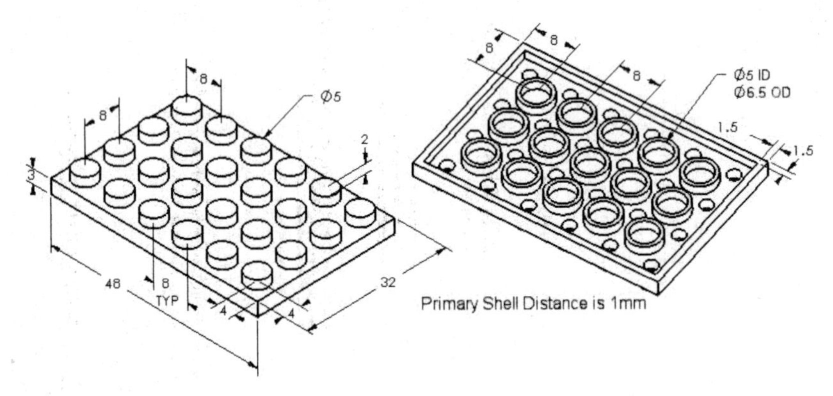

FIGURE 3.78

STEP 1

Create an extruded base feature on the Top Plane that has a length of 48mm and depth of 32mm. Use an extruded thickness distance of 3mm. The results of this operation are shown in Figure 3.79 on the left. Create a 5mm diameter cylinder on the top face of the plate. The center of the cylinder is located 4mm away from the horizontal and vertical edges. This cylinder is extruded a distance of 2mm. The results of this operation are shown in the following figure on the right.

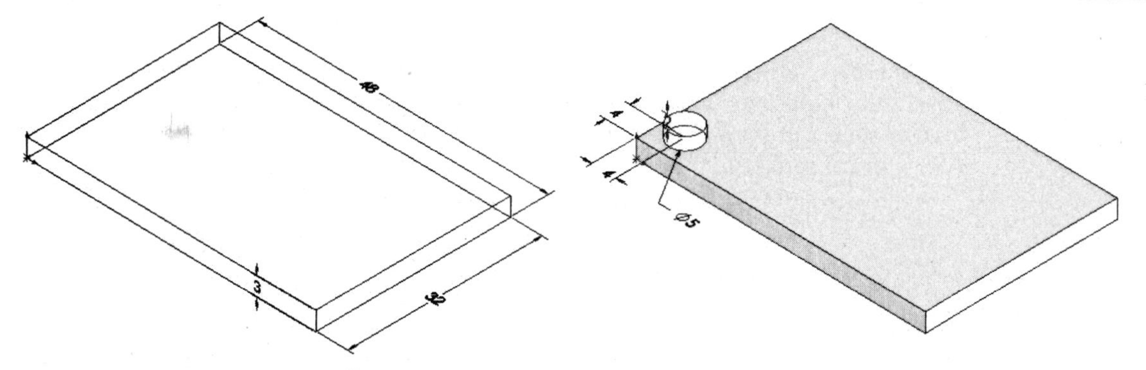

FIGURE 3.79

STEP 2

Create a linear pattern of the 5mm diameter cylinder. Click the bottom and right edges to provide the directions for the pattern. The number of instances (copies) for Direction 1 is 6 and for Direction 2, four instances are needed as shown in Figure 3.80 on the left. The results are illustrated in the following figure on the right.

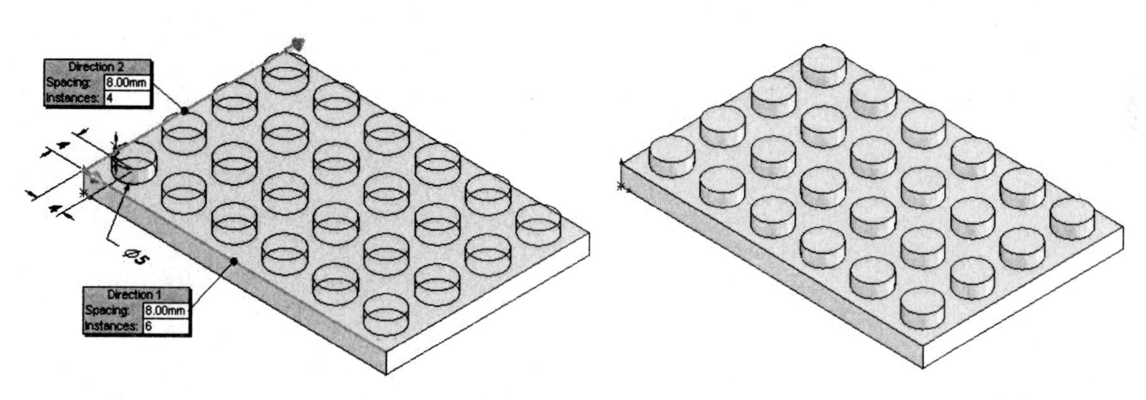

FIGURE 3.80

STEP 3

Rotate the part so you are looking underneath it. Create a shell feature with a wall thickness of 1mm. This initial wall thickness will create a thin wall based on the top of the plate and each individual cylinder. The sides of the rectangular extrusion, however, need to have a wall thickness of 1.50mm. This secondary wall thickness can be created by selecting the four side faces and entering the 1.50 distance under the Multi-thickness Settings area as shown in the following figure on the left. The results are illustrated on the right with the side walls created at a distance of 1.50 while the wall thickness of the other part areas measures 1mm.

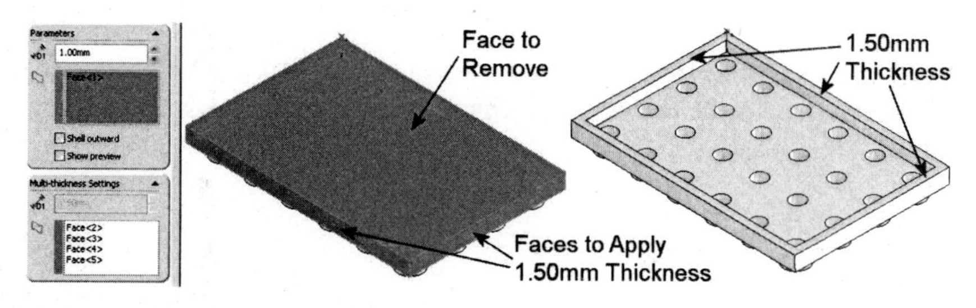

FIGURE 3.81

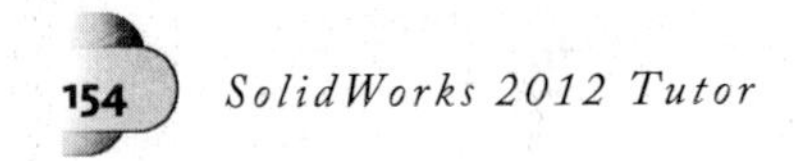

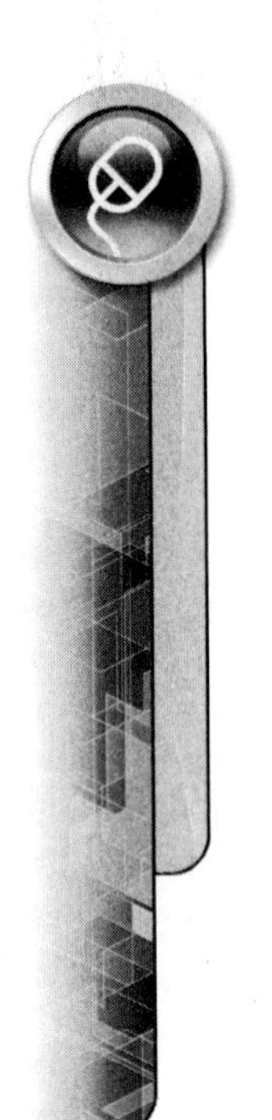

STEP 4

With the part still rotated to look underneath, make the bottom face the new sketch plane and create two circles as shown in Figure 3.82 on the left. Create an extruded boss as shown in the following figure on the right. The height of this boss should be even with the surface in the illustration.

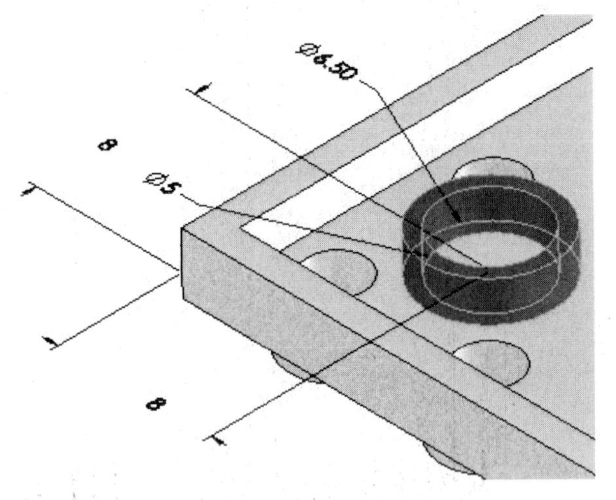

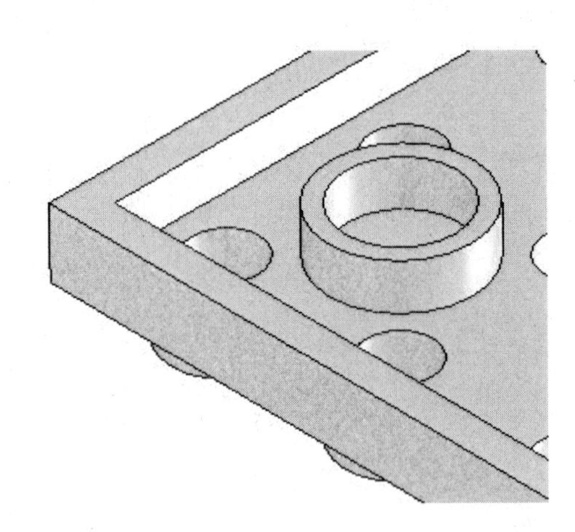

FIGURE 3.82

STEP 5

Create a linear pattern of the cylinder in the previous figure. Click edges to provide the directions for the pattern. The number of instances (copies) for Direction 1 is 5 and for Direction 2, three instances are needed as shown in Figure 3.83 on the left. The results are illustrated in the following figure on the right.

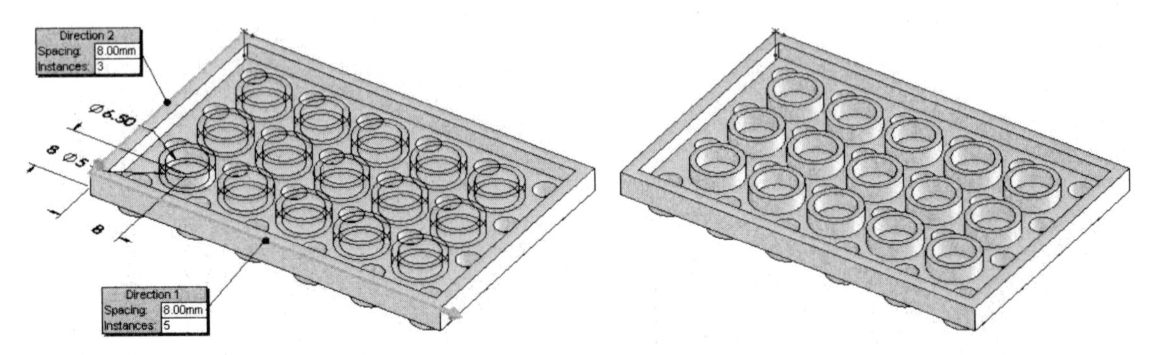

FIGURE 3.83

CREATING REVOLVED FEATURES

When a sketch is created and rotated around an axis, a revolved feature is formed. Figure 3.84 illustrates different part models where the first feature was created as a revolved feature.

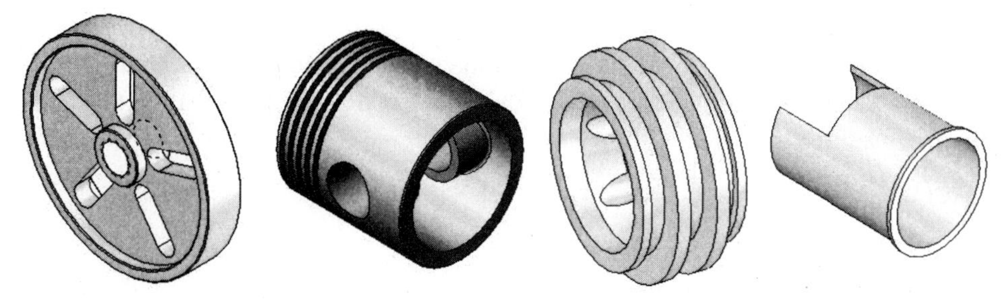

FIGURE 3.84

The Revolve tool can be selected by clicking on Revolved Boss/Base from the CommandManager as shown in Figure 3.85.

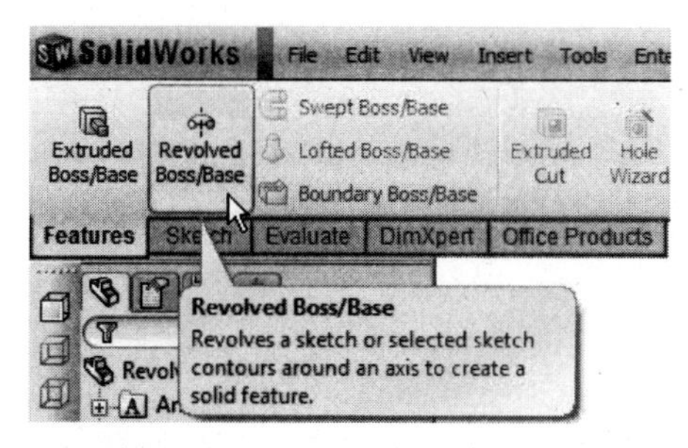

FIGURE 3.85

Figure 3.86 illustrates two techniques for dimensioning a sketch that is about to be revolved. The sketch on the left has a 50mm dimension that actually represents a diameter. However, this dimension shows only half of the required distance; this dimension would be similar to a radius distance. When creating revolved features, it is best to include a centerline in the sketch as shown in the following figure on the right. When dimensioning to the centerline and one of the edges to be revolved, a diametric (diameter) dimension can be created that better identifies the actual diameter value.

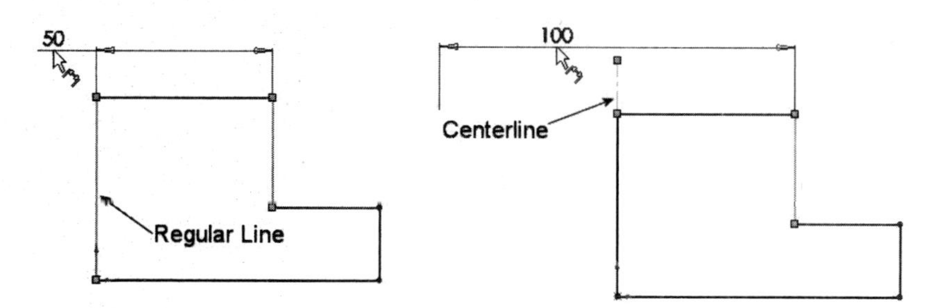

FIGURE 3.86

Create a new part file. This exercise is designed to create a sketch and convert the entities into a revolved boss/base feature.

TRY IT!

Create a new part file using metric units in millimeters. Sketch a series of lines to define the outline that will be revolved. Include a centerline in the sketch as shown in Figure 3.87. Add dimensions that actually represent diameters (100 and 160). The remaining height dimensions (20 and 50) are linear.

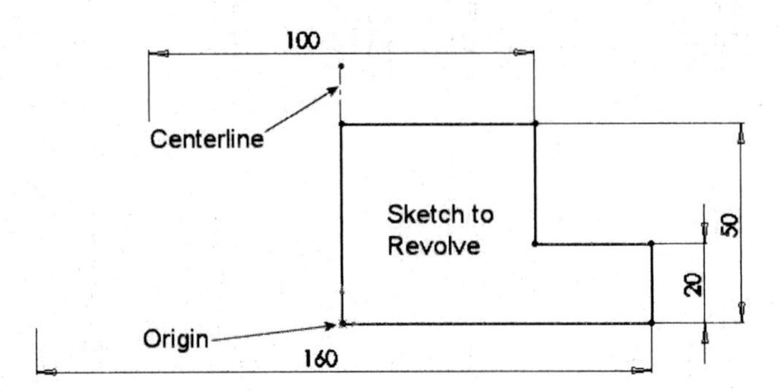

FIGURE 3.87

Click the Revolve Base/Boss tool. When the Revolve information displays in the PropertyManager as shown in Figure 3.88 on the left, click on the centerline as the axis of revolution. Since this is the first feature, the revolved feature will preview as shown in the following figure in the middle. Clicking the green OK checkmark will create the revolved feature shown in the following figure on the right.

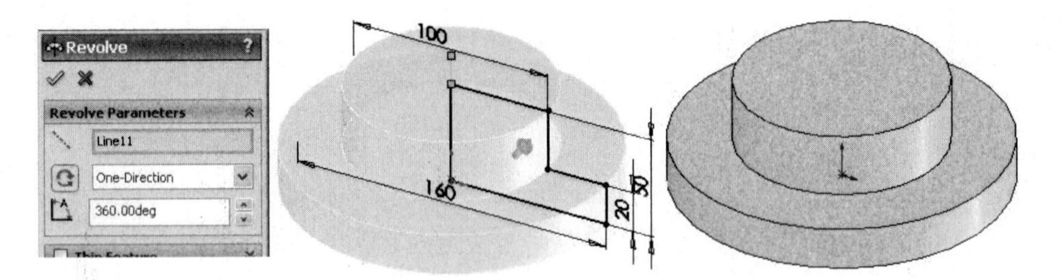

FIGURE 3.88

It is possible to include the creation of holes when creating revolved features. In the previous example, no central hole was created during the revolve operation. Illustrated in Figure 3.89 on the left is a sketch that has a 50mm dimension created from the common centerline. When forming the revolved feature, this void will generate a hole as shown in the following figure on the right.

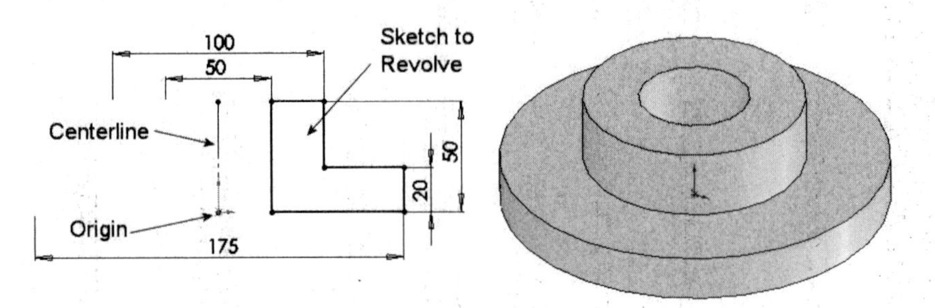

FIGURE 3.89

CREATING REVOLVED CUTS

As material can be added when creating revolved features, it can also be removed by creating a revolved cut. The Revolve Cut tool can be selected by clicking on Revolved Cut from the CommandManager as shown in Figure 3.90.

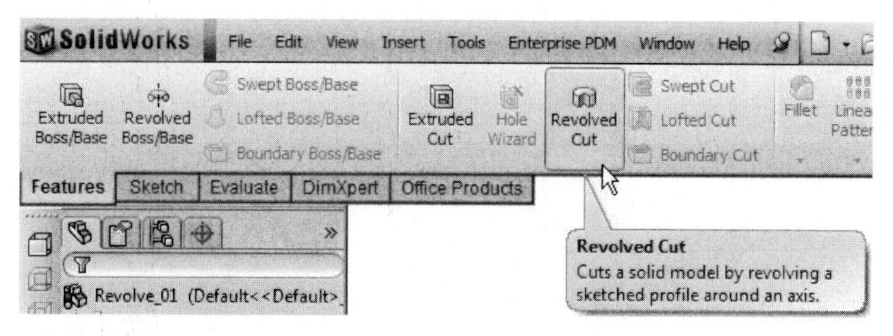

FIGURE 3.90

A semi-circular shaped cut will be made along the long cylindrical portion of the model as shown in Figure 3.91 on the left. The Right plane is made visible; then a new sketch plane is created on this plane as shown in the following figure on the right.

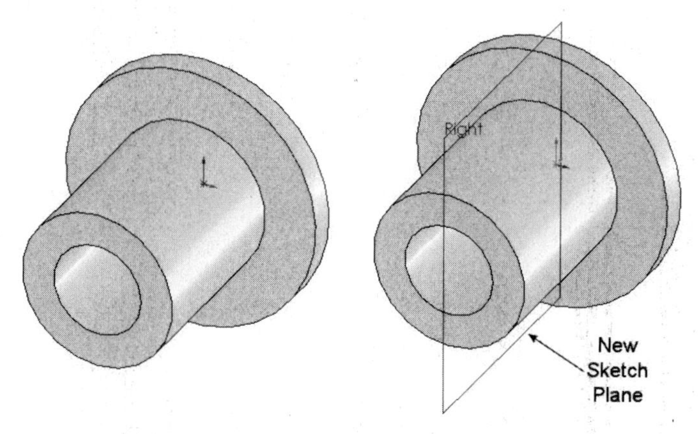

FIGURE 3.91

Change to a normal view and sketch a circle along the top edge of the model. Add dimensions and relations to fully constrain the sketch as shown in Figure 3.92.

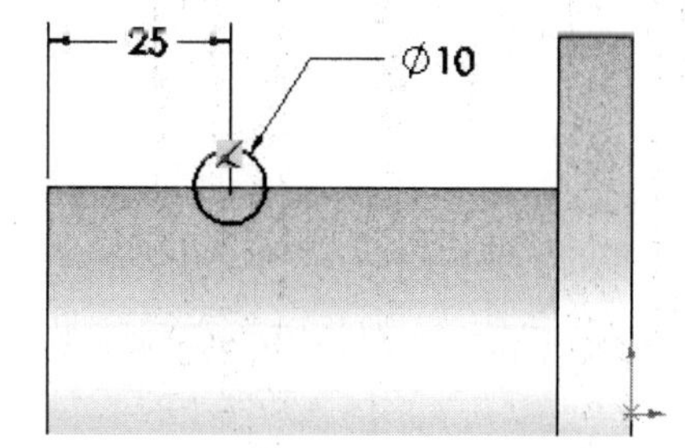

FIGURE 3.92

This type of cut requires an axis around which the sketch is revolved. An axis can easily be generated by clicking on Temporary Axes that is found under the View pull-down menu as shown in Figure 3.93 on the left. The temporary axis displays through the middle of the hole as shown in the following figure on the right. This entity will be used as the axis of revolution for the circle sketch.

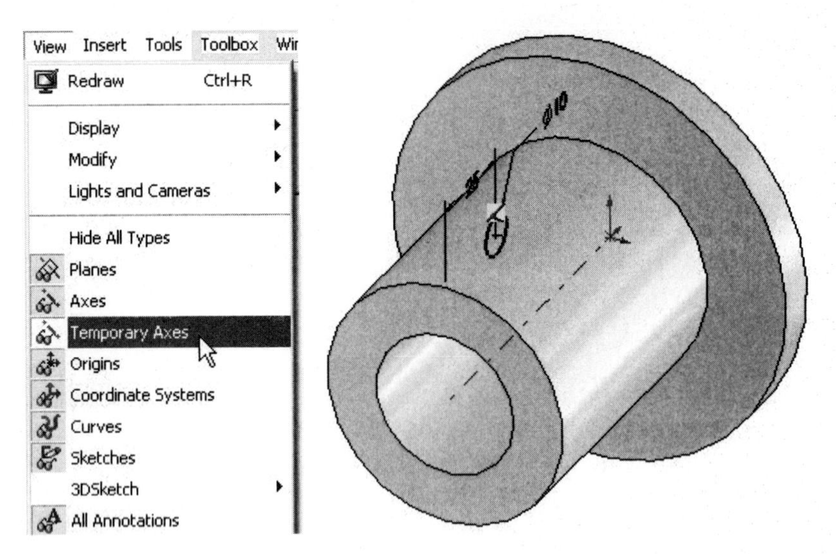

FIGURE 3.93

Clicking on the Cut Revolve tool activates the PropertyManager as shown in Figure 3.94 on the left. Select the centerline as the axis of rotation and notice the cut being previewed in the following figure in the middle. The result is illustrated in the following figure on the right with the creation of the semi-circular cut.

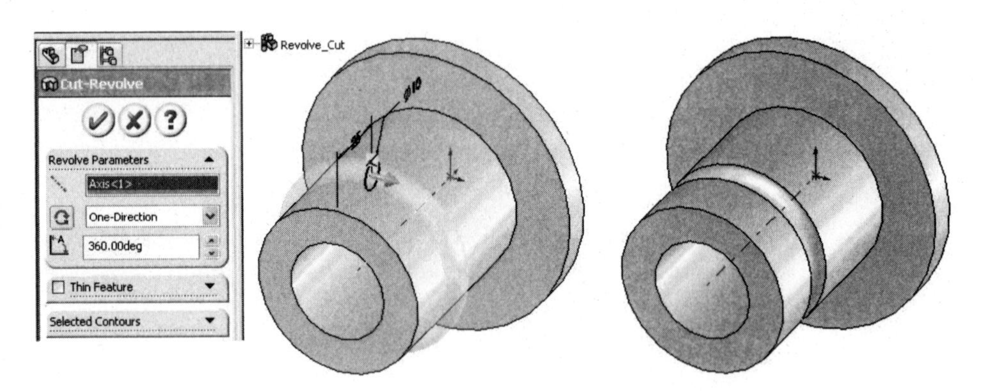

FIGURE 3.94

CREATING CIRCULAR PATTERNS

As with linear patterns, features can also be patterned in a circular direction as shown in the examples in Figure 3.95. Requirements include a valid axis by which to create the revolved feature, the number of instances (copies), and either an angle spacing in between each copy or an angle where the features being copied are equally spaced.

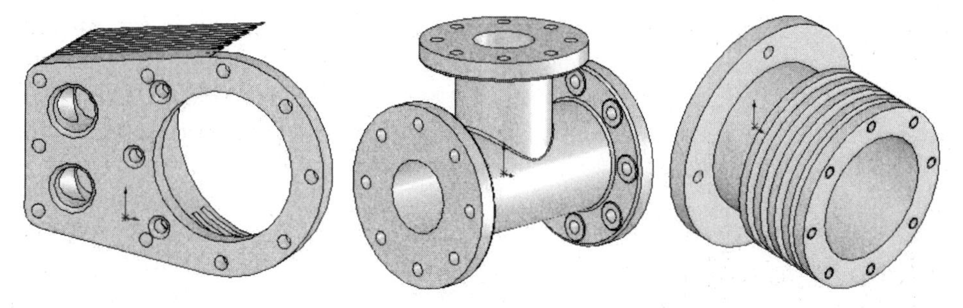

FIGURE 3.95

TRY IT!

This exercise is designed to create a circular pattern consisting of eight holes along a bolt circle. Construction geometry and relations will be used to create the first hole before creating the circular pattern.

Open the file Pattern_Circular as shown in Figure 3.96 on the left. Create a new sketch plane on the top of the base and sketch a circle as shown in the following figure on the right. An extruded cut operation will be performed on this circle creating a hole. This circle will then be duplicated in a circular pattern.

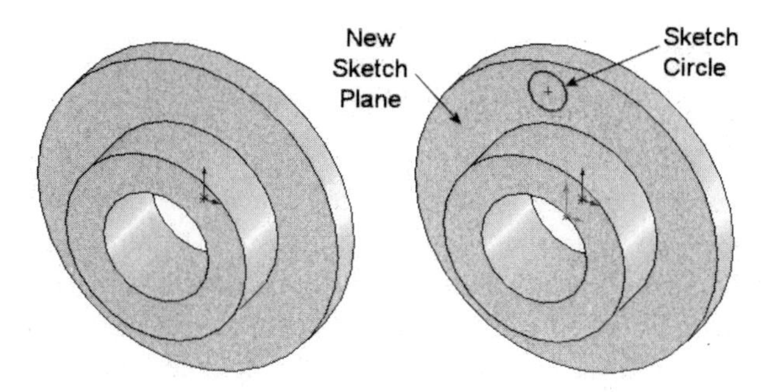

FIGURE 3.96

Before creating the pattern, the circle must first be properly positioned on the sketch for alignment purposes. Another circle is created with its center at the origin of the sketch. The circle intersects the center of the hole. So as not to interfere with any other features, this circle is converted into a construction entity. Make the center of the small circle vertical with the center origin of the sketch. Add diameter dimensions to the large centerline circle and small hole as shown in Figure 3.97 on the left. Switching back to an isometric view will display the model as shown in the following figure on the right.

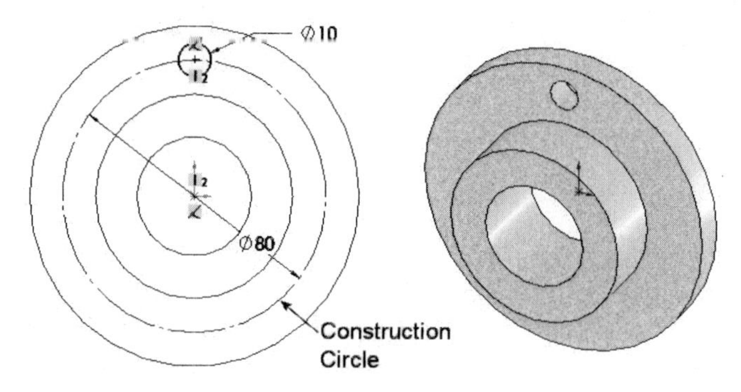

FIGURE 3.97

Before creating the circular pattern, an axis of rotation must first be present. All circular features automatically have axes at their centers; however, they are invisible by default. To turn on these temporary axes, click on the View pull-down menu and choose Temporary Axes; this will toggle on the display of all temporary axes for all circular features in the model as shown in Figure 3.98.

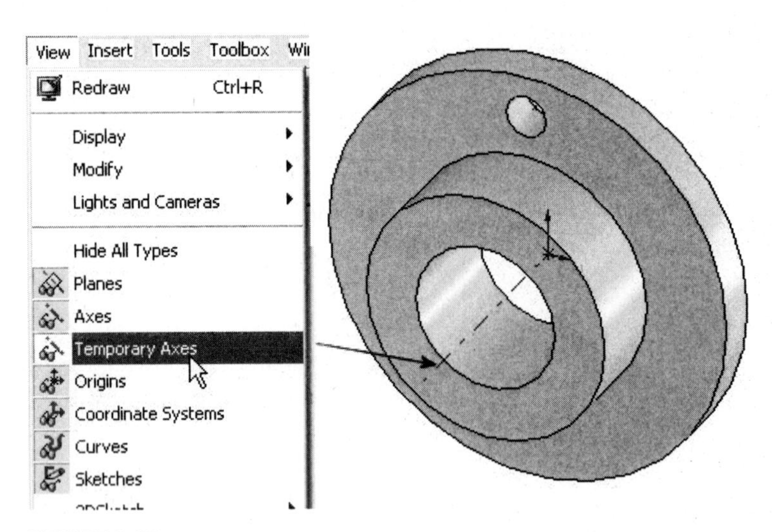

FIGURE 3.98

To create the circular pattern, click on this tool in the Features CommandManager. When the PropertyManager displays, click on the temporary axis as the pattern axis. Change the number of instances to 8. Keep the default angle set to 360°, which will rotate and copy the circle around the object. A preview will appear as shown in Figure 3.99 on the left. Click the green checkmark in the Feature Manager to create the hole pattern as in Figure 3.99 on the right.

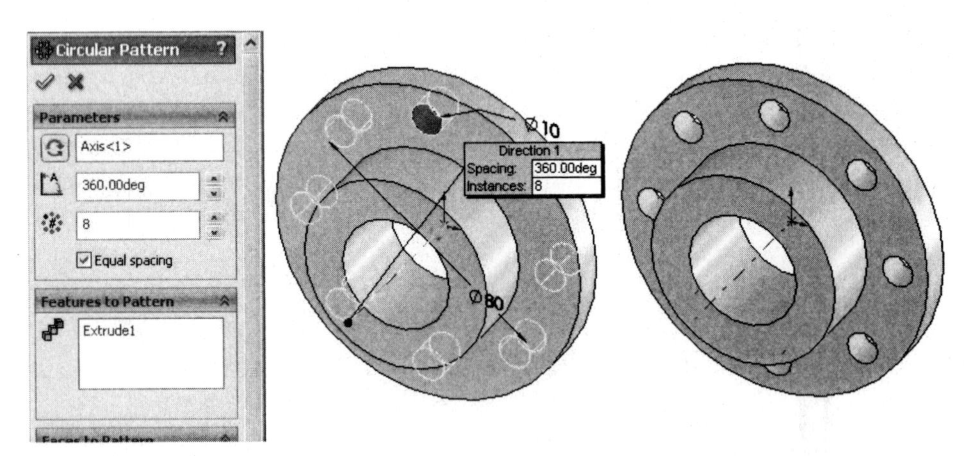

FIGURE 3.99

REVIEW QUESTIONS

1. What is the advantage of starting the sketch geometry at the origin? Disadvantage?
2. Can sketches be created on *any* plane?
3. Why do we rename feature names?
4. How can sketches be reentered in order to edit them?
5. Explain why the order of placing fillets matters.
6. When extruding a sketch, what are all of the options other than Blind?
7. What is unique about the Extrude Thin Feature command?
8. List the types of measurements that the Measure command can provide.
9. When revolving a sketch, if an existing line is not present to revolve about, what must be created?
10. What are the two types of patterns?

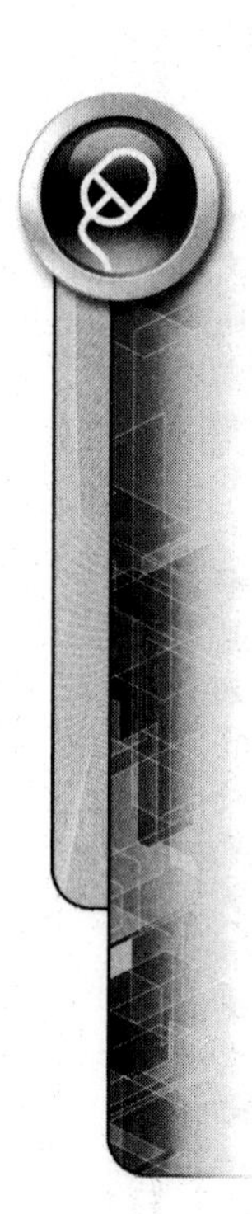

EXERCISES

3.1 a thru d

Begin each problem using a part template that has its units set to imperial (English). Use a grid spacing of 1" to construct these part models using SolidWorks sketch and feature tools. Perform a mass property calculation on each object and take note of the volume.

PROBLEM 3-1A (ENGLISH UNITS)

Use the Front Plane for the initial sketch.

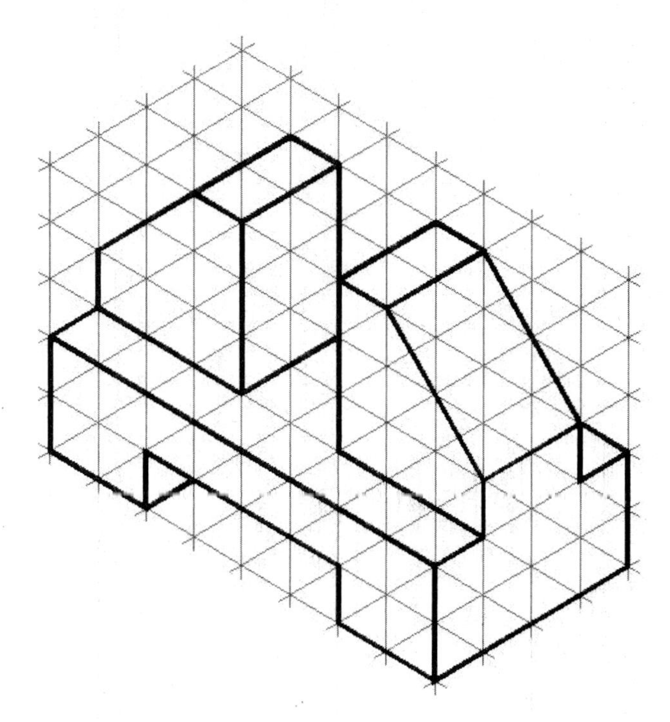

Volume

PROBLEM 3-1B (ENGLISH UNITS)

Use the Top Plane for the initial sketch.

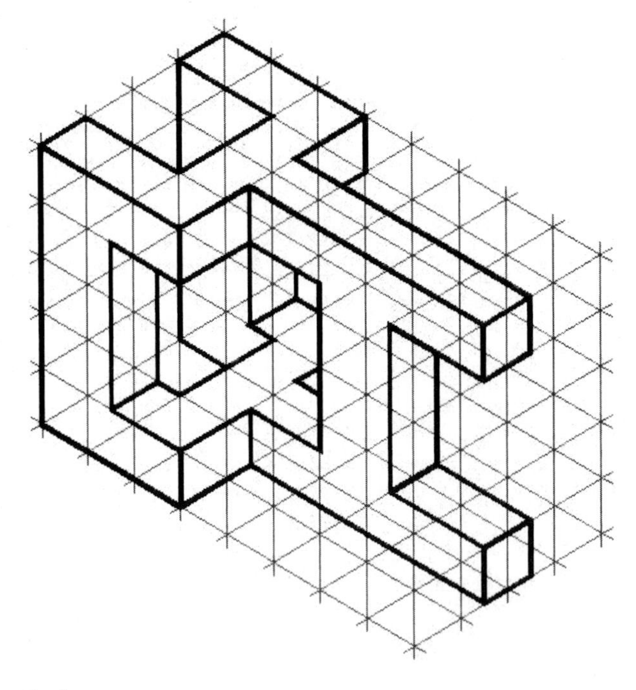

Volume________________________________

PROBLEM 3-1C (ENGLISH UNITS)

Use the Front Plane for the initial sketch.

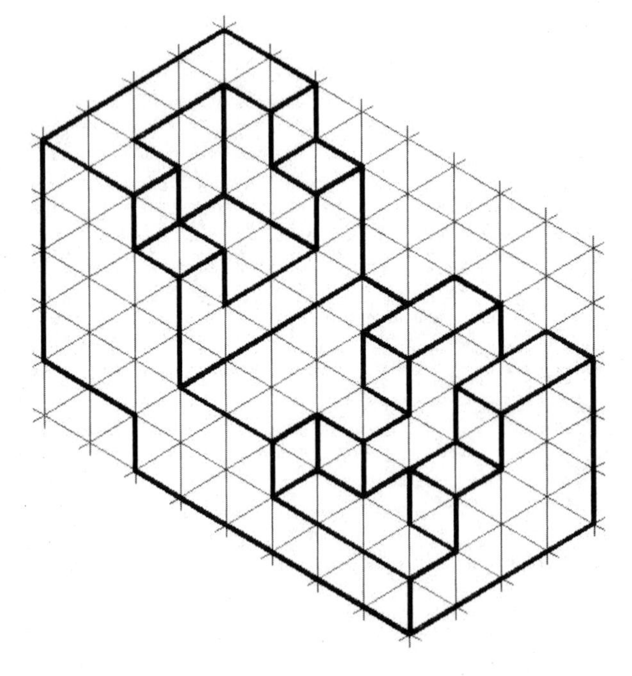

Volume________________________________

PROBLEM 3-1D (ENGLISH UNITS)

Use the Top Plane for the initial sketch.

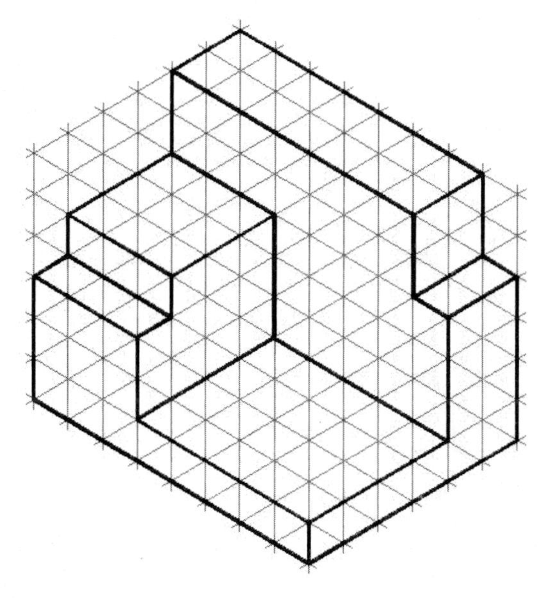

Volume________________________________

3.2 a thru d

Construct each problem using the appropriate part template. Perform a mass property calculation on each object and take note of the volume.

PROBLEM 3-2A (ENGLISH UNITS)

Use the Right Plane for the initial sketch.

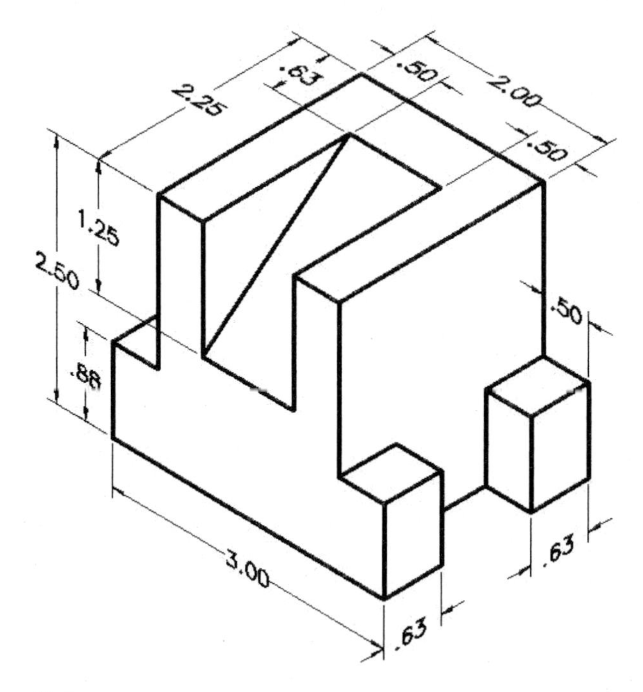

Volume________________________________

PROBLEM 3-2B (METRIC UNITS)

Use the Front Plane for the initial sketch.

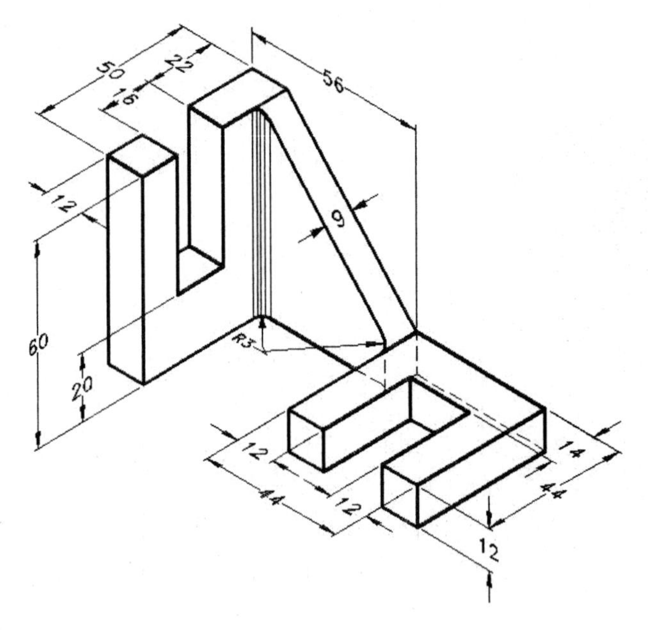

Volume________________________________

PROBLEM 3-2C (ENGLISH UNITS)

Use the Right Plane for the initial sketch.

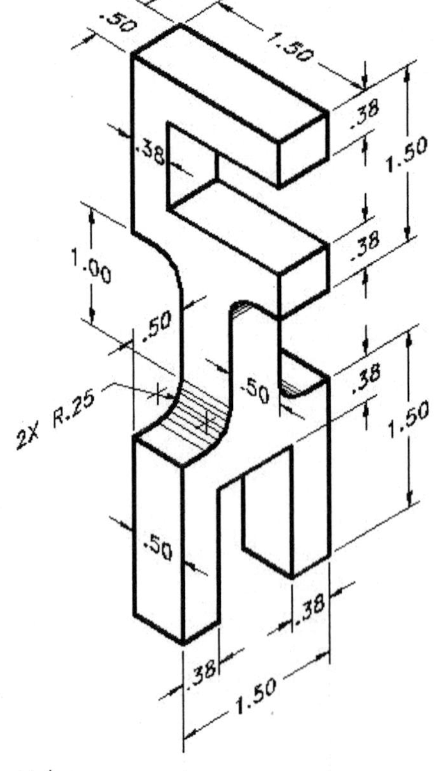

Volume________________________________

PROBLEM 3-2D (ENGLISH UNITS)

Use the Front Plane for the initial sketch.
The angled planes are parallel.

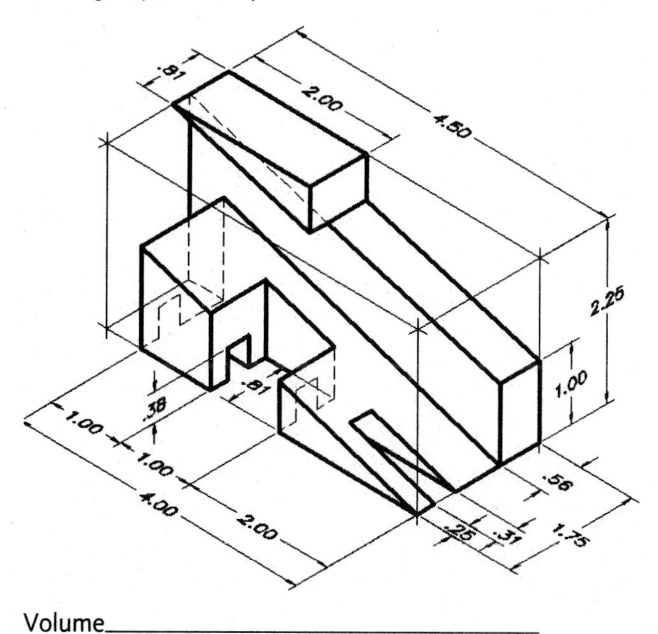

Volume________________________________

3.3 a thru d

Begin each problem using a part template that has its units set to imperial (English). Use a grid spacing of 1" to construct these part models using SolidWorks sketch and feature tools. Perform a mass property calculation on each object and take note of the volume.

PROBLEM 3-3A (ENGLISH UNITS)

Use the Top Plane for the initial sketch.
All hole diameters measure 1.25.

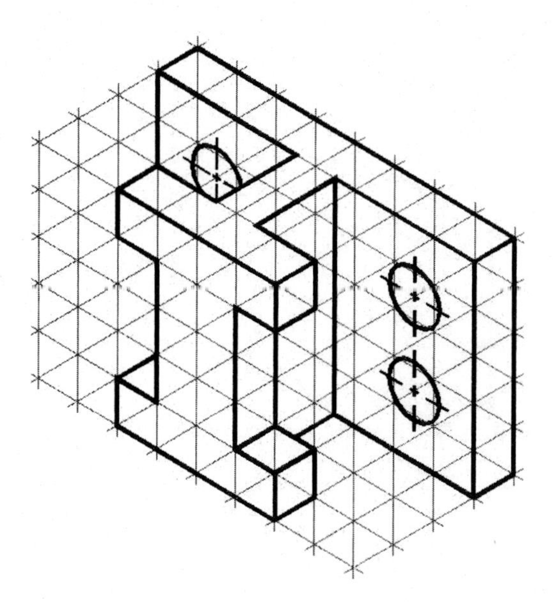

Volume________________________________

PROBLEM 3-3B (ENGLISH UNITS)

Use the Front Plane for the initial sketch.

The hole diameter measures 1.50.

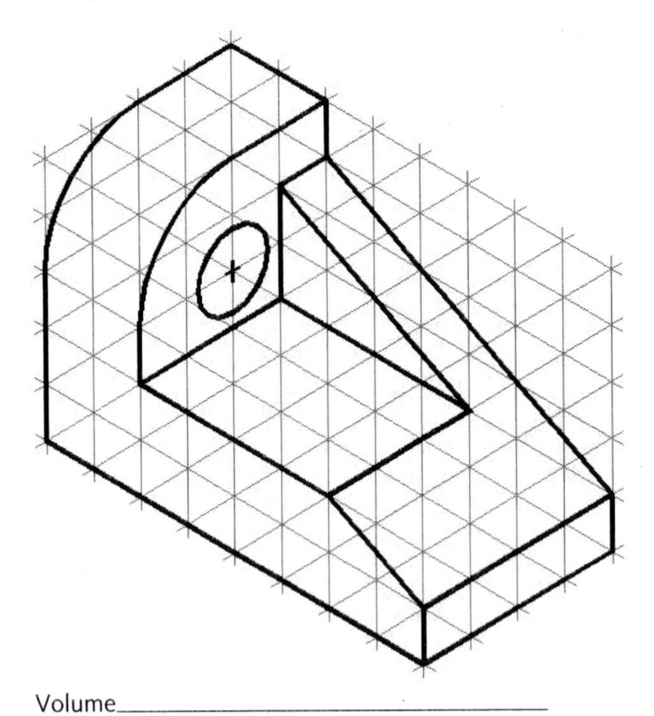

Volume______________________________

PROBLEM 3-3C (ENGLISH UNITS)

Use the Top Plane for the initial sketch.

The hole diameter measures 1.50.

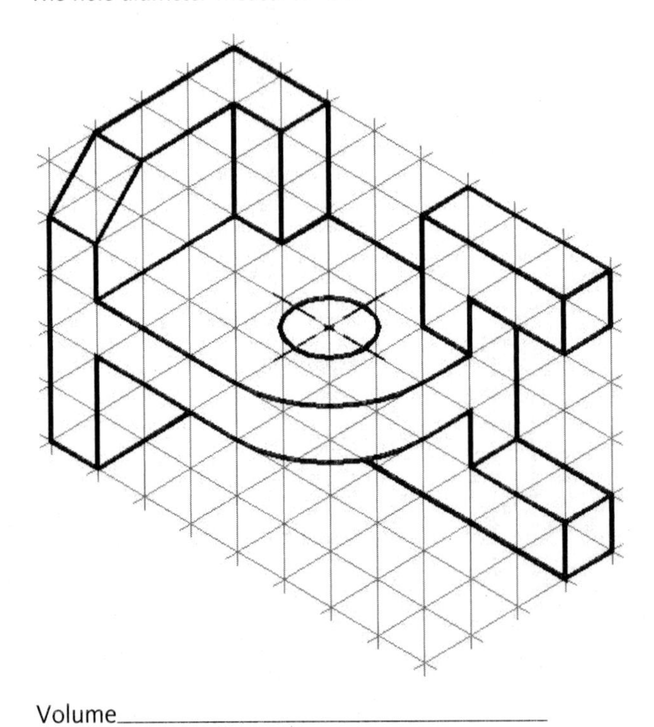

Volume______________________________

PROBLEM 3-3D (ENGLISH UNITS)

Use the Right Plane for the initial sketch.
The large hole diameter measures 2.00.
The small hole diameters measure .75 and are centered in the rectangle.

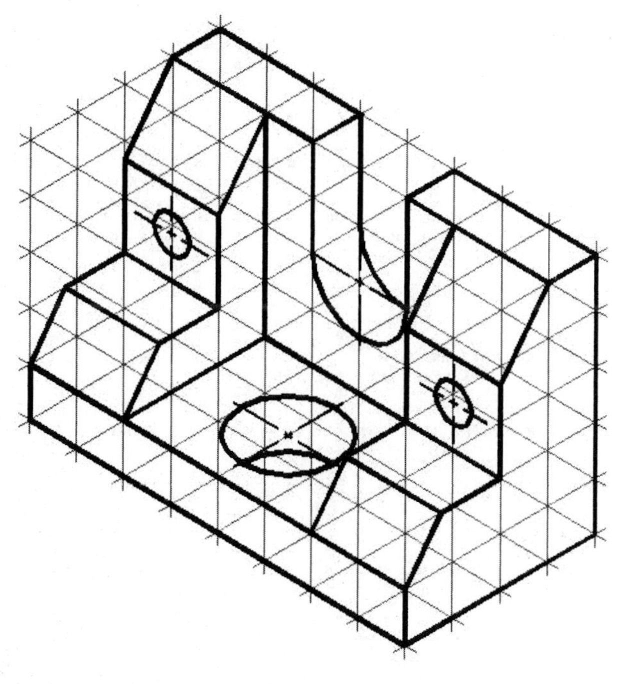

Volume________________________________

3.4 a thru e

Construct each problem using the appropriate part template. All holes should be created as separate features. Perform a mass property calculation on each object and take note of the volume.

PROBLEM 3-4A (ENGLISH UNITS)

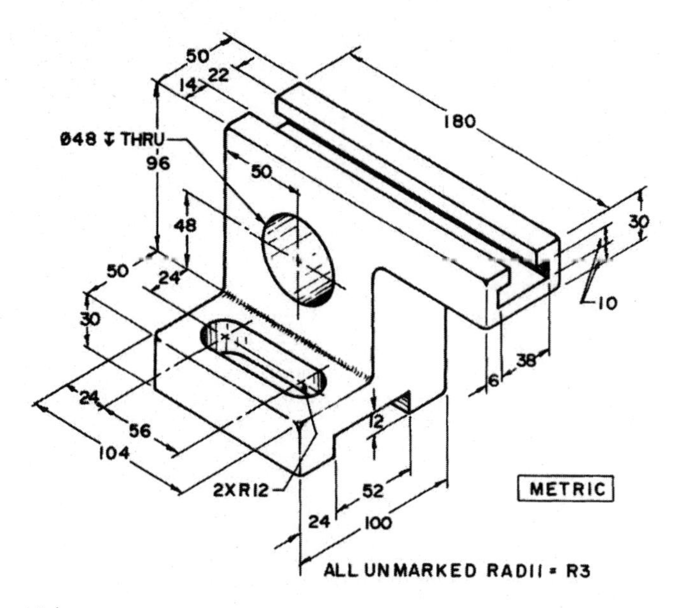

Volume________________________________

PROBLEM 3-4B (ENGLISH UNITS)

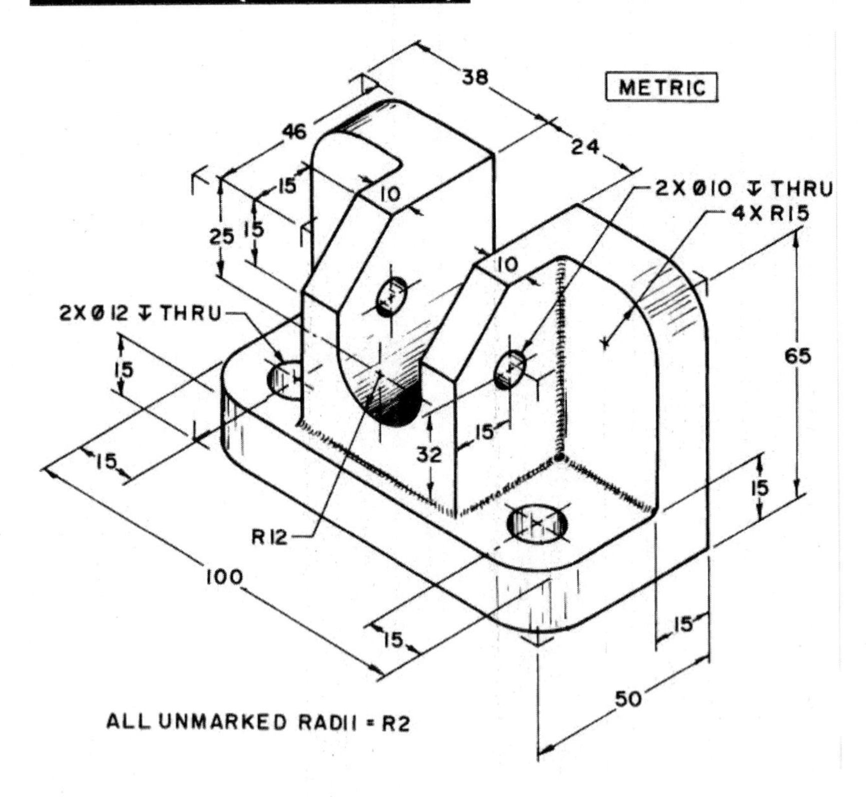

Volume_______________

PROBLEM 3-4C (ENGLISH UNITS)

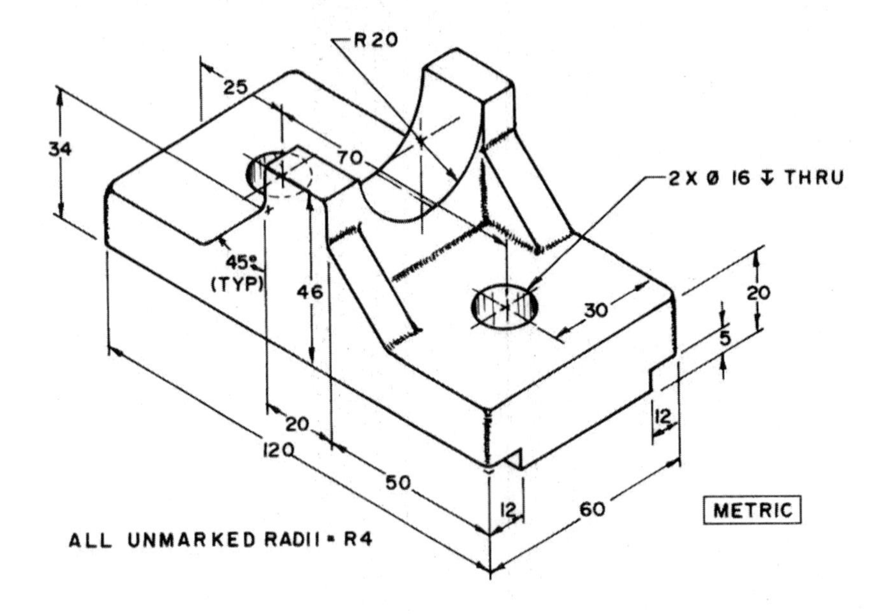

Volume_______________

PROBLEM 3-4D (ENGLISH UNITS)

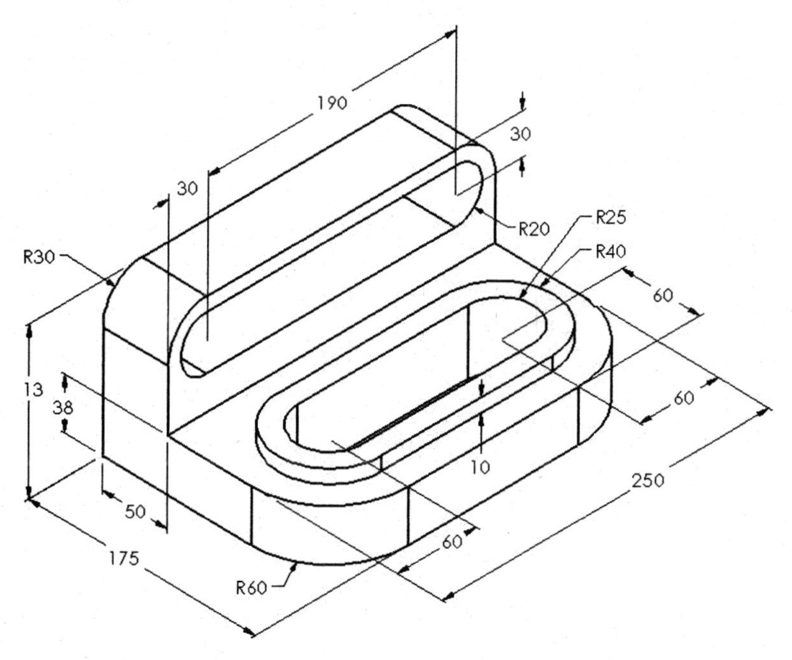

Volume____

PROBLEM 3-4E (ENGLISH UNITS)

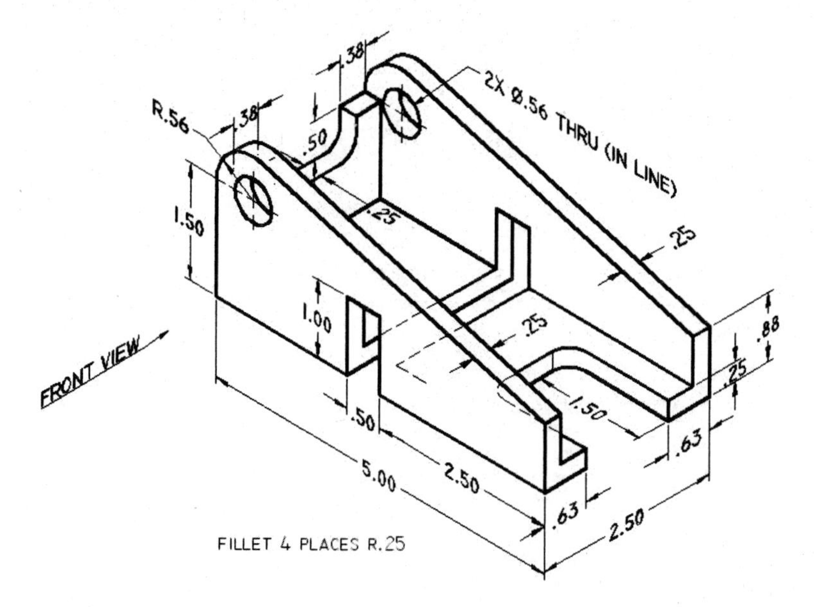

Volume____

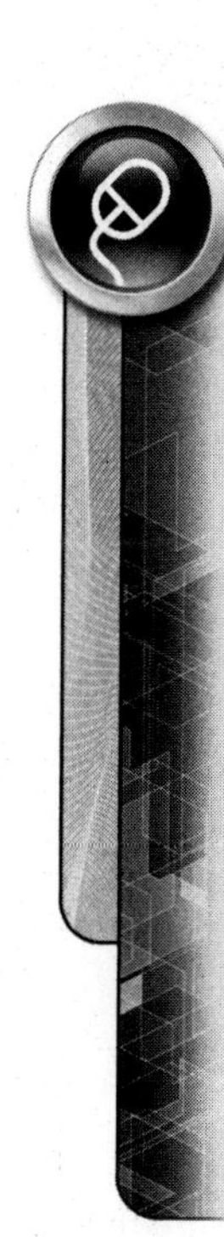

3.5 PATTERN PROBLEM: 251-W05-01.SLDPRT (METRIC)

Create only the base plate minus any holes with the initial sketch located at the center of the 70 radius arc. Extrude this shape a distance of 12.7 and add fillets to the four corners. Create a new sketch plane and cut the slot and 20 diameter hole. Create a new sketch plane, add one 7 diameter hole, and generate the pattern. Repeat this procedure for the remaining holes and hole patterns.

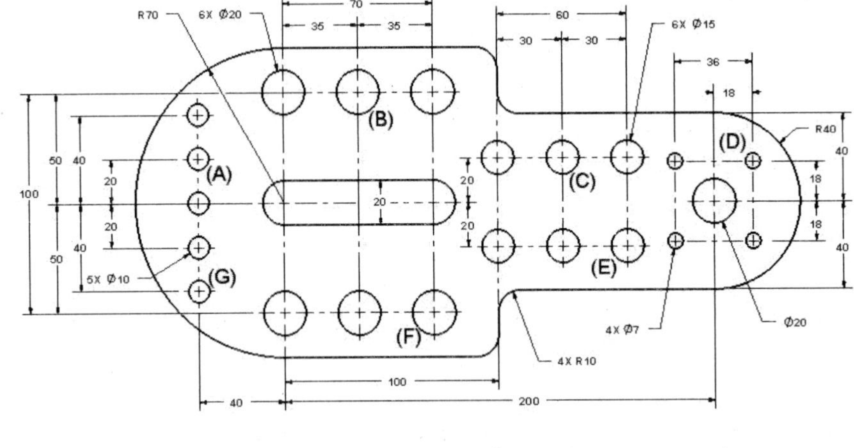

Arrange in a one page Word document with screen capture and mass properties. Also include the answers to the following questions:

What are the center-to-center distances between the following holes?

1. "A" to "B" ________________
2. "B" to "C" ________________
3. "C" to "D" ________________
4. "D" to "E" ________________
5. "E" to "F" ________________
6. "F" to "G" ________________
7. What is the surface area of the top of the plate? ________________
8. What is the perimeter of the top of the plate? ________________

3.6 PATTERN PROBLEM: 251-W05-02.SLDPRT (METRIC)

Create only the base plate minus any holes with the initial sketch located at the center of the 65 radius arc. Extrude this shape a distance of 12.7. Create a new sketch plane and cut the slot and 18 diameter hole. Create a new sketch plane, add one 7 diameter hole, and generate the pattern. Repeat this procedure for the remaining hole patterns.

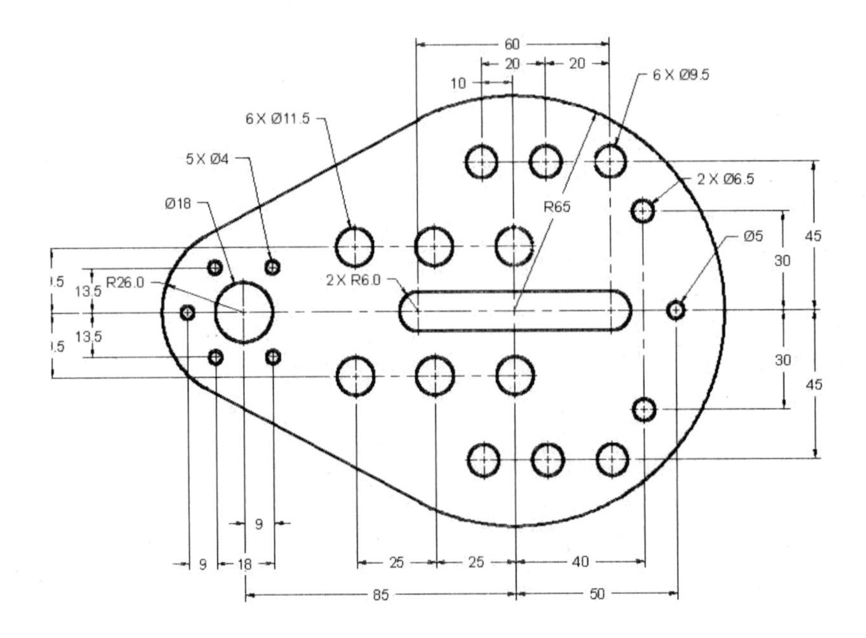

What are the center-to-center distances between the following holes?

1. "A" to "B" ________________
2. "B" to "C" ________________
3. "C" to "D" ________________
4. "D" to "E" ________________
5. "E" to "F" ________________
6. "F" to "G" ________________
7. What is the surface area of the top of the plate? ________________
8. What is the perimeter of the top of the plate? ________________

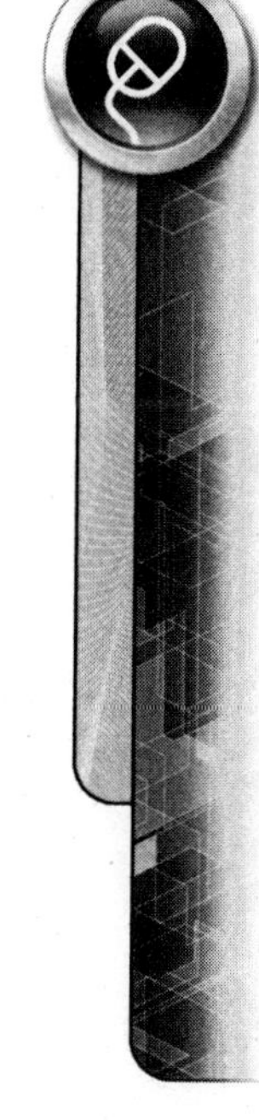

3.7 a thru e

Begin each problem using a part template that has its units set to metric (mm). Perform a mass property calculation on each object and take note of the volume.

PROBLEM 3-7A

1 × 6 Beam

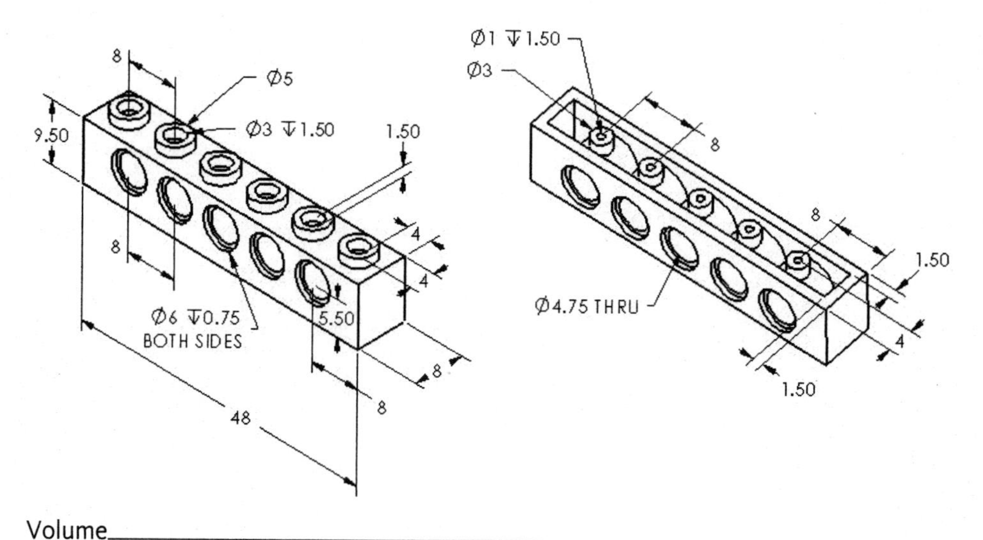

Volume________________________________

PROBLEM 3-7B

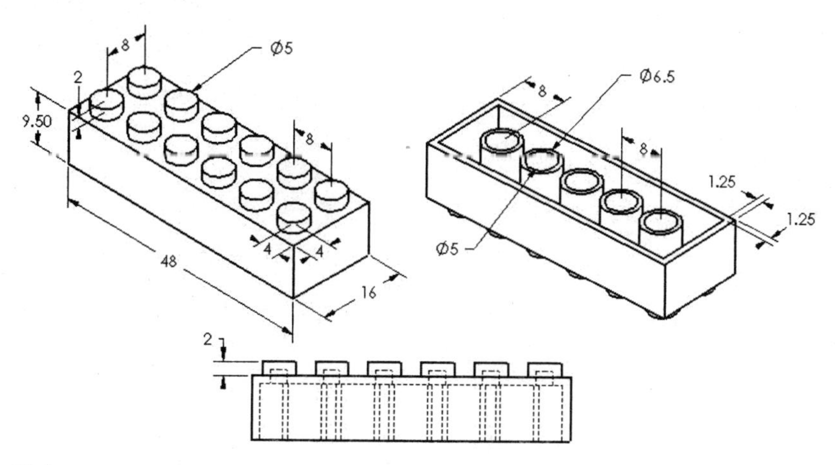

Volume________________________________

PROBLEM 3-7C

2 × 4 Plate Holes

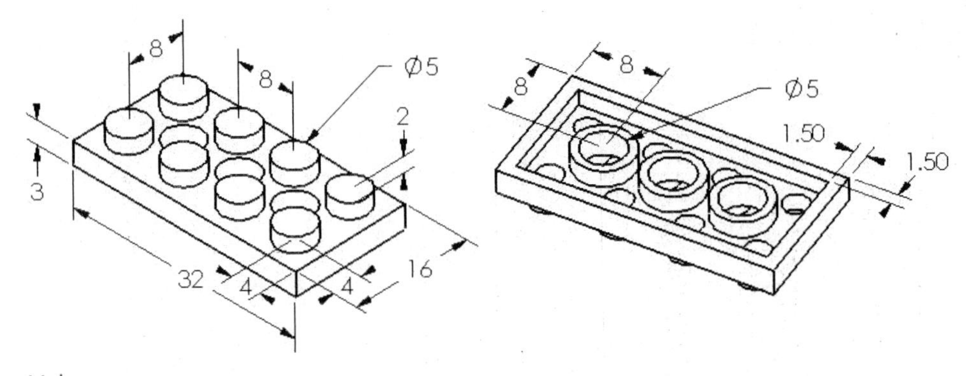

Volume______________________________

PROBLEM 3-7D

2 × 4 Plate

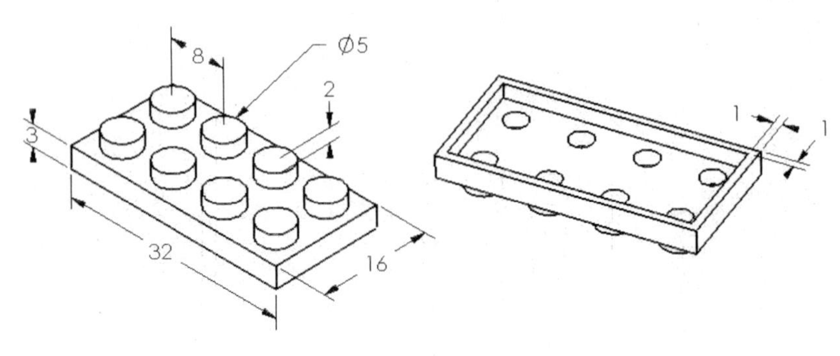

Volume______________________________

PROBLEM 3-7E

1 × 10 Plate

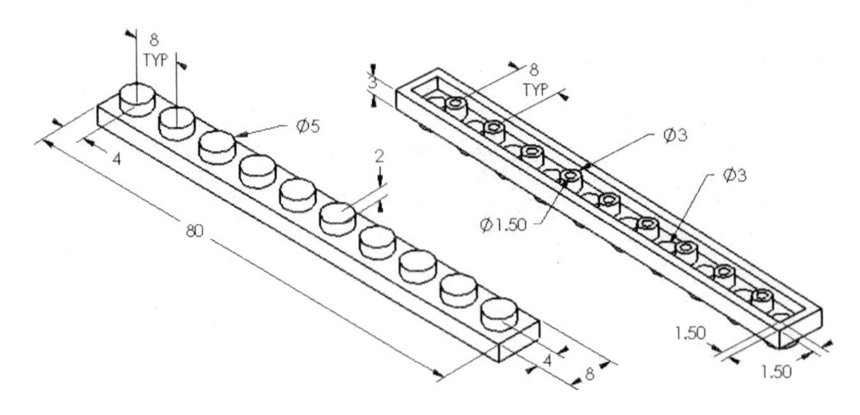

Volume______________________________

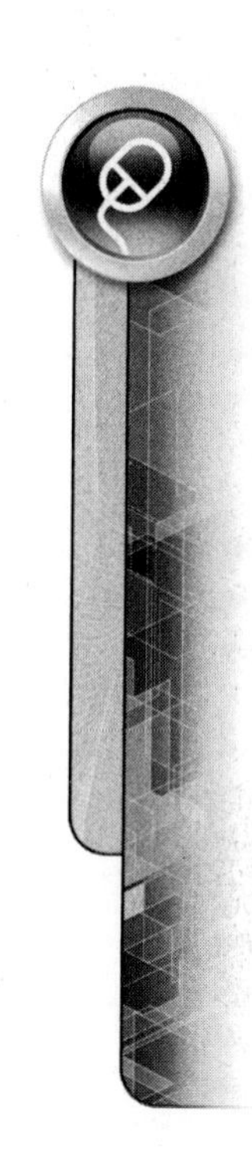

3.8 a thru c

For problems, use the same techniques outlined in the Plastic_Plate tutorial exercise.

PROBLEM 3–8A

4 × 4 Plate

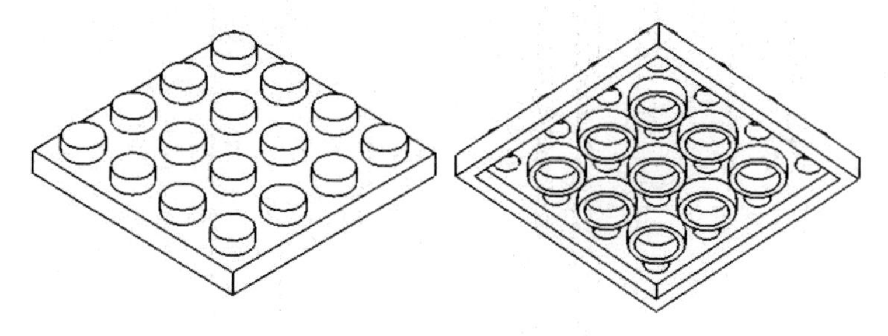

Volume____________________

PROBLEM 3–8B

4 × 12 Plate

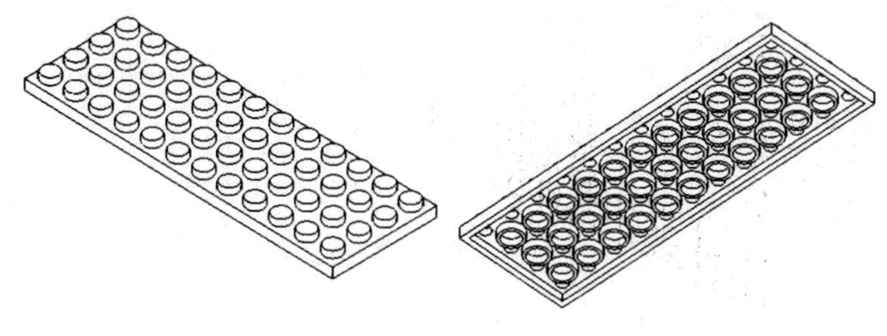

Volume____________________

PROBLEM 3–8C

6 × 16 Plate

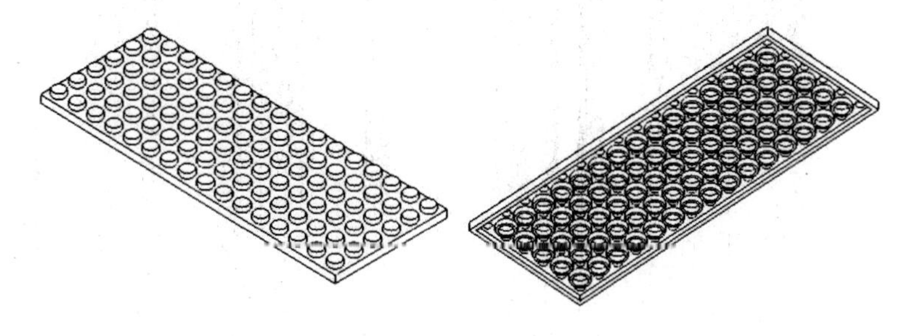

Volume____________________

3.9 a thru e

Begin each problem using a part template that has its units set to metric (mm). Set the dimension standard to ANSI. Construct these part models using SolidWorks sketch and revolved feature tools. Perform a mass property calculation on each object and take note of the volume.

PROBLEM 3-9A

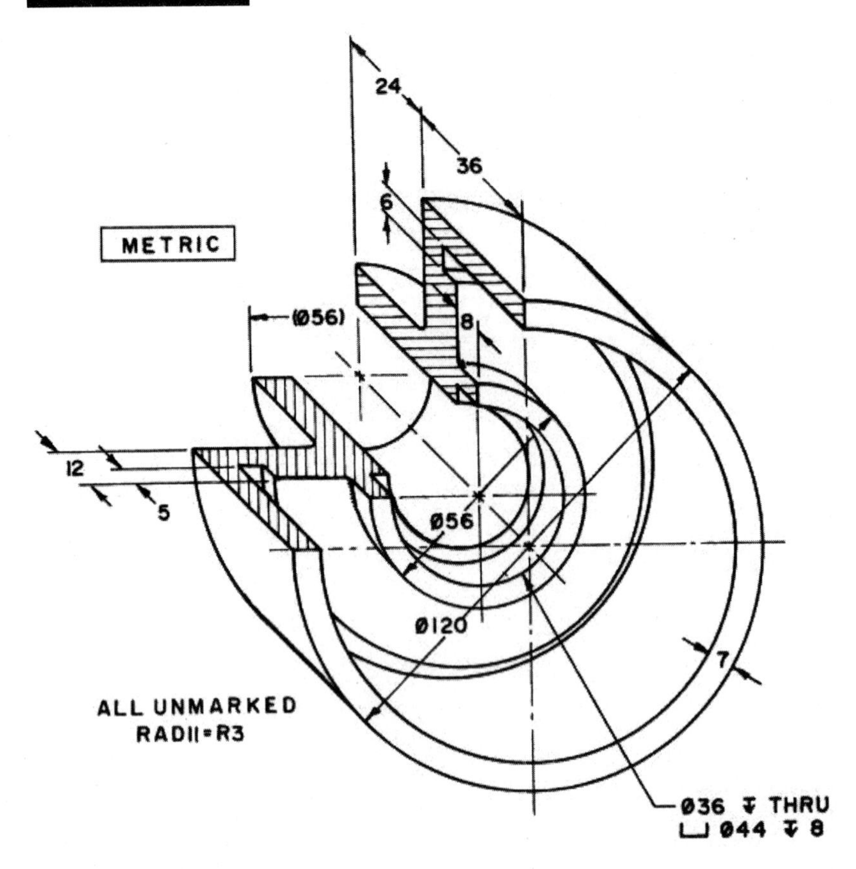

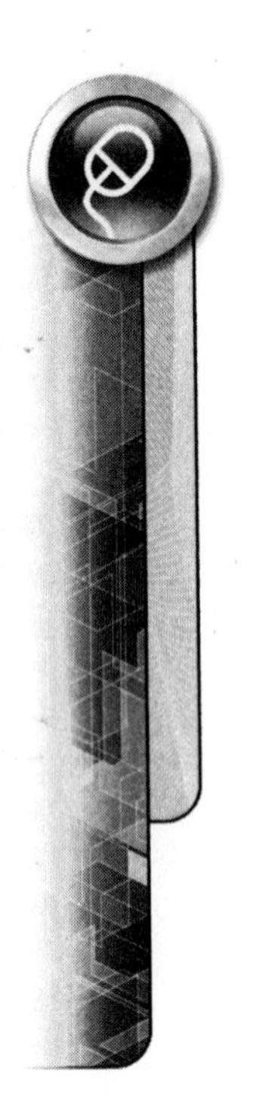

PROBLEM 3-9B

Use the Right Plane for the initial sketch.

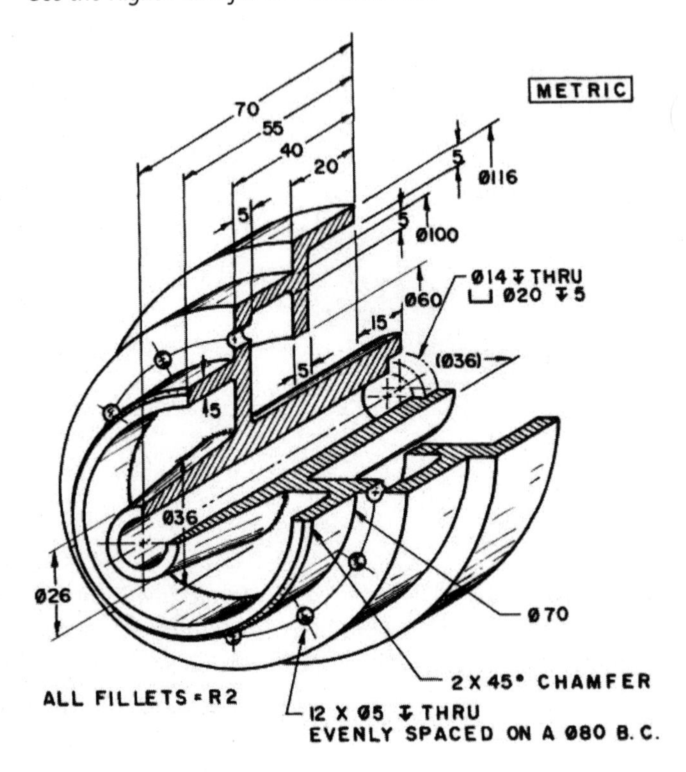

PROBLEM 3–9C

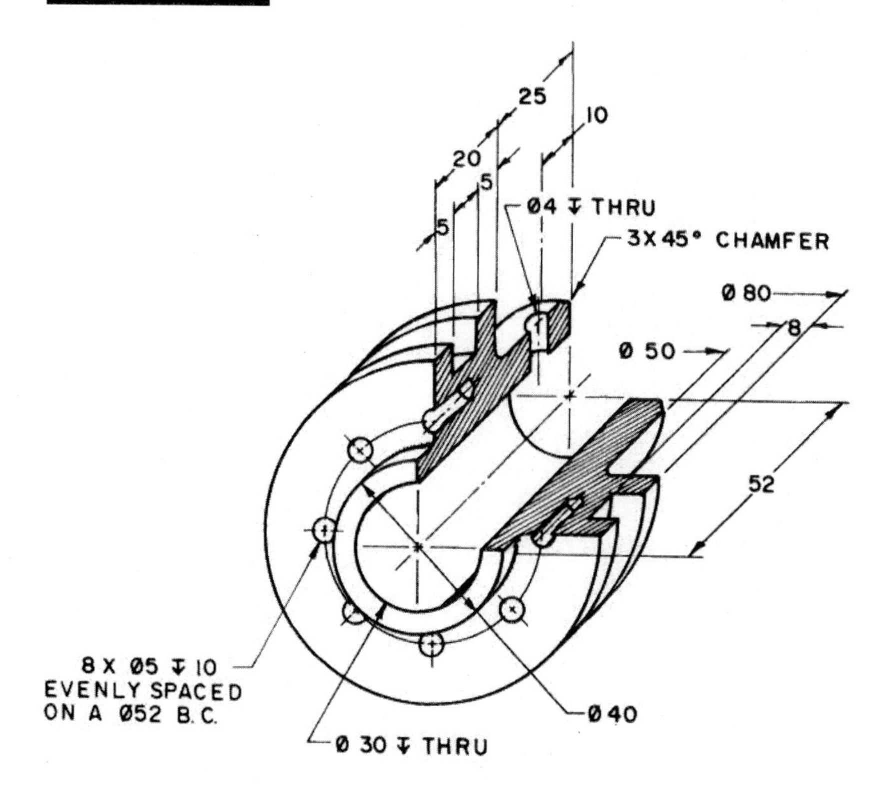

PROBLEM 3–9D

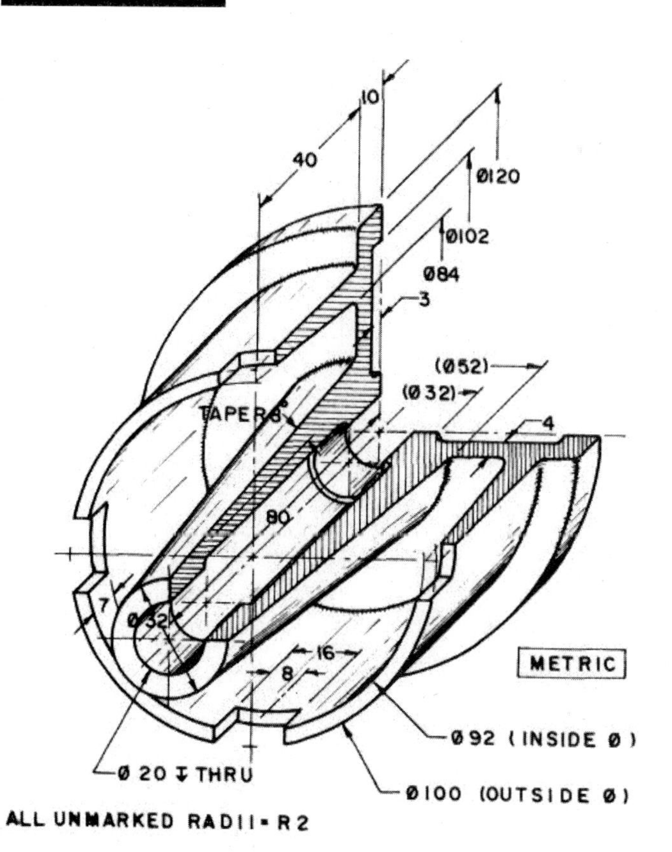

PROBLEM 3-9E

Use the Front Plane for the initial sketch.

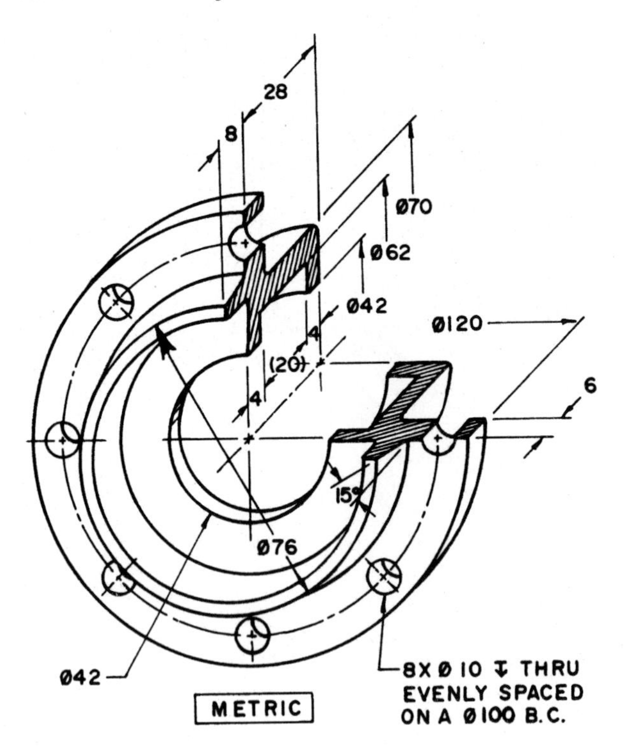

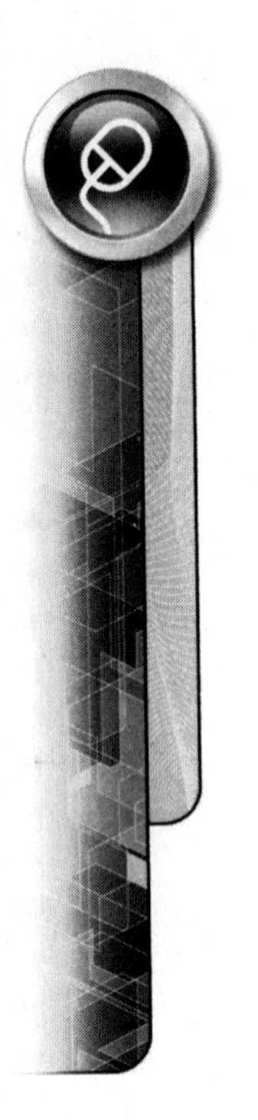

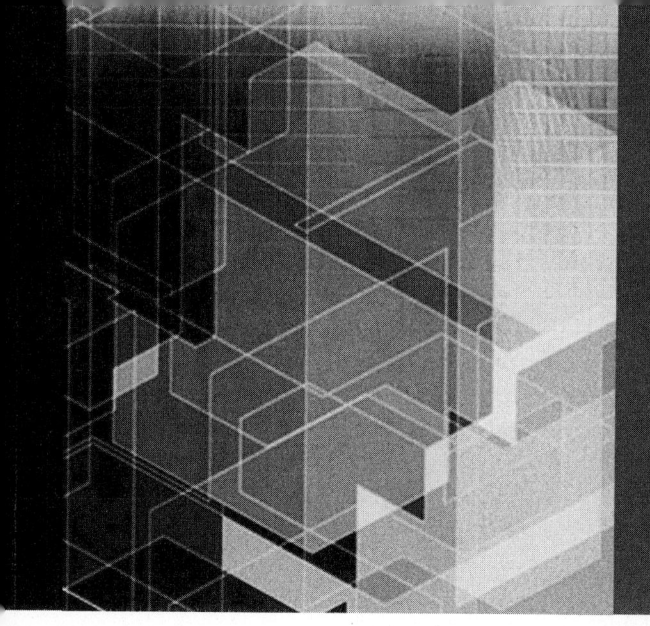

CHAPTER 4

Advanced Part Modeling

This chapter discusses various techniques and tools used to create part models and features in SolidWorks and includes the following topics:

- Sweep models
- Creating Reference Planes
- Sweep cut
- Defining and creating holes using the Hole Wizard
- Lofting models
- Creating ribs
- Creating drafts
- Embossing and engraving text
- Creating springs
- Adding dome features
- Mirroring solid parts
- Creating design tables
- Defining part material
- Introducing appearances

SWEEP BOSS/BASE

Performing a Sweep Boss/Base creates a solid part by extruding a profile sketch along a path. This is similar to an Extrude except an Extrude is performed in a straight line direction, whereas a Sweep can be swept across a complex path.

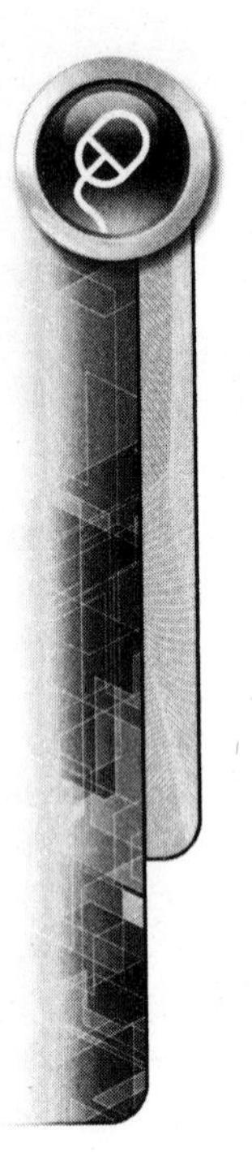

TUTORIAL EXERCISE: 04_SWEEP_BASIC.SLDPRT

Part Number: SWT-CH04-01.SLDPRT

Description: Sweep 1

Units: English

This tutorial exercise is designed to create a part model created by the Sweep Boss/Base command (Figure 4.1). Two sketches need to be created: the Profile Sketch and the Path Sketch. These sketches need to be accurately located from each other and the Origin will be used to accomplish this.

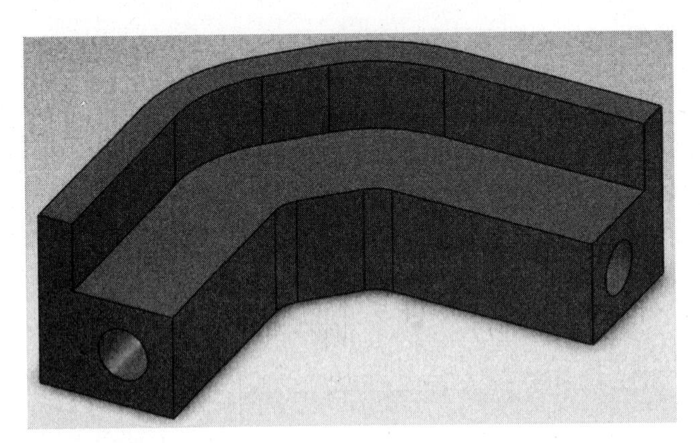

FIGURE 4.1

STEP 1

Begin a new part file, and click on the Front Plane. Then sketch the shape as shown in Figure 4.2. The origin should be located at the lower left corner of the sketch. All lines should be either horizontal or vertical. Also, add all dimensions to fully define the sketch, then exit the sketch.

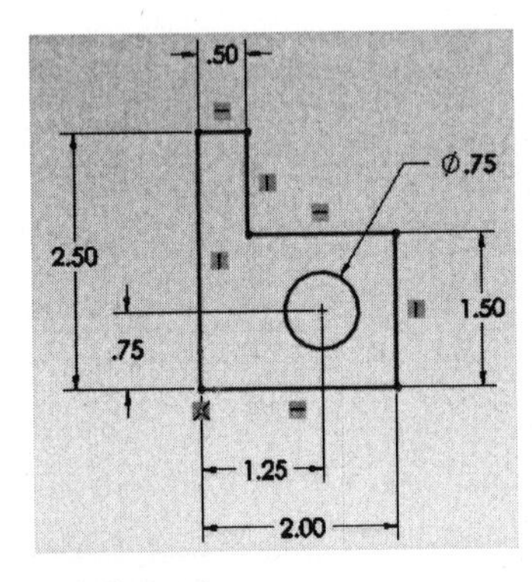

FIGURE 4.2

STEP 2

Start a new sketch and click on the Top Plane. Then sketch the shape as shown on the left in Figure 4.3. The origin should be located at the lower point of the sketch, and make sure the first line is vertical which will make it perpendicular to the sketch from Step 1.

Then add all dimensions to fully define the sketch, then add two Fillets of size R2.50 as shown on the right of the figure. Exit the sketch.

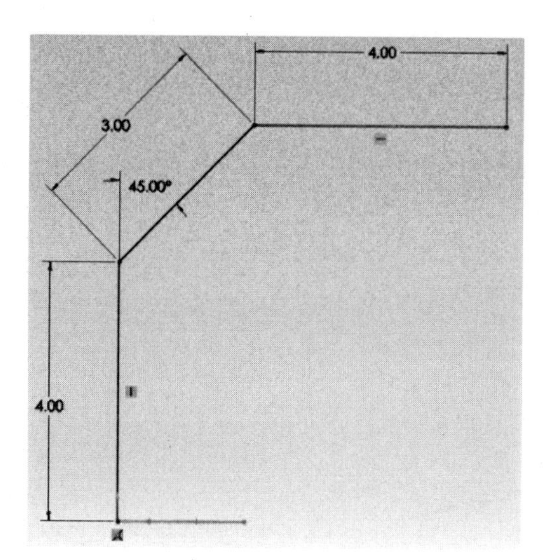

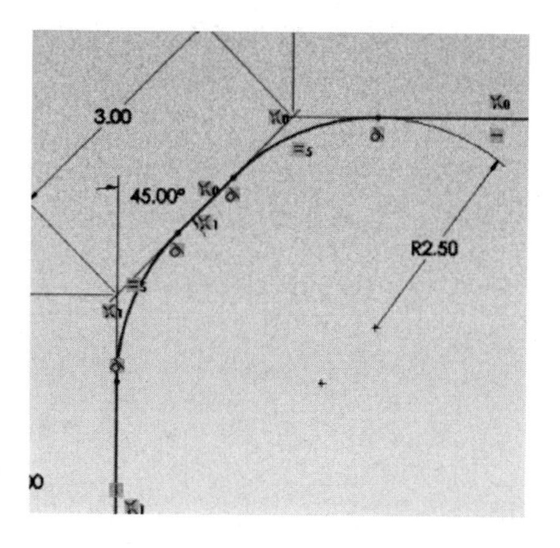

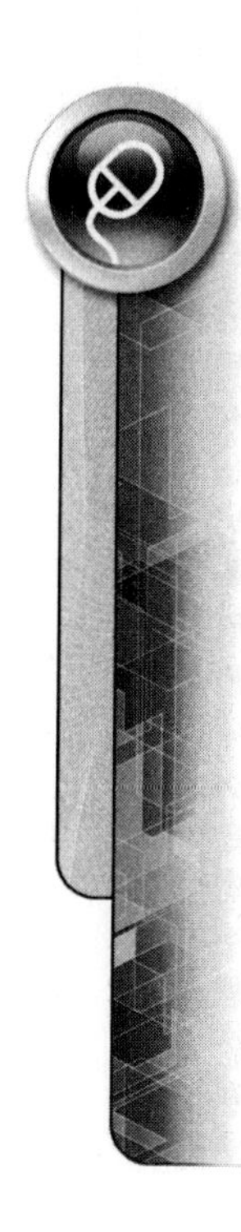

FIGURE 4.3

STEP 3

Using the Display menu, select the View Orientation pull-down and select Trimetric as shown in Figure 4.4. This should rotate the view allowing you to view both sketches in the Model Area.

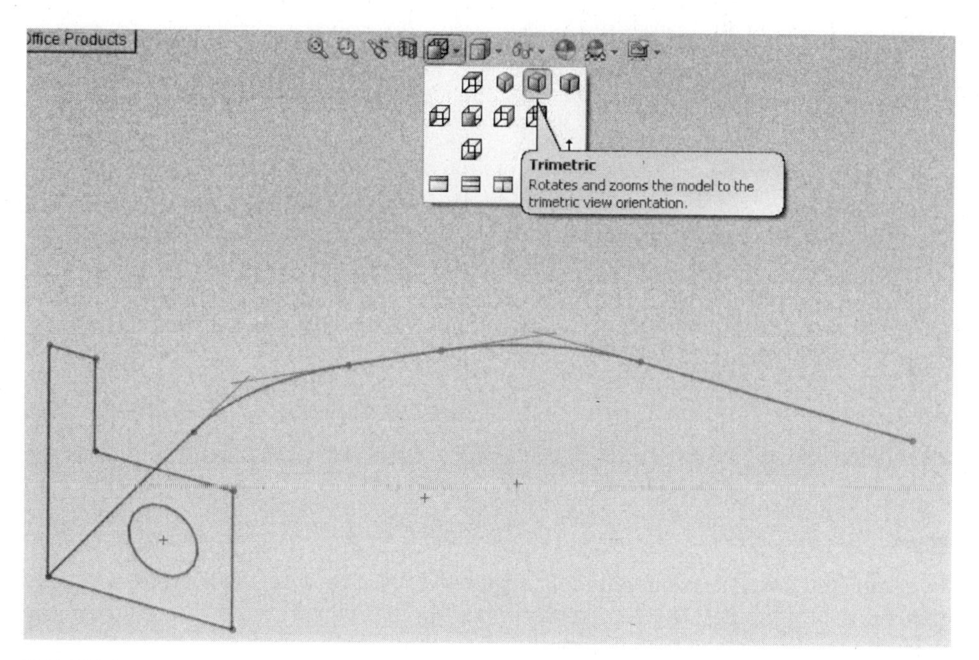

FIGURE 4.4

STEP 4

In the Feature Command Manager, select Sweep Boss/Base as shown on the left of Figure 4.5. The Feature Manager shows the Sweep options, and the Profile Sketch field

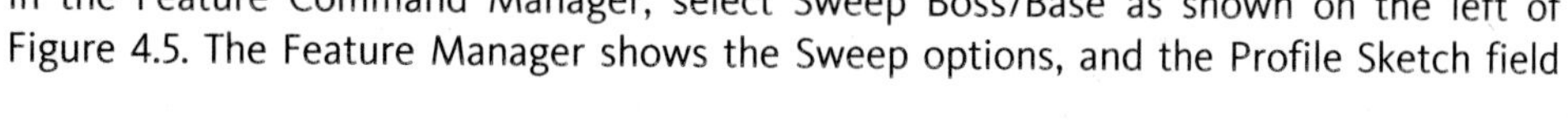

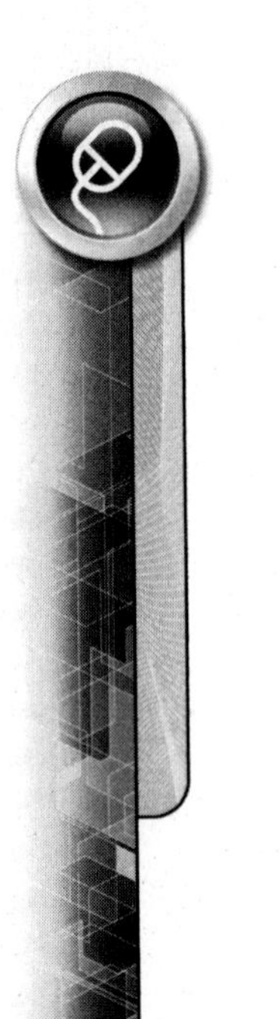

is highlighted in blue as seen in the figure on the right; select Sketch 1 by clicking on any line of the Sketch. Sketch 1 should highlight on the model as well as is called out in the Profile Sketch field.

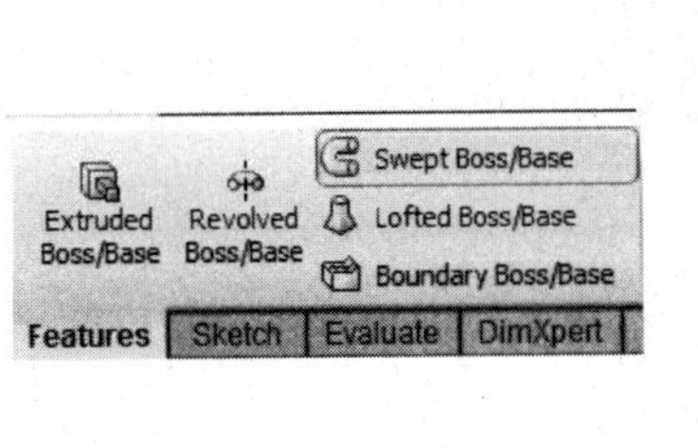

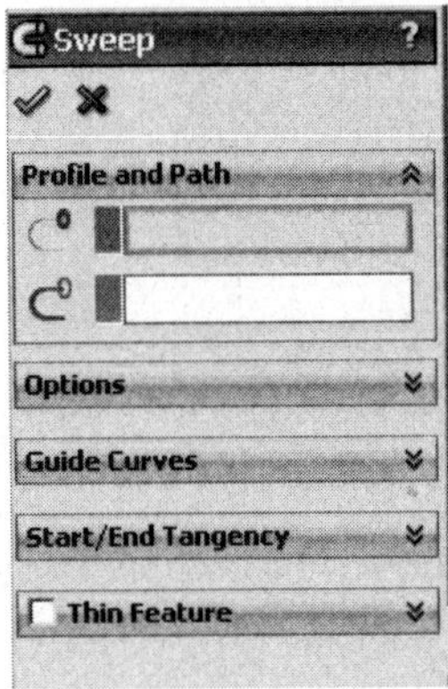

FIGURE 4.5

STEP 5

The Path Sketch field should now be highlighted in blue; select Sketch 2 by clicking on any line of the Sketch. The model should now look as shown in Figure 4.6. Click the checkmark to complete the command.

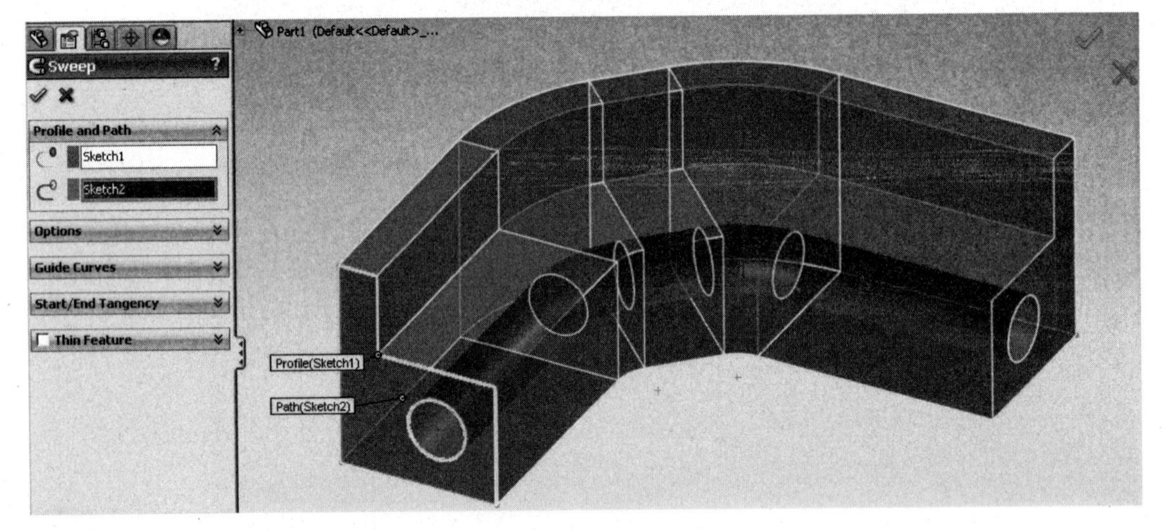

FIGURE 4.6

STEP 6

The completed Sweep is illustrated in Figure 4.7. Notice that the Feature Manager organizes the Sweep Boss/Base to be dependent on the two sketches. The sketches can be renamed as Profile and Path if desired.

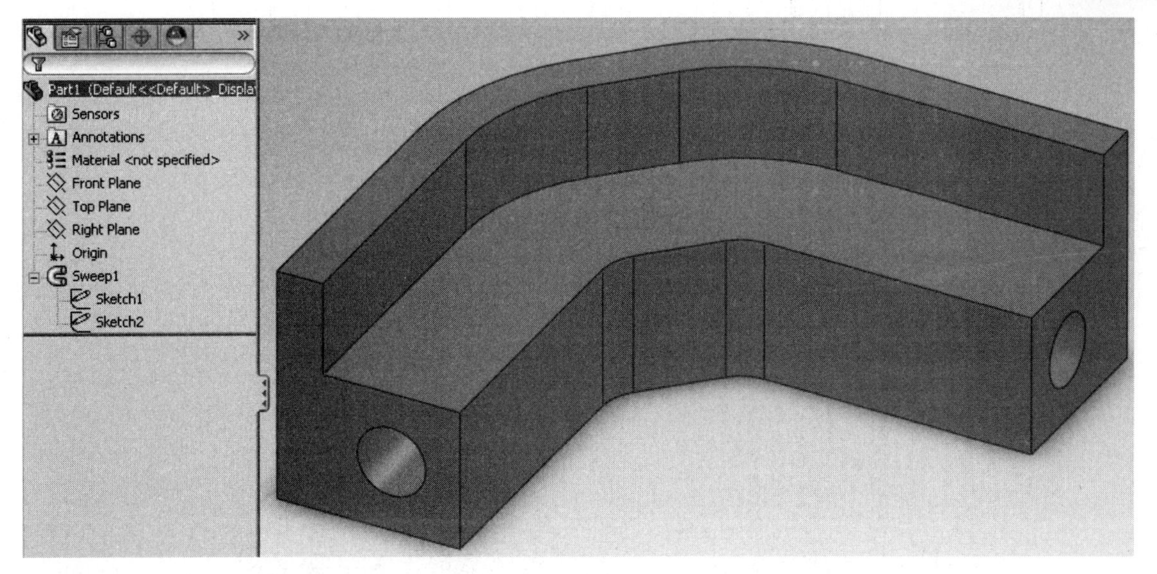

FIGURE 4.7

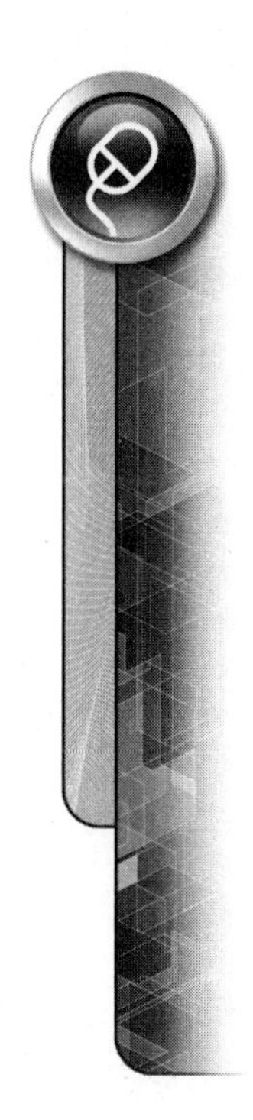

REFERENCE PLANES (SKETCHING PLANES)

When creating sketches, an existing two-dimensional plane is required to start. The sketch can be started on one of the three default phantom planes: front plane, top plane, and right plane, or you can use an existing face of the part. There are oftentimes where new phantom planes are required to create more complex geometry.

When additional planes are needed, the Reference Geometry Plane command is available. It is found on the Features ribbon as shown in Figure 4.8. The Reference Geometry Plane command is intuitive; that is, when a feature is selected, such as a plane, the Command Manager changes to options that are only relevant to create a plane from that feature.

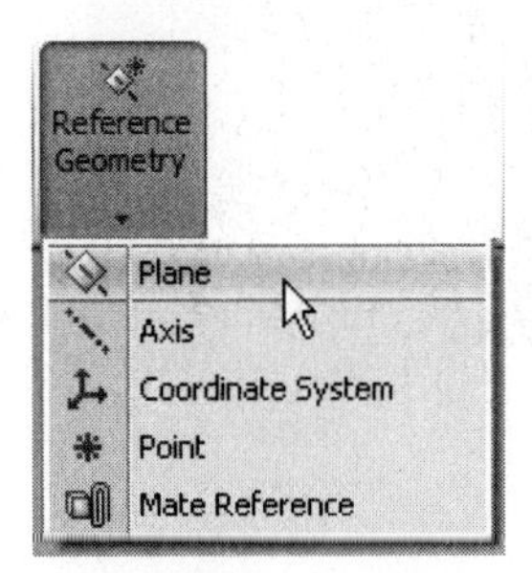

FIGURE 4.8

There are several types of Reference Planes that can be created. The major Reference Planes that can be created are:

TABLE 4.1

Reference Plane Type	Description	Visual
Offset	As an existing plane is selected, a plane is created at an offset of a known distance.	
Parallel to a face at a point	As a plane and a point are selected, a plane is created that is parallel to an existing face and offset at a distance identified by the point location.	
Perpendicular to a face at a point	As a plane and a point are selected, a plane is created that is perpendicular to the existing plane and is located by the point location.	
Midplane	As two existing planes are selected, a plane is created equidistantly between two parallel surfaces.	
Through three points	As three points are selected, a plane is created that passes through all three points.	

continued

TABLE 4.1 *Continued*

Reference Plane Type	Description	Visual
Through an edge and a point	As an edge and a point are selected, a plane is created that passes through the edge and the point.	
Through an edge at an angle to a plane	As an edge and a plane are selected, a plane is created that passes through an edge line and is referenced at an angle from an existing plane.	
Equal angled between two planes	As two nonparallel planes are selected, a plane is created that is located at the intersection of the two planes and is angled equally between the two planes.	

Tutorial examples of the various Reference Planes are incorporated in tutorial exercises.

SWEPT CUT

Performing a Swept Cut on a part removes material from the part. The material removed is defined similar to a Sweep Boss/Base by defining a Profile Sketch and a Path Sketch. A Sweep Cut is also similar to an Extrude Cut except an Extrude Cut is limited to removing material in a straight line while a Sweep Cut can remove material in a complex direction.

TUTORIAL EXERCISE: 04_SWEEP_CUT_BASIC.SLDPRT

Part Number: SWT-CH04-02.SLDPRT

Description: Sweep Cut 1

Units: English

This tutorial exercise is designed to create a part model that includes a feature created by the Sweep Cut Boss/Base command (Figure 4.9). After the initial block is created, two sketches need to be created: the Profile Sketch and the Path Sketch. These sketches need to be accurately located from each other and the Origin will be used to accomplish this.

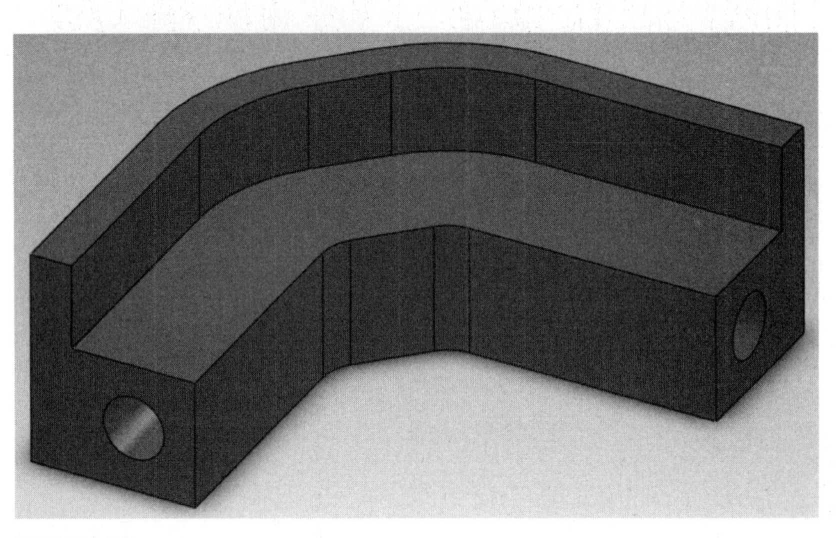

FIGURE 4.9

STEP 1

Begin a new part file and click on the Top Plane. Then sketch a 4.00 inch by 8.00 inch rectangle; the origin can be located at the lower left corner of the sketch. All lines should be either horizontal or vertical. After the sketch is fully defined, exit the sketch. Extrude the shape 1.00 inch high as seen in Figure 4.10.

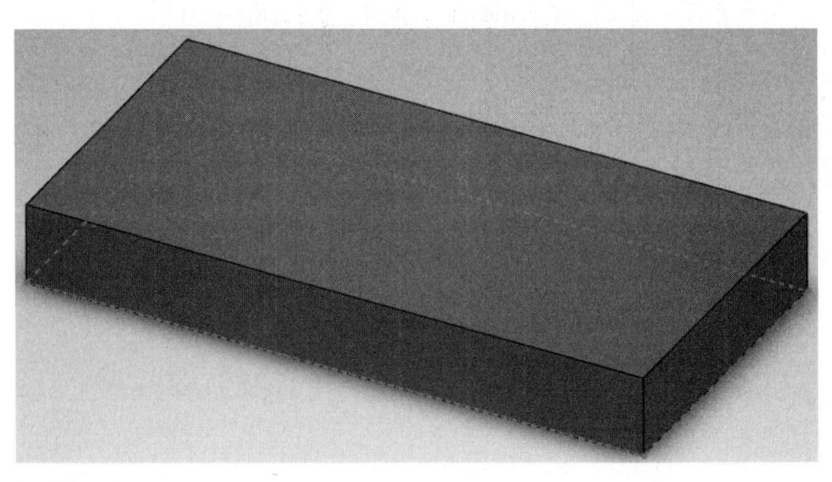

FIGURE 4.10

STEP 2

Add the sketch as shown in Figure 4.11 by starting a new sketch and selecting the Top Surface of the block. The view does not need to be rotated in order to define this sketch; SolidWorks can work in true 3D where the plane that is being used for a sketch does not have to be normal to the screen. Create a rectangle first and locate and size it with the four given dimensions, then add the fillets to complete the sketch. This will be the path sketch.

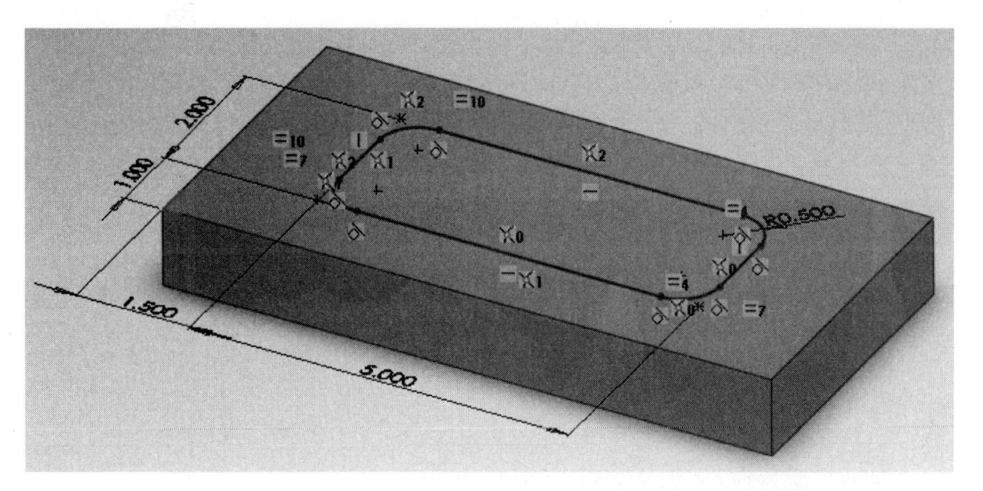

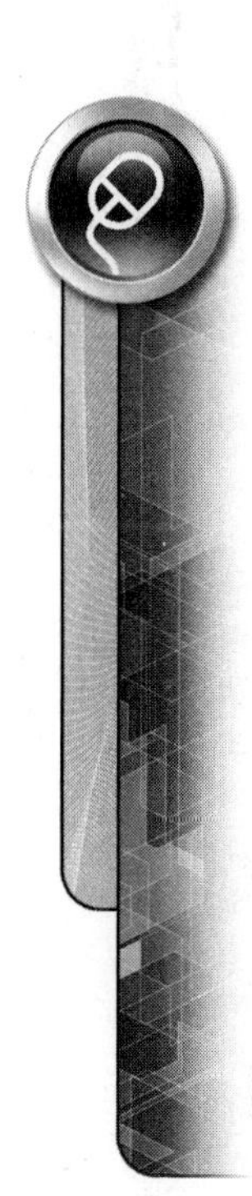

FIGURE 4.11

STEP 3

A plane perpendicular to the path sketch needs to be created that can be located anywhere along the path sketch. In this case, create a midpoint plane between the left and right sides of the block.

Select the Reference Geometry Plane command from the Features menu. Select the left face of the block, then select the right surface of the block; these two surfaces will be used to create a reference plane equidistant between them. Accept this plane by selecting OK (Figure 4.12).

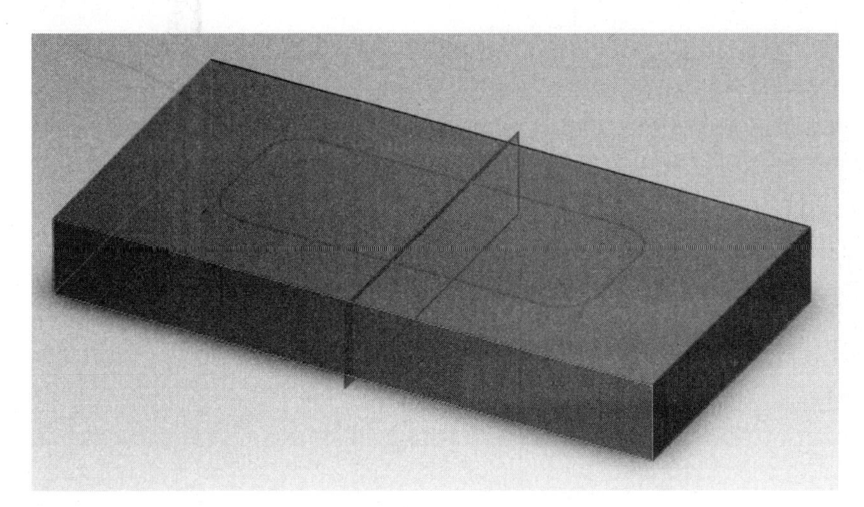

FIGURE 4.12

STEP 4

Now a circle that will represent the profile sketch needs to be created. Create a new sketch on the newly created midpoint plane.

Select the view orientation for a right side view. Notice in Figure 4.13 that the origin is in the lower left of the block in this plane. Also, there are points along the top surface that are representing points from the path sketch. Enter the sketch circle command and select the point on the top line that represents the path sketch and complete the circle. Dimension the diameter of the circle.

Notice that the sketch is fully defined with only one dimension. This is because the location of the circle is defined by the location of the point you selected on the path sketch. No further definition is necessary. Exit the sketch.

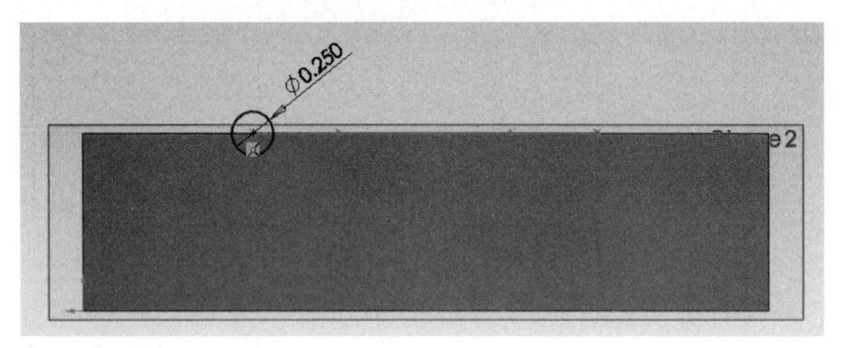

FIGURE 4.13

STEP 5

Change the view orientation to Trimetric and zoom in far enough to where you can see both sketches and they are selectable. Select the Swept Cut command off of the Features menu. While the profile sketch field is highlighted in the Feature Manager, select the sketch of the circle. Now the path sketch field is highlighted, so select the first sketch created. The preview of the Sweep should be represented as seen in Figure 4.14.

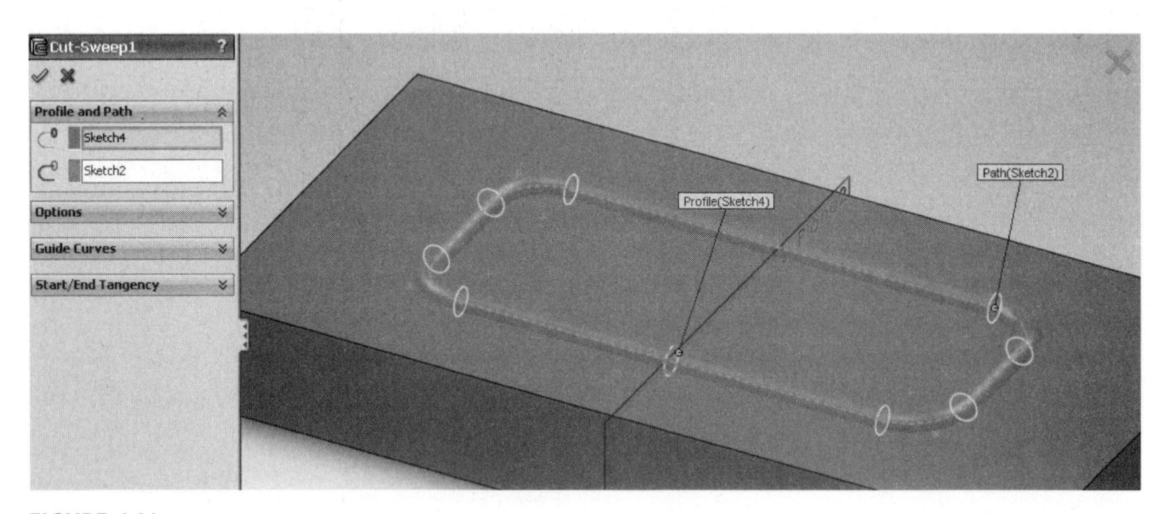

FIGURE 4.14

Accept the sweep by selecting OK. Here you can see that because only half of the profile circle was "inside" the block, the end result is a half-moon cut into the block (Figure 4.15).

FIGURE 4.15

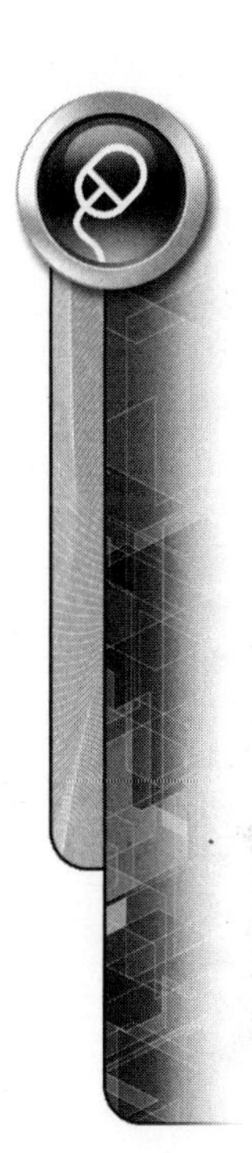

HOLE WIZARD

The Hole Wizard command is an advanced and efficient way of creating holes in a part.

The original method of creating a hole, for example, is using basic commands such as taking a sketch of a circle and extruding it through a part, and to make a counterbore, create a second sketch of a larger circle and extrude it partially into the part. And if this hole is to hold a fastener, standardized charts that give the fastener sizes need to be sought out and then the hole sizes need to be derived to be slightly larger than the fastener as well as the depth of the counterbore. The Hole Wizard eliminates the majority of these steps.

The Hole Wizard command is driven mostly by the Feature Manager. While on the Type tab of the Feature Manager, define specifically the type of hole desired. In most cases the type of fastener that will be inserted into the hole is selected from a list and all dimensions are predefined except the depth, which is derived by a selected End Condition. The second Feature Manager tab, which is called Positions, is then used to locate the hole by defining locating dimensions.

The Hole Wizard command has an exhaustive amount of predefined holes. They are available for ANSI and ISO series as well as several other standard series. There are also holes defined for counterbore, countersink, straight tap, and tapered tap. They are also predefined for dozens of styles of fastener heads. And in the rare case where the conditions desired are not available, the option of creating a Legacy Hole is available, which simply means all of the dimensions are defined by the user.

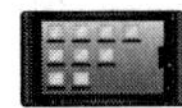

TUTORIAL EXERCISE: 04_HOLE_WIZARD_BASIC.SLDPRT

Part Number: SWT-CH04-03.SLDPRT

Description: Hole Wizard 1

Units: English

This tutorial exercise is designed to create a counterbore hole for a known fastener size (a ½" socket head cap screw). Using the Hole Wizard command will demonstrate the ease of creating a hole designed around any type of fastener (Figure 4.16).

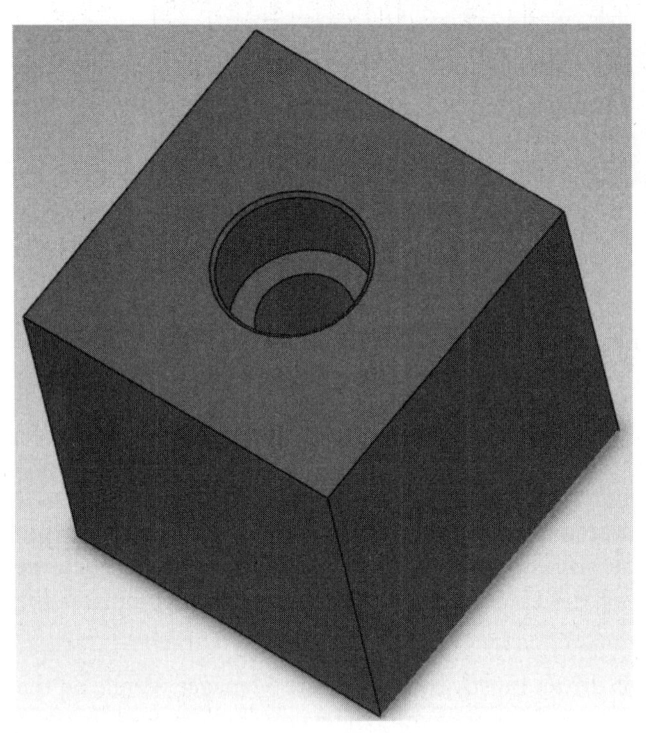

FIGURE 4.16

STEP 1

Create a file and set the units to inches (IPS). Using the sketch and then extrude process, create a 2.00 × 2.00 × 2.00 block. Select the Hole Wizard command from the Features menu (Figure 4.17).

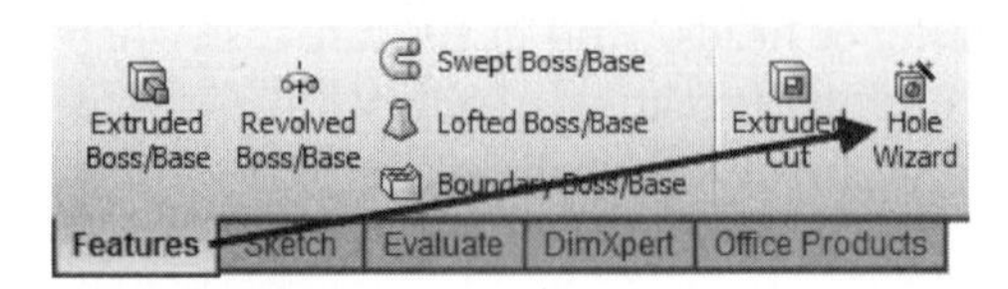

FIGURE 4.17

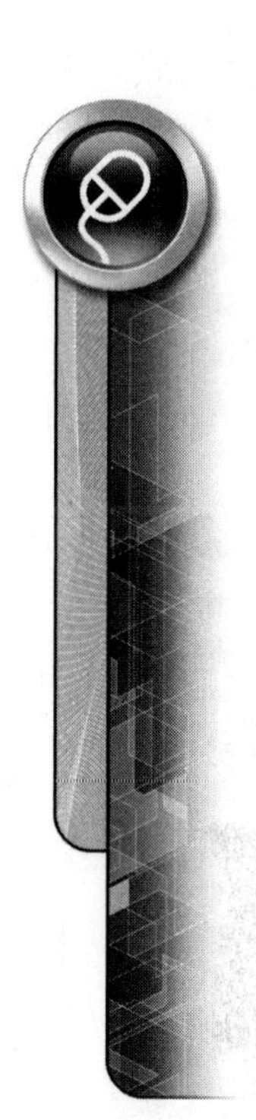

STEP 2

Define the type of fastener the hole is being designed to receive. In the Hole Wizard Feature Manager shown in Figure 4.18, select the Counterbore icon. Also select the ANSI Inch option in the Standard field and select Socket Head Cap Screw in the Type field.

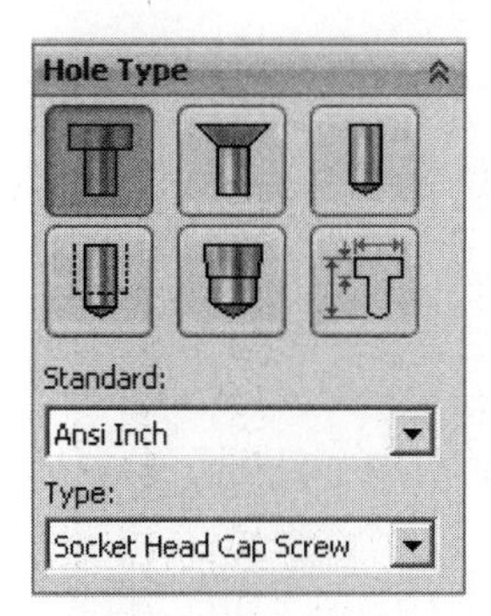

FIGURE 4.18

STEP 3

Define the specifications of the hole and the end condition as seen in Figure 4.19. In the Feature Manager select ½ for the size and Normal for the fit. Also select Through All for the End Condition.

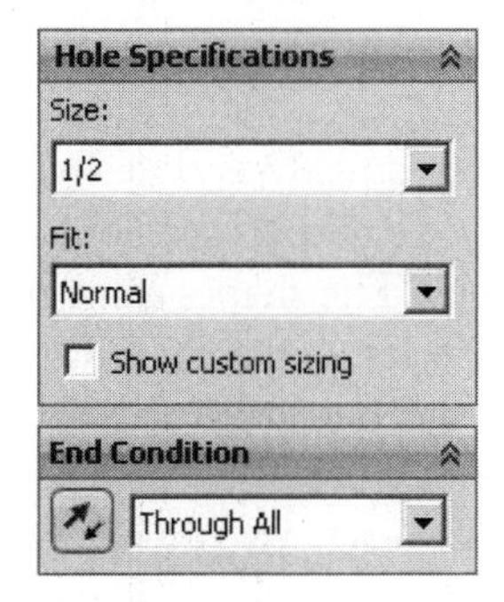

FIGURE 4.19

STEP 4

In the Feature Manager, select the Positions tab. This allows the placement and location of the hole. Select the top face of the block; you will see a preview of the fastener inserting at this surface. Use smart dimension to locate the hole in two directions; it is best to select the center point of the hole as well as edges of the part. Place the dimensions as shown in Figure 4.20. Accept the hole by selecting OK.

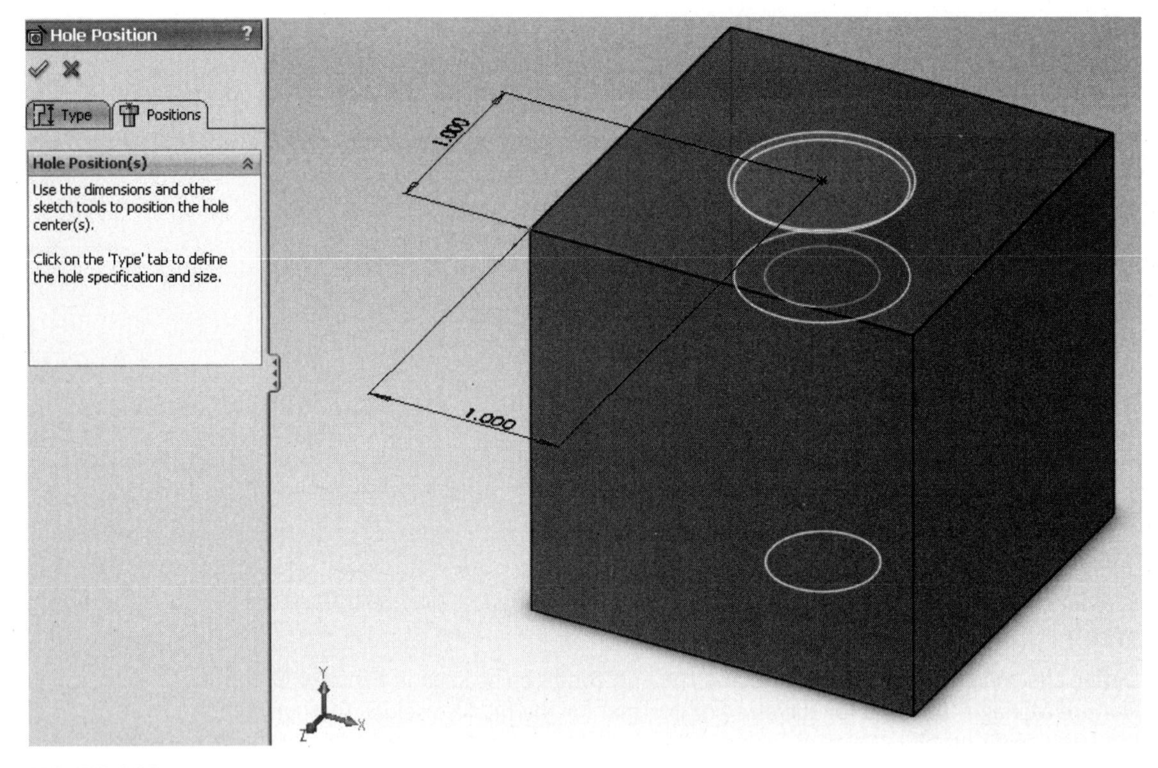

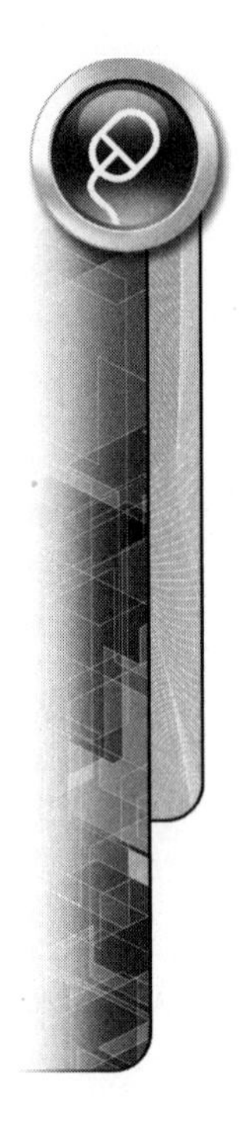

FIGURE 4.20

LOFT BOSS/BASE

The Loft command allows the construction of complex shapes quickly and easily. The Loft command creates solid objects by transitioning from one two-dimensional shape to another, each shape being created as individual sketches. The shapes can be of any type as long as they are closed shapes such as rectangles, circles, polygons, slots, ellipses, complex shapes and also including points. The shapes can also be directly aligned with each other, or they can also be offset. For even more complex parts the loft can be twisted. Figure 4.21 shows several examples of lofts. In each case there are two sketches created to make the parts.

FIGURE 4.21

TUTORIAL EXERCISE: 04_LOFT.SLDPRT

Part Number: SWT-CH04-04.SLDPRT

Description: Loft 1

Units: English

This tutorial exercise is designed to create a part created by the Loft command. Once the two sketches are created, a Loft will be created using the two sketches (Figure 4.22).

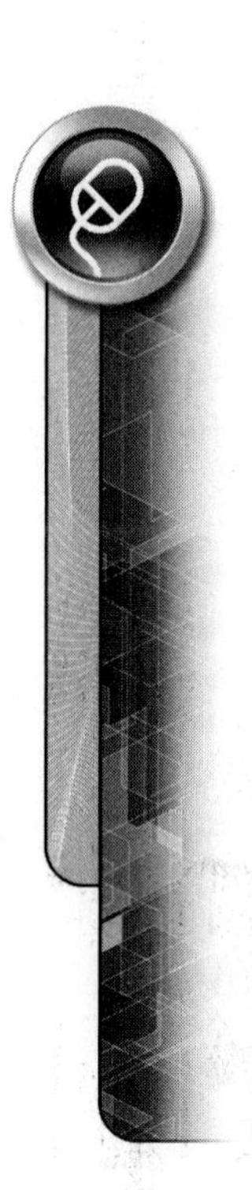

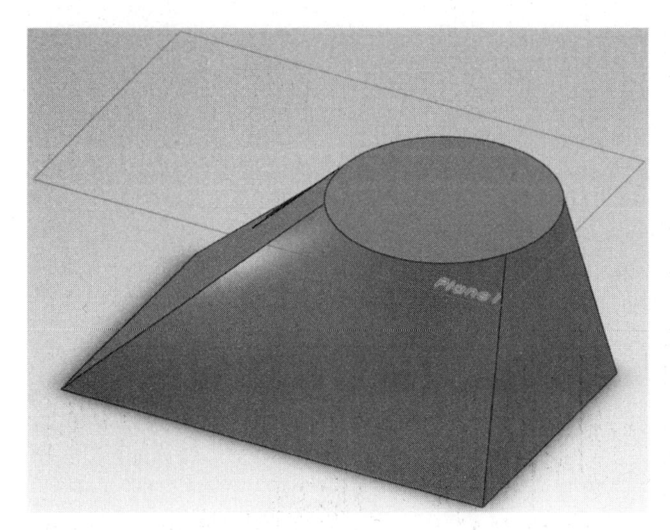

FIGURE 4.22

STEP 1

Create a file and set the units to metric (MMGS). On the Top Plane create a sketch of a rectangle. Dimension the sketch as seen in Figure 4.23. When it is fully defined, exit the sketch.

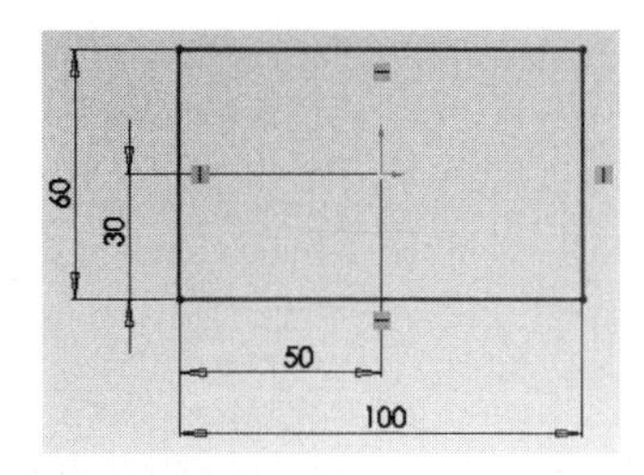

FIGURE 4.23

STEP 2

Create a plane by going to the Features Manager, pulling down the Reference Geometry icon, and selecting the Plane command as shown in Figure 4.24.

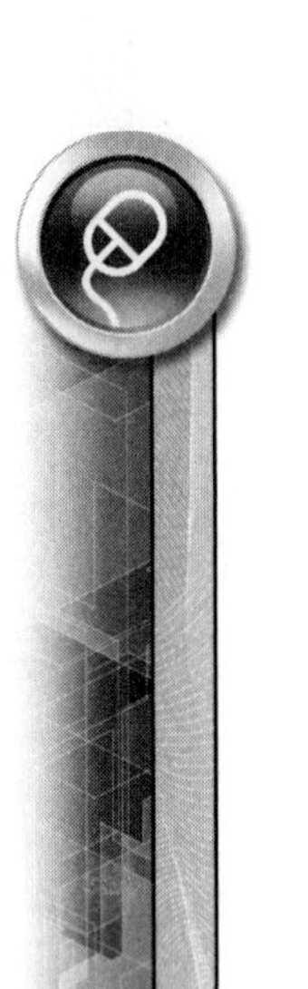

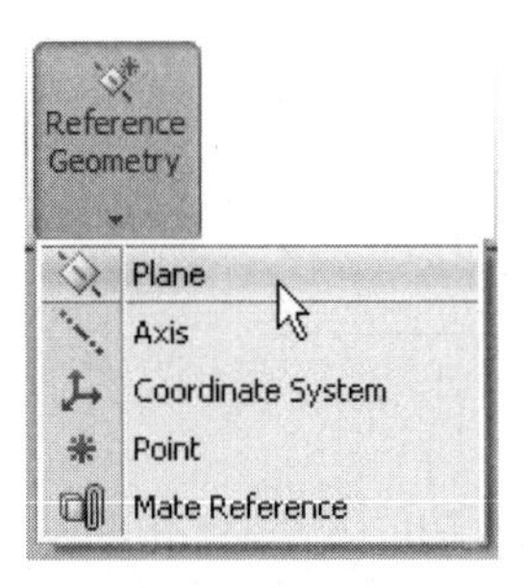

FIGURE 4.24

Select the Top Plane as a reference for a new plane to be created from. In the Command Manager, type in a distance of 50.00mm. A new plane should be shown located above the Top Plane (if it is beneath, select Flip) (Figure 4.25).

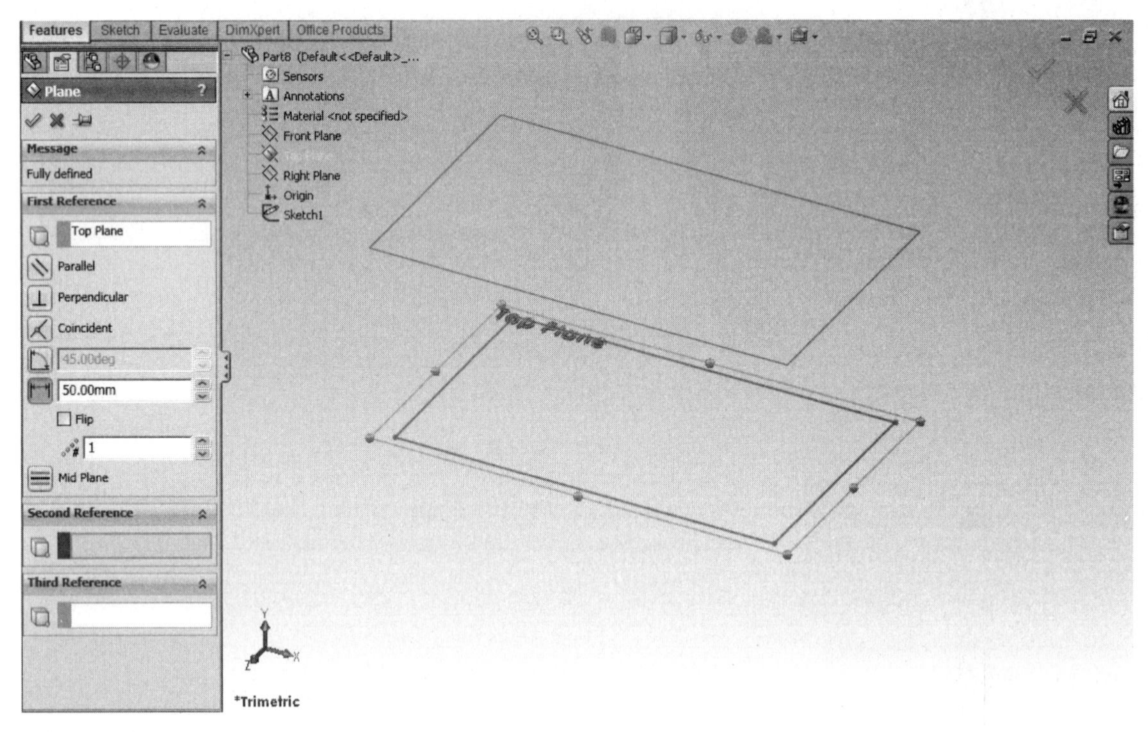

FIGURE 4.25

STEP 3

Create a sketch on the newly created plane by selecting the Sketch command and then selecting the new plane. Create the sketch using the dimensions found in Figure 4.26.

While defining this sketch a horizontal relation needs to be added between the origin and the center point of the circle. Do this by first dropping down the Display/Delete Relations icon from the Sketch ribbon and selecting the Add Relations command as seen in Figure 4.26. Then select the center point of the circle and the then origin. Once this is done the Command Manager has three relations to choose from. Select the Horizontal relation and accept the relation by checking OK. Exit the sketch and change the view orientation to an isometric view.

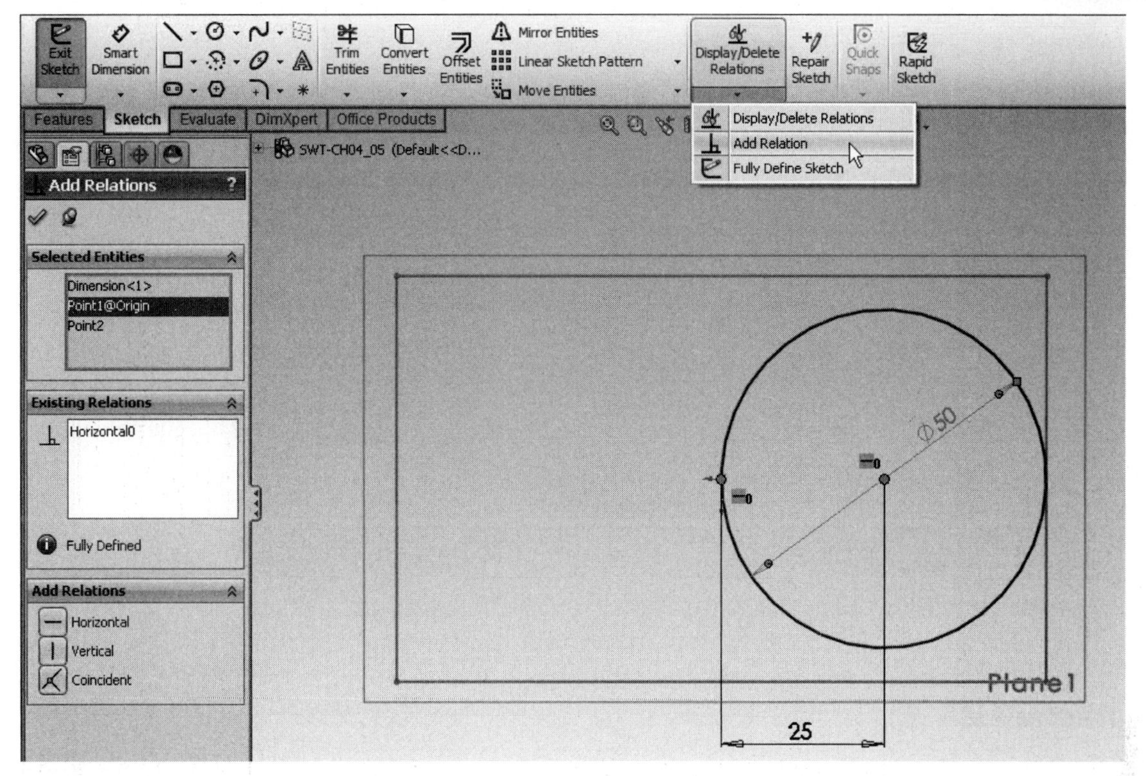

FIGURE 4.26

STEP 4

Now select the Lofted Boss/Base command from the Features ribbon as shown in Figure 4.27.

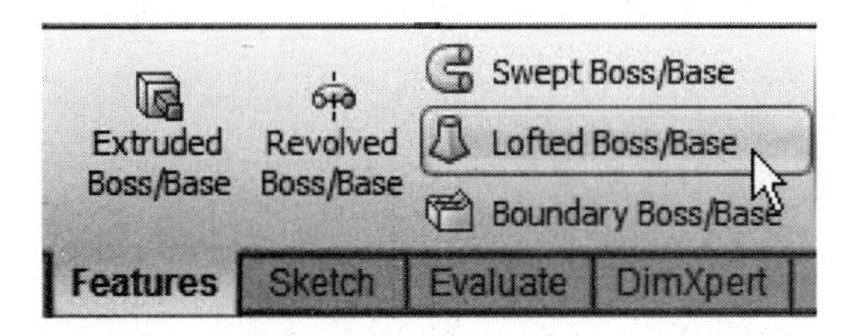

FIGURE 4.27

In the Feature Manager while the Profiles area is highlighted in blue, select sketch 1 then select sketch 2. The loft should be visually represented as shown (Figure 4.28). Accept the loft by selecting OK.

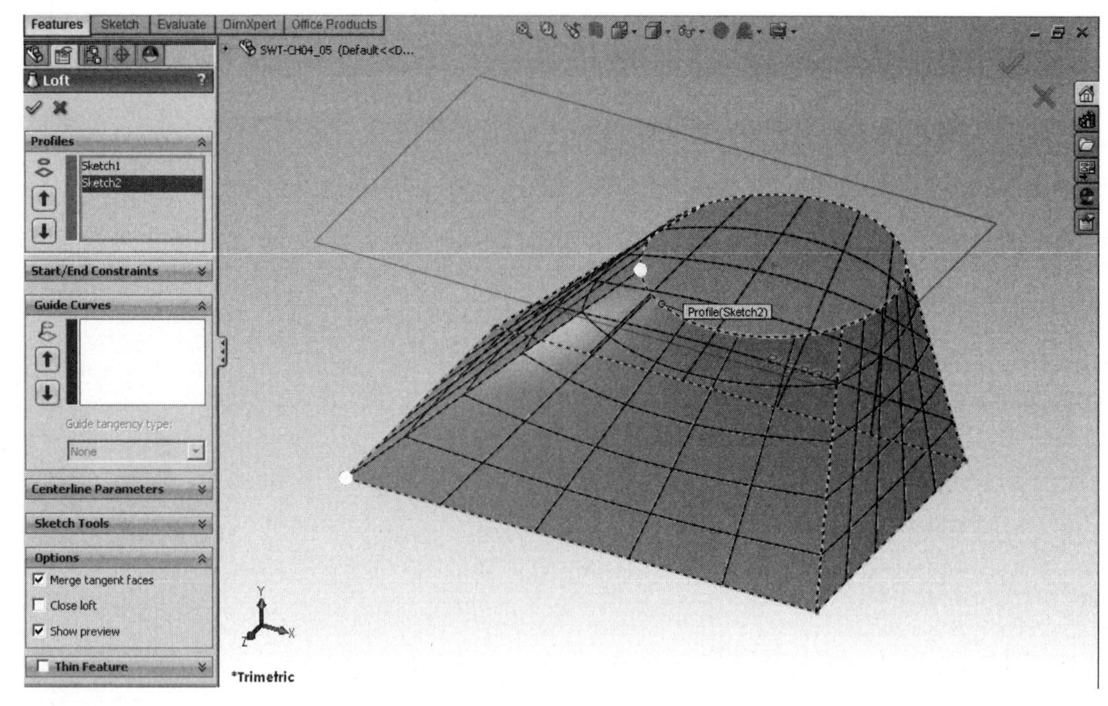

FIGURE 4.28

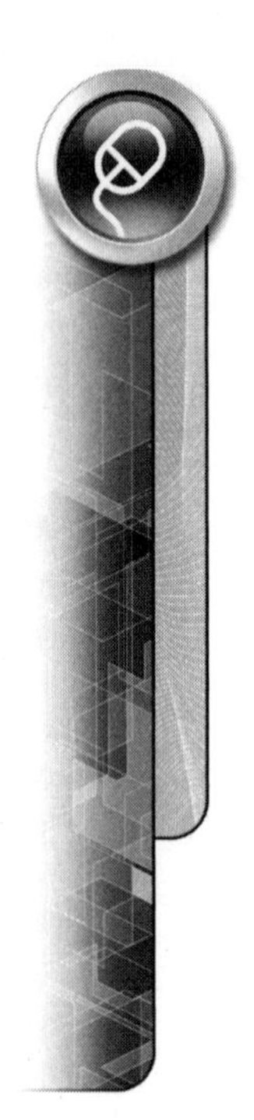

RIBS

A rib is used in mechanical design as reinforcement for any structure whether it is a boss, wall, or other structure that needs additional support for strength. Ribs are used often in casting and molding designs for cast iron or plastics.

The Rib command is a specialty extrude command where instead of having to create a closed shape to extrude, a line or a group of lines are used to create the extrude. A specified thickness and extrusion direction are given to the line(s).

Figure 4.29 shows an internal boss inside a walled structure where the boss needs a rib for support. A sketch of an individual line is shown on the top surface of the part.

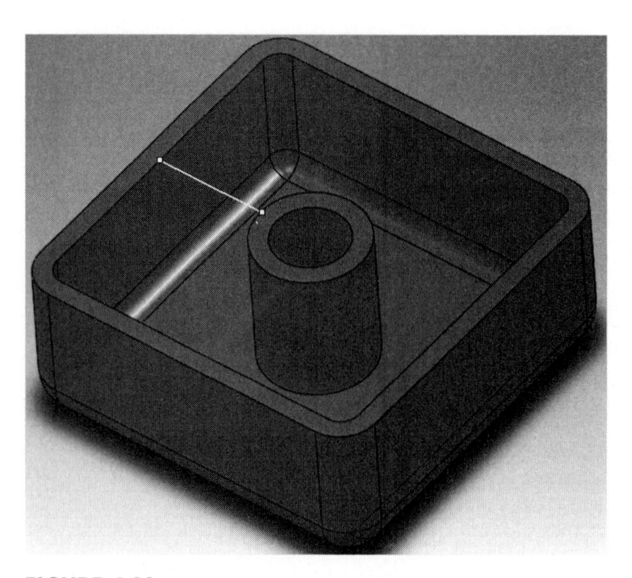

FIGURE 4.29

Figure 4.30 shows that the line was selected while in the Rib command. Notice in the Feature Manager that the thickness can be set to either offset to the left, right, or be centered on the line, in this case centered is selected. Also notice the extrude direction setting. The line can either be extruded parallel to the sketch or normal (perpendicular) to the sketch; in most cases the Rib command will only function in one of these directions. In this case normal to the sketch is selected since the sketch plane lies flat across the top of the part.

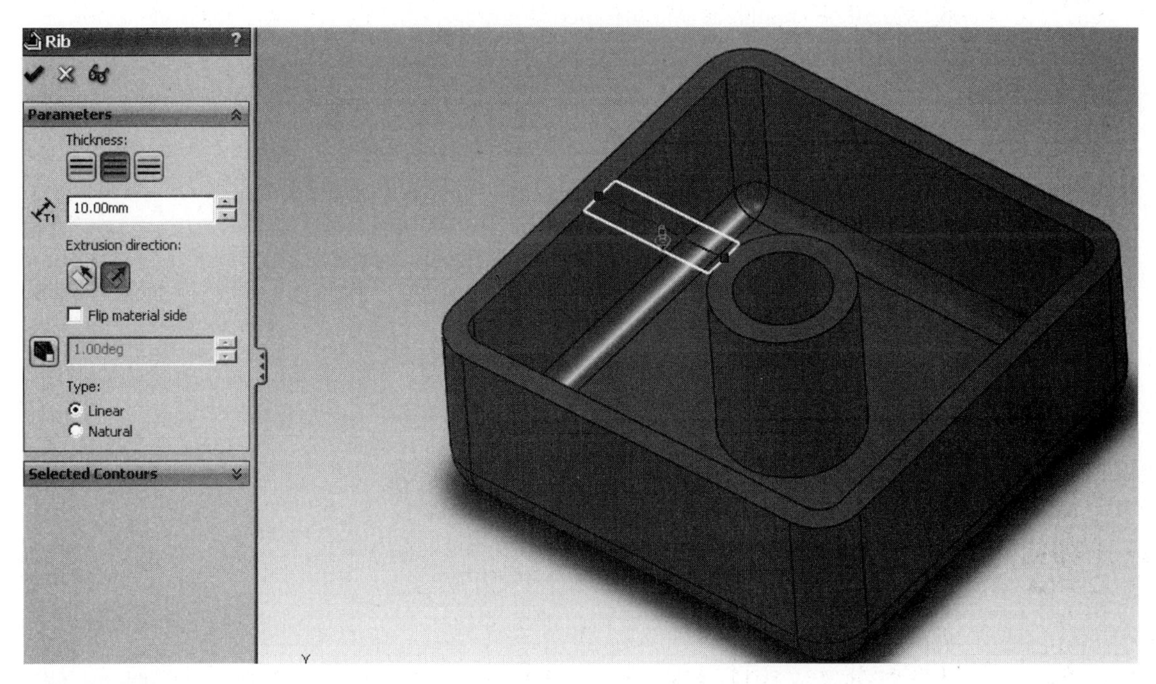

FIGURE 4.30

Figure 4.31 shows the complete rib on the left. The rib is projected down until it finds a barrier (the floor) as well as side barriers (wall and boss). In the right of the figure it shows the same example with four lines ribbed simultaneously.

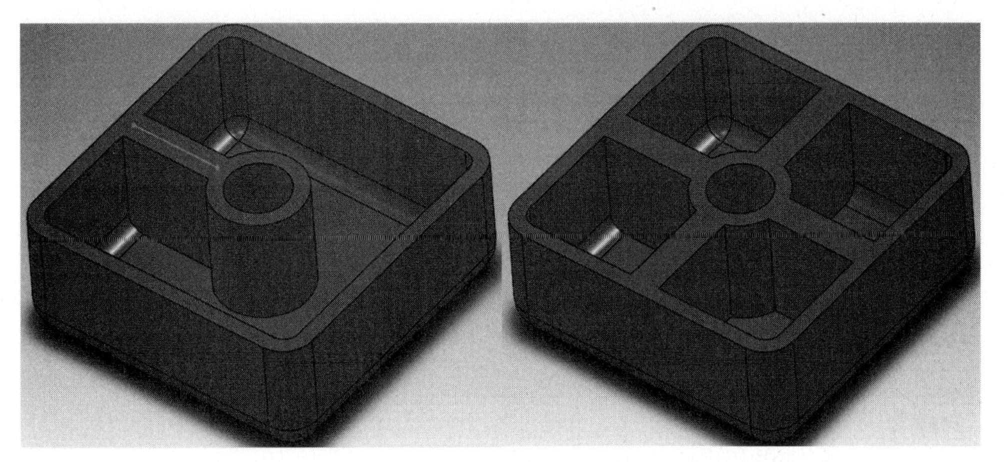

FIGURE 4.31

TUTORIAL EXERCISE: 04_RIB.SLDPRT

Part Number: SWT-CH04-05.SLDPRT

Description: Rib 1

Units: Metric (MMGS)

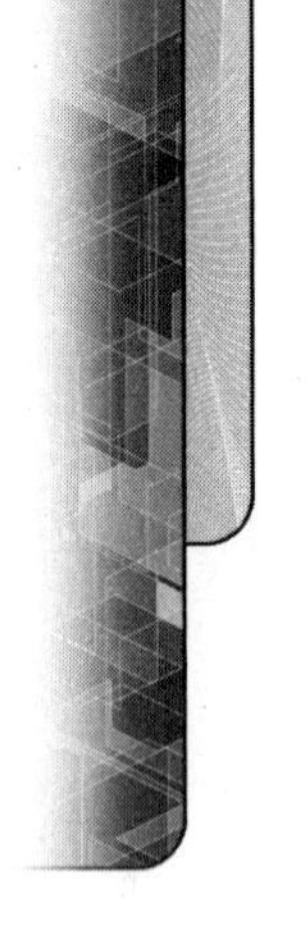

This tutorial exercise is designed to introduce creating a part with multiple ribs. The ribs will all be created at one instance (Figure 4.32).

FIGURE 4.32

STEP 1

Start by creating a sketch on the Top Plane with the dimensions shown in Figure 4.33. Note that the location of the origin is at the lower left corner of the rectangle. Exit the sketch when it is fully defined.

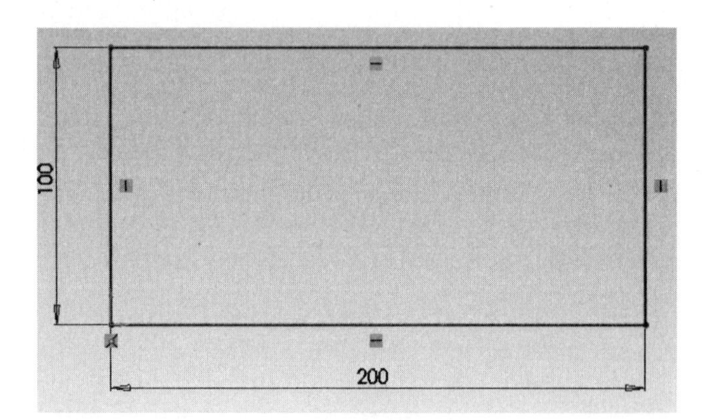

FIGURE 4.33

STEP 2

Select the Extrude command and extrude the part by 30mm. Exit the Extrude command when complete (Figure 4.34).

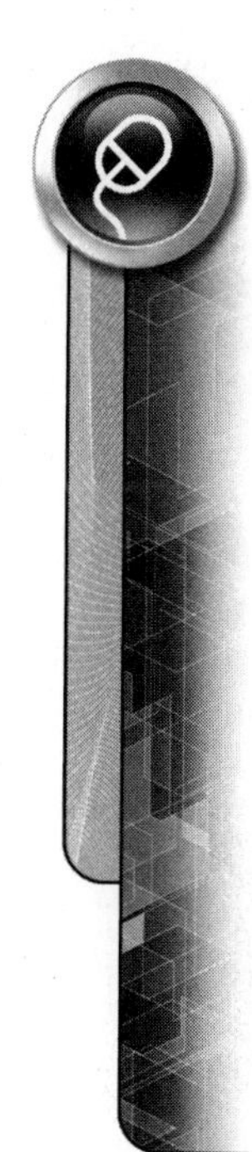

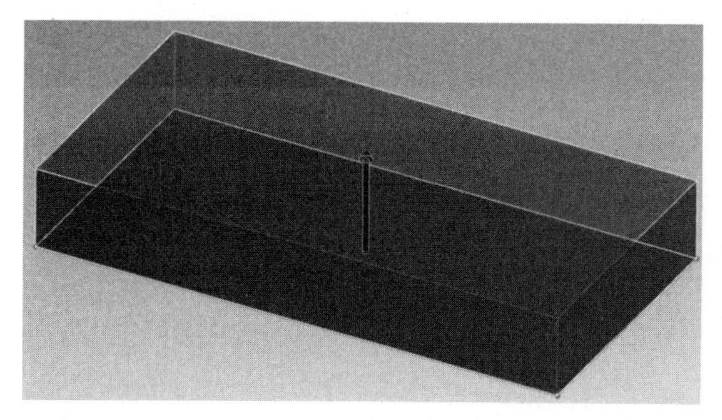

FIGURE 4.34

STEP 3

Create a sketch located on the top surface of the block and define it by using the dimensions in Figure 4.35. When drawing the rectangle, be sure to select the left and right edges of the block. Exit the sketch command when the sketch is fully defined.

FIGURE 4.35

STEP 4

Using the Extrude Cut command, extrude this sketch into the part at a depth of 25mm as shown (Figure 4.36).

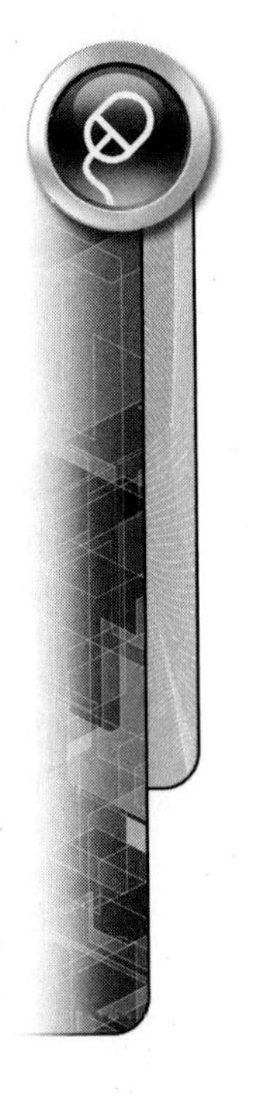

FIGURE 4.36

STEP 5

Select the Fillet command from the Feature Command Manager. Create two inside radii of 10mm as shown in Figure 4.37.

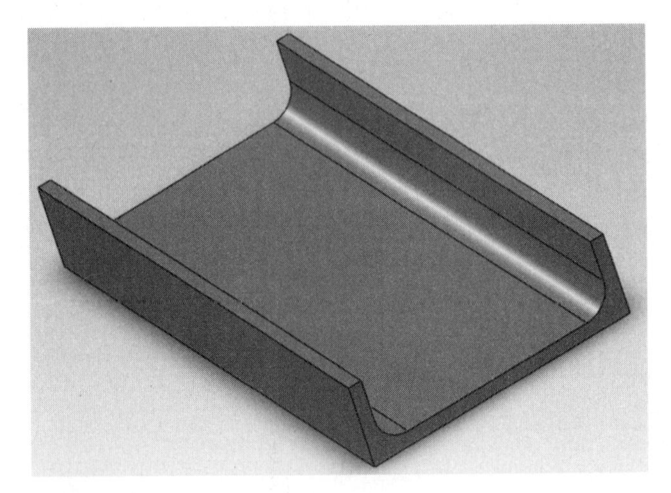

FIGURE 4.37

STEP 6

Select the Fillet command from the Feature Command Manager. Create two outside radii of 15mm as shown (Figure 4.38).

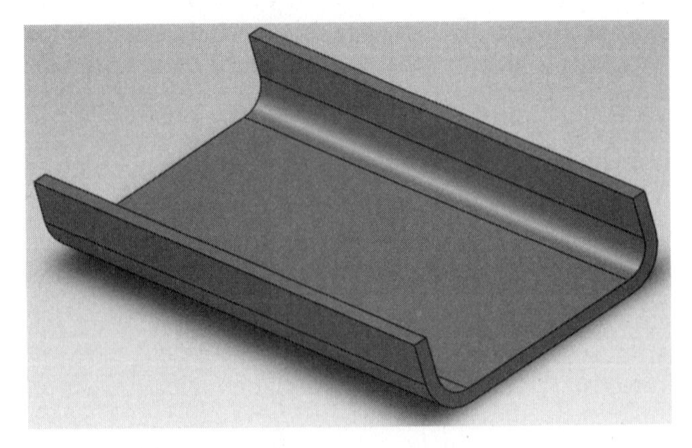

FIGURE 4.38

STEP 7

Create another sketch on the top surface of the part (select one of the two long narrow surfaces to locate the sketch plane). Create the two parallel lines first by selecting the appropriate edges as shown in Figure 4.39. Create the longer line last by selecting the midpoints of the first two lines of the sketch; this will define the third line. Add the two dimensions of 25mm and 150mm to fully define the sketch. Exit the sketch.

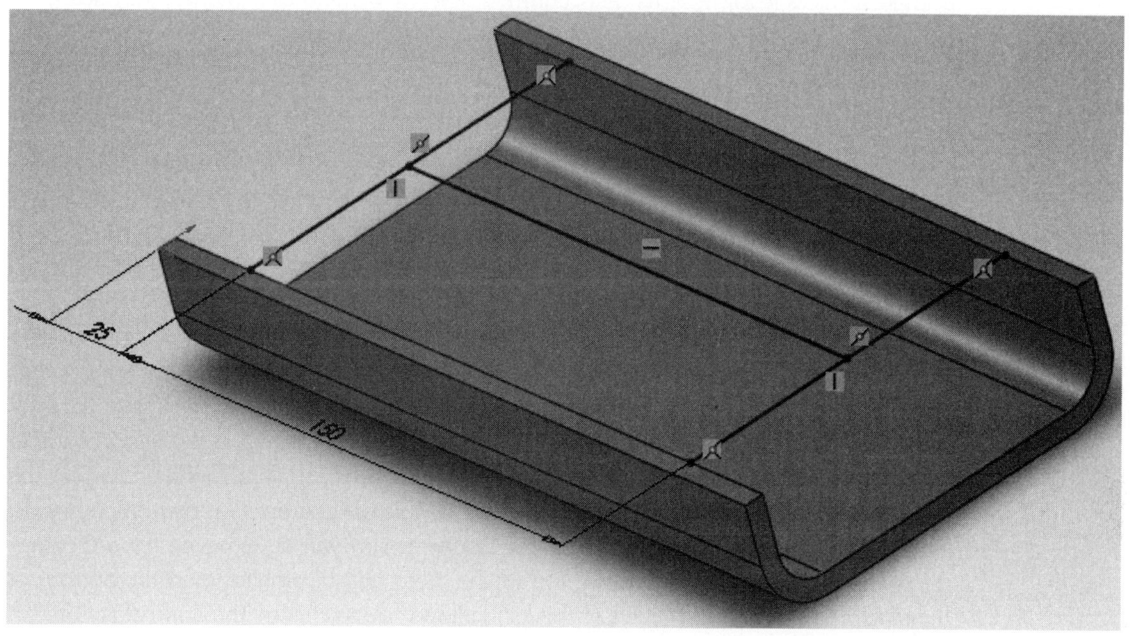

FIGURE 4.39

STEP 8

Select the Rib command from the Feature Command Manager. If prompted to select an existing sketch, select the newly created sketch by selecting one of the lines. Set the Thickness method to "Both Sides." Set the thickness to 5mm. Accept the rib by selecting the green checkmark. The part is complete (Figure 4.40).

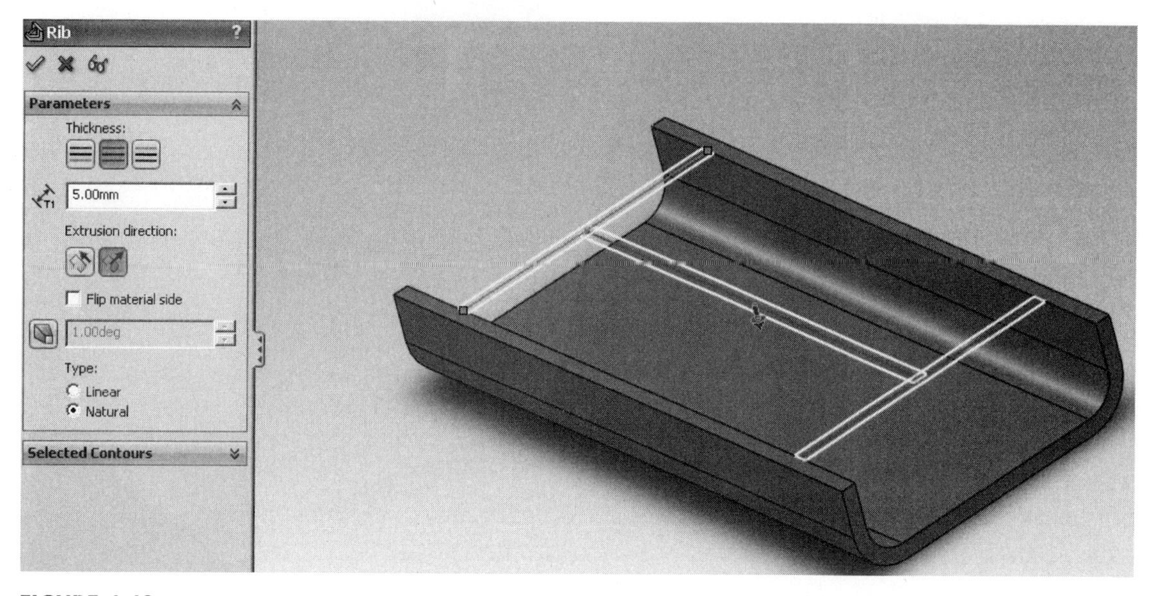

FIGURE 4.40

CREATING DRAFTS

The Draft command is used to give a draft taper to a surface or multiple surfaces. Draft surfaces are common in many designs particularly if a part is being manufactured by casting or molding as are most cast iron and plastic components. Surfaces are required to be designed with a draft, typically of 1° or 1.5°, to allow the part to be easily released from the manufacturing mold or die when it is completed. Another reason for creating drafted surfaces is for designs that require a tight fit with an easy method of release.

Activate the Draft command located on the Feature Command Manager.

Draft Methods

There are two methods to create drafts using the Draft command, manually or using the DraftXpert. Both methods use virtually the same variables to define the draft. Through this example we will be using the DraftXpert method.

DraftXpert

The DraftXpert should be the default method when entering the Draft command. The Property Manager should be manipulated top to bottom. First set the desired draft angle typically 1° to 1.5°, but to visualize an example a larger angle may be used. When the pink field which is the Direction of Pull is highlighted, select a surface that is perpendicular to the surfaces that will be tapered. When the surface is selected, an arrow representing the draw direction is placed on the surface. In this example the gray arrow is pointing upwards, which represents the direction the die (or casting mold) will be removed from the part; therefore the result will be that the top of the block will be smaller than the bottom of the block. The button to the left of this field allows you to change the draw direction.

Once the Pull surface is selected, the Faces to Draft field will be highlighted in blue. Any number of surfaces can be selected as long as they are perpendicular to the Pull surface. In this example all four sides around the part are selected as Draft Faces as seen in Figure 4.41.

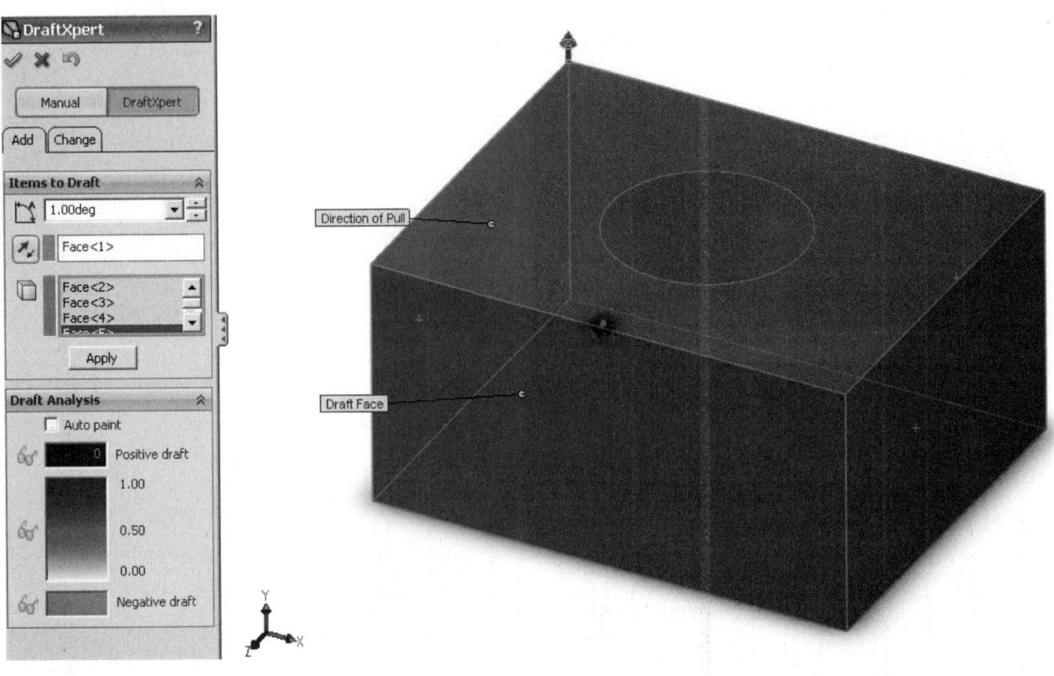

FIGURE 4.41

At this point the draft can be applied to the part by selecting the Apply button found on the Property Manager. Prior to this a color-coded visual representation of the drafts can be viewed by checking the Auto paint box as shown in Figure 4.42. The direction of pull remains magenta and any surfaces that are not tapered are represented in yellow. Any tapered surfaces are represented on a gradient scale from 0° to the maximum taper angle. In this example all surfaces are tapered at the maximum angle and therefore are all represented in blue.

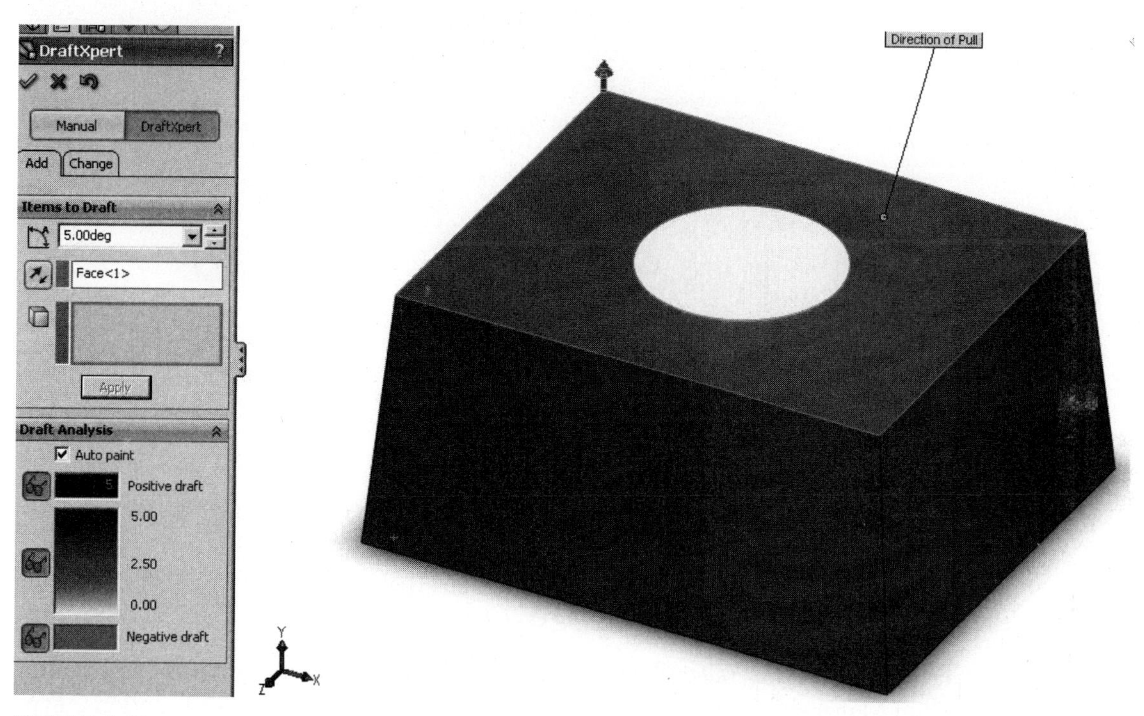

FIGURE 4.42

EMBOSSING

Embossing text on a part is a method of extruding where the extrusion is of text where each character is an enclosed shape. Most products have some form of embossing (or engraving) whether it is the manufacturer's name or logo, a warning note, a part number, or a model number.

Create any solid part with a face large enough to hold an embossing as seen in Figure 4.43. In order to emboss, a sketch is required on the desired surface and this sketch should only contain text on it and any other closed shapes to be included in the embossing. When selecting a font, all of the characters must be closed shapes. The majority of available fonts in SolidWorks have been designed to be enclosed shapes when placed in a sketch.

FIGURE 4.43

Text can be placed either along a curve (a straight line or curved object) or by selecting a location point. If a curve or point has yet to be identified, the text will be located at the origin until it is relocated to the desired location. Once the desired text is located and the font settings are set, exit the sketch.

To emboss the text, select the Extruded Boss/Base command from the Feature Command Manager. The embossing can be created using any of the extrude options. Typically an embossing is created using a blind distance of a low profile, in this case .03". Once the extrude settings are complete, accept the settings to complete the embossing as seen in the right of Figure 4.44.

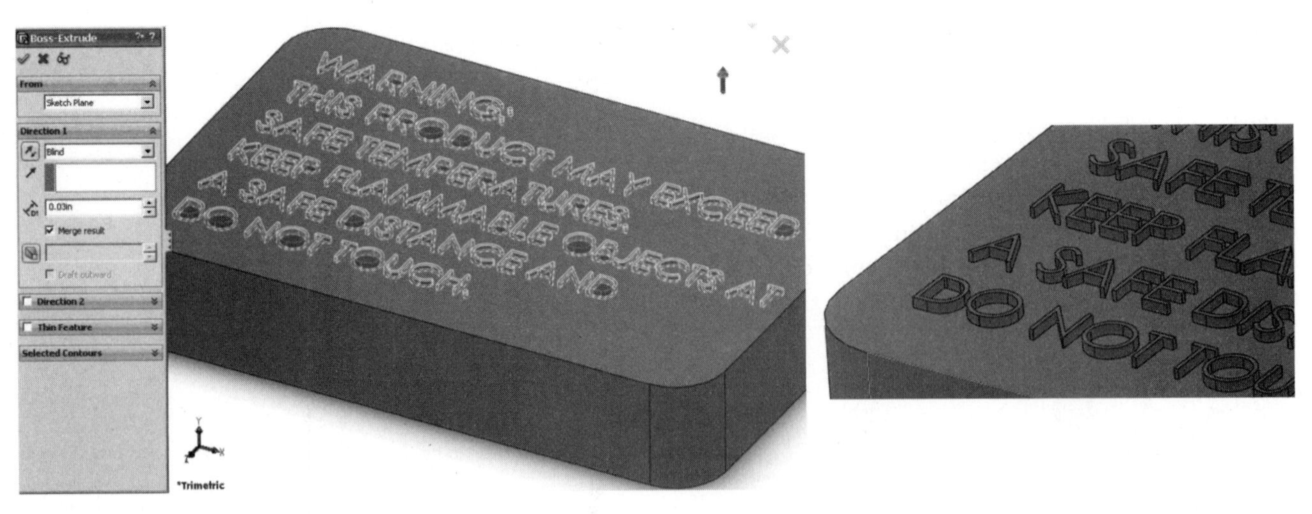

FIGURE 4.44

ENGRAVING

Engraving a part is a method of extrude cutting where the cut is in the shape of text. Engraving is a common method of adding text and logos to a part because the method removes material which is common in manufacturing processes.

TUTORIAL EXERCISE: 04_ENGRAVE.SLDPRT

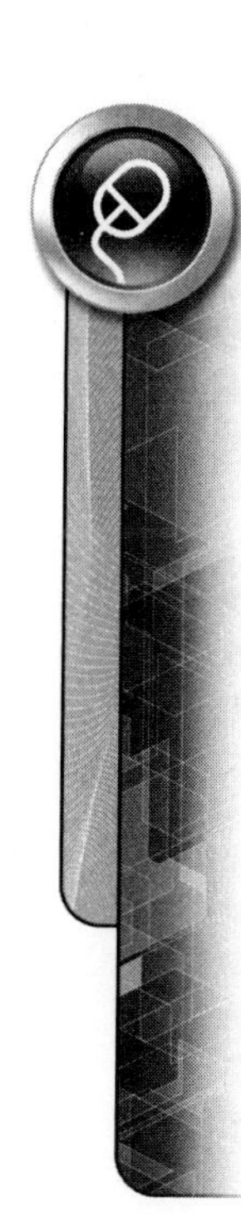

Part Number: SWT-CH04-06.SLDPRT

Description: Engrave 1

Units: English (IPS)

This tutorial exercise is designed to introduce engraving a part. This exercise follows the same process for embossing except an Extruded Cut is used instead of an Extruded Boss/Base (Figure 4.45).

FIGURE 4.45

STEP 1

Create a sketch on the top plane with the dimensions shown in Figure 4-46. Create the rectangle first and locate the lower left corner at the origin. Add the sketch fillets after the rectangle has been fully defined. Exit the sketch and extrude it 0.048" thick (18 gage steel).

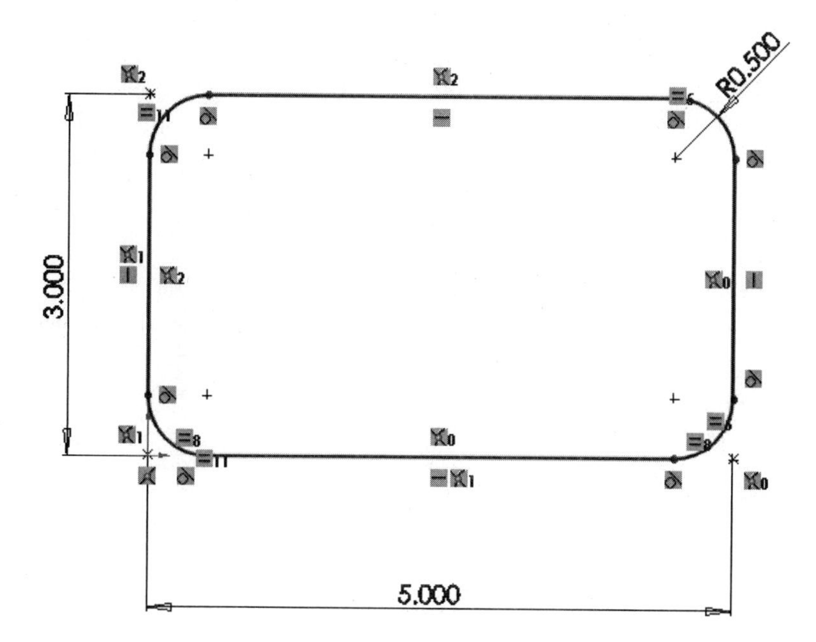

FIGURE 4.46

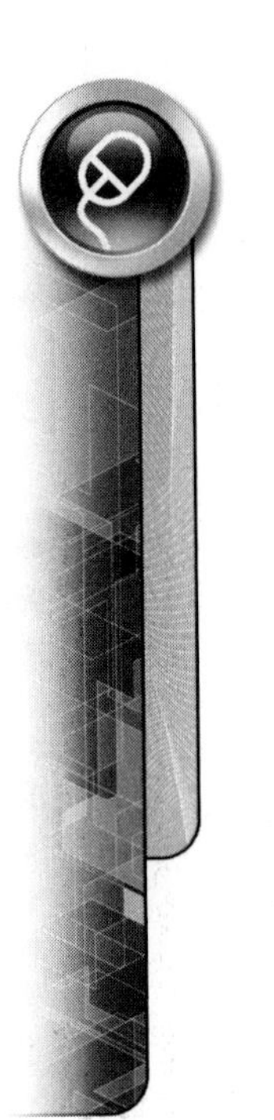

STEP 2

Create a new sketch on the top surface of the plate. Type in the text as shown in Figure 4.47. Locate the text on the plate by selecting a point in the upper left area of the plate as shown on the right of the figure. The text placement point (justification point) is the lower left corner of the top line of text.

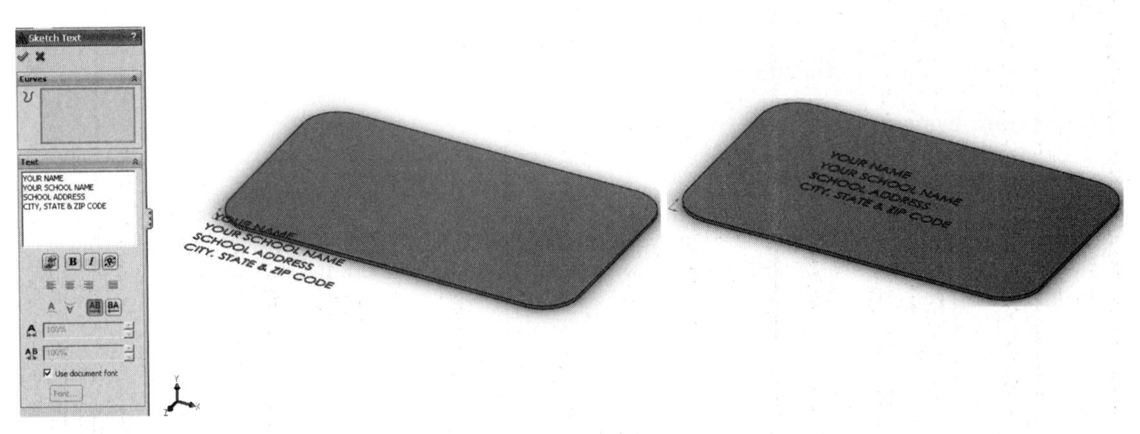

FIGURE 4.47

STEP 3

Uncheck the "Use Document Font" box and select the font button. The font, font style, and text height can all be modified using this tool. Change the Font to Engravers MT. Change the Font Style to Bold. And change the height to a unit height of 0.1875". Accept the new settings by selecting the OK button (Figure 4.48). Exit the sketch.

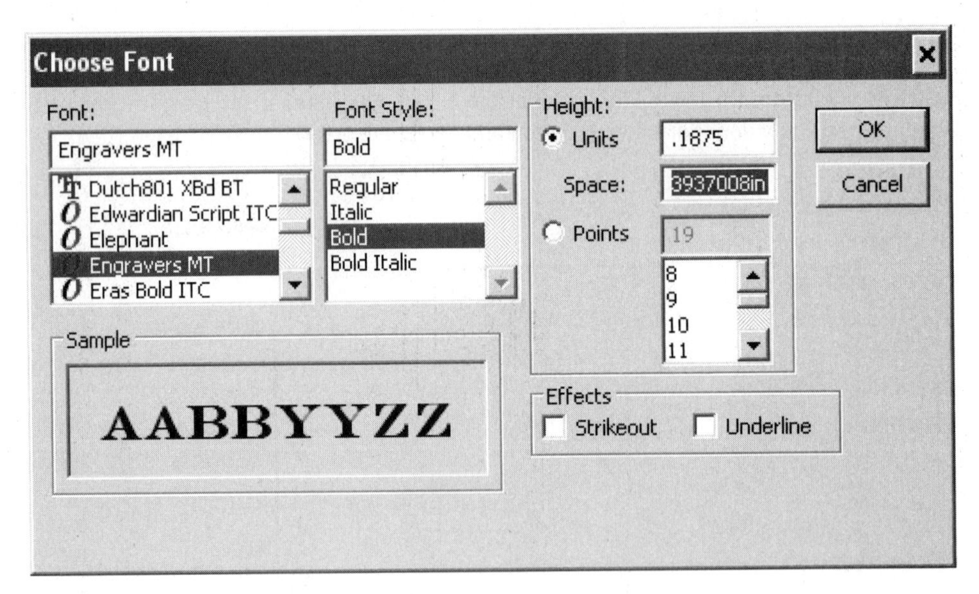

FIGURE 4.48

STEP 4

To create an engraving of the text, select the Extruded Cut command. Select the text sketch for extrusion. Choose the Extrude method of Through All. Complete extruding the text by completing the command (Figure 4.49).

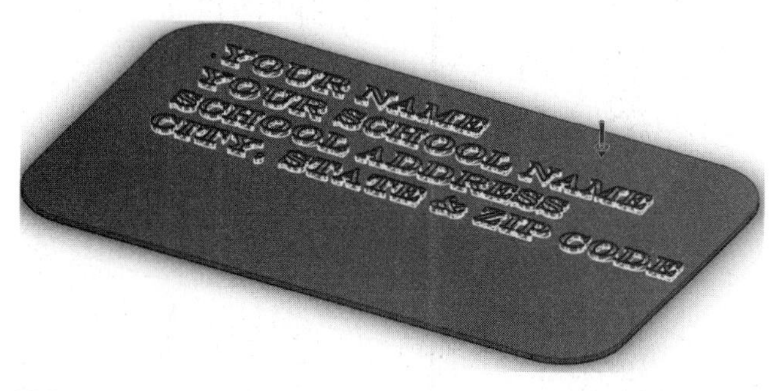

FIGURE 4.49

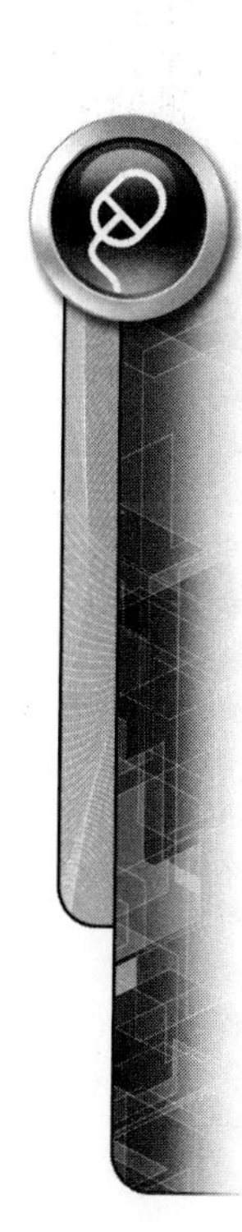

SPRINGS

Springs are a common mechanical device regularly found in mechanical design. Standard mechanical compression spring can be created by creating a helix which is a simple curve command and then sweeping a sketch of a circle along the helix.

Helix

The helix command creates a three dimensional spiral, where the spacing between each rotation, called the pitch, can be constant or variable. In order to create a helix, a sketch of a circle representing the diameter of the helix is required.

TUTORIAL EXERCISE: 04_SPRING.SLDPRT

Part Number: SWT-CH04-07.SLDPRT

Description: Spring 1

Units: Metric (MMGS)

This tutorial exercise is designed to show how to create a simple spring as shown in Figure 4-54. Springs can be shown in several ways in assembly and typically in a static position; for this reason the working length must be known and modeled to, not the free length of the spring. In assemblies where motion is required, the spring component is often suppressed to allow for the dynamic motion. The spring part file will not compress or extend in the solid modeling assembly.

STEP 1

On the Top Plane, create a circle of 50mm in diameter. Once the sketch is fully defined, exit the sketch (Figure 4.50).

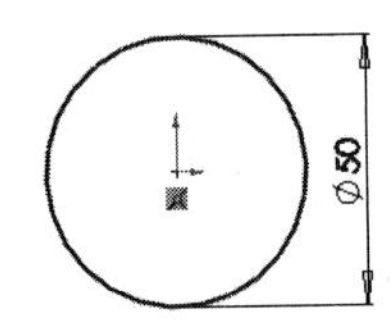

FIGURE 4.50

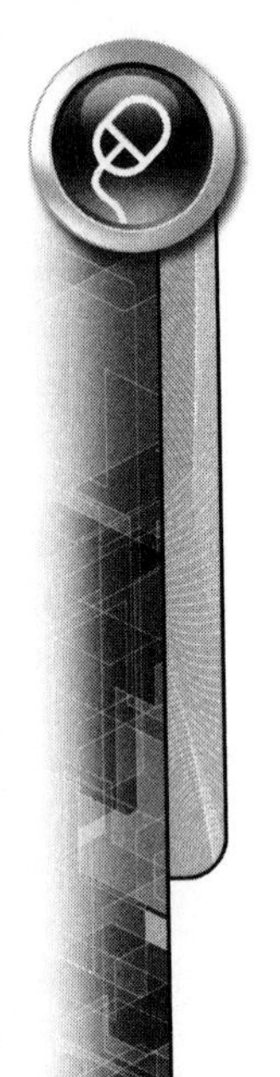

STEP 2

Select the Helix/Spiral command from the Insert pull-down menu in the Curve pull out as shown in Figure 4.51. The Helix/Spiral should automatically select the sketch of the circle; if it does not, select the circle.

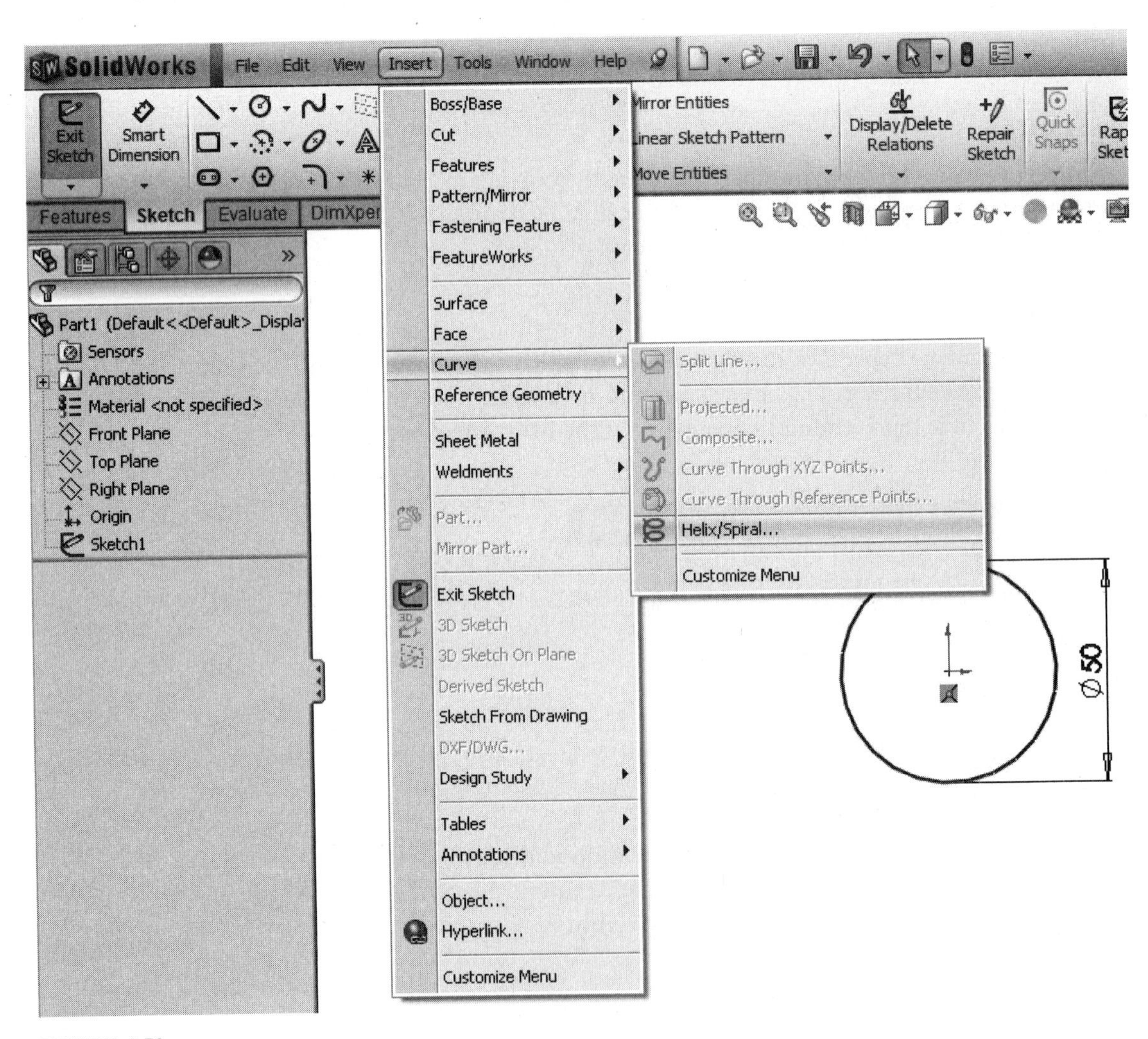

FIGURE 4.51

STEP 3

The spring can be defined using a number of variables; select the Pitch and Revolutions method. The pitch can be a constant or a variable pitch, where we should select constant pitch. Define the rest of the parameters as 10mm for a pitch, 8 revolutions, a starting angle of 0.0°, and a clockwise helix. Accept the helix by selecting OK (Figure 4.52).

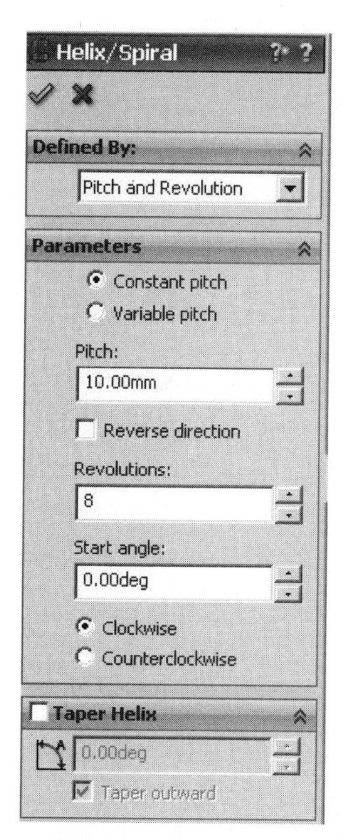

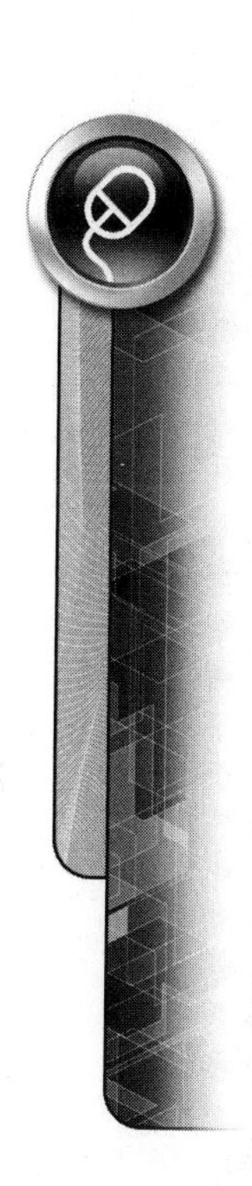

FIGURE 4.52

STEP 4

The location of the starting point of the helix is critical to creating the spring; since 0.0° was used the starting point is normal to (perpendicular to) the right plane. Change the View Orientation to Trimetric, then create a new sketch on the right plane. The sketch should be a circle horizontal to the origin and locate it 25mm from the origin and have a diameter of 4mm. Exit the sketch when it is fully defined (Figure 4.53).

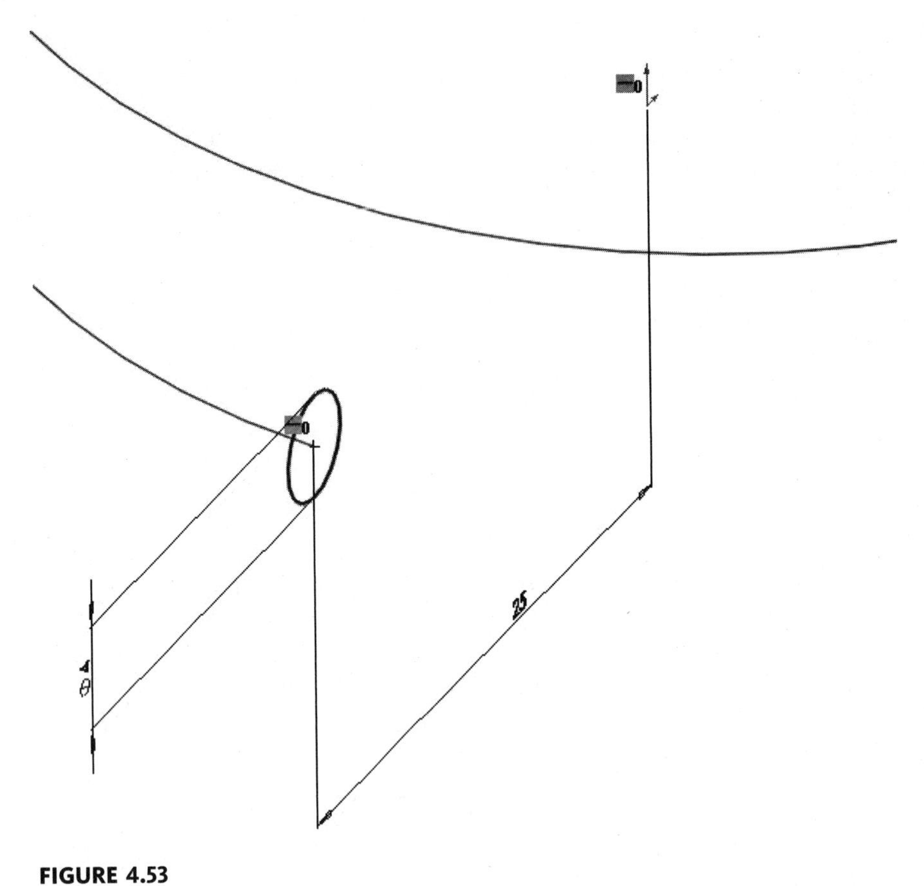

FIGURE 4.53

STEP 5

Select the Swept Boss/Base command from the Feature Manager. Select the sketch of the circle for the profile sketch and select the helix for the path sketch. Complete the spring by selecting OK (Figure 4.54).

FIGURE 4.54

CREATING DOMES

Spherical shapes are common in mechanical designs, everywhere from ball pens to pressure vessels. The Dome command is used to create spherical shapes under a number of constraints. The dome is created from any existing face whether it is round or any other closed shape. Upon entering the Dome command, select the face from which it will be derived and the dome will project in a direction normal to the surface by default. A visual preview representation of the dome is shown on the model as seen in Figure 4.55. The preview representation will dynamically change as settings are modified in the Feature Manager.

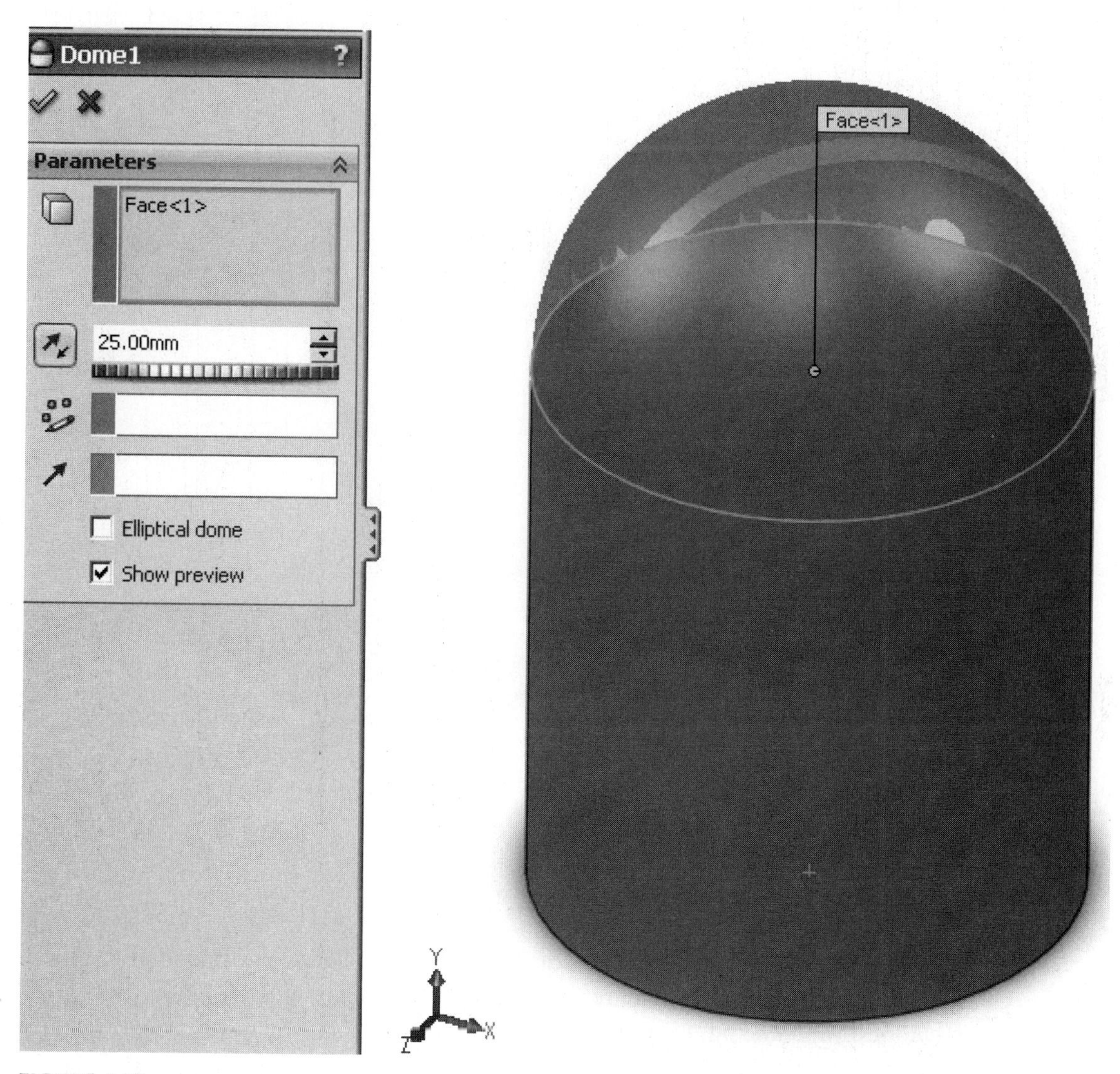

FIGURE 4.55

Dome Direction

Most commonly spherical ends are convex or protruding although the Dome command can create concave surfaces as well. The dome will be represented convex, which is the default setting.

In order to change the dome to concave, simply select the Reverse Direction icon .

If a dome direction that is not normal to the selected face is desired, then the direction arrow field can be selected and a defined line can be selected to define the desired angle. In Figure 4.56, the dome follows an angle described by one of the 45° edge lines of the chamfer. It can be seen that the apex of the dome is at a 45° degree angle from the center of the face.

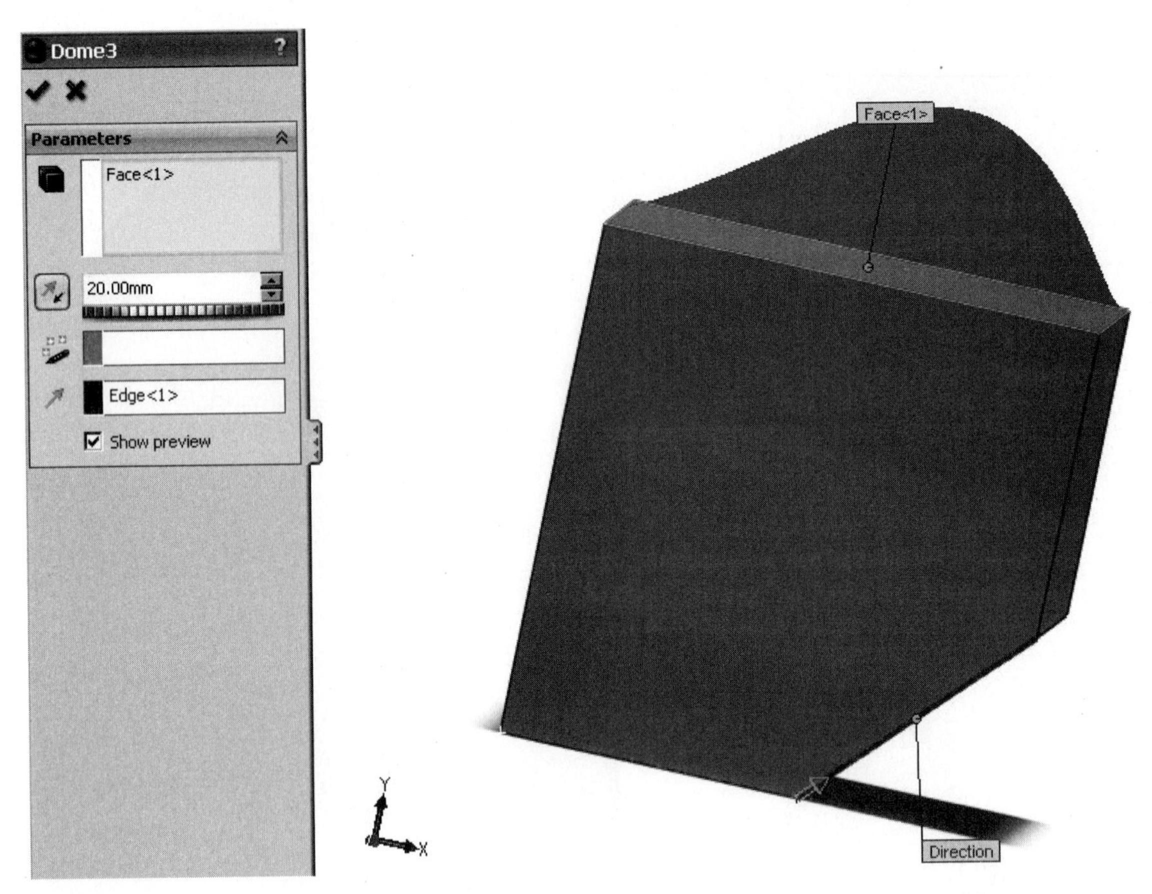

FIGURE 4.56

Elliptical Dome

The dome can be created in an elliptical manner instead of a constant radius by selecting the Elliptical Dome parameter. The example on the left side of Figure 4.57 is elliptical while the right is a constant radius.

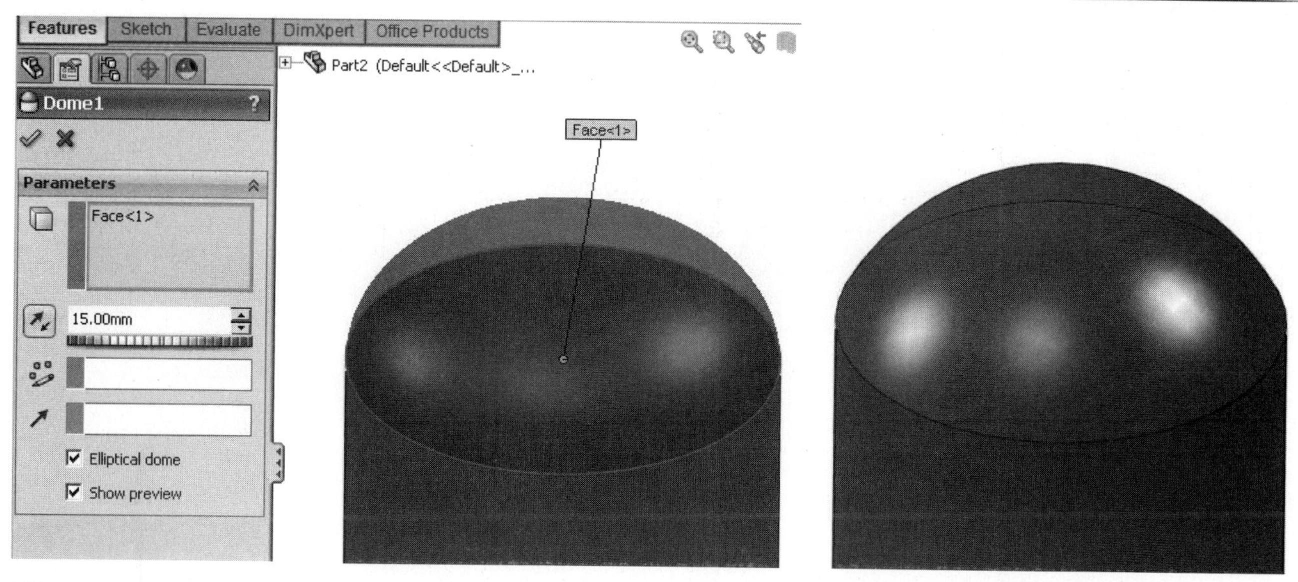

FIGURE 4.57

MIRRORING SOLIDS

Mirroring geometry can be accomplished in a sketch, but oftentimes it is more efficient to mirror a solid part instead of mirroring a sketch or group of sketches before the sketches are manipulated into a solid part; this is preferred because of the ease of editing features.

The Mirror command can be found on the Feature Command Manager. Once a part is created, it can be mirrored across any existing planar face or reference plane.

The following part in Figure 4.58 is created with two features, first is an extrusion of the complex outside shape, then a separate hole was added.

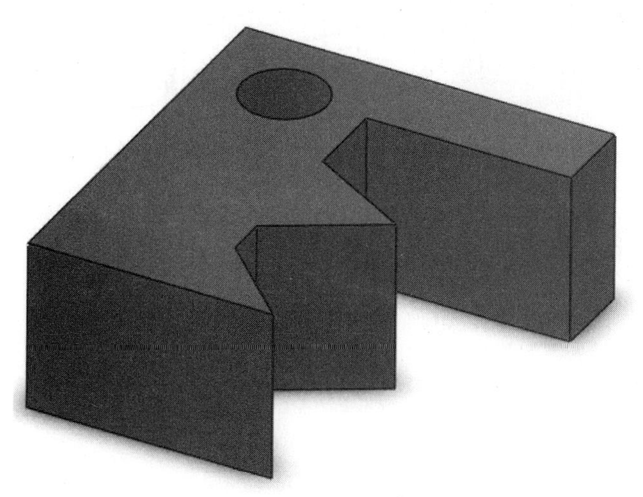

FIGURE 4.58

The first field in the feature manager is to select the plane in which you want to mirror the part across, so a planar face or plane should be selected. The second field is Features to Mirror where all of the features desired to mirror are selected. The example in

Figure 4.59 shows only the selection of the Boss Extrude and not the hole and on the right of the figure it can be seen that only the body was mirrored.

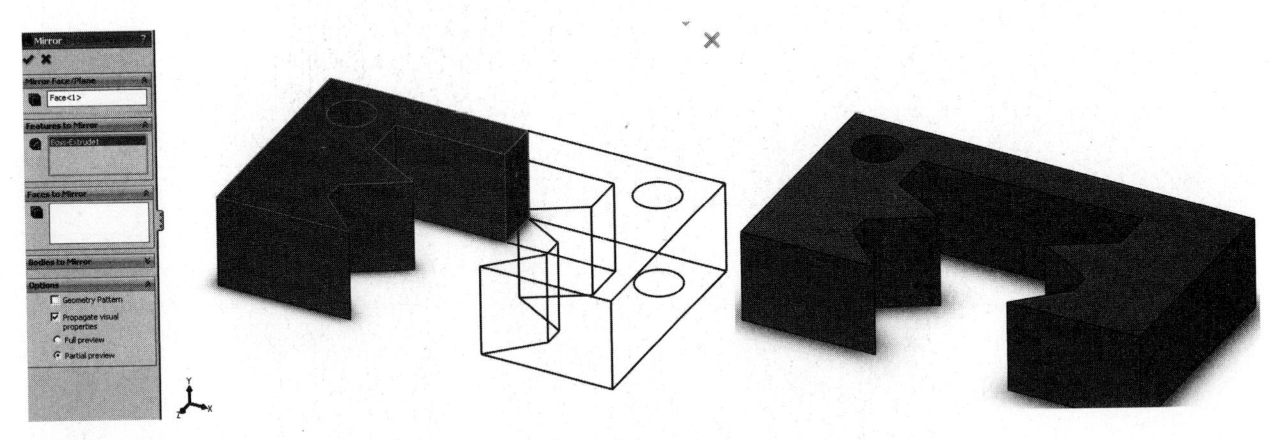

FIGURE 4.59

Disjoined Mirror

When the mirror is complete, the result is to be a single part. If the existing volume and added volume are not connected by a face or more, the pieces will be disjoined and as a result no mirror will be completed. In Figure 4.60, a preview of a disjoined mirror is shown, but upon completion of the command the second solid will not be created and a warning will be shown in the Feature Manager.

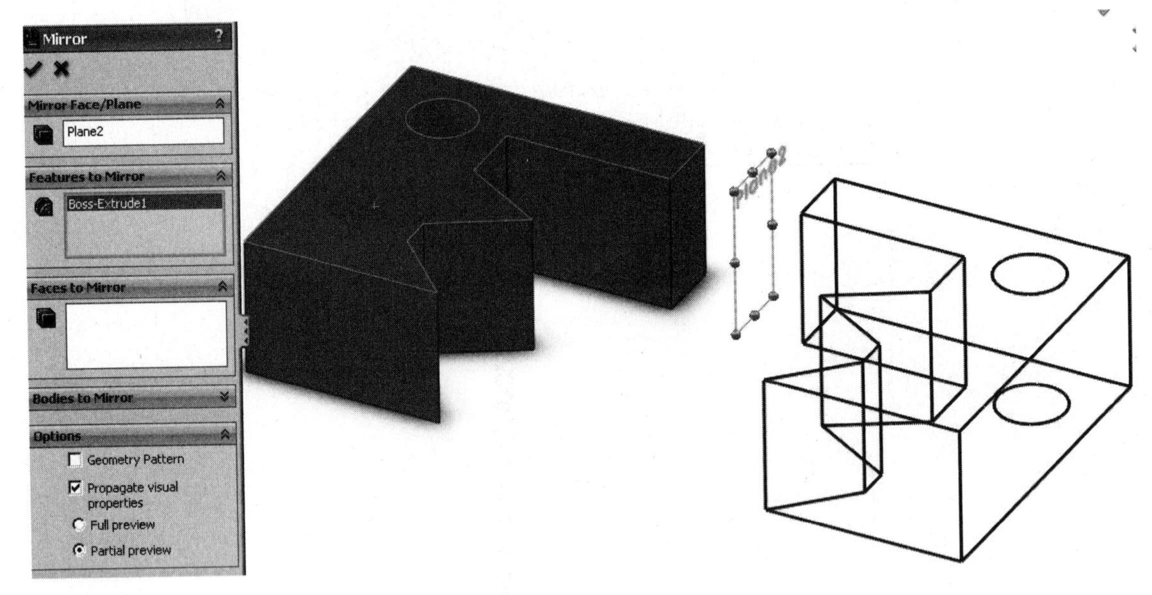

FIGURE 4.60

Mirroring with Overlapping Volume

When a mirror plane that would result in the overlapping of geometry is selected, the command will proceed by correcting the result of one solid piece that bounds all of the volume of the original shape as well as the added volume as shown (Figure 4.61).

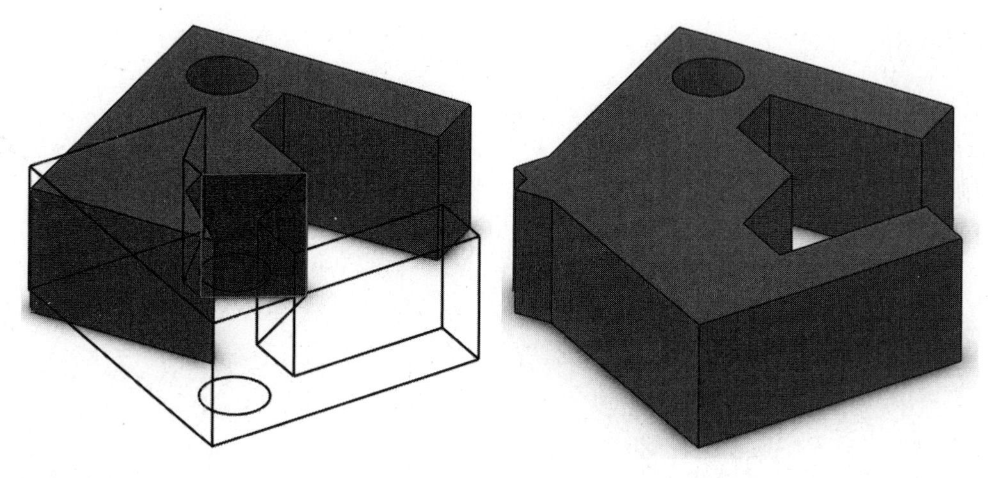

FIGURE 4.61

DESIGN TABLES

Having some spreadsheet experience is desired in order to use this function.

NOTE

Oftentimes with a design, there may be a desire of several variations of the same shape. Examples of this may be a paper clip, fishing hook, or a structural shape such as an "I" beam where the design parameters do not change from one size to the next but the actual dimensional sizes do change. SolidWorks has many tables of products available in the Toolbox, but if you are creating a new design where you are creating the design parameters in a part file, a Design Table can be set up to allow variations in size. This is done in conjunction with a spreadsheet embedded into the part file. There is no limit as to the number of variables used in the spreadsheet.

In this example we will create a design table for a series of simple flat washers. A flat washer has three dimensions, the inside diameter, the outside diameter, and the thickness. Other than the variation in sizes, the design of the washer itself remains the same, such as the two diameters are always concentric and in the same plane and the thickness is always perpendicular to that plane.

Variable Names

When creating a sketch of the two diameters, take note of how each dimension is assigned a variable name; for example, the first dimension placed on your sketch is identified as D1 and the second dimension is identified as D2; this can be seen in the Primary Value field of the Feature Manager. Each variable, such as a dimension or a feature size, is assigned a name (Figure 4.62).

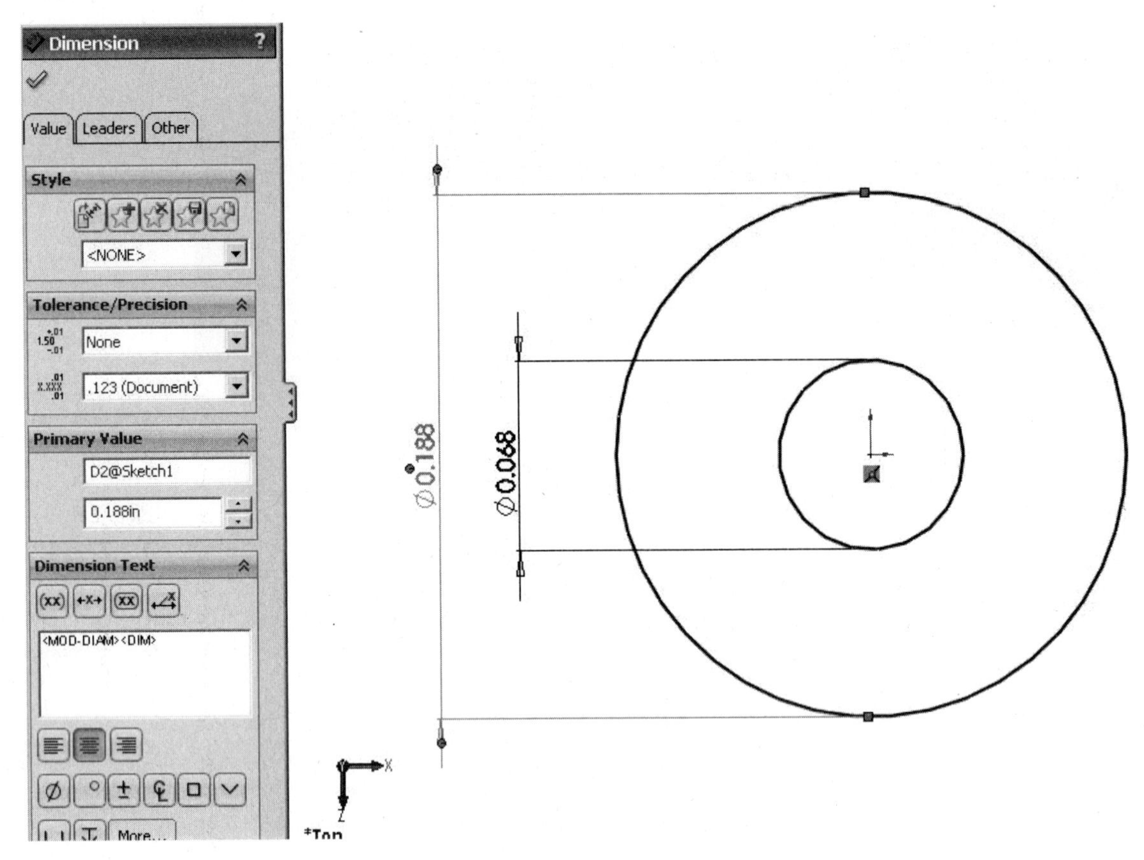

FIGURE 4.62

Feature names can be changed by selecting the Primary Value field and typing over a new name. Be sure to keep the @ symbol and the remaining information. In this case change "D2@Sketch1" to "OD@Sketch1" (Figure 4.63)

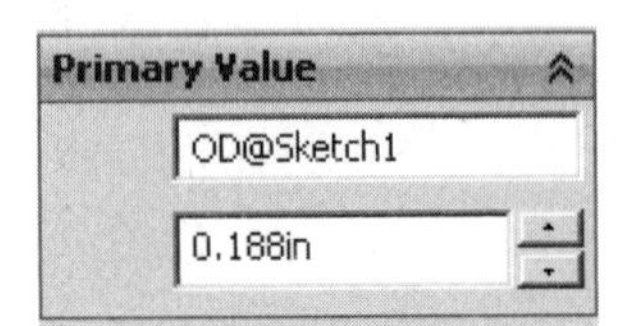

FIGURE 4.63

Creating the Design Table

Once the sketch is extruded to a thickness, the part is designed and the design table is ready to be set up. It is good practice to save the file at this point before continuing on. Enter the Design Table command by going to the Insert pull-down menu, selecting Tables, and then selecting Design Tables (Figure 4.64).

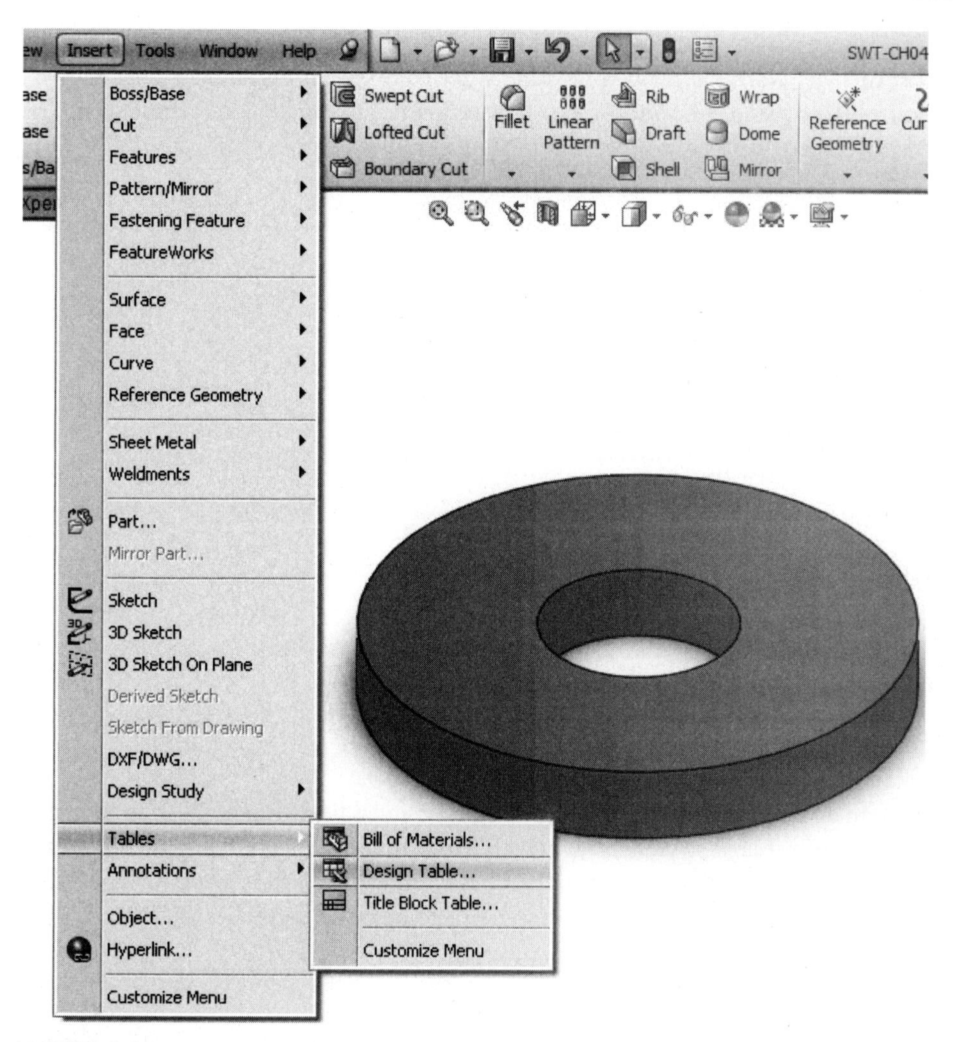

FIGURE 4.64

In the Design Table Feature Manager, select Auto-create and then select OK. A box appears that allows the selection of desired variables from the part file to include in the Design Table. Only variables that are going to change should be included. In this case all three variables should be selected by using the mouse selection button (Figure 4.65).

FIGURE 4.65

A spreadsheet now appears on the part file. The left column is unlabeled and can be labeled such as "Nominal Size." The other three columns are also labeled, being given the full variable name. The advantage of giving names to dimensions can be seen here especially if there were even more columns of variables (Figure 4.66).

	A	B	C	D	E	F	G
1	Design Table for: SWT-CH04-DTable01						
2		ID@Sketch1	OD@Sketch1	D1@Boss-Extrude1			
3	Default	0.068	0.188	0.025			
4							
5							
6							
7							
8							
9							
10							

Sheet1

FIGURE 4.66

Names can be assigned to each row and the dimensional fields can be filled in to match the various sizes desired as seen in Figure 4.67. The functionality of filling in the spreadsheet cells is the same as using any spreadsheet. Once the spreadsheet is completed, it can be saved and exited. The part file also needs to be saved before it can be used.

	A	B	C	D	E	F
2		ID@Sketch1	OD@Sketch1	D1@Boss-Extrude1		
3	Default	0.068	0.188	0.025		
4	No.0	0.068	0.188	0.025		
5	No.1	0.084	0.219	0.025		
6	No.2	0.094	0.25	0.032		
7	No.3	0.109	0.312	0.032		
8	No.4	0.125	0.375	0.04		
9	No.5	0.141	0.406	0.04		
10	No.6	0.156	0.438	0.04		
11	No.8	0.188	0.5	0.04		
12	No.10	0.203	0.562	0.04		
13	No.12	0.234	0.625	0.063		
14	.25	0.281	0.734	0.063		
15	.3125	0.344	0.875	0.063		
16	.375	0.406	1	0.063		
17	.4375	0.469	1.125	0.063		
18	.5	0.531	1.25	0.1		
19	.5625	0.594	1.469	0.1		
20	.625	0.656	1.75	0.1		
21	.75	0.812	2	0.1		

FIGURE 4.67

Using the Design Table

Once the table information is saved, any of the design iterations can be used. Typically they are inserted into assembly files; therefore insertion into files will be covered in Chapter 6.

DEFINING PART MATERIAL

In order to take advantage of engineering computation-based commands or software, the parts being created need to have a material specified. For example, weight calculations or strength computations need the part to have specific material properties in order to complete the calculations. Applying a material to any part file is necessary and easy to do.

Applying a Material to a Part File

While in a part file, material can be applied by right-clicking on the Material setting found in the Feature Manager tree. There are several options to choose from in this pull-out menu, including the selection of ten common materials; selecting "Edit Material" gives you a more thorough list of materials to choose from as well as additional options (Figure 4.68).

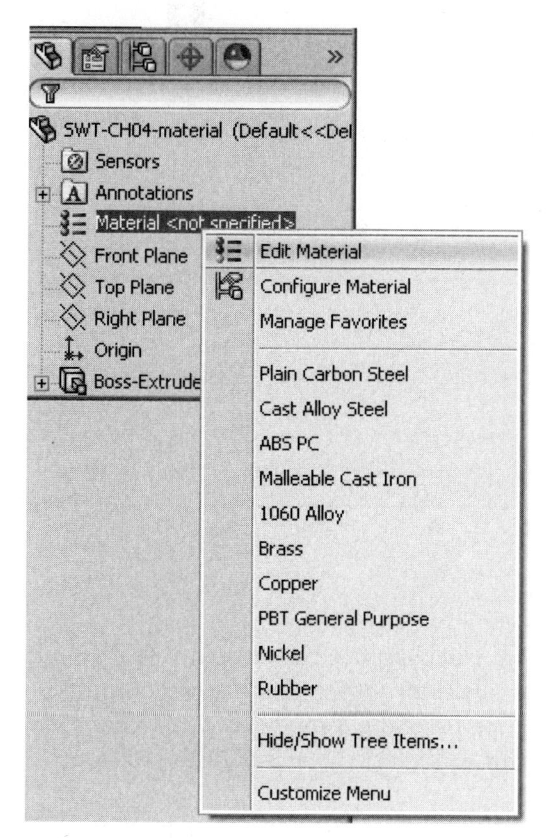

FIGURE 4.68

After selecting Edit Material, the Material dialog box appears. Along the left side of the dialog box, there are over two hundred fifty common industrial materials listed and categorized in major groups. To select the desired material, simply click on the folder describing the general type of material (e.g., steel, iron, plastic) and select the specific type of material inside that folder. When complete, select the Apply button and all of the documented material properties will be applied to the part (Figure 4.69).

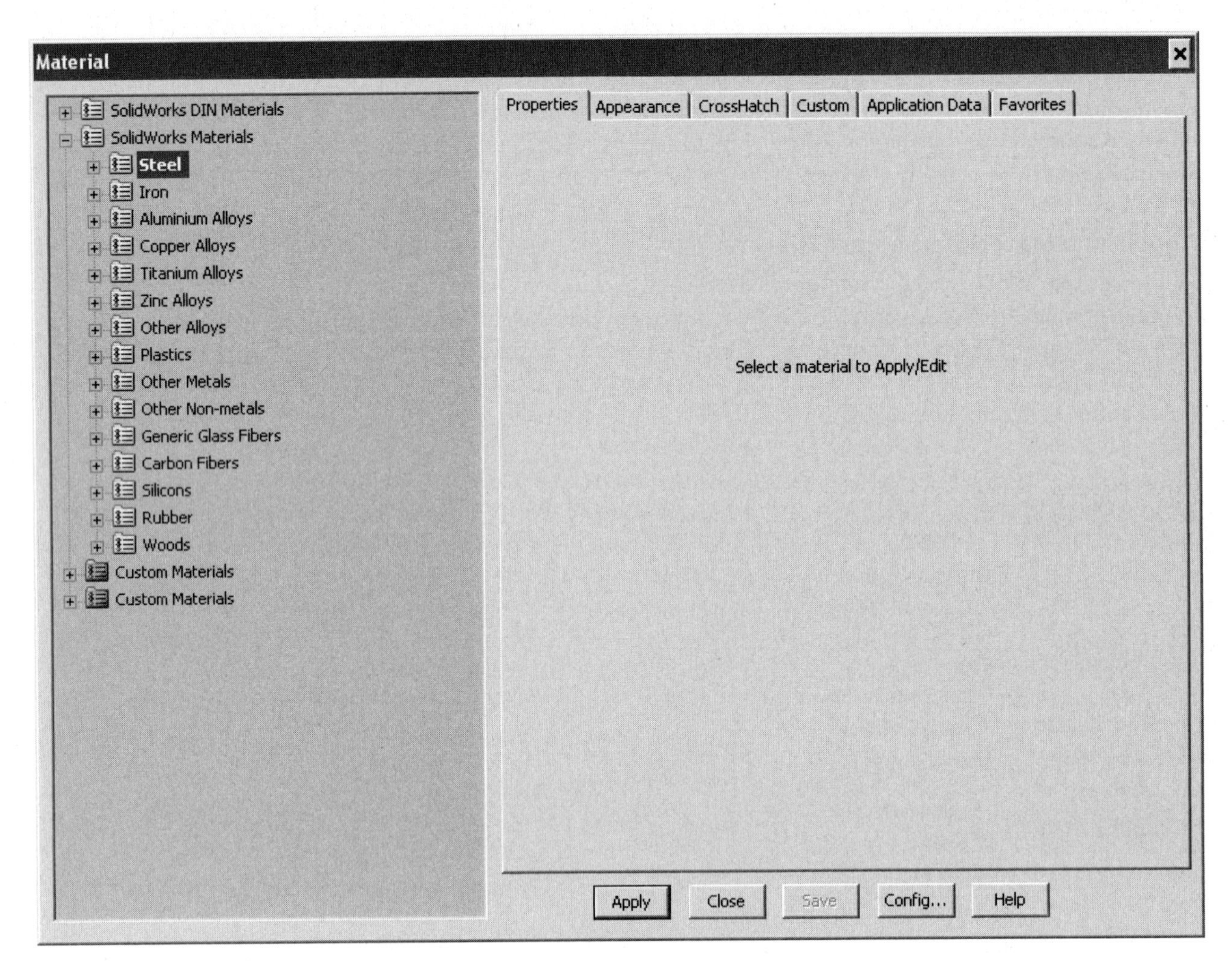

FIGURE 4.69

Each material selection has a set of known material properties attributed to it as seen in Figure 4.70. These material properties allow any engineering calculations imbedded in analysis software that use these properties to be completed. Oftentimes some of the properties are not available and this can affect the ability of the software to complete some analysis.

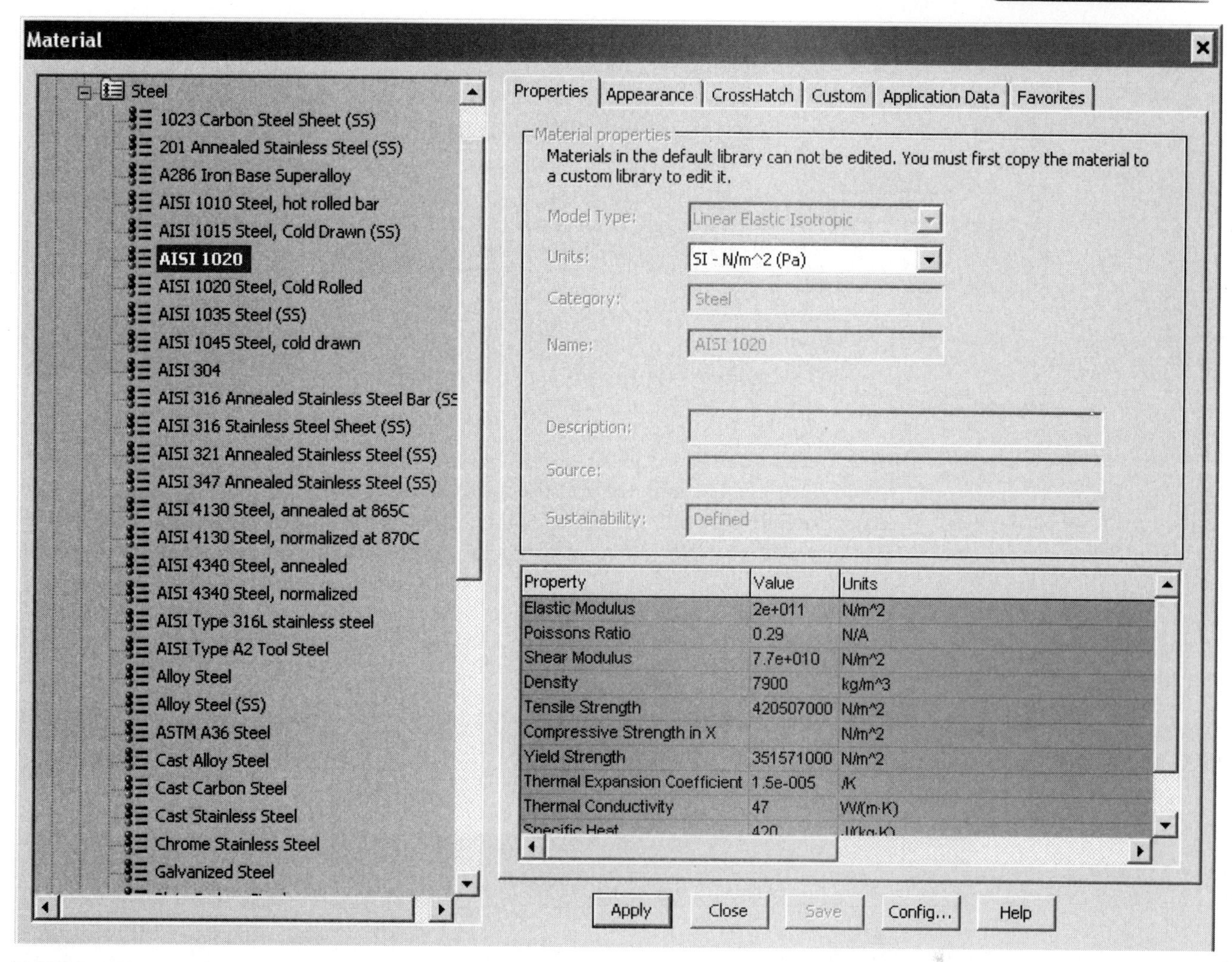

FIGURE 4.70

Creating a Custom Material

There may be times when the material desired is not available in the material list provided by SolidWorks. If the material is known to have similar properties to another material, selecting the other material can be used for estimated results, but this is not a recommended practice. The material list in the dialog box has a Custom Materials option. The majority if not all of the listed properties should be known to the designer as found in external publications or more likely from the manufacturer of the material.

For example, if the desired material is AISI5060 high carbon steel, it can be seen as not being a selectable option in the current material list. In the materials library under Custom Materials look to see if Steel is an option. If not the new custom category must be created by right-clicking on the Custom Materials and selecting New Category. Name the new category "Steel" and hit enter. Then scroll up into the existing material list and right-click on an existing similar material such as 4340 steel and select Copy. Next, right-click on the newly created Steel category and select Paste. This will create a new material under the Steel category with the name and material properties of the copied material, but the name and properties are all editable (Figure 4.71).

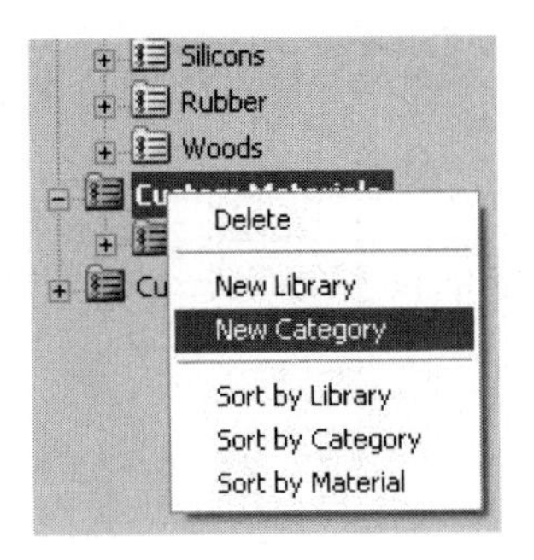

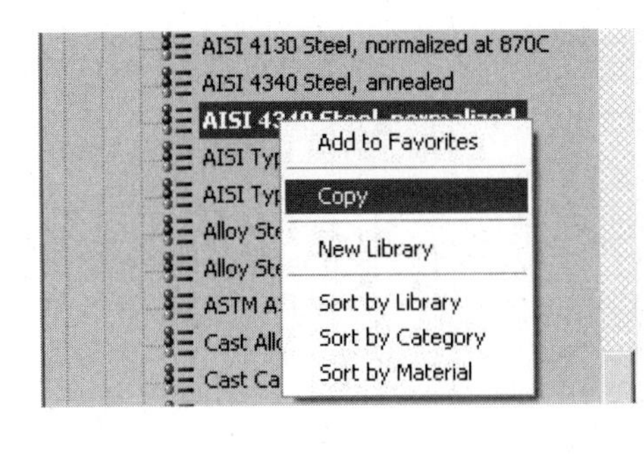

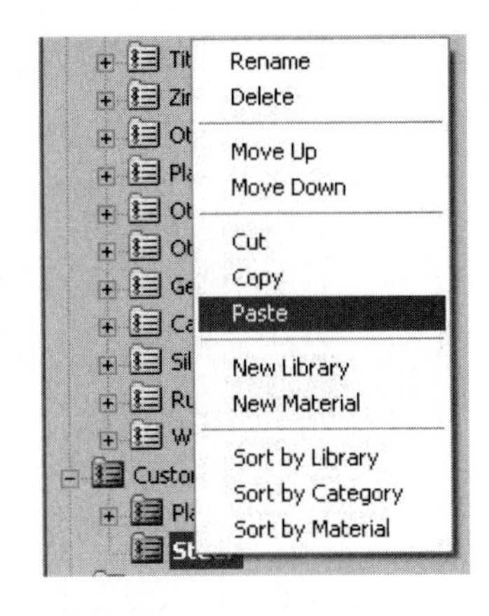

FIGURE 4.71

Next, right-click on the name of the material and change it. All of the material properties are fully editable; therefore, change these to the known properties. Since a similar material was selected to modify, in some cases the value does not need to be changed.

If this new material is desired for the existing part, select the Apply button. In order to permanently save the material on your computer, select the Save button.

For additional materials, there are Internet sources, some are free and some can be paid for, that are downloadable onto your computer (Figure 4.72).

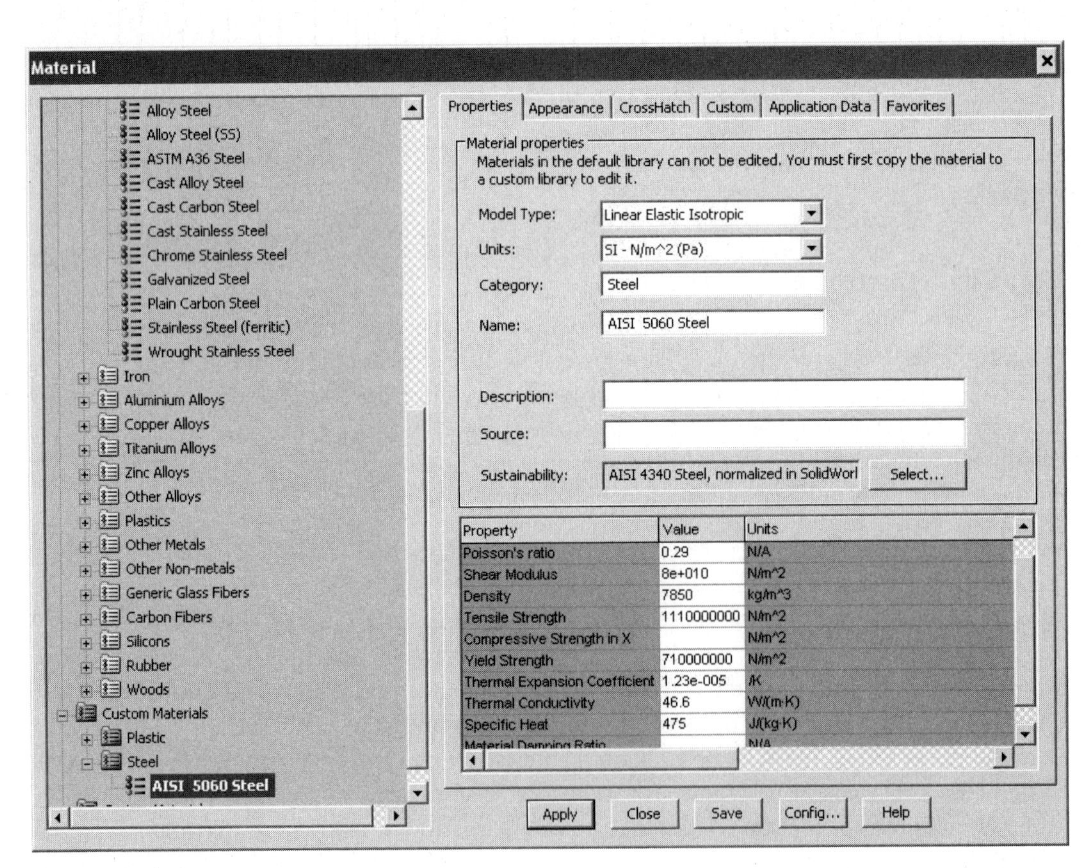

FIGURE 4.72

The appearance of a newly created material can be modified by selecting the Appearance tab. It is best to select color and effects that most closely look like the material (Figure 4.73).

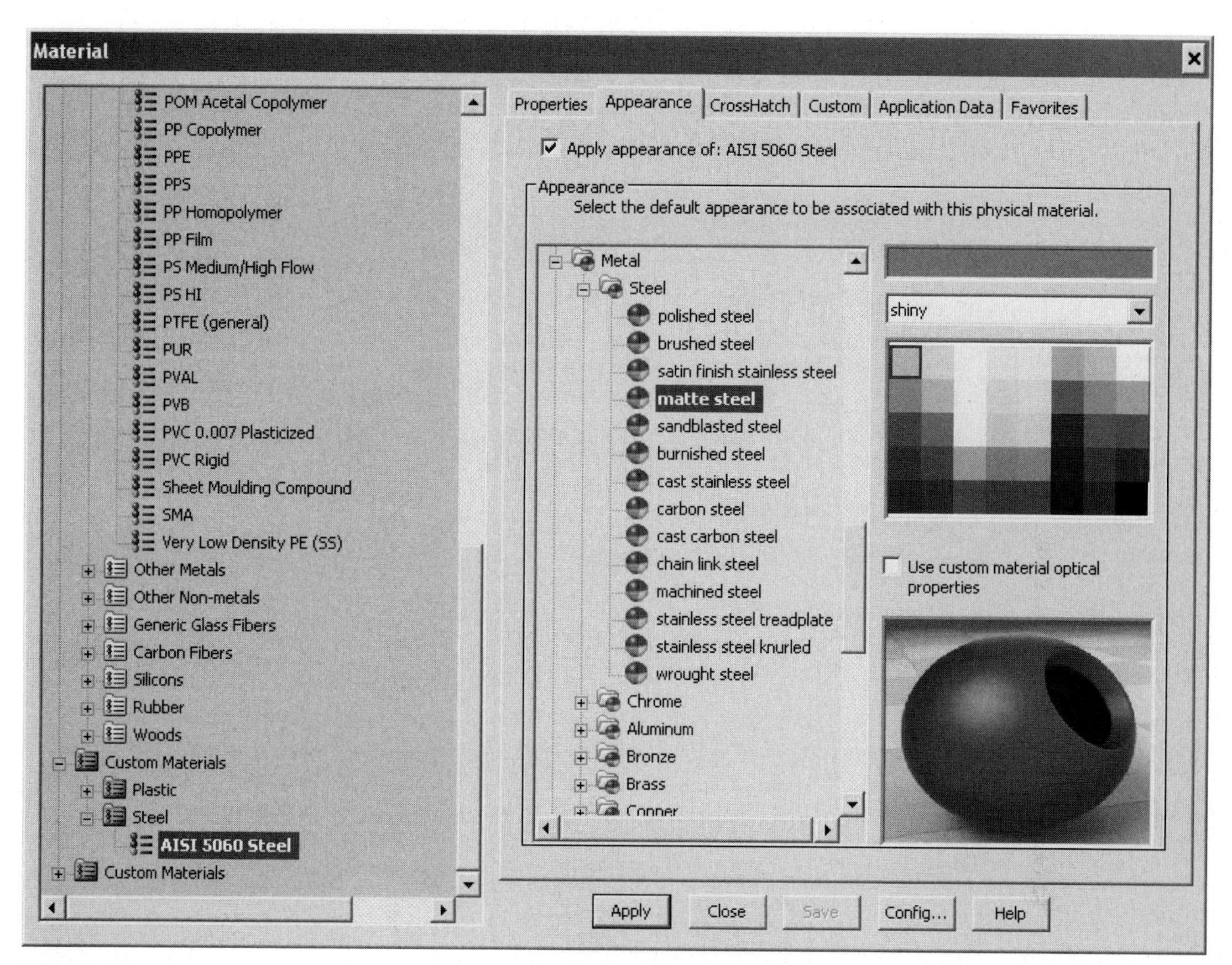

FIGURE 4.73

APPEARANCE

Oftentimes a part's appearance is not that of the raw material; for example, the part may be finished to a fine polish, roughed, or painted. Once the material is set, the appearance can be changed for a part. This is modified using the Edit Appearance command found on the Embedded Views Display Menu (Figure 4.74).

FIGURE 4.74

The Appearance command is quite powerful and can change the appearance of a part to look realistic depending on the level of effort put into this command. Initially under the basic setting of the Appearance command, either the whole part or individual surfaces or features can be visually modified by selecting the appropriate icon in the Selected Geometry field. It is most common to stay with the default of editing the whole part.

In the Color field the color can be selected from the default forty colors available in the swatch selection area, or a new color can be selected using the RGB (or HSV) color palette. In the color field, the appearance can be set to standard, dull, shiny, or transparent as well (Figure 4.75).

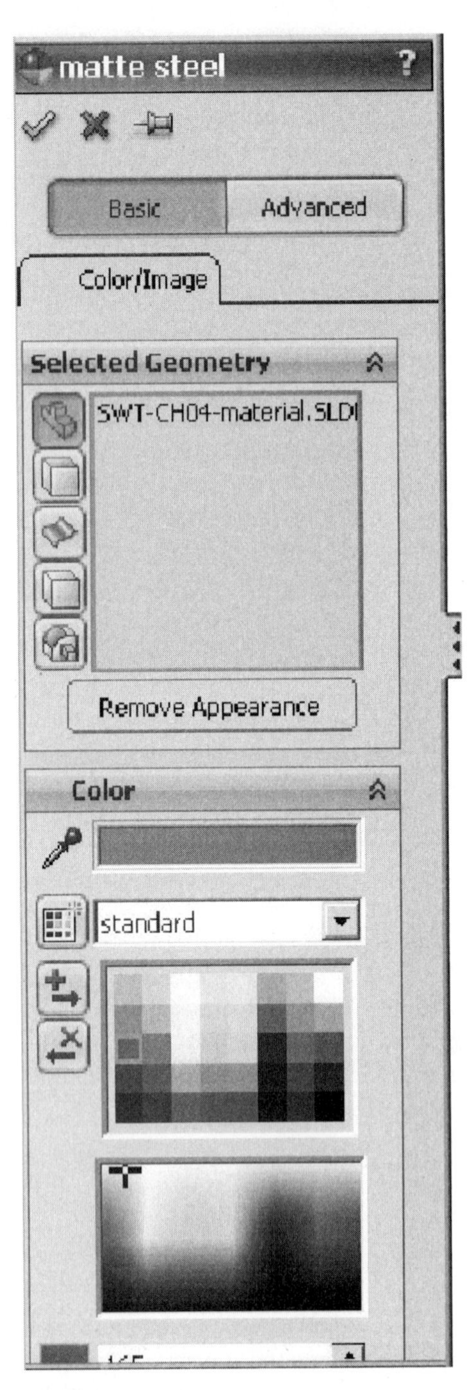

FIGURE 4.75

The Appearance command has an advanced button where the part can be further enhanced. Illumination can be added to the part where advanced topics such as diffusion, specular, reflection, transparency, and luminosity can be varied. Many variations of surface finish can also be applied and the mapping can also be modified. These topics are advanced graphics topics (Figure 4.76).

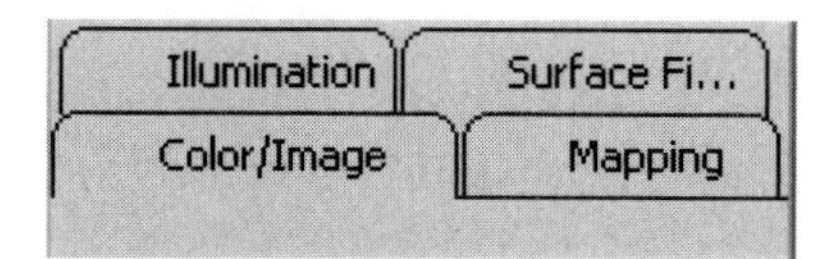

FIGURE 4.76

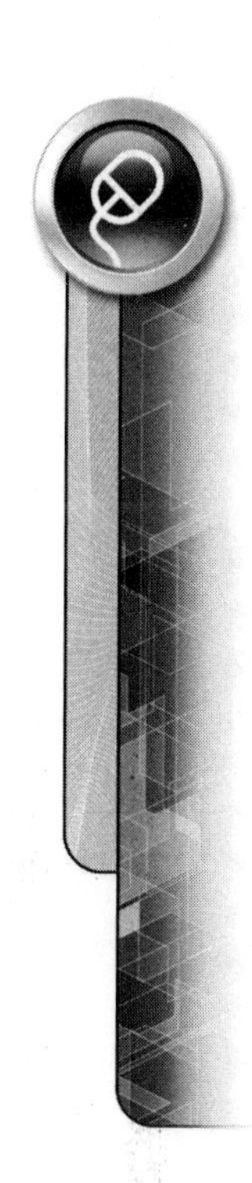

REVIEW QUESTIONS

1. Why are sweeps used instead of an extrude?
2. Why would a new reference plane need to be created?
3. How has the Hole Wizard made creating holes more efficient?
4. Define Loft.
5. Why are ribs used in a design?
6. How does a Rib differ from an extrude?
7. When are Drafts commonly used in design?
8. What primary commands are used for an Emboss? Engrave?
9. What is the advantage of Design Tables?
10. How is the material of a part defined?

EXERCISES

4.1

Create the following part using the Sweep command.

Hint: After creating the body using a sweep, create a sketch on the face of the left end of the part. Create two horizontal lines that are .03125 from the origin and a total of .0625 from each other. Next, create Reference Planes that pass through the lines and are 15 degrees from the Top Plane (one plane through each line) so the planes are 30 degrees from each other. Using the Insert → Cut → With Surface command, cut back the ends to each plane (make use of the Flip Side option if needed).

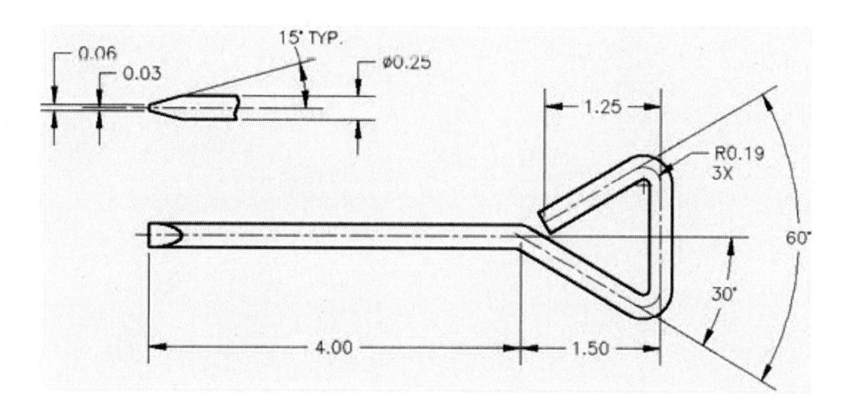

4.2

Create the following part. The main shaft should be created using the Sweep command. The bend radius is 1.00".

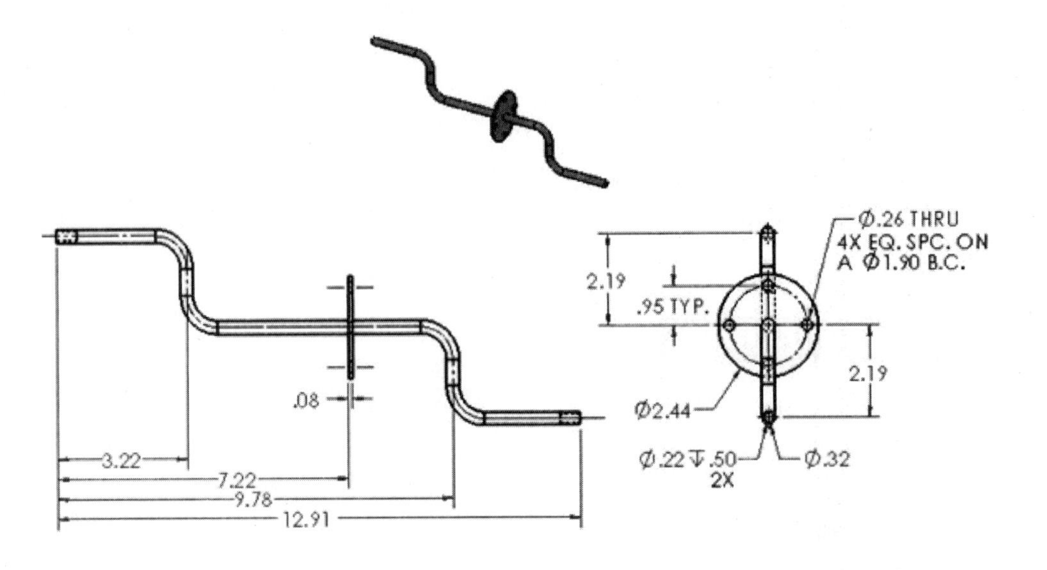

4.3

Create the following part. Use the Dome command to create the end of the handle. The bend radius is 10mm.

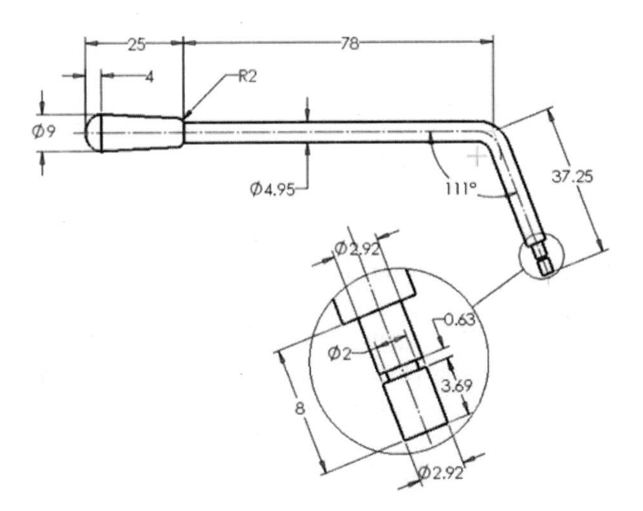

4.4

Create the following part with an inclined face.

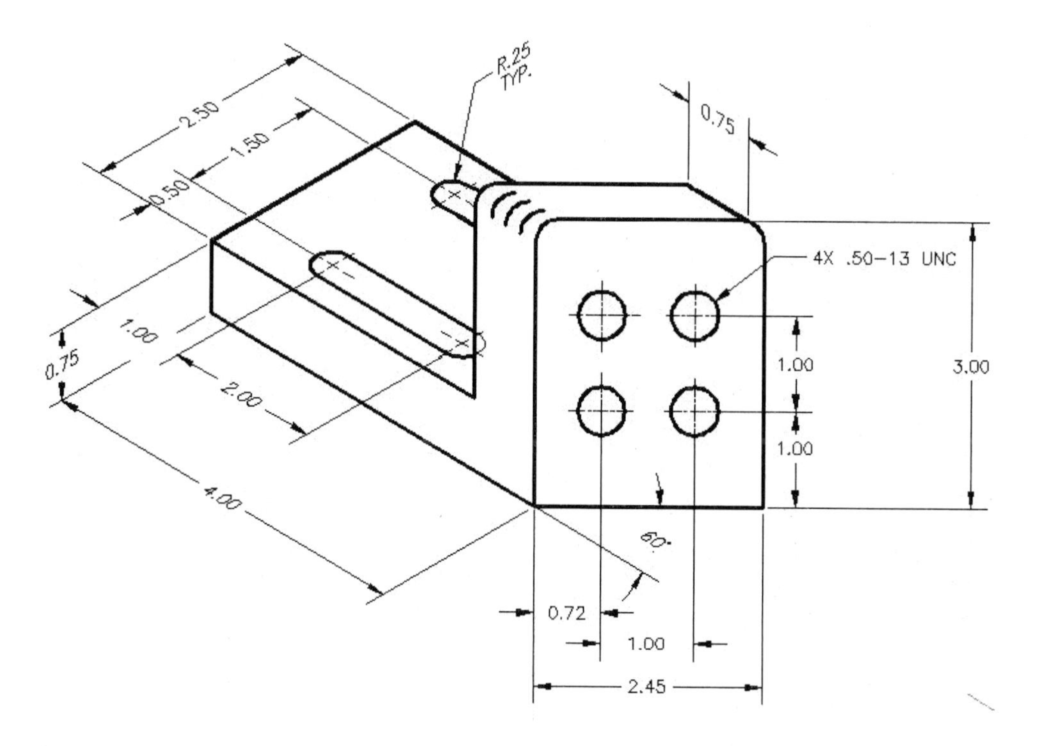

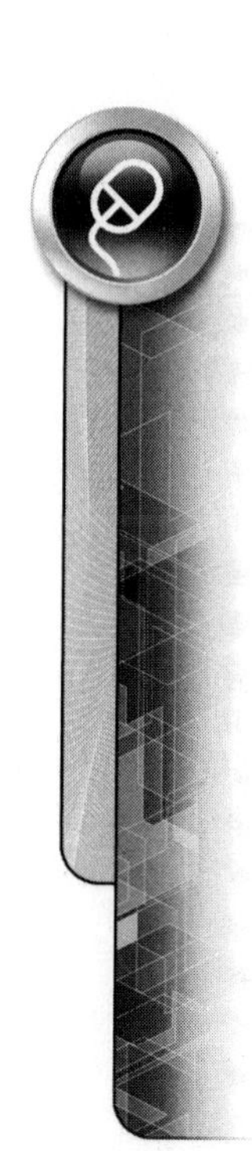

4.5

Create the following part with an inclined surface.

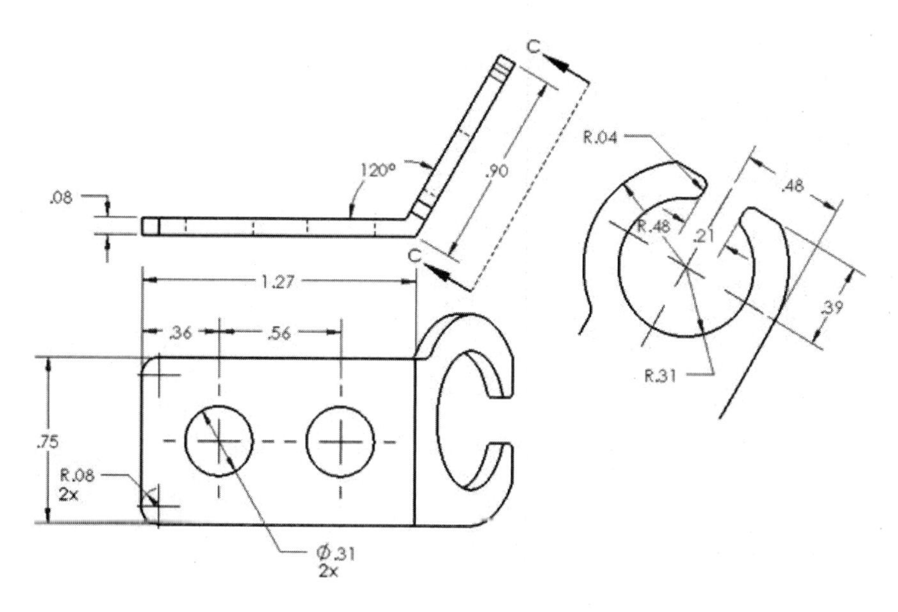

4.6

Create the following part. Use the Hole Wizard to create all of the holes.

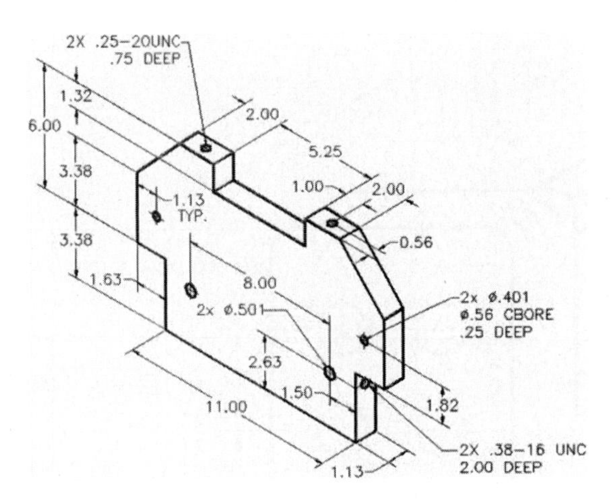

4.7

Create the following lofted part.

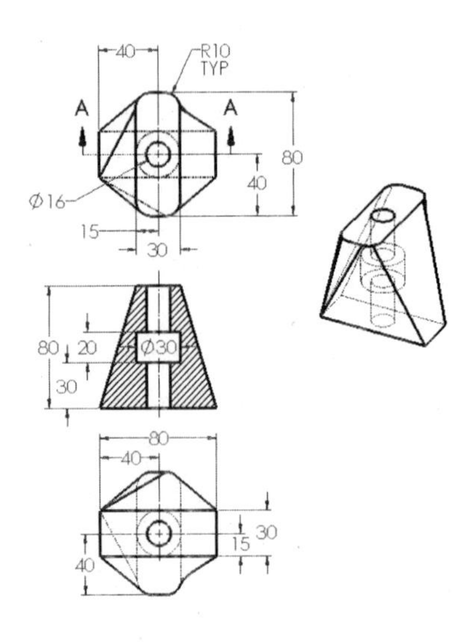

4.8 a & b

Create part files for the following springs.

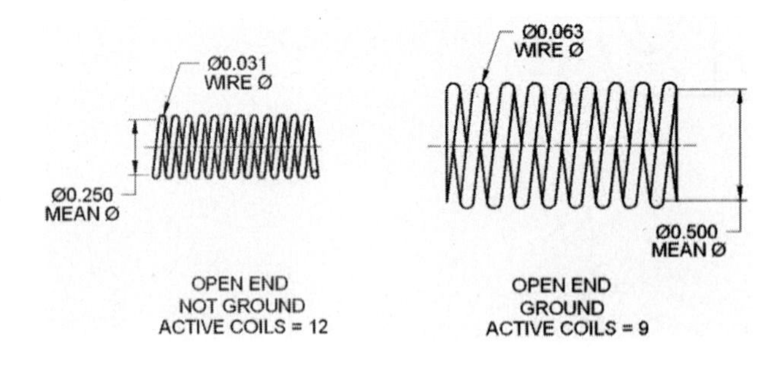

4.9

Create the following part. Use the Rib command to create the four ribs supporting the center boss.

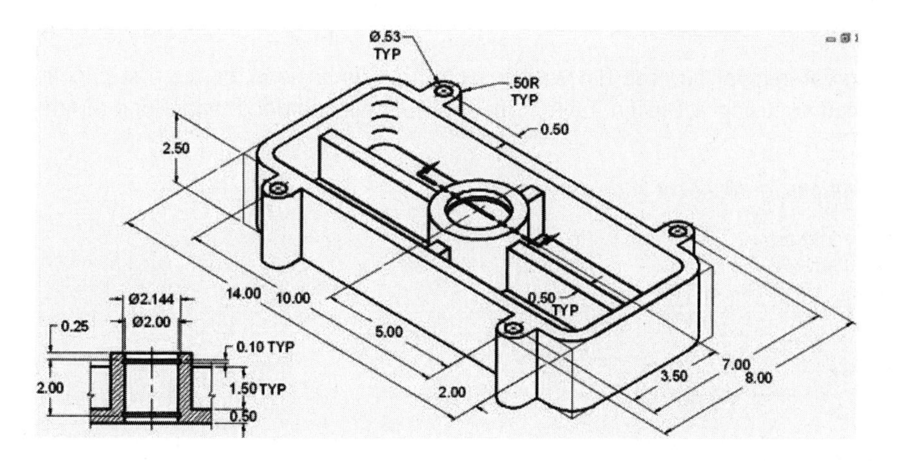

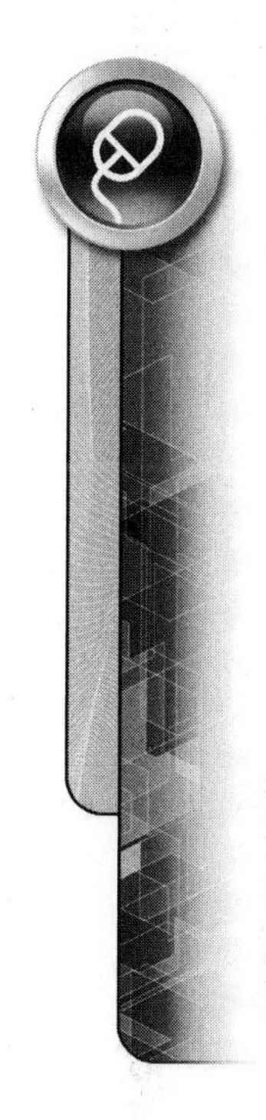

4.10

Create the initial general purpose flat washer using the dimensions for the first part in the table. Then by using a Design Table, create all of the additional variations of the part.

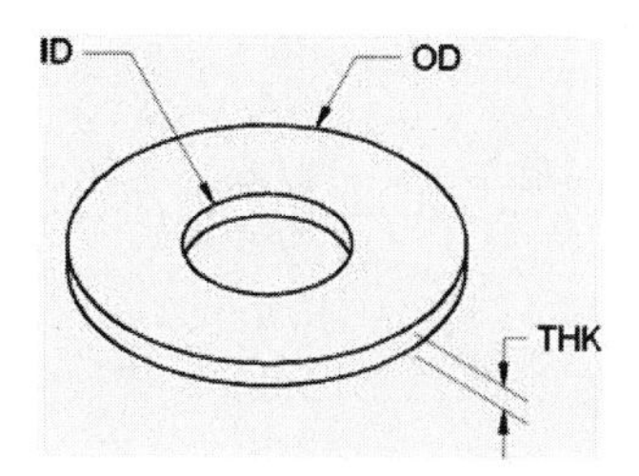

Screw Size	ID	OD	Thk.
#10	1/4	9/16	0.040
1/4	5/16	47/64	0.065
5/16	3/8	7/8	0.085
3/8	7/16	1	0.085
7/16	1/2	1 1/4	0.085
1/2	9/16	1 3/8	0.110
9/16	5/8	1 15/32	0.110
5/8	11/16	1 3/4	0.130
3/4	13/16	2	0.150
7/8	15/16	2 1/4	0.165
1	1 1/16	2 1/2	0.165
1 1/8	1 1/4	2 3/4	0.165
1 1/4	1 3/8	3	0.165
1 3/8	1 1/2	3 1/4	0.185
1 1/2	1 5/8	3 1/2	0.185
1 5/8	1 3/4	3 3/4	0.185
1 3/4	1 7/8	4	0.185
2	2 1/8	4 1/2	0.185
2 1/4	2 3/8	4 3/4	0.22
2 1/2	2 5/8	5	0.245
3	3 1/8	5 1/2	0.285

4.11

Create the initial general purpose flat washer using the dimensions for the first part in the table. Then by using a Design Table, create all of the additional variations of the part.

Hint: Use equations for all of the radii dimensions.

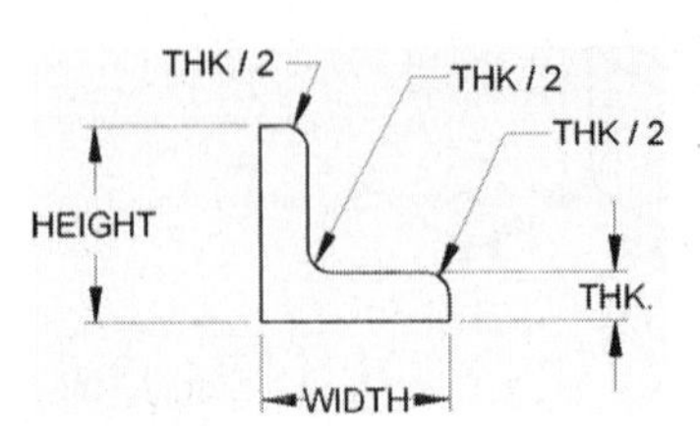

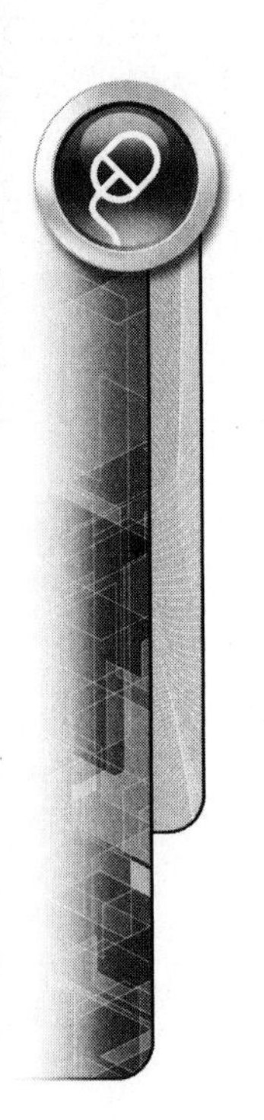

Width	Height	Thk.	Width	Height	Thk.
1/2	1/2	1/8	2 1/2	1 1/2	3/16
3/4	3/4	1/8	2 1/2	1 1/2	1/4
1	1	1/8	2 1/2	1 1/2	5/16
1	1	3/16	2 1/2	2	3/16
1	1	1/4	2 1/2	2	1/4
1 1/4	1 1/4	1/8	2 1/2	2	5/16
1 1/4	1 1/4	3/16	2 1/2	2	3/8
1 1/4	1 1/4	1/4	2 1/2	2 1/2	3/16
1 1/2	1 1/2	1/8	2 1/2	2 1/2	1/4
1 1/2	1 1/2	3/16	2 1/2	2 1/2	5/16
1 1/2	1 1/2	1/4	2 1/2	2 1/2	3/8
1 3/4	1 3/4	1/8	2 1/2	2 1/2	1/2
1 3/4	1 3/4	3/16	3	2	3/16
1 3/4	1 3/4	1/4	3	2	1/4
2	1 1/2	1/8	3	2	5/16
2	1 1/2	3/16	3	2	3/8
2	1 1/2	1/4	3	2	1/2
2	2	1/8	3	2 1/2	3/16
2	2	3/16	3	2 1/2	1/4
2	2	1/4	3	2 1/2	5/16
2	2	5/16	3	2 1/2	3/8
2	2	3/8	3	2 1/2	1/2

4.12

Select a series of components that is repeatable with the same design parameters for different sizes. Part catalogs have many components to choose from. With the approval of the instructor, create the initial part then by using a Design Table, and create all of the additional variations of the part.

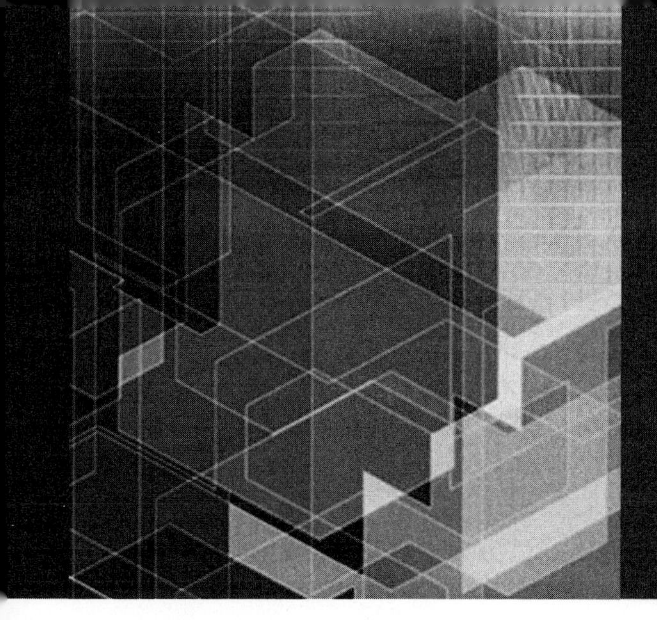

CHAPTER 5

Generating 2D Part Drawings

This chapter discusses various techniques and tools used to create 2D part drawings in SolidWorks and includes the following topics:

- Creating a new drawing file and setting up the drawing sheet
- Modifying drawing settings
- Filling in the titleblock
- Importing a part file into a drawing file
- Creating centerlines and centermarks
- Placing dimensions on a drawing
- Moving views
- Creating auxiliary views
- Creating section views
- Creating detail views
- Placing annotations (notes, surface finishes, GD&T)
- Placing a revision block
- Plotting a drawing

THE DESIGN PROCESS

In past 2D drafting practices, orthographic projection drawings were one of the first steps and an integral part of the design process. With solid modeling, creating ANSI standard drawings is one of the last steps. The majority of the designing is done during the modeling stage.

The historic design process is:

1. Initial concept is conceived
2. Ideas are conceptualized and sketched
3. Design is refined by drafting
4. Prototype part(s) is manufactured

5. Prototype parts are tested for strength, durability, safety, and life
6. Decision: Does the design meet the desired requirements?
   If yes, continue.
   If no, go back to step 3.
7. Design and fabricate equipment to manufacture the product
8. Manufacture

Many companies still use this design process today. Oftentimes, the design to decision process (steps 3 through 6) can be repeated several times. This results in a long development timeline due to the time taken to redesign drawings as well as fabricating prototype parts. This timeline is accentuated when we are manufacturing parts that require molds to be fabricated.

The modern design process is:

1. Initial concept is conceived
2. Communicate with customer and develop "Design Input Requirements"
3. Ideas are conceptualized and sketched
4. Solid parts and assemblies are modeled
5. Virtually test the parts for strength, durability, safety, and life
6. Rapid prototype
7. Decision: Does the design meet the desired requirements?
   If yes, continue.
   If no, go back to step 3.
8. Create drawings to document the part and tolerances
9. Design and fabricate equipment to manufacture the product
10. Manufacture

As modern engineering segues into the modern processes we will realize an improved design process speed increase of 3X or greater.

FILE ASSOCIATION

The drawing files are associated to the original part/assembly file. That means that when changes are made to the part/assembly file, the changes are automatically represented in the drawing file.

Automatic is an ambiguous term though. While the geometry changes may be reflected, and dimensions may change in size, any additional features or features that have been removed will result in additional work on the part of the designer to update the drawing with new dimensions. Also, if the design intent changes, the method of dimensioning may also change.

THE NEED FOR DRAFTING

The sole purpose for drafting working drawings is to communicate. The designers (engineer, designer, drafter, and architect) need to communicate detailed information to the fabricators (machinist, assembly person, welder, constructor, mold maker, inspector, etc.). With the advent of Solid Modeling, much of this information is communicated automatically in the modern design process. The size and shape description of the part is often automatically transferred electronically using the designed models.

It is believed by many modern designers that the drafting process may become obsolete due to the fact that many modeled part files are sent directly to manufacturing equipment such as Programmable CNC's, Wire EDM Machines, or Stereo Lithography Machines. Additionally, the modeled part files may be directly sent to a manufacturing designer, such as a mold designer, and the part file is used to develop the mold.

So why are drafted drawings needed? There are two major reasons. The first is to communicate the design intent. The dimensioning method used to define the part communicates to the manufacturer what dimensions are critical to maintain.

Secondly, in a closely related need, tolerances need to be communicated. To date, there is no easy way to communicate the tolerances of dimensions and the tolerances of surface finishes, as well as many other design issues that are put in notes.

These are the reasons the drafting step of the design process is necessary for some time to come. The need to describe shape description of a part has lessened, but there are still critical details that require a drawing.

Although there are currently few other methods that are reliable within Solid Modeling programs to communicate this type of information, it can't be assumed that these methods will never be available.

If and when these methods are realized, there is a possibility that the drafting step in the design process will become obsolete. When this happens, the role of the pure drafter will become obsolete.

CREATING A NEW DRAWING FILE AND SETTING UP A DRAWING SHEET

The basic process of creating a drawing for a part is to create a new blank drawing file and then import the necessary views from an existing saved part file; therefore, the part file and drawing file are separate files which are associated to each other.

To create a new drawing file, select the "New…" command from the "File" pull-down menu and the three file options of "Part," "Assembly," and "Drawing" appear. Select "Drawing" and the Sheet Format/Size window will appear.

In the left window panel there are dozens of standard sheet sizes to select from. Select a sheet size from the list and then click on OK to create a pre-sized sheet with a default titleblock and border. If a sheet size is desired and not listed, the Custom Sheet Size button can be selected (Figure 5.1).

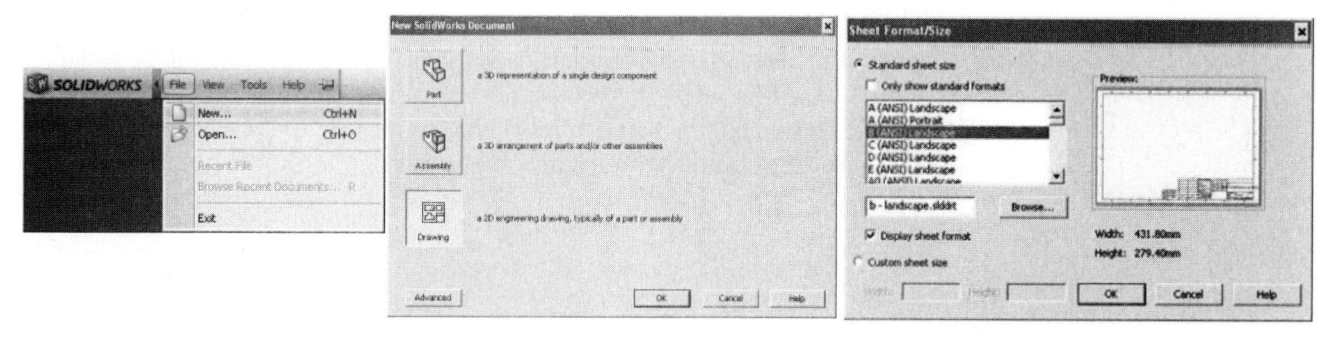

FIGURE 5.1

The sheet size can be changed at any time by right-clicking on Sheet Format1 in the Feature Manager tree. The same list of choices of sizes including custom size will be available.

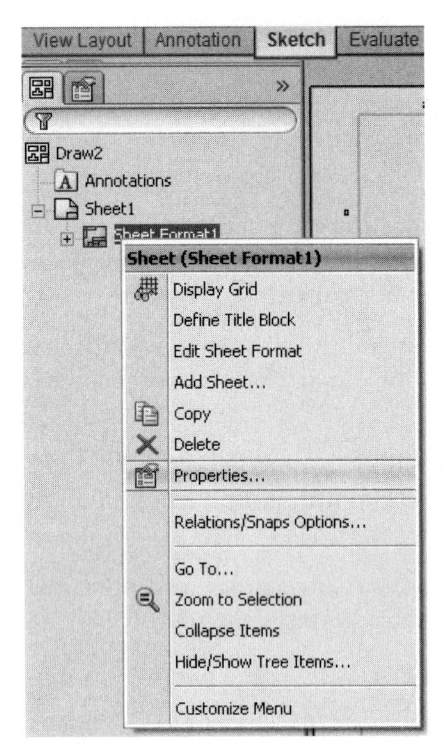

FIGURE 5.2

MODIFYING DRAWING SETTINGS

There are several drawing settings that can be changed easily by right-clicking Sheet Manager1 as shown in Figure 5.2. The Sheet Properties window will appear that has additional settings besides Sheet Format and Size.

In the upper left of the window is the name of the sheet, the default being Sheet1. The overall scale of the drawing can also be changed where the default is set to 1:1.

The type of projection can also be changed between first angle and third angle projection. First angle projection, being the ISO standard projection, is more common in Europe. Third angle projection is the standard for ASME and British Standard and is primarily used in North America and Great Britain. *It is critical that the proper type of projection is set.*

When creating new non-orthographic drawing views, the first view will be assigned the label "A" as it is set as the default setting. When creating multiple datums in GD&T dimensioning, the first datum will be designated "A" followed by B, C, etc. as "A" is set as the default starting letter (Figure 5.3).

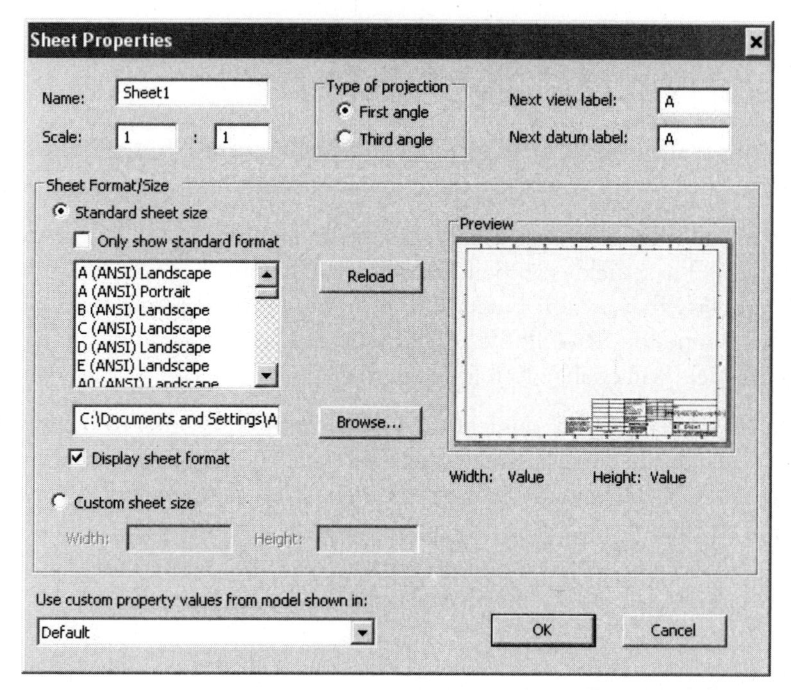

FIGURE 5.3

FILLING IN THE TITLEBLOCK

One of the first things to do on a drawing is to fill in the titleblock. In order to do this, right-click on the Sheet Format1 in the Feature Manager and select the Edit Sheet Format option. You will notice that the titleblock visually changes where the lines turn blue as well as various predefined fields are present in many areas on the titleblock but are poorly represented as you can see on the right side of Figure 5.4. At the bottom of the figure the fields can be seen as black boxes with Xs in them; use this figure to assist in locating them until you become more familiar with the locations.

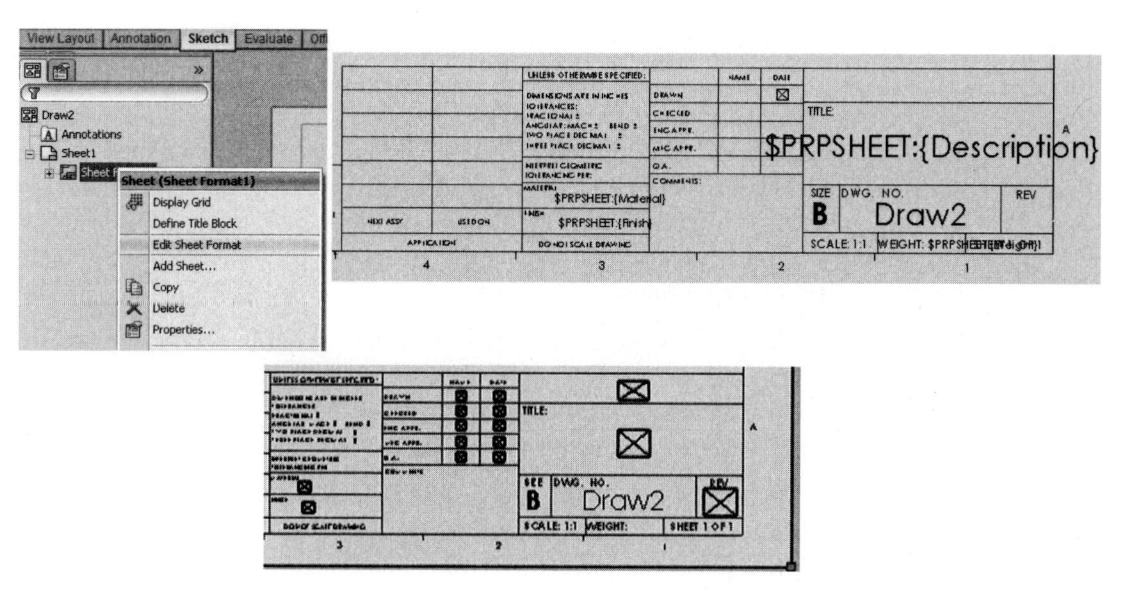

FIGURE 5.4

In order to fill in any of the fields, put the cursor over one and double-click it. Fill in the field and select OK to accept it. All fields should be filled in using capital letters and numeric values.

There are several fields that appear to be filled out with default settings. In some cases the defaults are logical text such as the scale and number of sheets.

Other fields, such as the title, drawing number, Material, and Finish, are linked to other document values. These fields can be manually changed but they will lose any association to the property they are attributed to. In order to keep the associativity of the drawing number, name the drawing file as described in Chapter 1 and when the drawing is saved, this field will change to match the file name.

Blank fields can be linked to existing document values. For example, in order to change the drawing date to match the creation date of the file, which is a typical custom in industry, first select the field. Then in the Feature Manager, select the Link to Property button. In the Link to Property window, select the property to link to, in this case "SW-Created Date," select the Short Date option, and deselect the Show Time field. Accept the settings and the date will be inserted into the date field (Figure 5.5).

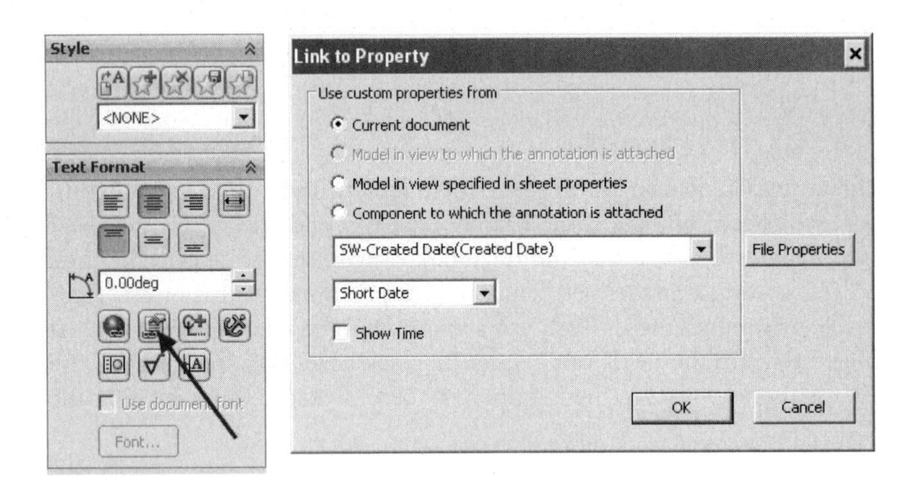

FIGURE 5.5

SETTING THE DIMENSIONING STANDARD

Dimensioning standards vary throughout the world; the two major dimensioning standards concerns in the United State are ANSI (American National Standards Institute) and ISO (International Organization of Standardization). The ability to change the settings between these two as well as to other more obscure standards can be changed in the Options.

In the Tools pull-down menu, Options can be selected. Within Options, there are two tabs, Systems Options and Document Properties. Under Document Properties, the first selection is Drafting Standard. The standard default is ISO. If another is desired such as ANSI, select in the dropdown menu then select OK.

Changing the Drafting Standard will change all of the default settings appropriately with respect to text heights, dimension geometry sizes (e.g., arrow sizes), and any other standards that are controlled per governing body (Figure 5.6).

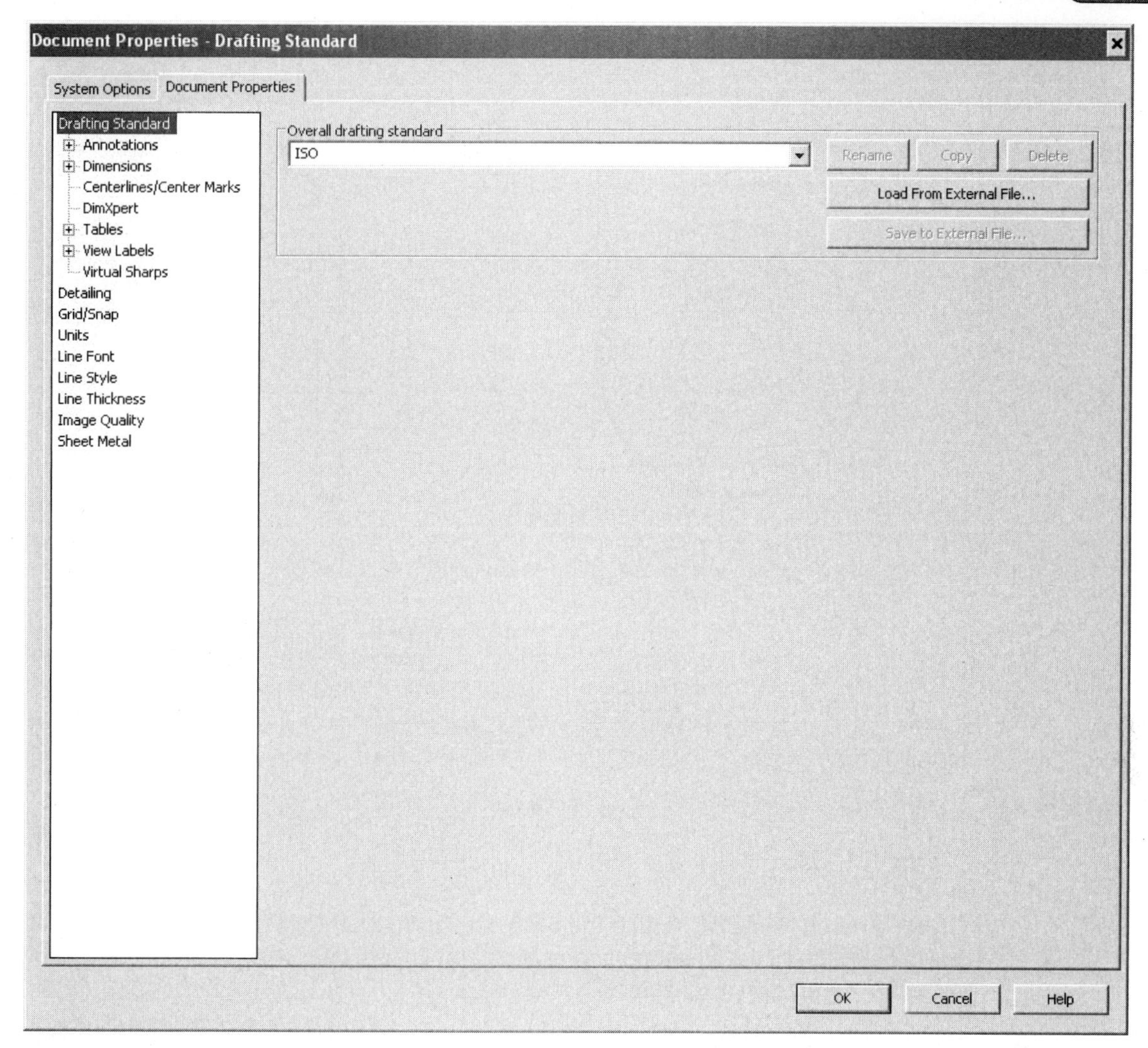

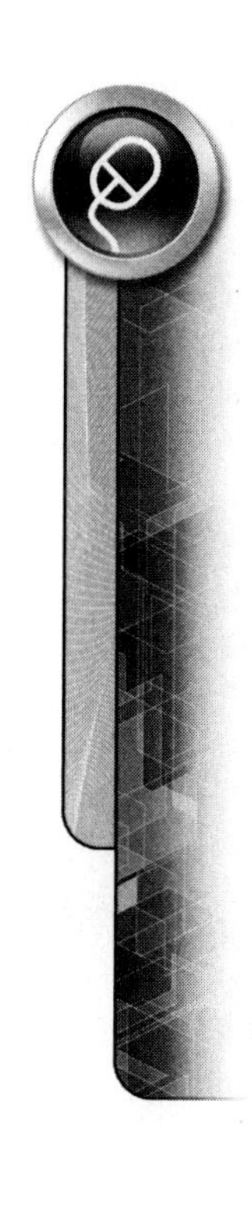

FIGURE 5.6

IMPORTING A PART FILE INTO A DRAWING FILE

One of the most important parts of a drawing is to represent the shape of the part (called shape description) using multiple orthographic views. The views generally should be represented with object, hidden, and center lines. The views should be projected from each other and they should be at a scale that clearly represents every feature.

The process of creating views in SolidWorks is to bring in one base view where all other views are projected from. In order to place the first view, a knowledge of which views are available in the original part file is needed. Generally speaking, one of the six primary orthographic views is selected. The first view placed is not necessarily the primary front view.

TUTORIAL EXERCISE: 05_CREATING_DRAWING.SLDPRT

Part Number: SWT-CH05-01.SLDPRT

Description: Creating Drawing 1

Units: English

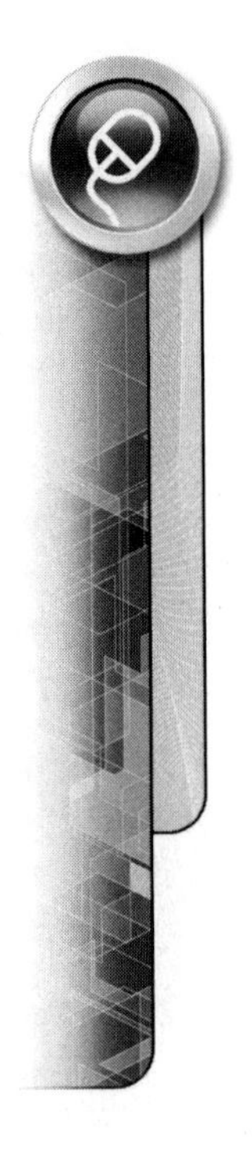

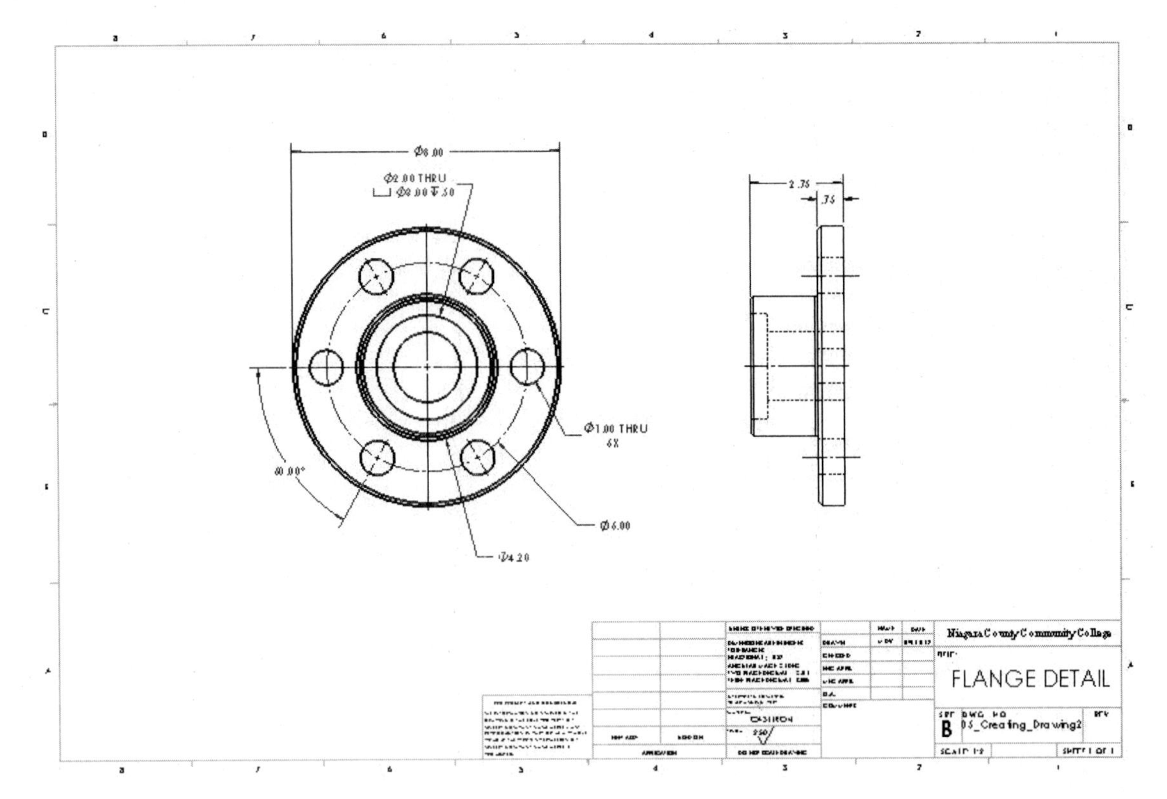

FIGURE 5.7

STEP 1

Create a new drawing file setting the paper size as B(ANSI) Landscape. Also ensure that the units setting in document properties be set for English units (IPS). Also set the type of projection to third angle (Figure 5.7).

When the feature manager shows Model View, exit by selecting the red X. Although importing the part file can be done at this time, we will address it at a future step (Figure 5.8).

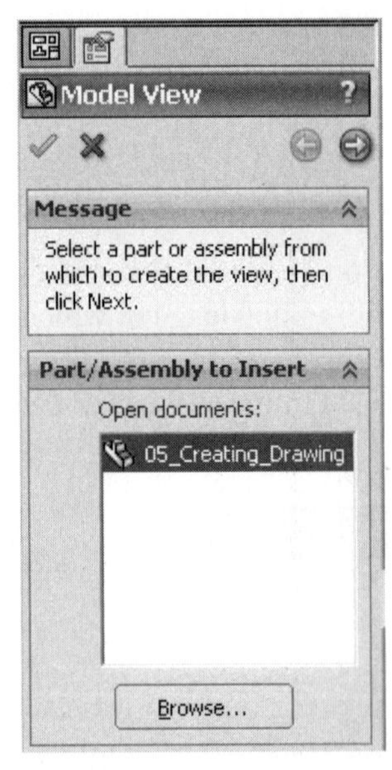

FIGURE 5.8

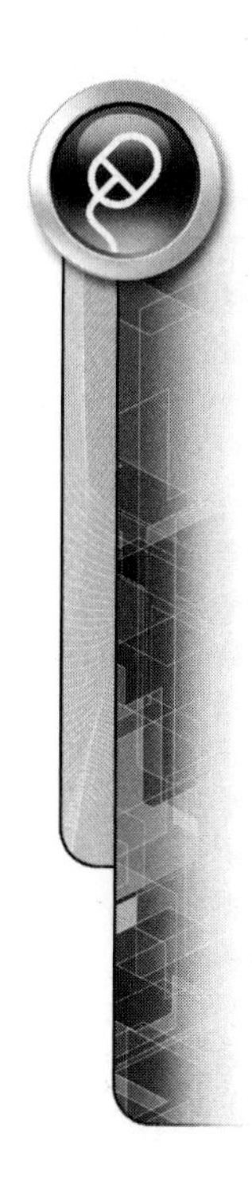

STEP 2

Fill in the titleblock as shown in Figure 5.9. The drawing title should be called Flange Detail. Put your school's name in the top right field, your initials in the name drawn field, and set the date to match the creation date of the drawing. Fill in as much of the titleblock as possible.

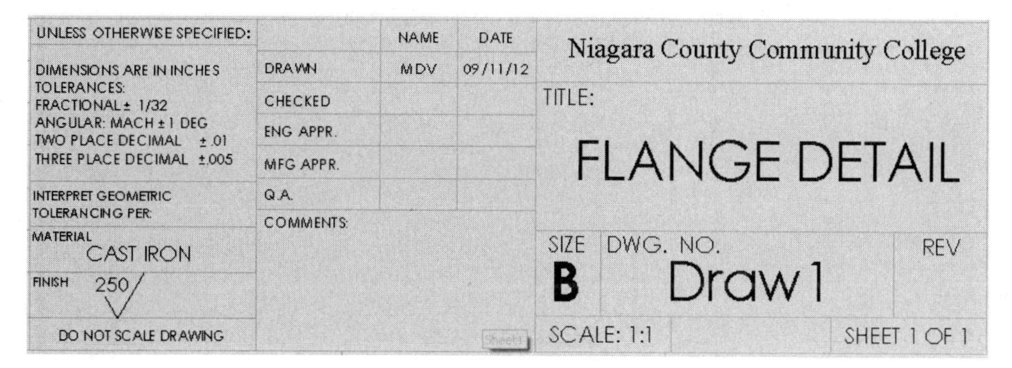

FIGURE 5.9

STEP 3

Select the View Layout tab and select the Model View command. There are two ways to import a file into the drawing file. First, a list of all currently open files is listed in the Open Documents window. If the part you are seeking to draft is in this list, select it here. If the file you are seeking is currently closed, select the browse button. An Open window, which will allow you to navigate through the computer drives and folders to find the desired file, will appear (Figure 5.10).

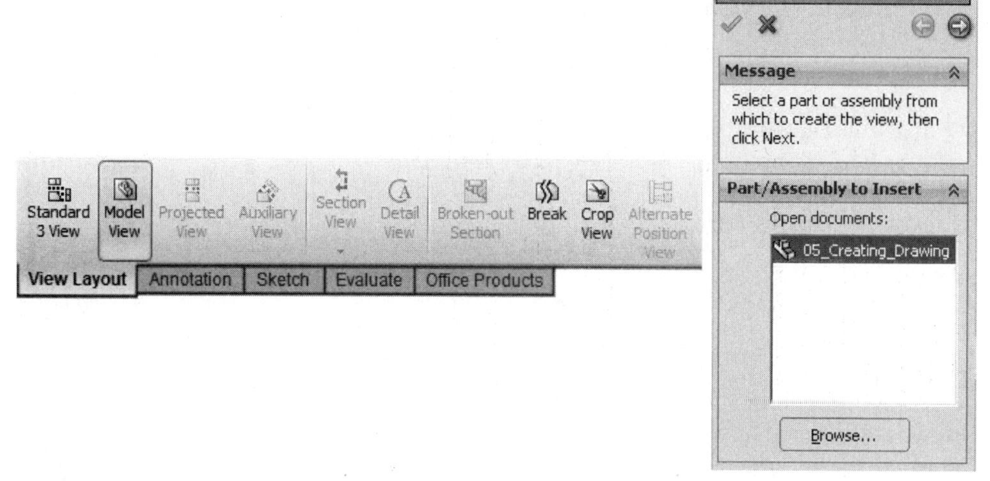

FIGURE 5.10

STEP 4

Once the file is selected, the Model View Feature Manager shows a sequence of settings for the first view to be placed on the drawing. Under the Orientation section, select the appropriate view, the top view icon in this case. Under the Display Style section, select the Hidden Lines Visible icon.

The scale can also be selected by selecting Use Sheet Scale (when a second scale is needed, a custom scale can be set). Lastly, the dimension type should be set to Projected. This forces the dimension to read the true 2D distance on the sheet instead of the actual selected points on the part which may be at different depths into the sheet.

Only after all of these settings are set should the view be placed on the drawing. The view is represented as a rectangle representing the outer edges of the view. Place the view on the drawing where appropriate. If the view is placed in an undesirable area, it may be moved easily (Figure 5.11).

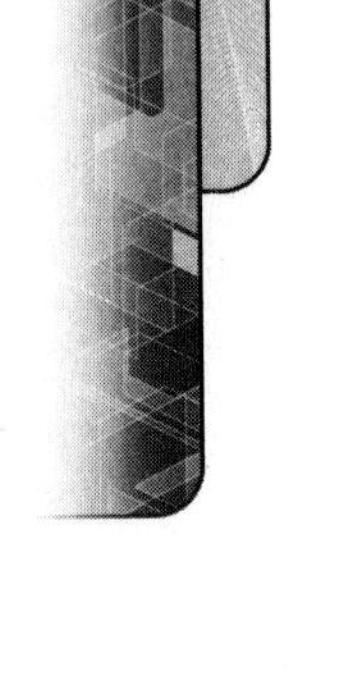

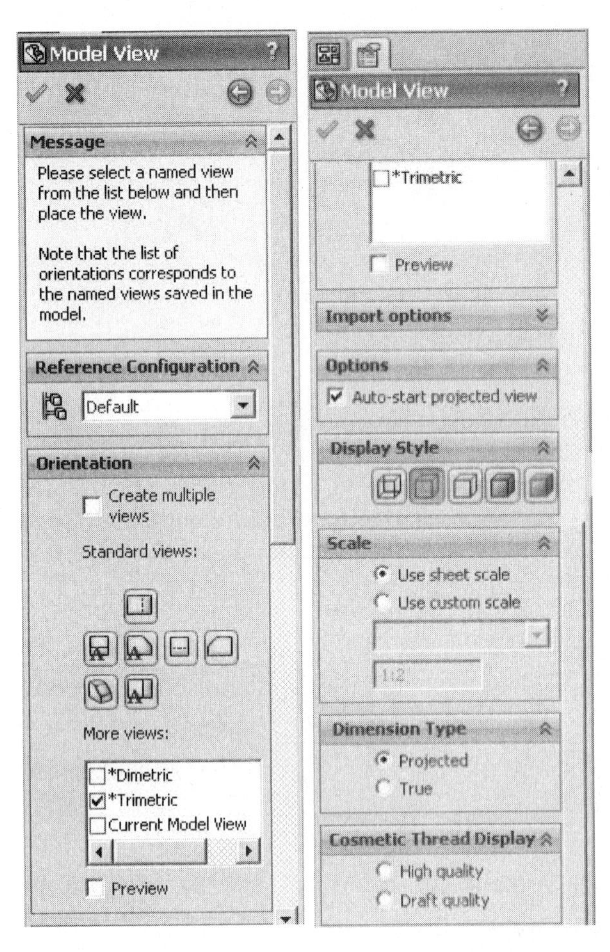

FIGURE 5.11

STEP 5

Place the view on the drawing sheet. You should have a preplanned idea as to how many views are required as well as generally where you want them located on the sheet. This drawing should have two views, located horizontally from each other; therefore, place this first primary view centered and slightly to the left. The view will appear as a rectangle that represents the size of the view. Once the view is placed, the detail can be seen (Figure 5.12).

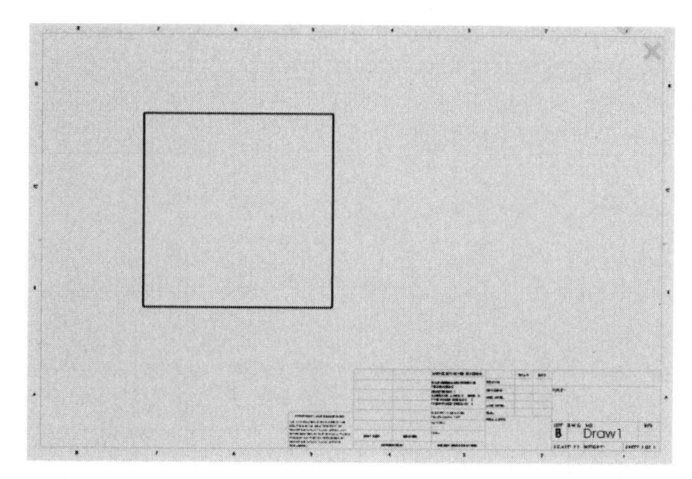

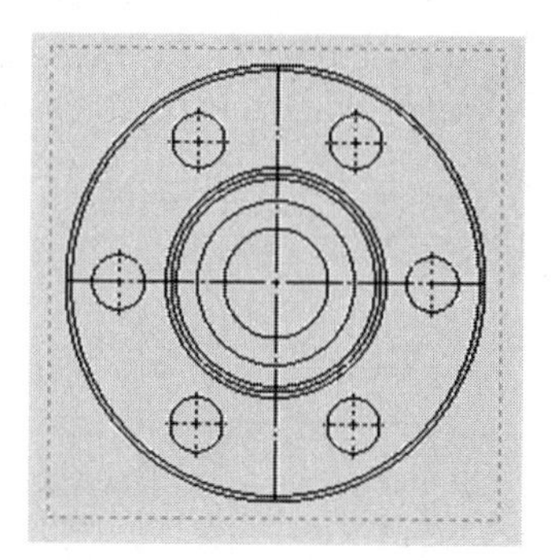

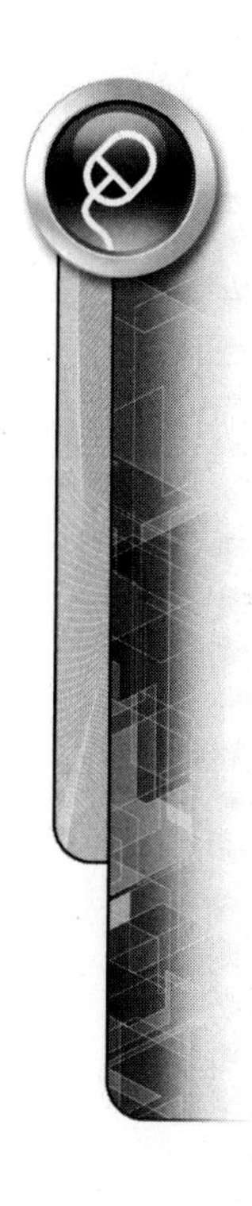

FIGURE 5.12

STEP 6

Now we place the second view. In the View Layout ribbon, select the Projected View command. As the cursor is moved around the drawing, the new view will dynamically change to any of the primary projection views (top, bottom, left, right, as well as isometrics) depending on where your cursor is with respect to the first view. Move your cursor on the drawing to the area where you want to place the second view and select the location, in this case to the right.

Additional views can be quickly placed if desired, but this drawing will be a two-view drawing; therefore, accept the view by selecting OK (Figure 5.13).

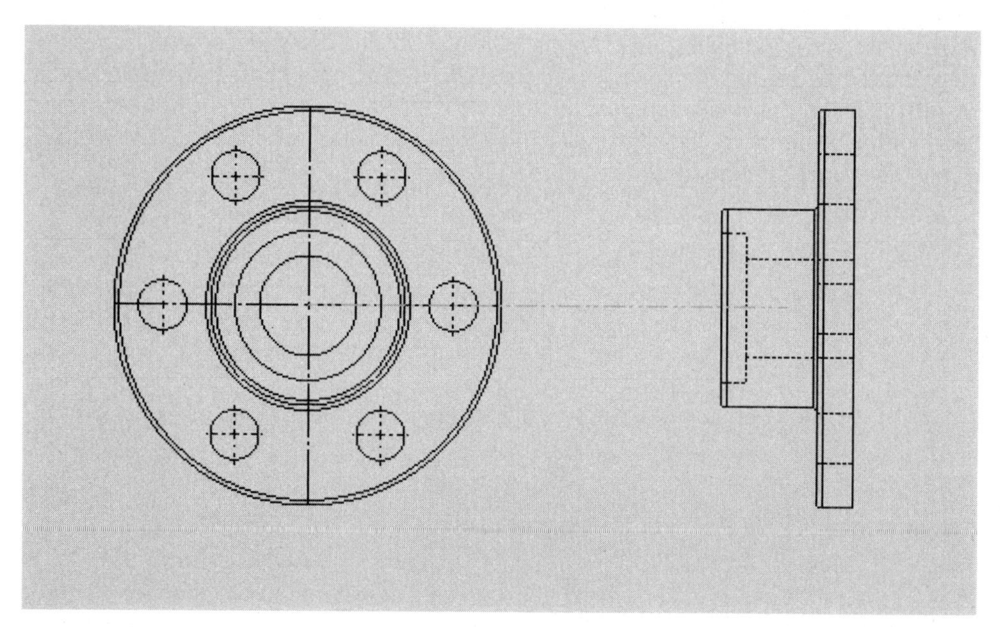

FIGURE 5.13

STEP 7

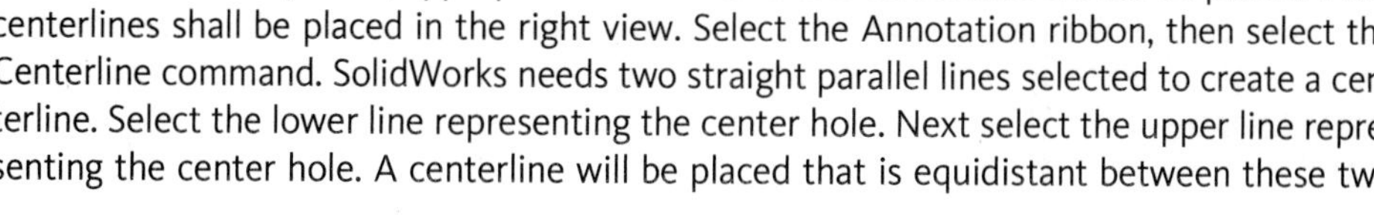

Once all views are placed, appropriate centermarks and centerlines should be placed. First, centerlines shall be placed in the right view. Select the Annotation ribbon, then select the Centerline command. SolidWorks needs two straight parallel lines selected to create a centerline. Select the lower line representing the center hole. Next select the upper line representing the center hole. A centerline will be placed that is equidistant between these two lines and the centerline will extend past the lengths of the two lines. Since the centerline does not properly represent a complete centerline on the left because it does not extend past the whole view, select the line and drag the blue endpoint outside the view. Create any additional centerlines needed in the right side view (Figure 5.14).

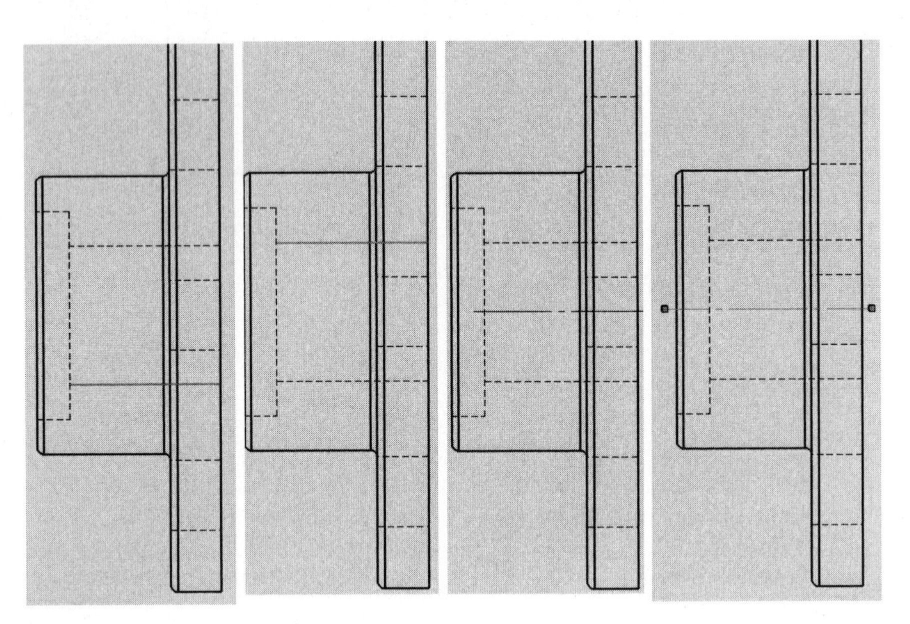

FIGURE 5.14

STEP 8

Notice in the front view where SolidWorks automatically places a centermark for every hole in the view. Although this is normally helpful, in this case the centermarks can be represented better for the six-hole pattern. Before creating a centermark pattern that meets technical drawing standards, the existing centermarks need to be deleted. Select each of the centermarks individually and hit the delete button on the keyboard for each. Only the overall centermark should remain (Figure 5.15).

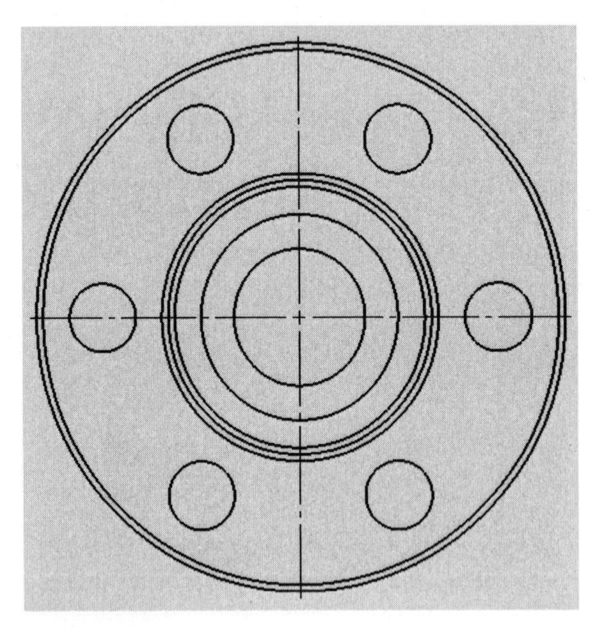

FIGURE 5.15

On the Annotation ribbon, select the Centermark command. In the Feature Manager, there is a section for Manual Insert Options. Among these options are the ability to create single centermarks, linear centermarks (a rectangular pattern), or circular centermarks. Select the Circular Center Mark option. Select four of the six holes, the two upper holes and the two lower holes. The left and right holes do not need to be selected because there is already a centerline horizontally through these two holes. The result will be a base circle centerline that passes through the center of all six holes as well as four radial centerlines that pass through the four holes which needed them (Figure 5.16).

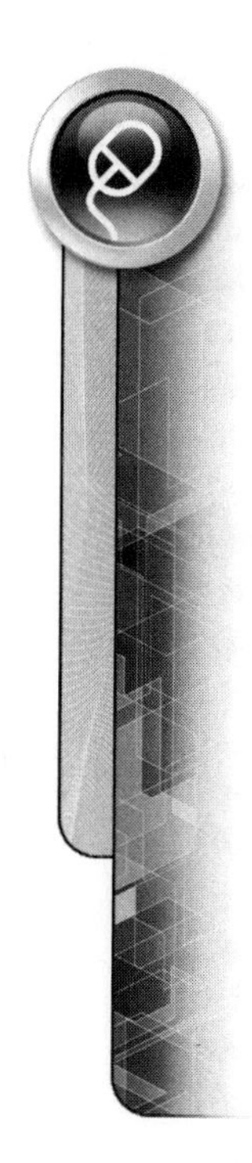

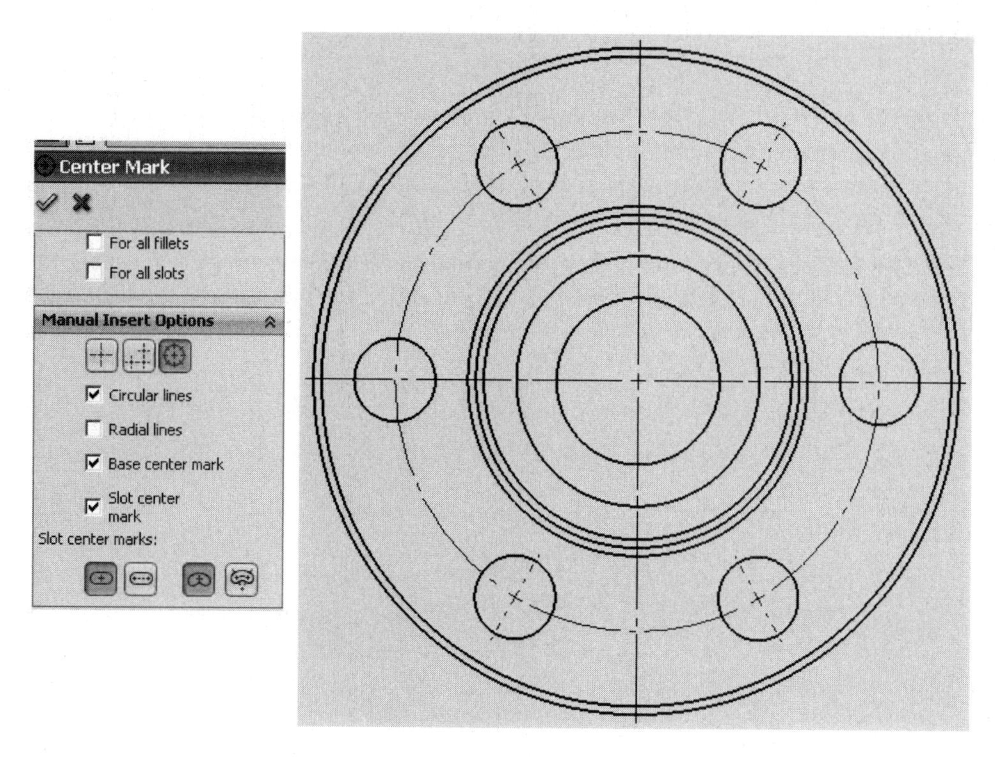

FIGURE 5.16

STEP 9

Next is to dimension the drawing. The Smart Dimension is an intuitive command meaning that depending on what geometry is selected decides what type of dimension will be placed. Many types of dimensions, including linear horizontal, linear vertical, linear inclined, angle, radius, and diameter dimensions, can be placed using Smart Dimension.

On the Annotation ribbon, select the Smart Dimension command. In the right side view select the far right line near the top that represents the bottom of the part, then select the vertical line representing the top of the main base that is to the left. As the cursor is moved upward, notice that the dimension "snaps" to various heights. These heights are predefined in the dimension settings to assist in locating the dimensions the appropriate distance from each other. Place this dimension located at the first snapping height and center the text within the dimension lines. Accept the dimension by selecting OK. Next, place the 2.75 dimension above the .75 dimension (Figure 5.17).

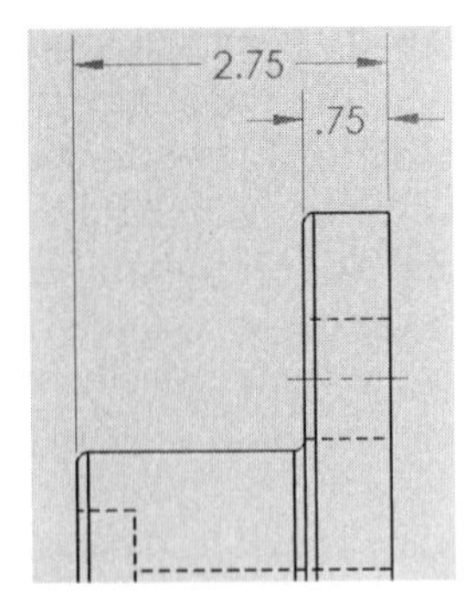

FIGURE 5.17

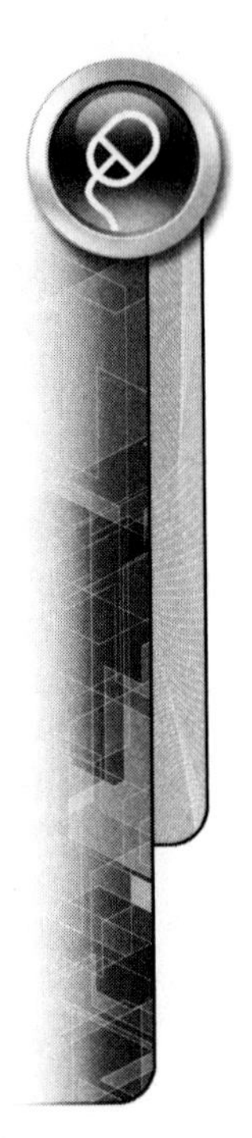

If holes on part files are created using the Hole Wizard, the Hole Callout command can be used to place dimensions efficiently using all of the appropriate standard symbols. Oftentimes holes made separately using the Extrude command will combine into standard callout using the Hole Callout command also.

Select Hole Callout command on the Annotation ribbon. In the front view, select the three-inch counterbore circle and drag the dimension off the part. Accept the dimension by selecting OK (Figure 5.18).

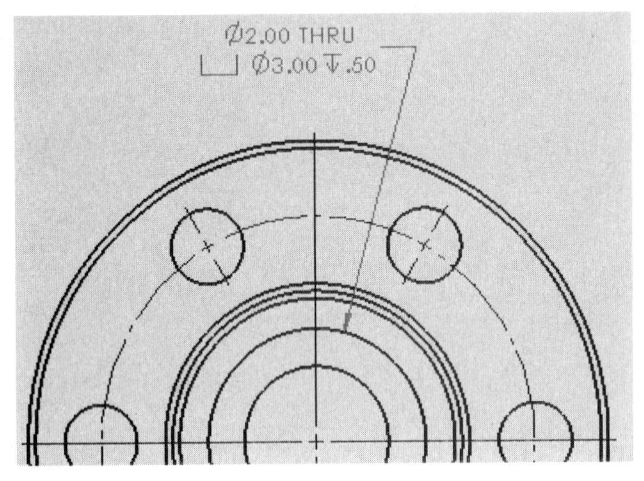

FIGURE 5.18

Select Smart Dimension again and select the outer diameter that represents the overall diameter of the part. Locate the dimension in an appropriate location. Notice that a linear dimension or a leader dimension can be placed (Figure 5.19).

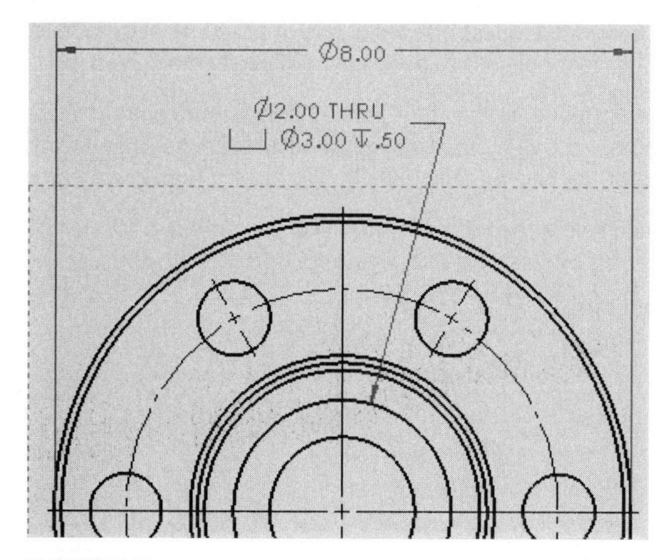

FIGURE 5.19

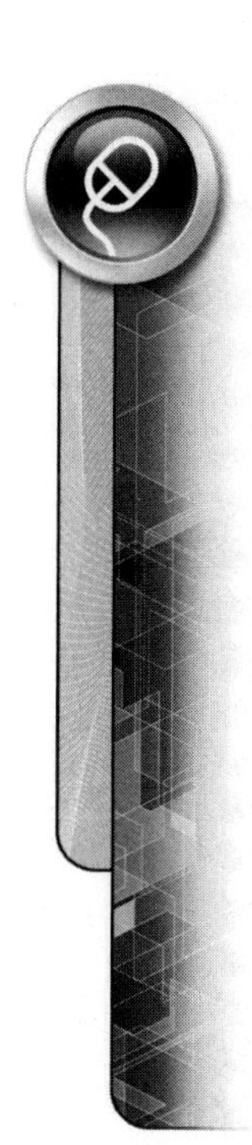

While in Hole Callout, select one of the holes in the circular pattern. After placing the location of the dimension, modify the Dimension Text window in the Feature Manager. When the text field is selected, SolidWorks warns that the associativity of the dimension may be affected; it will not be affected as long as the initial text is not modified. After the existing text in the text window, hit enter then type 6X. Accept the dimension by selecting OK (Figure 5.20).

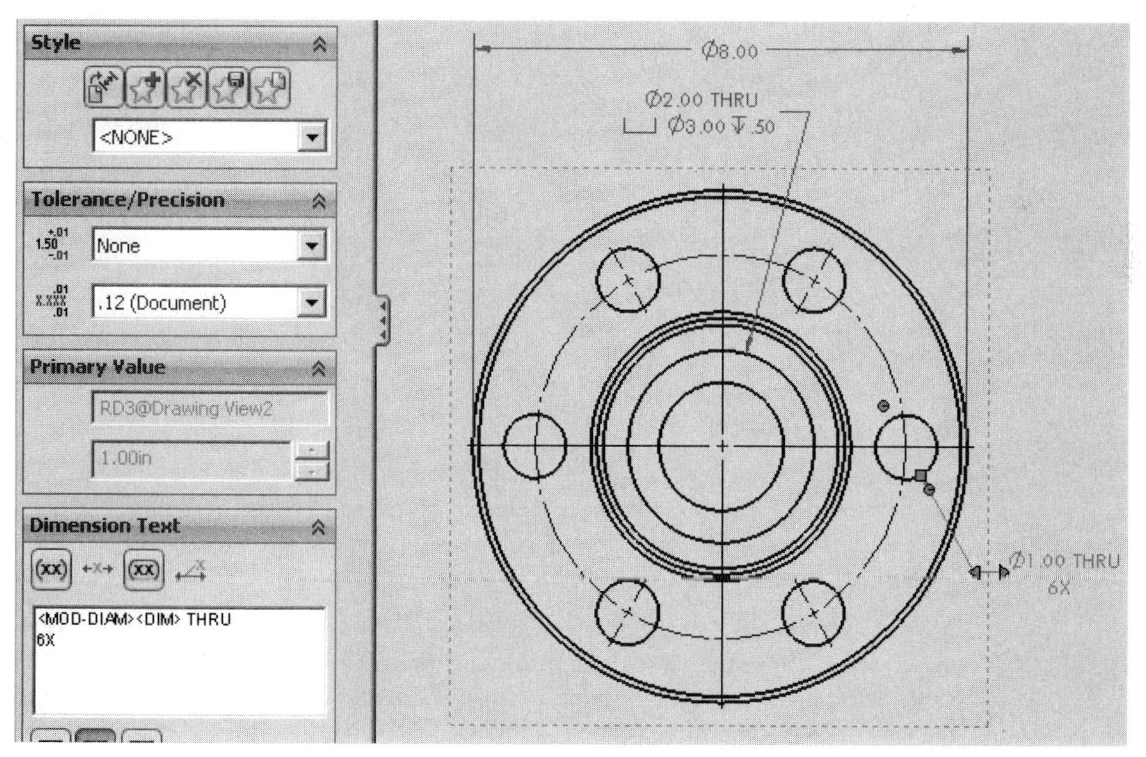

FIGURE 5.20

Select Smart Dimension and then select the base circle for the hole pattern and place the dimension. If the dimension is specified as a radius, right-click on the dimension and select "Display as a Diameter." Accept the dimension by selecting OK.

While in the Smart Dimension Command, select one radial line of a circular pattern hole, then select a radial line of the adjacent hole. Place the angular dimension off the part.

In the Smart Dimension pull-down menu, select the Chamfer Dimension command. Select the edge of a chamfer in the right view, then select an adjacent edge, then locate the dimension off the part. Accept the part by selecting OK. Do this for both chamfers.

Select Smart Dimension and then select the radius located in the right view. Locate the dimension off the part and accept it by selecting OK (Figure 5.21).

The part is fully dimensioned. Save the completed file.

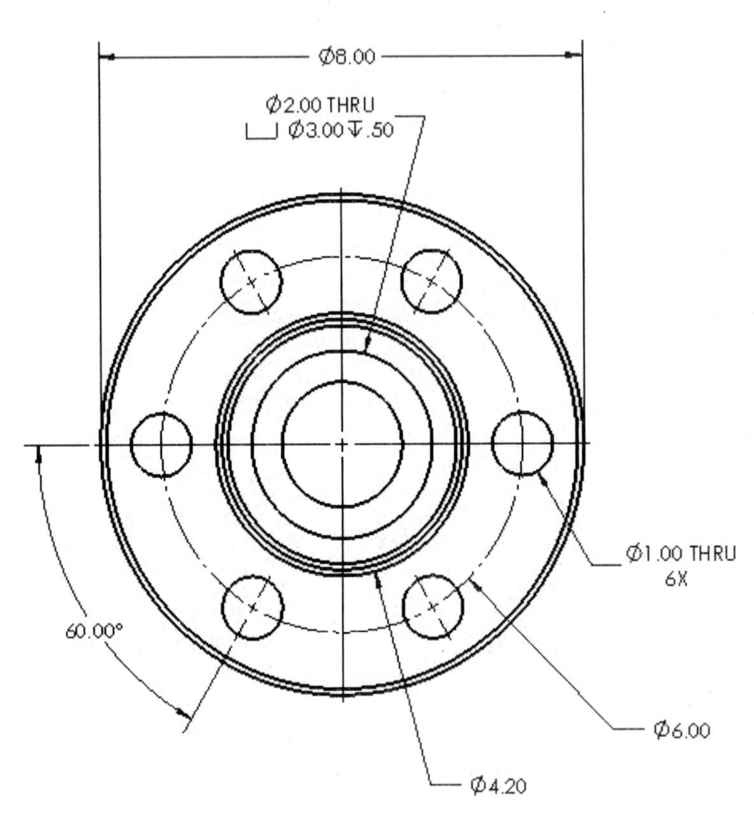

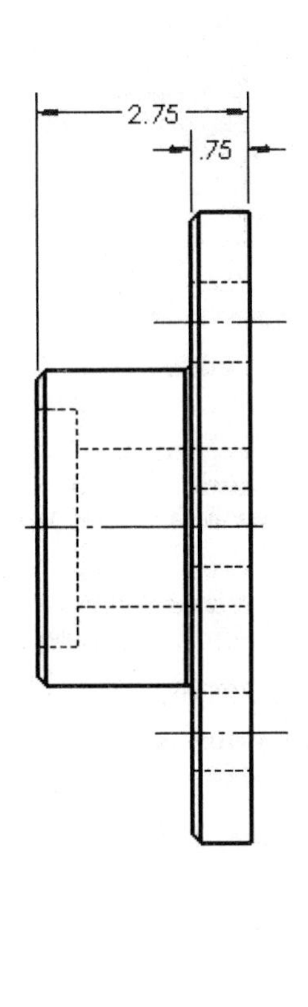

FIGURE 5.21

MOVING VIEWS

Oftentimes when creating a drawing the location of the views may not be ideal. This can be caused by a number of issues including the need to add move views, the need for additional room between views for dimensions, or the need to aesthetically balance the drawing.

Moving views within drawings is an easy and intuitive task. While not in any command, select the view that is desired to move by grabbing the dashed border of the view. Select and hold the mouse button while simultaneously moving the mouse to the new desired location.

When moving views, all of the dimensions attributed to the view will move along with it. Also all projections to all other views are maintained; the other views will move along with the view being moved.

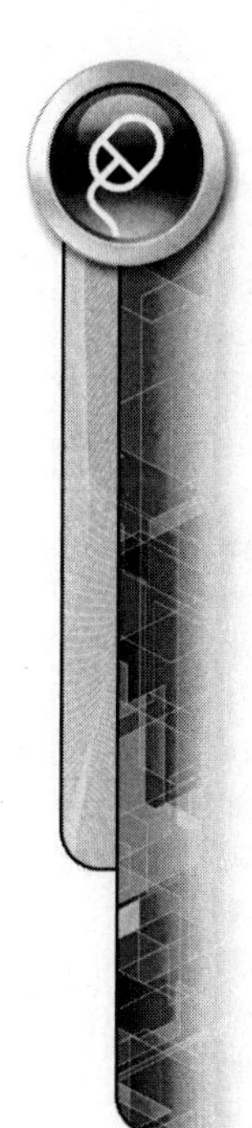

CREATING AUXILIARY VIEWS

When creating drawings, all features are required to be dimensioned in a True View, that is, a view in which the feature is shown in true shape and size and not distorted. For surfaces and features that are vertical or horizontal, they are simply in true view in one of the primary orthographic views, but when a feature is located on an inclined plane, an auxiliary view is required to dimension the size and location of that feature.

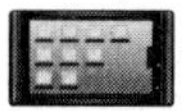

TUTORIAL EXERCISE: 05_AUXVIEW.SLDPRT

Part Number: SWT-CH05-auxview01.SLDPRT; SWT-CH05-auxview01.slddrw

Description: Auxiliary View

Units: English

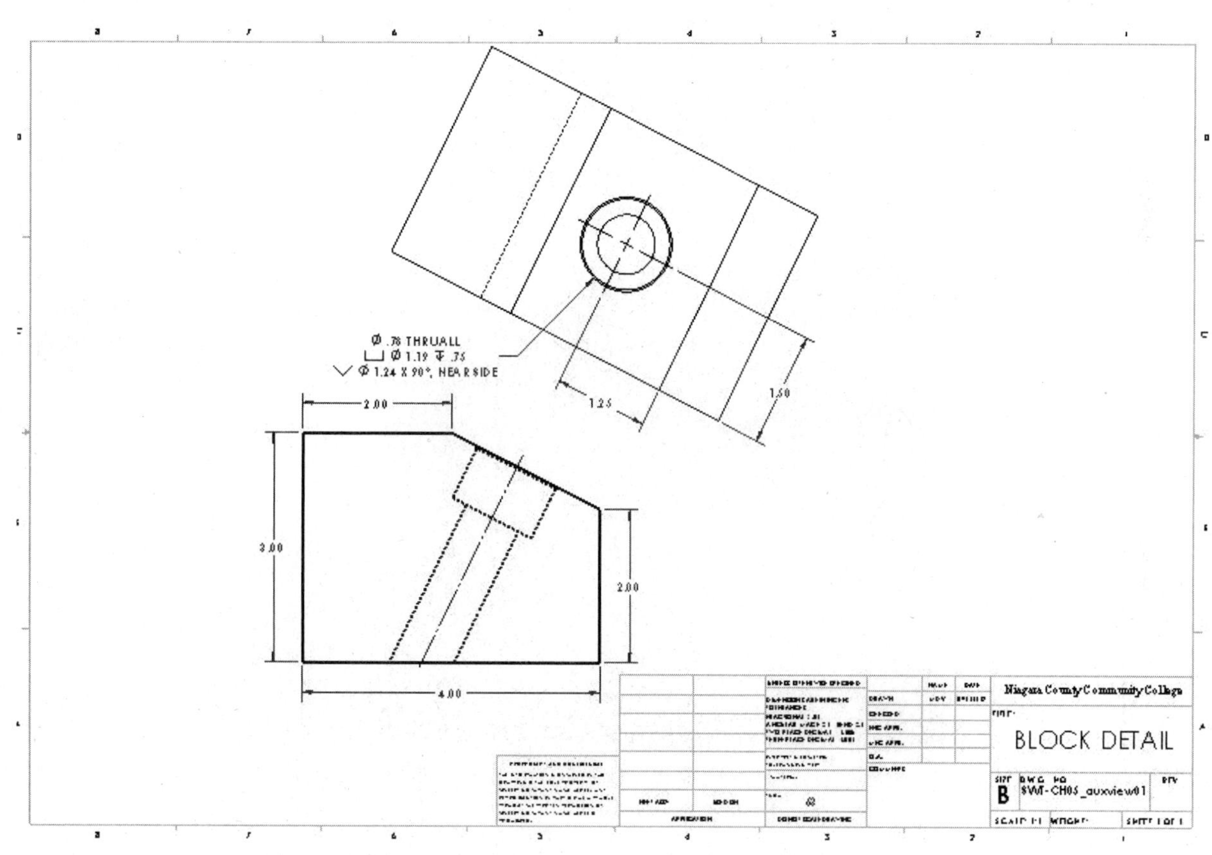

FIGURE 5.22

In this example, there is one inclined plane with a counterbore hole located on and perpendicular to that plane. The appropriate view to dimension the size and location of that hole is in an auxiliary view (Figure 5.22).

STEP 1

Create the following sketch and extrude it 3.00 inches deep (Figure 5.23).

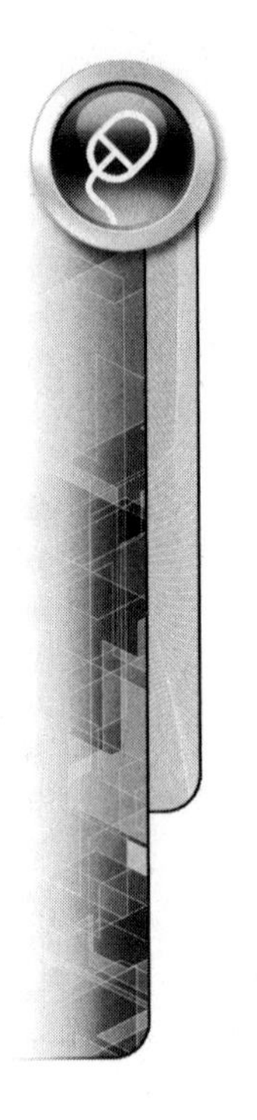

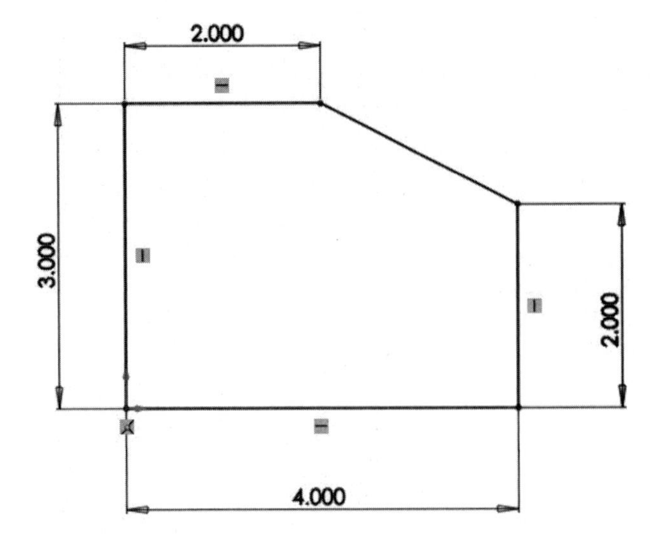

FIGURE 5.23

STEP 2

Place a counterbore hole for a ¾-inch socket head cap screw on the inclined plane surface using the hole wizard. Using Smart Dimensions, locate the hole using the dimensions shown in Figure 5.24. Accept the Hole Wizard command. Save the part file.

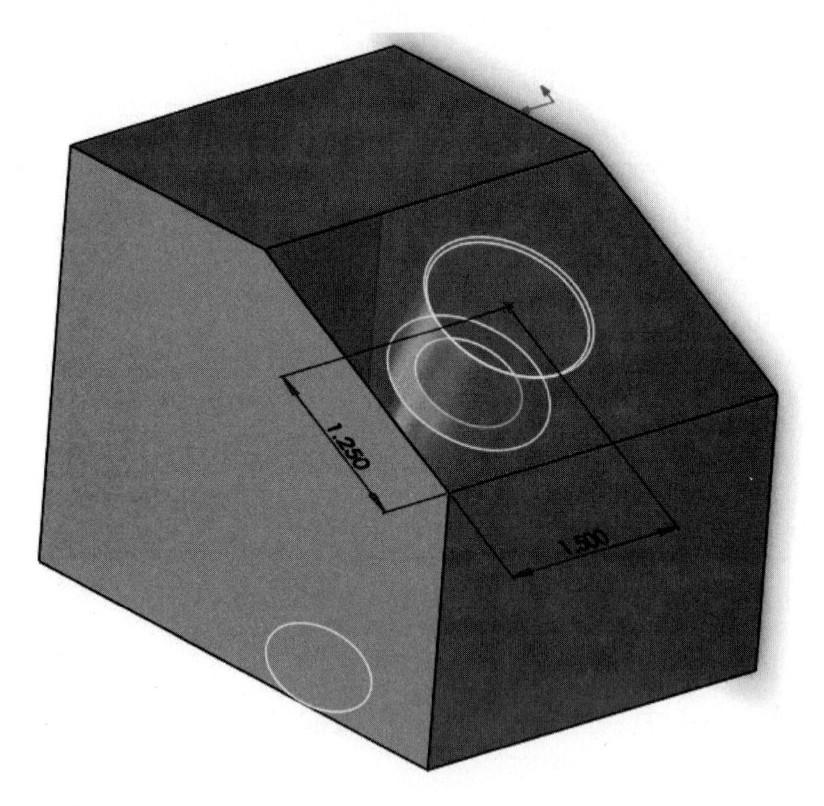

FIGURE 5.24

STEP 3

Create a new drawing file of the B(ANSI) Landscape size. Ensure that the Sheet Format property of the Type of Projection is set to Third Angle. Also make sure the drawing units are set properly to IPS in the Options menu.

Insert the top view of the part file, making sure that the Hidden Line Visible Display Style is selected. Locate the view in the lower left area of the paper sheet. Accept the view by selecting OK (Figure 5.25).

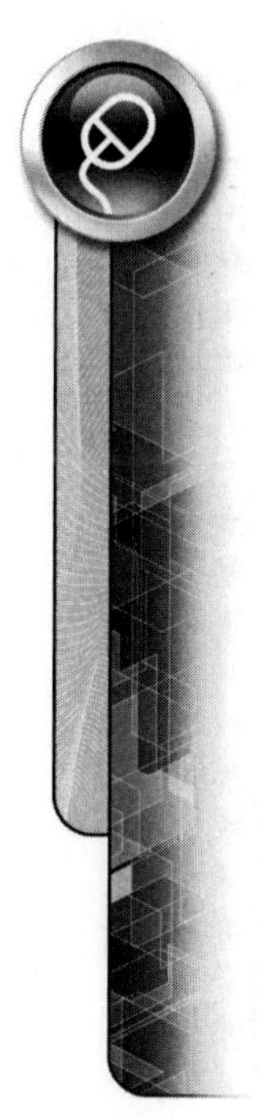

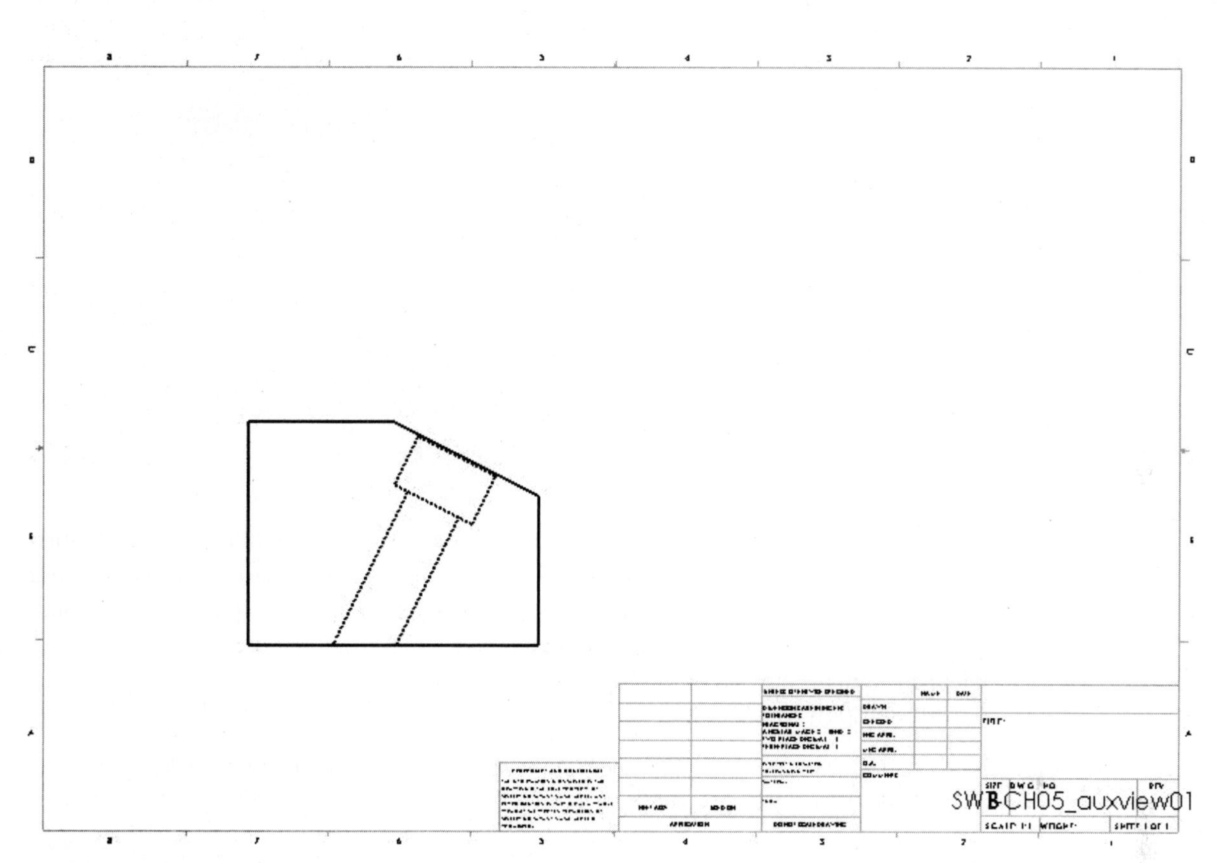

FIGURE 5.25

STEP 4

On the View Layout menu, select the Auxiliary View command. The Feature Manager notes to "Please select a reference edge to continue." The Reference Edge is the hinge edge for the auxiliary view to rotate from. The view will be rotated perpendicular to this line. Typically, an inclined line is selected.

Select the inclined line (the angled line) from the front view and place the auxiliary view off this line on the drawing (Figure 5.26).

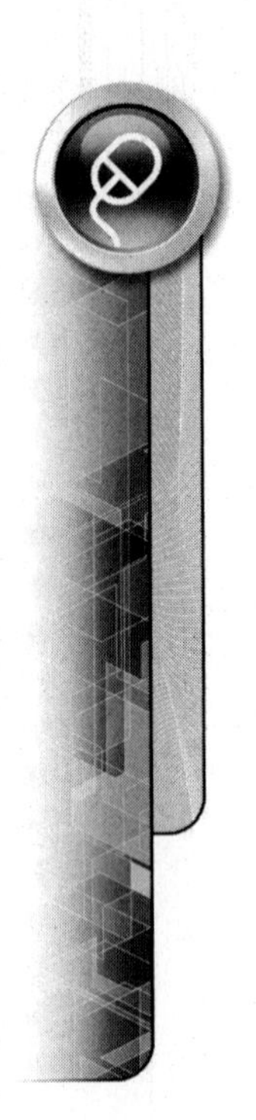

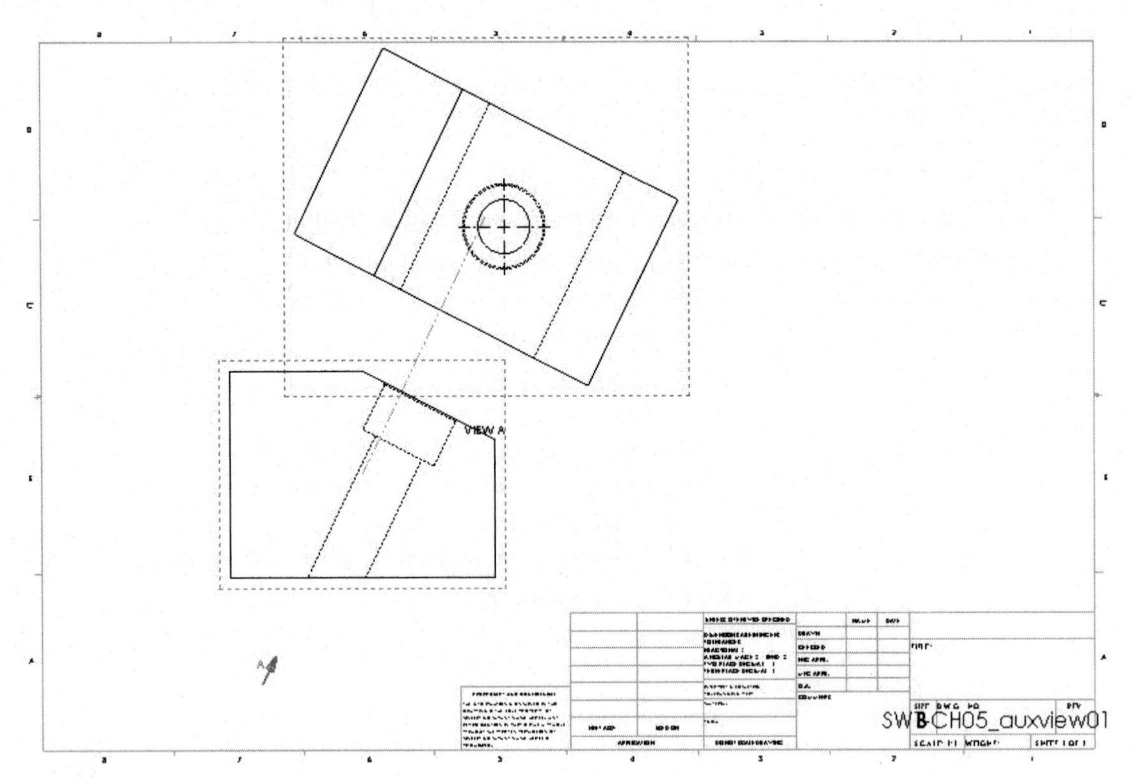

FIGURE 5.26

STEP 5

Before dimensioning, the views can be relocated to ensure there is room for all of the dimensions. Place a centerline in the front view to reflect the axis of the hole. Also, delete the centermark from the auxiliary view and place a new centermark using the Annotation Centermark command; this will place a centermark rotated that is aligned with the view itself (Figure 5.27).

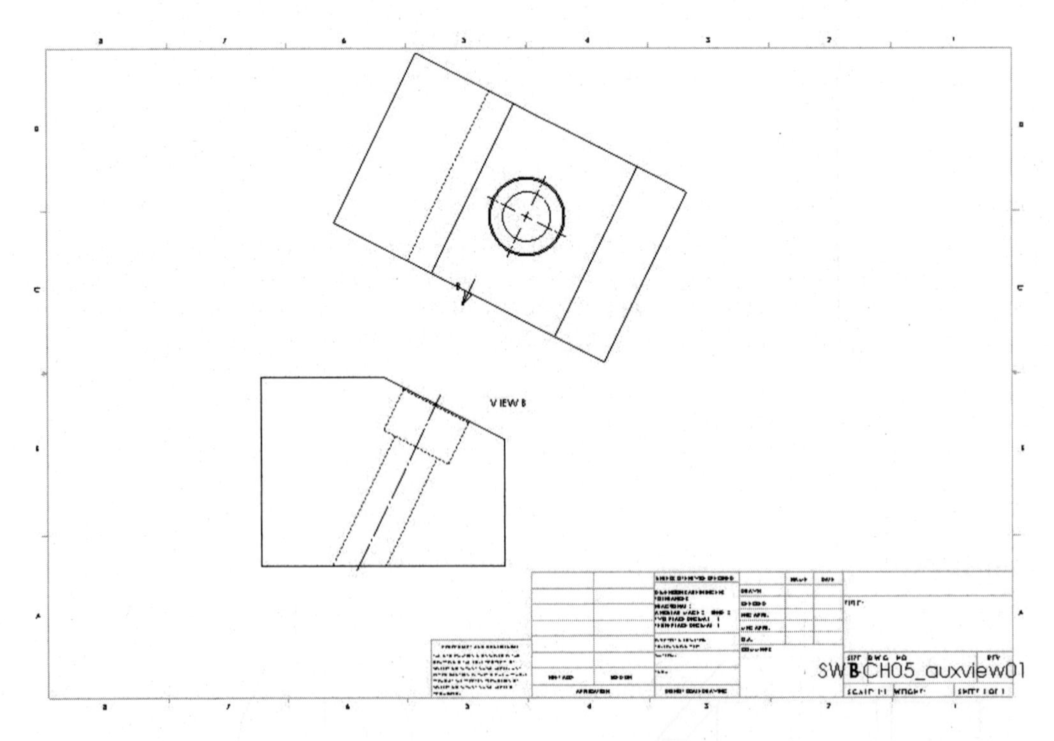

FIGURE 5.27

STEP 6

Any unwanted view labeling can be deleted. If the views are in projection labels such as "VIEW A," and view arrows can be deleted.

Next is to place the dimensions. Make sure that the Drafting Standards are set for ANSI Standards in the Document Properties Options. Using the Annotation Smart Dimension command, place both location dimensions for the hole. Try to locate the dimensions between the views when possible. Using the Hole Callout command on the Annotation toolbar, dimension the hole. Place the callout between the views if possible (Figure 5.28).

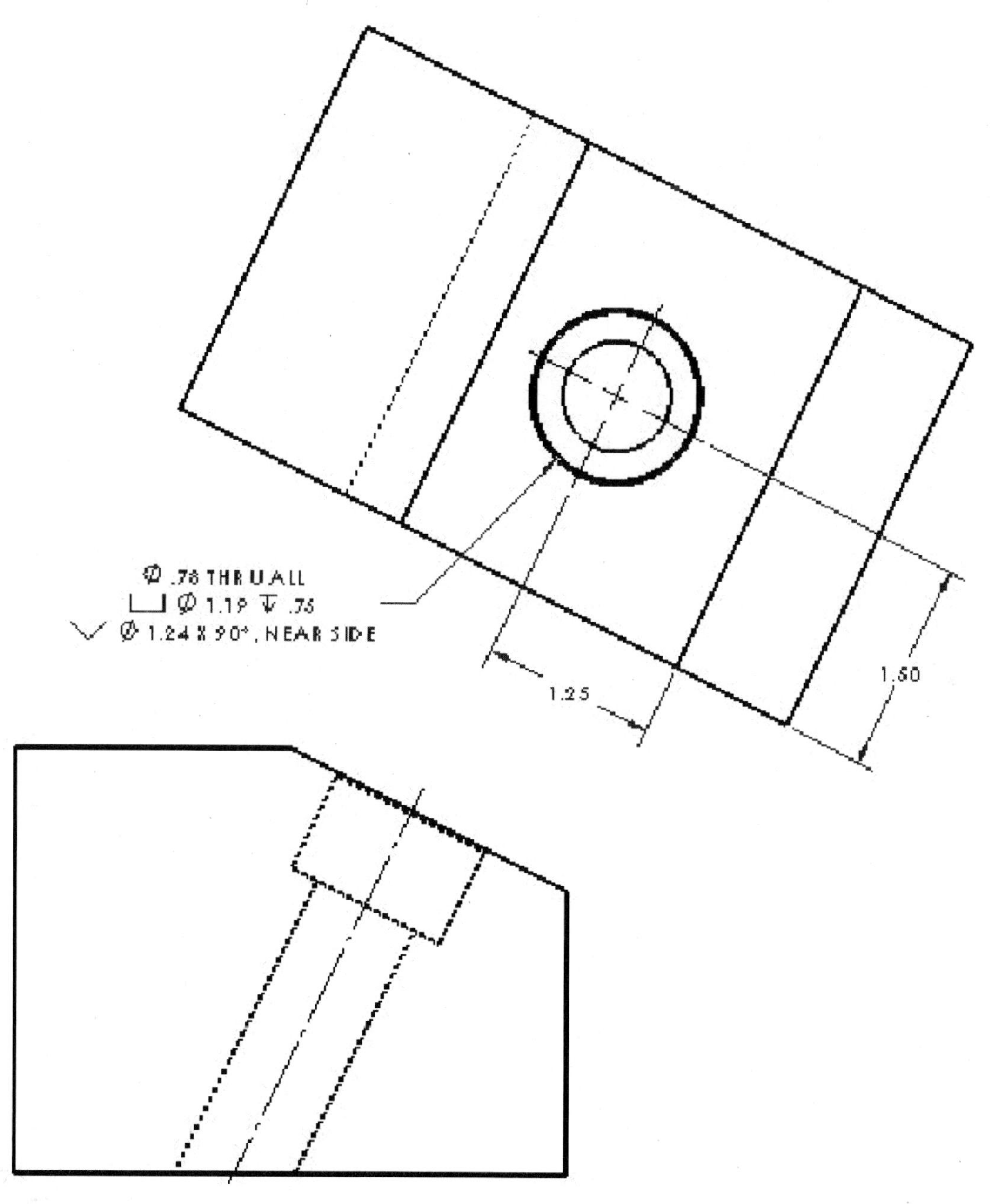

FIGURE 5.28

STEP 7

Complete the drawing by finishing all necessary dimensions as well as complete filling in the titleblock. Save both files, the part file and the drawing file. Add any additional notes as needed.

CREATING SECTION VIEWS

All features of a part on a detail drawing need to be dimensioned in order to fully manufacture the part. It is incorrect to dimension any hidden lines or hidden views; all features need to be shown in object line in at least one view. Often with parts there are internal features that cannot be dimensioned in any of the primary views or in auxiliary views; therefore, section views need to be created in order to expose the internal features and show them as object lines so they can be dimensioned.

TUTORIAL EXERCISE: 05_SECTIONVIEW.SLDPRT

Part Number: SWT-CH05-sectionview01.SLDPRT; SWT-CH05-sectionview01.slddrw

Description: Section View

Units: English

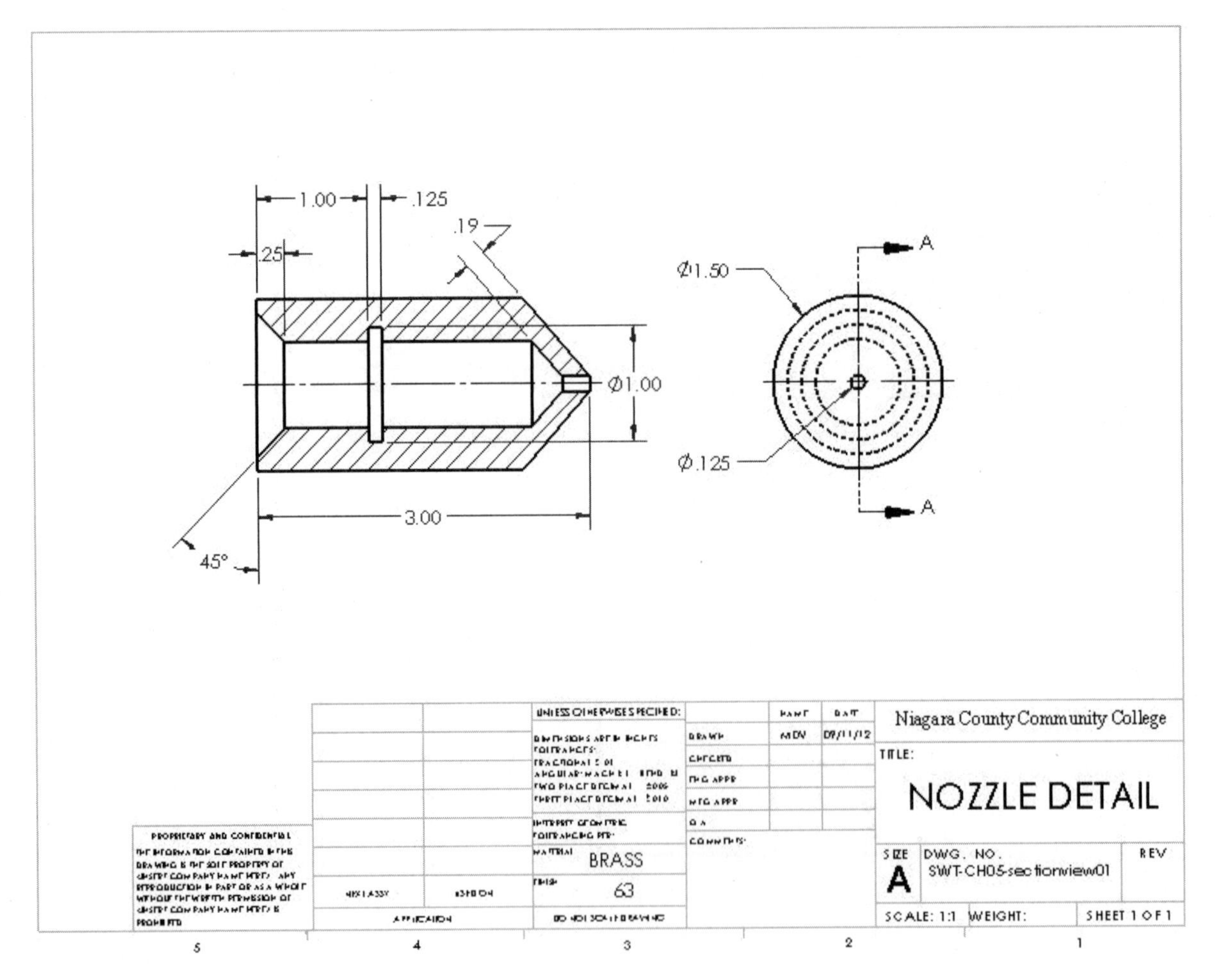

FIGURE 5.29

STEP 1

Create the following sketch on the top view including the axis that passes through the origin (Figure 5.29). Exit the sketch (Figure 5.30).

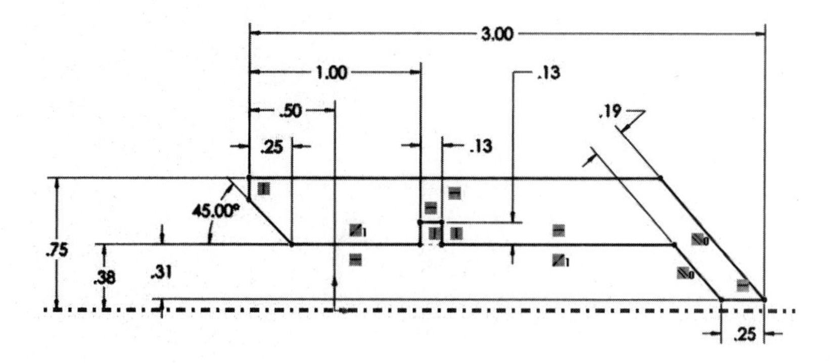

FIGURE 5.30

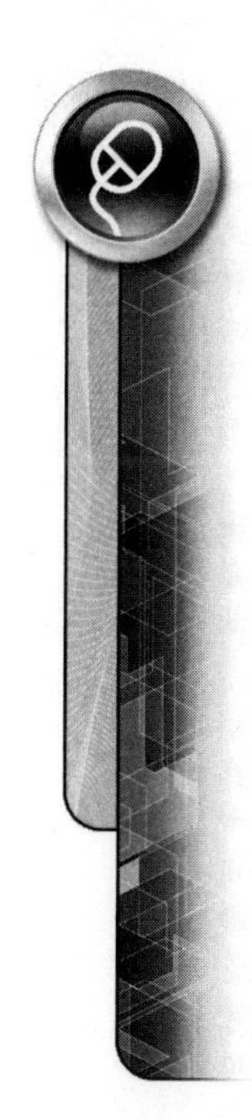

STEP 2

Revolve the sketch around the axis and save the part (Figure 5.31).

FIGURE 5.31

STEP 3

Create a new drawing file using the A(ANSI) Landscape template. Make sure the units are set to IPS, the Sheet Properties is set to Third Angle Projection, and the Dimension Standard is set to ANSI.

Using the Model View command on the Insert toolbar, insert the right side view and set it to the right side of the sheet as shown in Figure 5.32. Make sure the Display Style is set to Hidden Lines Visible.

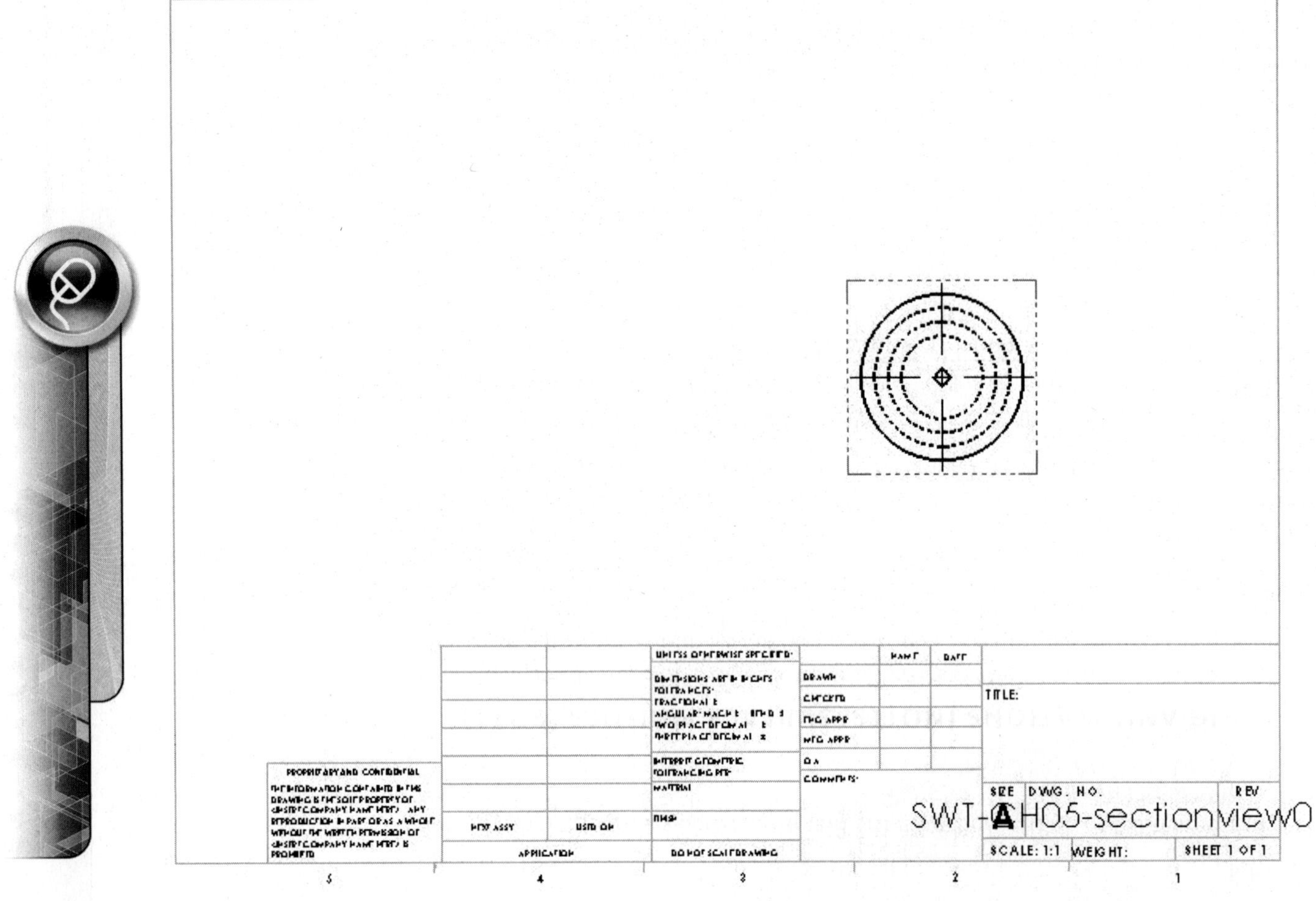

FIGURE 5.32

STEP 4

On the View Layout toolbar, select the Section View command. Sketch a line on the drawing view that represents the cutting plane line for the section view. Object snapping points can be used, so hover over the top quadrant of the outer most circle and move the cursor straight up so an alignment line (blue dashed) is visible ensuring that the cutting line will pass through the exact center of the part and select a point a distance above the point. Move the cursor below the view and select a second point to ensure the line passes all the way through the part.

Drag the newly created section view to the left and place it at a reasonable distance from the right side view. The line-of-sight arrows should be pointing to the right; if they are not, select the "Flip direction" option for the Section Line in the Feature Manager. Since hidden lines are not supposed to be shown in section views, make sure the Display Style is set to Hidden Lines Removed before the view is placed (Figure 5.33).

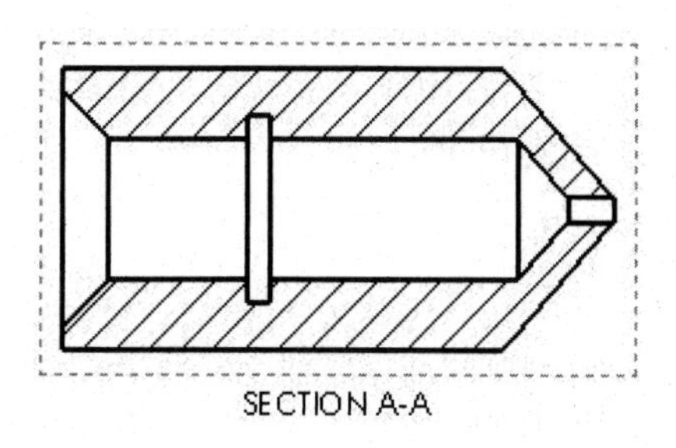

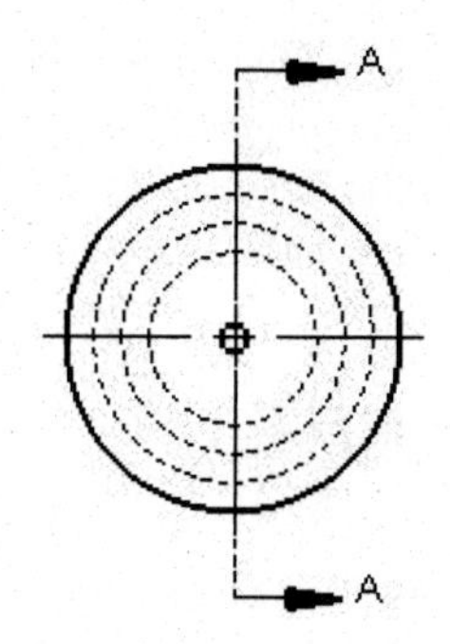

FIGURE 5.33

STEP 5

If the views are in projection, the section view labeling is unnecessary and can be removed. Also place a centerline through the part in the left view. Add the necessary centerline in the side view and fully dimension the part.

The number of decimal places can be changed as per dimension. When placing the three place dimensions, first place the dimension at its default two decimal places at the desired location (Figure 5.34).

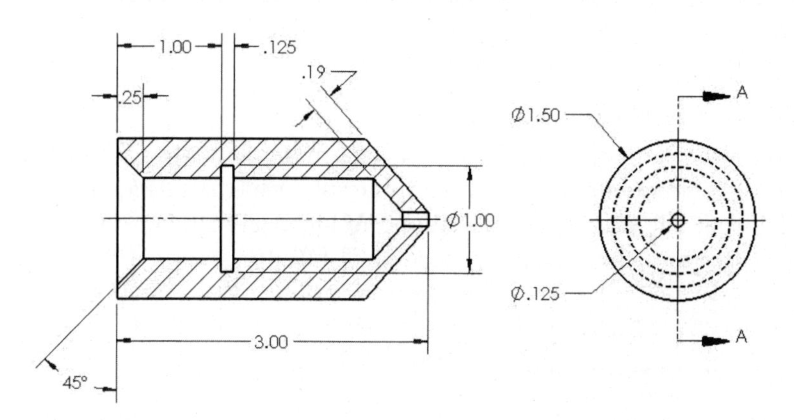

FIGURE 5.34

The Feature Manager has several settings that can be changed before moving on to the next dimension. By changing the Tolerance/Precision settings, tolerances can be added using a number of methods such as:

- None - leaves no tolerance
- Basic - creates a GD&T basic dimension
- Bilateral - creates a tolerance that is different in the positive and the negative
- Limit - creates a tolerance shown in the high/low format
- Symmetric - creates a tolerance that is the same in the positive and the negative
- MIN - creates the suffix "MIN"
- MAX - creates the suffix "MAX"
- Fit - creates a standardized fit
- Fit with a tolerance - creates a standardized fit and a tolerance
- Fit (tolerance only) - creates a tolerance from a standardized fit

The number of decimal places of a dimension can also be changed from anywhere between no decimal places to eight decimal places. The number of decimal places can also be independently set but be aware that there should be the same number of decimal places in the tolerance as they are in the dimension (Figure 5.35).

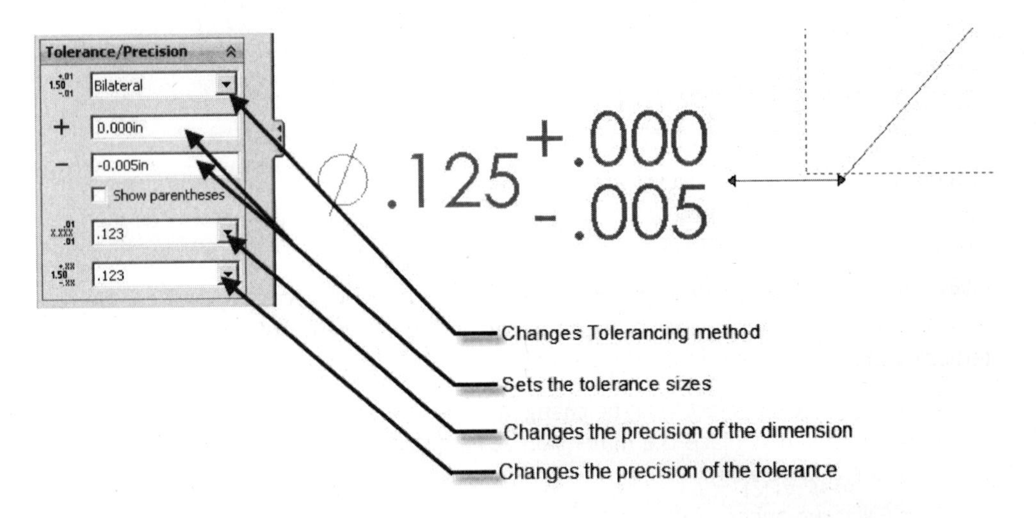

FIGURE 5.35

CREATING DETAIL VIEWS

Occasionally there are scenarios where drawings require more than one scale. The drafter should select a scale for a drawing where all of the detail that is required is shown and dimensioned; one scale may not suffice for all details though. For parts where there are large features combined with very small features, the drawing may require methods to show multiple scales of the same part. One of these methods is to create a Detail View.

A Detail View allows the drafter to accent a small section of the drawing and scale it up in order to better see the detail and dimension to it. Take for example the shaft shown in Figure 5.36. It is a lengthy shaft that has several small retaining ring grooves machined into it.

FIGURE 5.36

When creating the drawing for the shaft, first insert the side and end views in at full scale. At this scale certain details can be dimensioned, such as the overall sizes (length and diameter). Also, the location of the ring grooves can also be located. The size of the rings though is too small visually to see clearly and therefore should not be dimensioned at full scale (Figure 5.37).

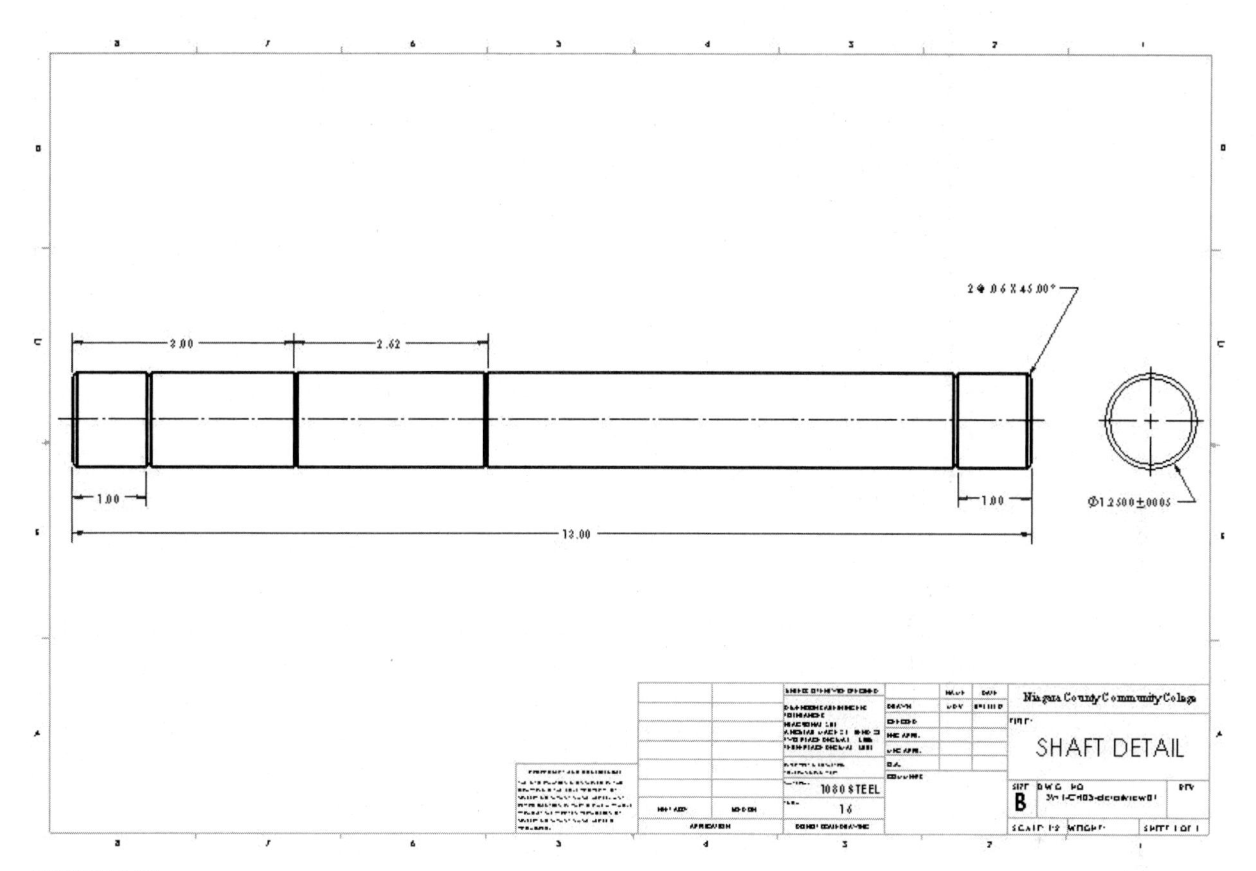

FIGURE 5.37

The Detail View command is found on the View Layout menu. When selected, the first action is to create a circle around part of the full scale drawing in which more detail is desired. The circle is placed with two points; select a center point, then select a point on its diameter. Make sure that all areas that need dimensions are included. The geometry of the part within the circle will then be copied at a different scale; the default is 2X. There are also other view controls such as "View Display" for displaying hidden lines. Another setting will create or remove a connecting line between the two circles.

When dimensioning the enlarged view, the dimensions reflect the actual size of the part. All other typical dimensional settings are available (Figure 5.38).

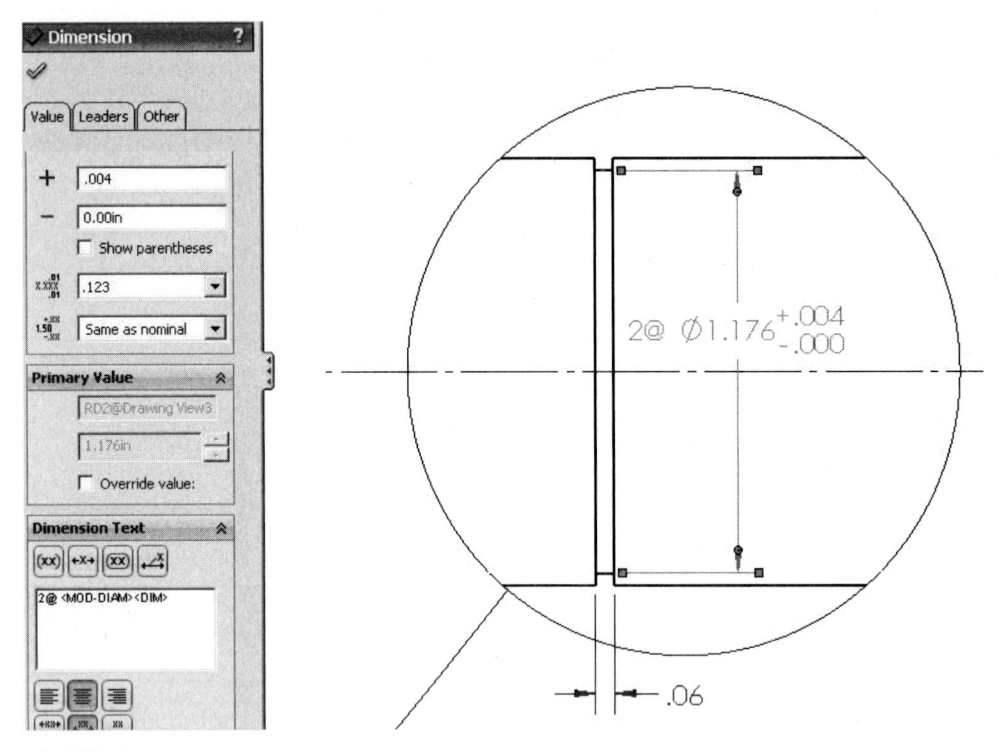

FIGURE 5.38

Views can then be rearranged to balance the drawing (Figure 5.39).

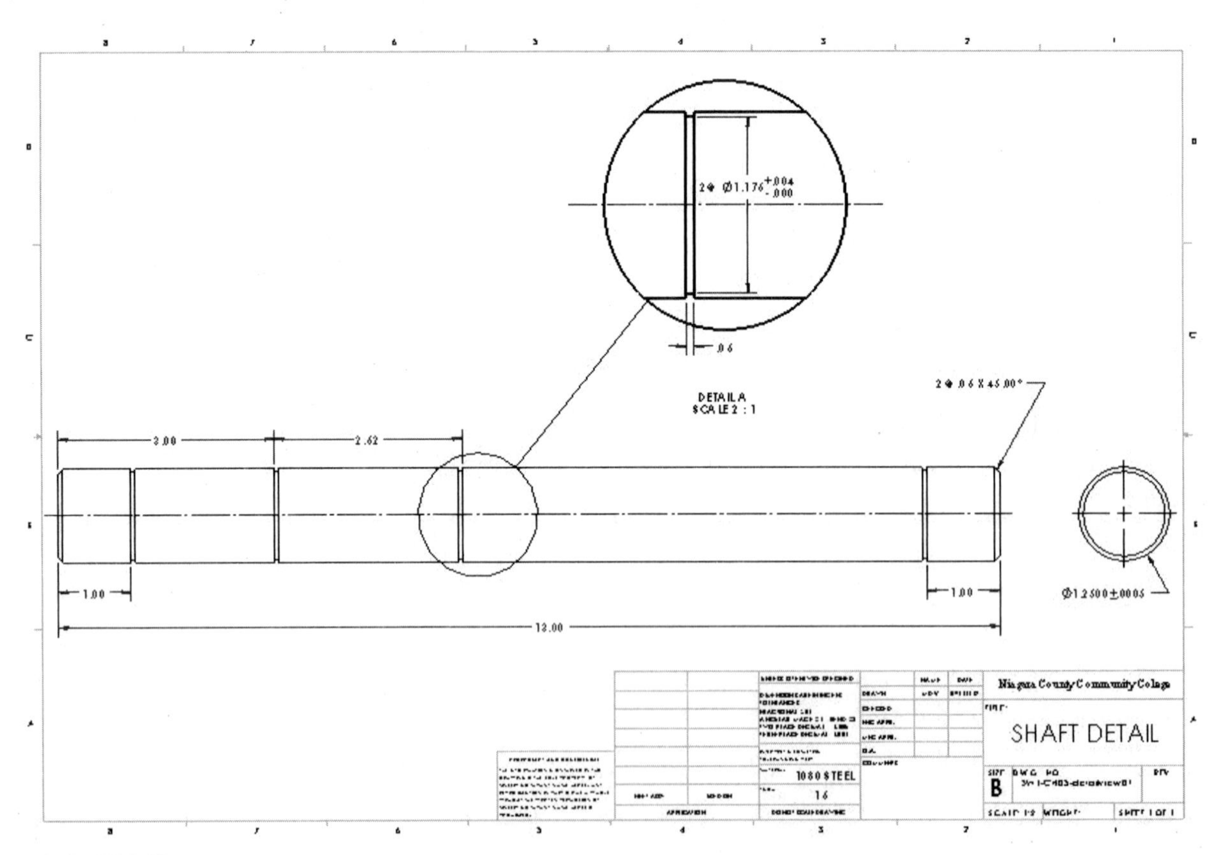

FIGURE 5.39

BROKEN-OUT SECTION VIEWS

When there is a substantial amount of internal features that need to be exposed for dimensioning, full sections and half sections are typically used. If there is only a small part of internal features that need to be exposed, a Broken-Out section is used.

The Broken-out sections command is found on the View Layout toolbar. When selected, it prompts to create a closed spline around the area desired to be exposed. Then the depth of the section is requested. Once the view is accepted, the internal features are ready for dimensioning (Figure 5.40).

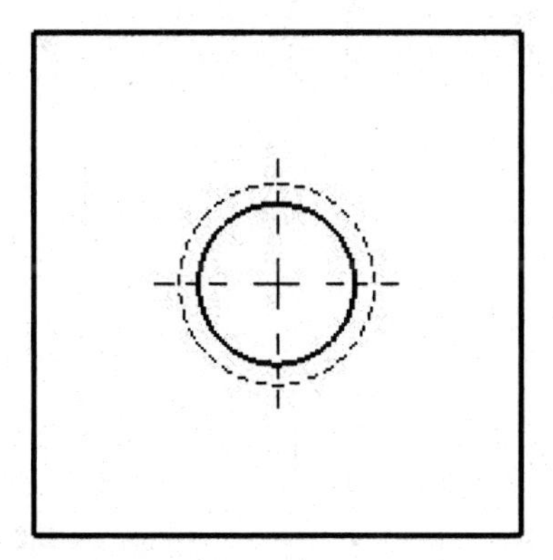

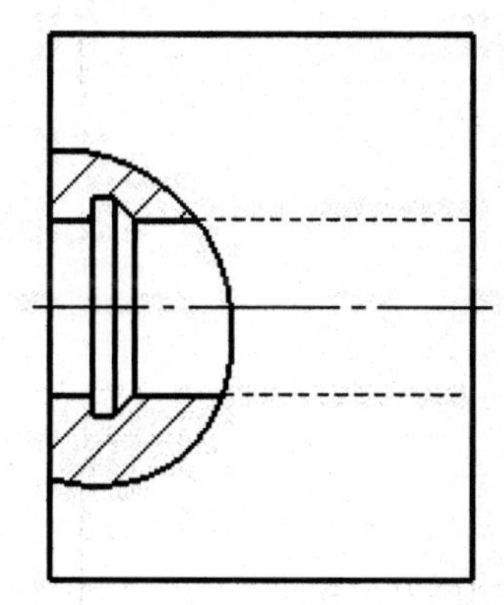

FIGURE 5.40

ROTATING VIEWS

Often when one of the default views is brought in as the first view on a drawing, the orientation is not what is desired. This can be changed before any other views are placed. First right-click and activate the pull-out menu for Zoom/Pan/Rotate, then select Rotate View. The ability to rotate the view to 90° or to any other angle is available.

It is recommended to do this only with the first primary view although it can be done at any time with any view.

PLACING ANNOTATIONS (NOTES, SURFACE FINISHES, GD&T)

In addition to dimensioning using standard dimensions, additional details, such as additional notes, surface finishes, geometric tolerances, and datum references, may need to be added to the drawing.

Placing Notes

Notes can be placed anywhere within the border of a drawing. Most notes are contained within the titleblock and additional notes can be added to the drawing either with or without a leader.

The Note command is found on the Annotation toolbar. When selected, the Feature Manager displays all of the possible settings (Figure 5.41).

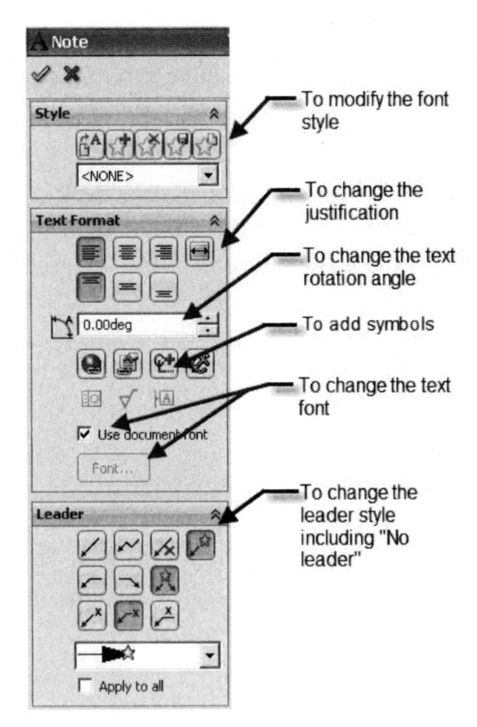

FIGURE 5.41

Before typing the desired text, select the location of the text on the drawing. A text field will appear to allow typing the text afterward. If the leader style is set to the default Auto Leader, a leader will appear if the cursor is near geometry and will disappear if the cursor is in an open area of the drawing.

Placing Surface Finishes

The Surface Finish command can be found on the Annotations toolbar. The surface finish callout controls the surface texture of any surface. The surface finish can be a general note controlling all of the part surfaces with one callout or the surface finish symbol can be placed individually on specific surfaces (Figure 5.42).

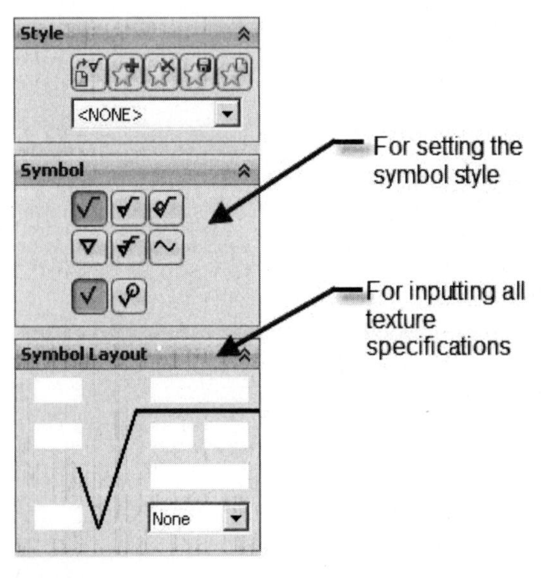

FIGURE 5.42

A typical surface finish callout will have a basic symbol and one finish control size but the command also allows for additional information to be added to the symbol as well as has the ability to change the symbol for various conditions.

As shown in Figure 5.42, the surface finish mark can be put on a leader, placed on an extension line, or placed on the part itself. It is best to locate the surface finish mark on the part itself. Only if the symbol cannot be placed on the object line, then it can be put on an extension line. Placing the symbol on a leader should be used only as a last resort.

In order to meet standards, when the finish mark symbol is placed on a line, it should lie in a manner that the symbol is located outside of the part body. SolidWorks does not always place it in this manner automatically. If the rotation is incorrect, it should be rotated using the angle setting in the Feature Manager. Notice that as the symbol is rotated, the finish size rotates appropriately (Figure 5.43).

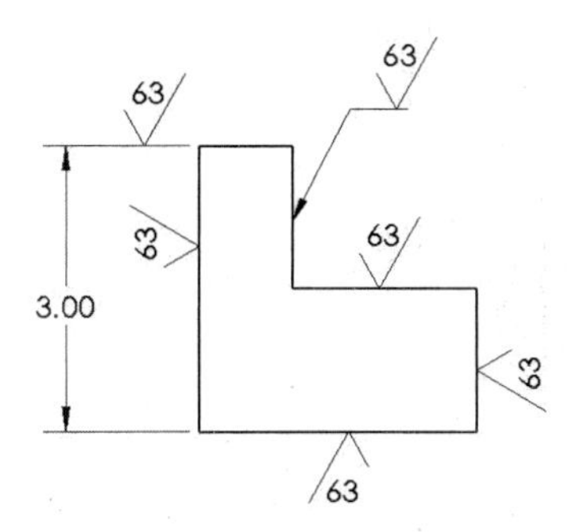

FIGURE 5.43

Placing Geometric Tolerances

Geometric dimensioning tolerances can be placed on drawings by placing feature control frames (FCF). FCFs can be placed by using the Geometric Tolerance command found on the Annotation toolbar.

When the command is selected, the Properties menu appears as shown in Figure 5.44. All fourteen tolerance symbols are available. Material conditions can be added to the tolerances at any time by selecting the icons for them. Tolerance sizes as well as datum references are to be keyed in by using the keyboard.

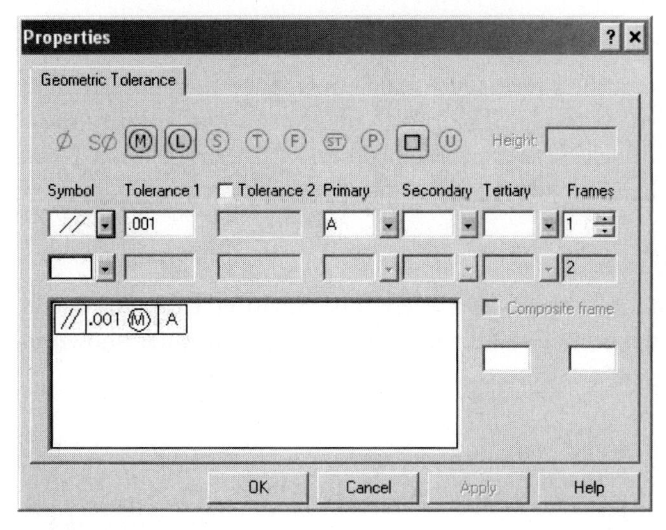

FIGURE 5.44

Once the FCF is created, it can be placed on the drawing with or without a leader. When Auto leader is selected, a leader will appear if the cursor is near geometry. It is recommended to place FCFs onto extension lines primarily. Use leaders only if no other option is available (Figure 5.45).

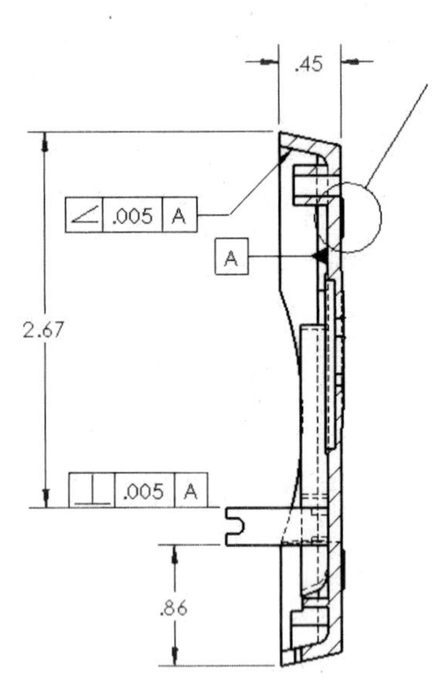

FIGURE 5.45

Placing Datums

The Datum command is found on the Annotation toolbar. Any feature can be identified as a datum as the drafter sees fit. The Datum command will automatically start with datum "A" first then sequentially go through the alphabet. The datum callout can be changed though in the Feature Manager by changing the label setting.

It is recommended to place the datum symbol directly on the object lines where possible. If it is not possible to place it on the object, then placing the symbol on an extension line representing the feature is acceptable. Placing datum symbols on leaders should be used only as a last resort.

PLACING A REVISION BLOCK

When drawings are to be modified formally in an industrial setting, the changes should be documented on the drawing. The most common way to identify these changes is with a Revision Block. It is a two-step process: first the revision block header needs to be placed, then individual revision blocks can be added.

Inserting a Revision Header

The revision header is a table and is easily inserted. The Revision Table command is found in the Insert pull-down menu and the Tables pull-out menu as shown in Figure 5.46.

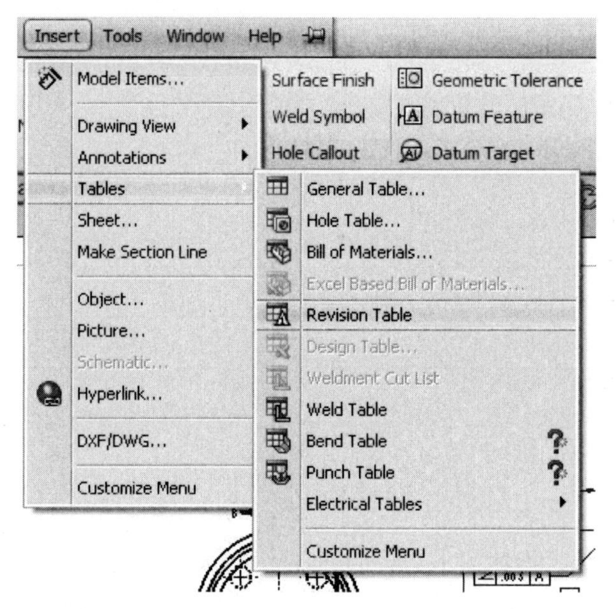

FIGURE 5.46

When the Revision Table command is selected, the Command Toolbar has several options that can be set. First, the Table Template should be kept as the default "standard revision block" unless the company or institution for which the drawing is being created has made custom tables (Figure 5.47).

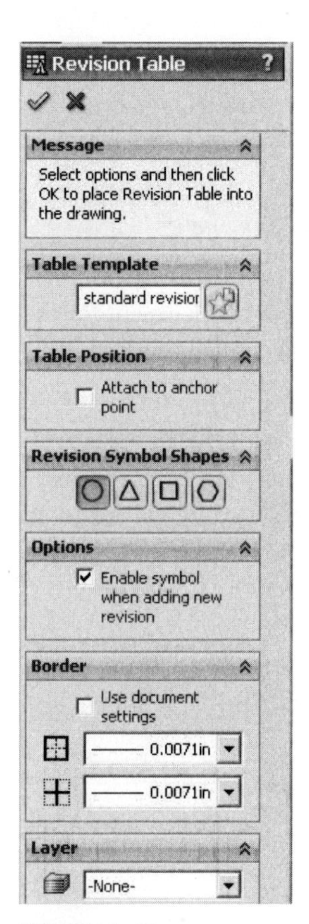

FIGURE 5.47

Next, the Table Position "Attach to anchor point" should be left unchecked; the block will be placed in the default position which is the upper right corner of the border. Once all desired settings are set, accept the header by selecting OK.

The revision header can be used for all paper sizes (Figure 5.48).

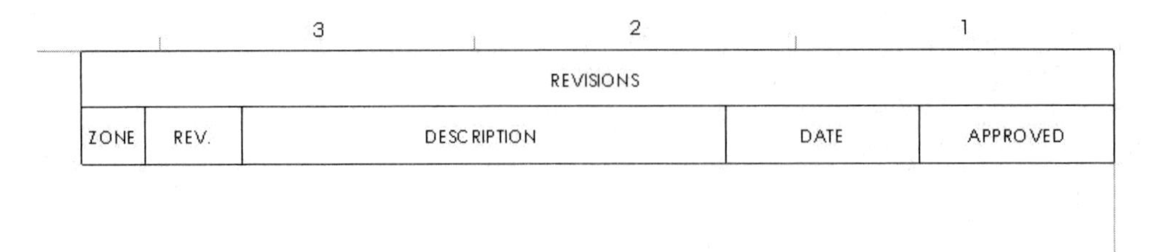

FIGURE 5.48

Inserting a Revision Block

Whenever a revision is to be added to a drawing, right-click anywhere on the revision header, go to the pull-out command "Revisions" and select "Add Revision" (Figure 5.49).

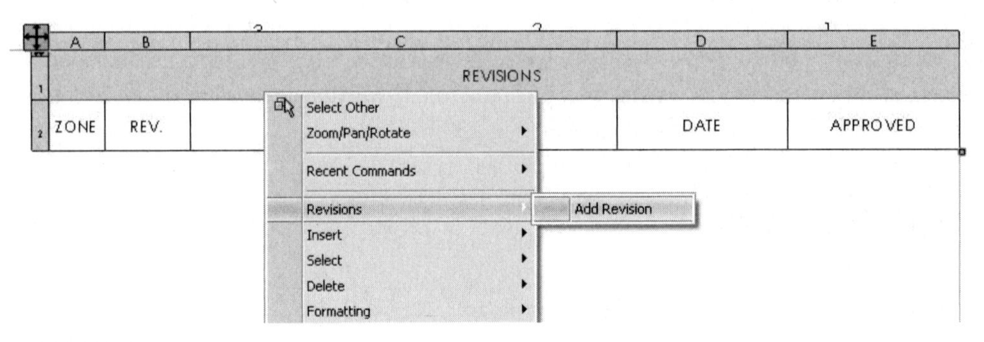

FIGURE 5.49

When the Add Revision is selected, two things appear; first, a new blank row of revision information appears below the revision header. All fields are blank except for the revision letter and the date field which has the current date. Second, a revision indicator is attached to the cursor.

The next step is to place the revision indicator on the drawing and close to the area that is being revised. The revision indicator will automatically have the revision letter (or number) indicated inside of it. If there are multiple changes being recorded for this revision, the revision indicator can be placed in as many locations as desired (Figure 5.50).

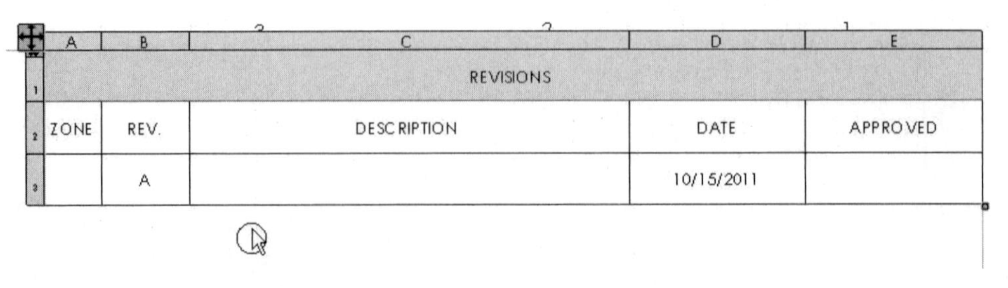

FIGURE 5.50

In order to add information to the revision block, select the field desired (e.g., Description). Multiple lines can be added using spaces as line separations. Jumping from field to field can be accomplished using the tab button (Figure 5.51).

REVISIONS				
ZONE	REV.	DESCRIPTION	DATE	APPROVED
C5 C5 B6 D3	A	DATUM A WAS ADDED .005 PERPENDICULAR TOLERANCE WAS ADDED .005 ANGULAR TOLERANCE WAS ADDED 63 MICROINCH SURFACE TEXTURE WAS ADDED	10/15/2011	MDV

FIGURE 5.51

CREATING SCALED DRAWINGS

Very often the desired views for a particular drawing do not fit on the selected paper size. The paper size could always be changed to a larger size, but even then there is a limitation in sizes. So, in order to place larger views on a drawing, the views may need to be scaled down. These are called reduced scales.

In the case of very small parts, in order to see all of the detail and dimension to it, views may need to be scaled up. These are called enlarged scales.

In both cases there is a limited amount of specified scales that can be used. The most common reduced mechanical scales are Full Scale (1 = 1), Half Scale (.5 = 1), Quarter Scale (.25 = 1), Tenth Scale (.1 = 1), and Twentieth Scale (.05 = 1). The most common enlarged scales are Double Scale (2 = 1), Four Times Scale (4 = 1), Ten Times Scale (10 = 1), and Twenty Times Scale (20 = 1). Other scales outside of this list can also be used.

When the primary views of a part are inserted onto a drawing, it quickly becomes apparent whether the views will fit on the specified sheet size. When placing the first view, SolidWorks automatically selects a scale based on the size of the view being inserted; this scale is called the Sheet Scale. This scale may be acceptable or it may not. The true scale to use is only known by the drafter for only the drafter knows how many views are appropriate to fully represent the part or assembly.

To calculate the appropriate scale, the drafter would typically roughly sketch out the views needed and compare the total height of the views as well as the width of the views with the height and width of available drawing space on the sheet.

Once the desired scale is calculated, the scale for the first view inserted on the drawing sheet can be selected. The desired scale must be known prior to inserting the first model view. In the Feature Manager, the software selected scale is indicated by the grayed out scale represented when "Use sheet scale" is selected. If this is the desired scale, then place the view on the drawing. If a different scale is desired, select the "Use custom scale" button in the Feature Manager and change the desired scale prior to placing the first view (Figure 5.52).

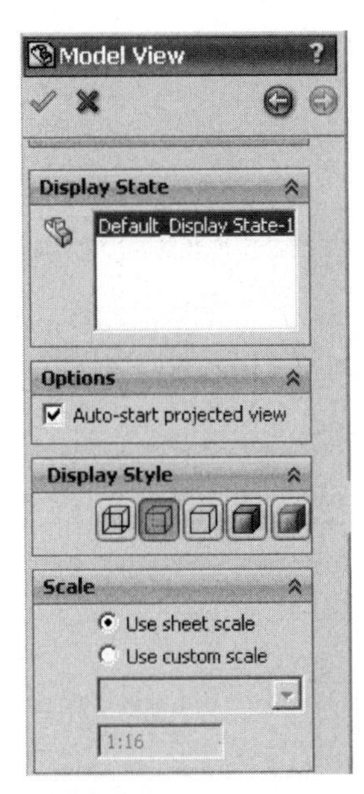

FIGURE 5.52

Once the first view is placed, the scale used is reflected in the Scale field in the title-block. The other views can now be placed and then the part can be dimensioned.

Since the paper size has not changed, and the views were scaled down (or up) to fit on the paper sheet, the dimension scale of dimension text heights and arrow sizes does not need to be changed. Also, the actual dimension values will now reflect the original part sizes. No other settings need to be modified (Figure 5.53).

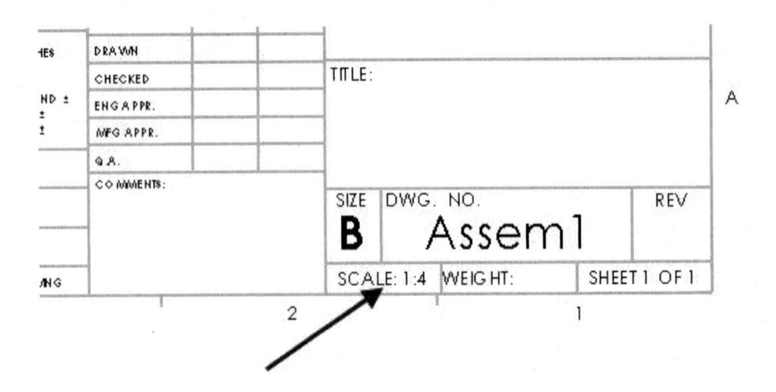

FIGURE 5.53

PLOTTING A DRAWING

Once a drawing is complete, it may be plotted to create paper copies of it to share with other members involved in either the design or manufacturing of the part. Drawings should meet the standards required for drawings, ANSI and/or ISO.

The drawings should be plotted to the scale identified, all lines and text should be black, and all standards with respect to line weights should be reflected.

To start plotting, both the File pull-down menu and Print can be selected, or the Print icon in the SolidWorks toolbar can be selected (Figure 5.54).

FIGURE 5.54

Once the Print command is selected, the Print menu appears. The first thing to select is the desired printer to create the print because the print driver may change many of the settings imbedded in the print menu (Figure 5.55).

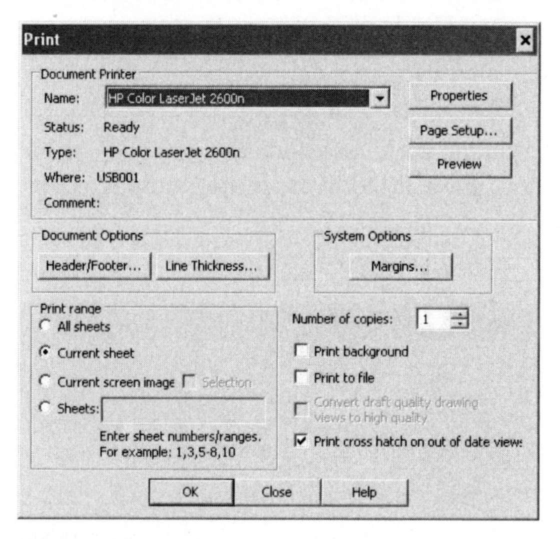

FIGURE 5.55

Selecting Paper Size

Once the printer is selected, then the paper size should be chosen. In order to select the paper size, the Properties button needs to be selected on the Print menu. The default tab should be the first tab called Paper/Quality. Half way down the menu is the "Size is:" field where all sizes of paper are listed for the selected printer. The desired paper size should be selected here then accepted by clicking the OK button (Figure 5.56).

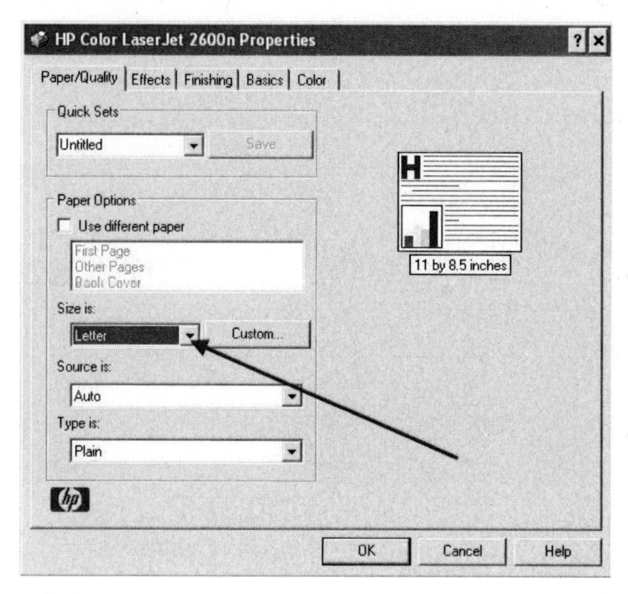

FIGURE 5.56

Defining the Line Weights

Once the paper size is chosen, the line thicknesses should be checked. In order to view and/or modify the line thicknesses on the print, the Line Thickness button needs to be selected on the Print menu. The line thicknesses can be set in Document Properties prior to plotting or can be changed at the time of plotting.

The software defines the size of each line on a drawing while it is being created. It will put each entity (line, circle, arc, text, etc.) into one of the default size categories of Thin, Normal, Thick, Thick(2), etc. as the list of sizes can be seen in Figure 5.57. Object lines are assigned a normal thickness. Hidden lines, center lines, text, dimensions, and hatch are all assigned a thin thickness. The other categories are custom thicknesses and individual lines can be assigned different thicknesses if desired.

In order for the object lines to visually meet the thickness standards it should be changed to 0.02in thick. Once the line settings are changed to the desired settings, accept them by clicking on OK.

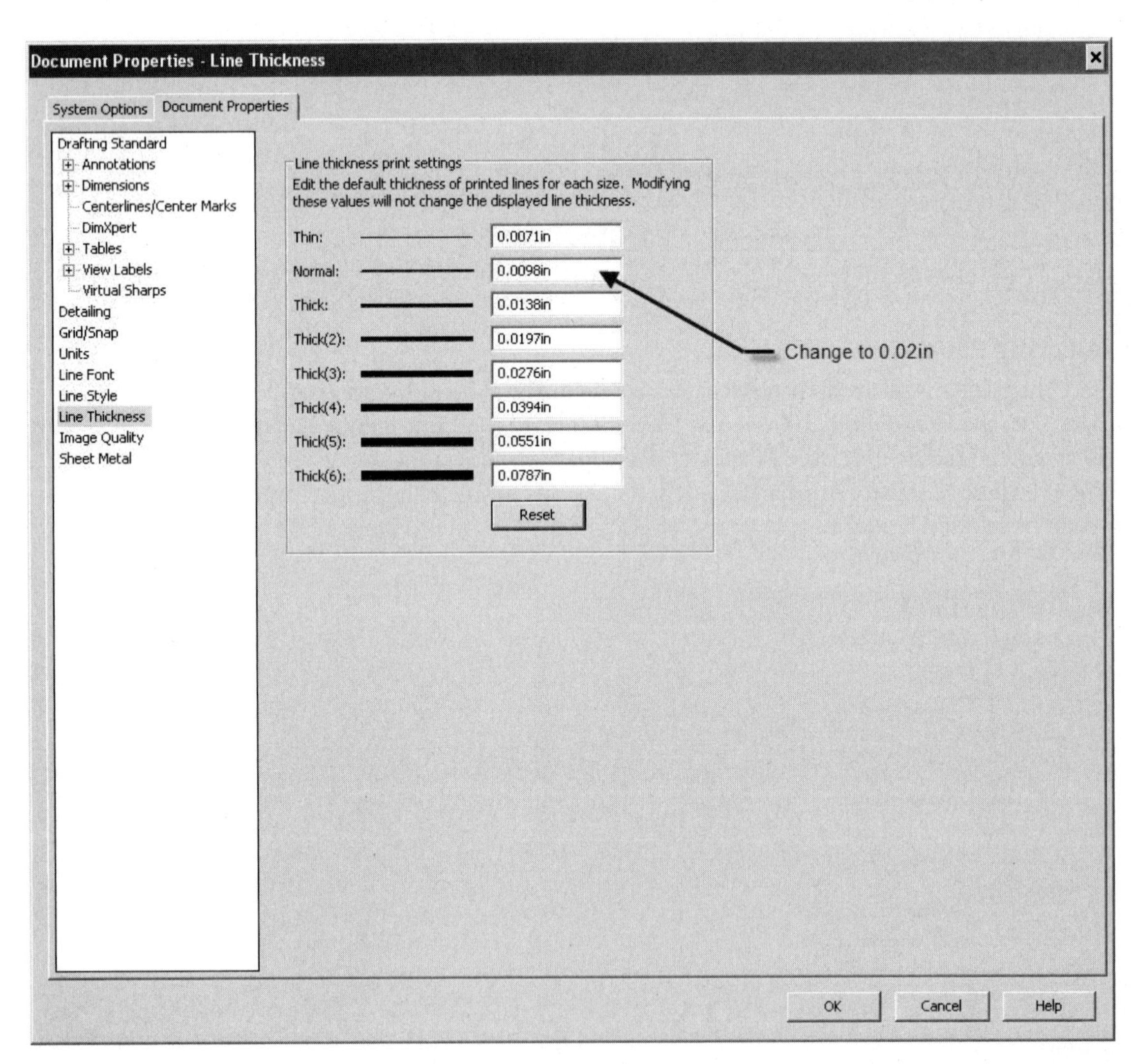

FIGURE 5.57

Finalizing the Plot

Once the paper size and the line thicknesses are set, the plot is ready. Always complete a plot preview to ensure that scale, line thicknesses, and everything else look acceptable. Once everything is set as desired, click on OK to send the plot to the printer.

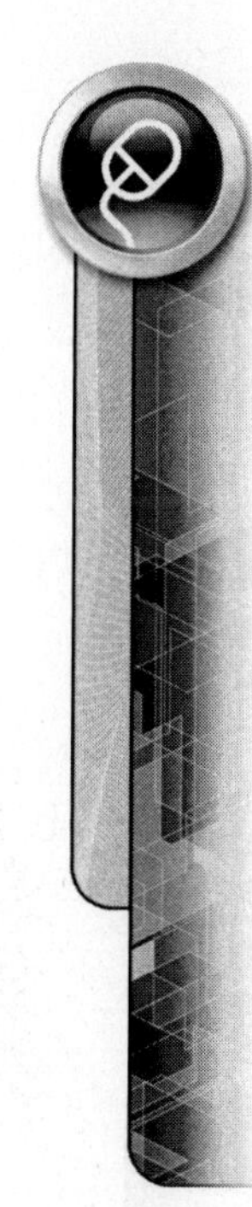

REVIEW QUESTIONS

1. At what step was drafting in the historic design process?
2. At what step is drafting in the modern design process?
3. Explain the change in the role of drafting in the design process.
4. If features are changed in a part file, what effect does it have on any drawings that reference that part?
5. What type of projection is required on all ANSI drawings?
6. When moving a view on a drawing to a more desired location, do the dimensions placed on that view move along with the view?
7. When moving a view on a drawing to a more desired location, do the adjacent views move with it to remain in projection?
8. What is the purpose of creating an auxiliary view?
9. What is the purpose of creating a section view?
10. What are the preferred line thicknesses for each type of line?

EXERCISES

5.1

Create a drawing for English units assignment 2.1.C. Select a drawing size as assigned by the instructor. Choose the appropriate scale with the assistance of SolidWorks' automatic scaling. Make sure to add all necessary centerlines and centermarks. Completely fill in the titleblock and fully dimension the part.

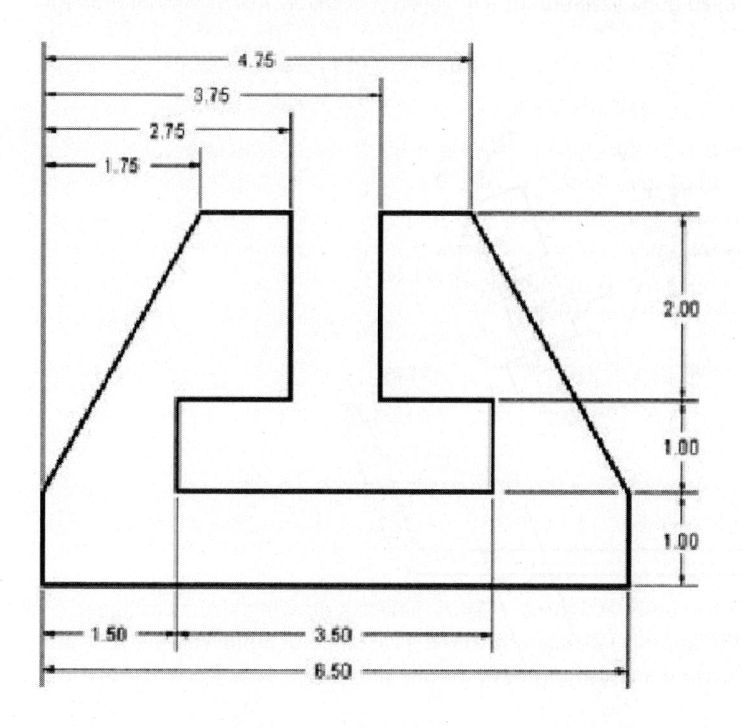

5.2

Create a drawing for the Metric units assignment 2.3.C. Select a drawing size as assigned by the instructor. Choose the appropriate scale with the assistance of SolidWorks' automatic scaling. Make sure to add all necessary centerlines and centermarks. Completely fill in the titleblock and fully dimension the part.

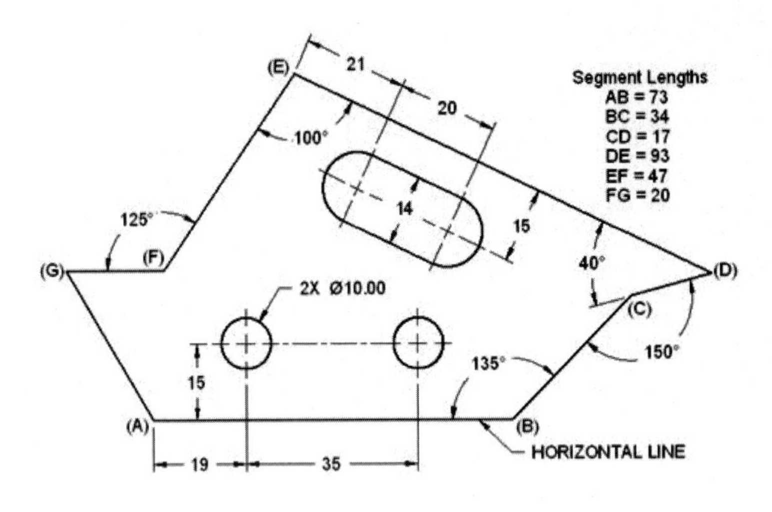

5.3

Create a drawing for English units assignment 2.4.D. Select a drawing size as assigned by the instructor. Choose the appropriate scale with the assistance of SolidWorks' automatic scaling. Make sure to add all necessary centerlines and centermarks. Completely fill in the titleblock and fully dimension the part.

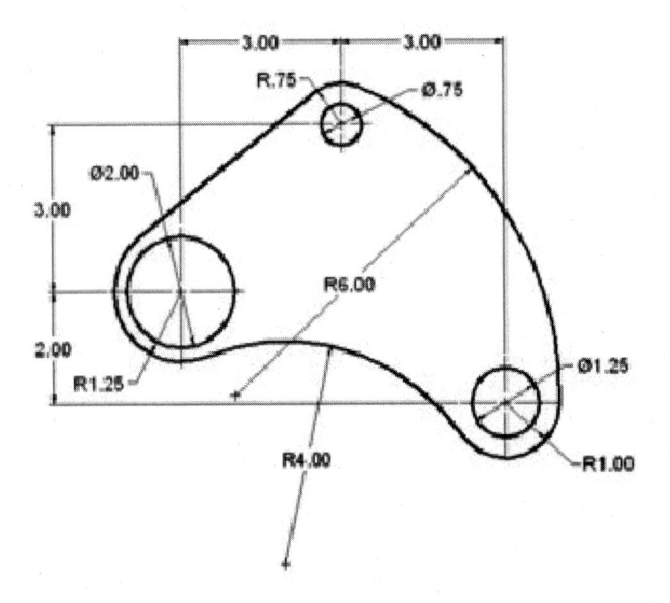

5.4

Create a drawing for Metric units assignment 3.4.A. Select a drawing size as assigned by the instructor. Choose the appropriate scale with the assistance of SolidWorks' automatic scaling. Select the appropriate amount of views to fully define the part. Make sure to add all necessary centerlines and centermarks. Completely fill in the titleblock and fully dimension the part.

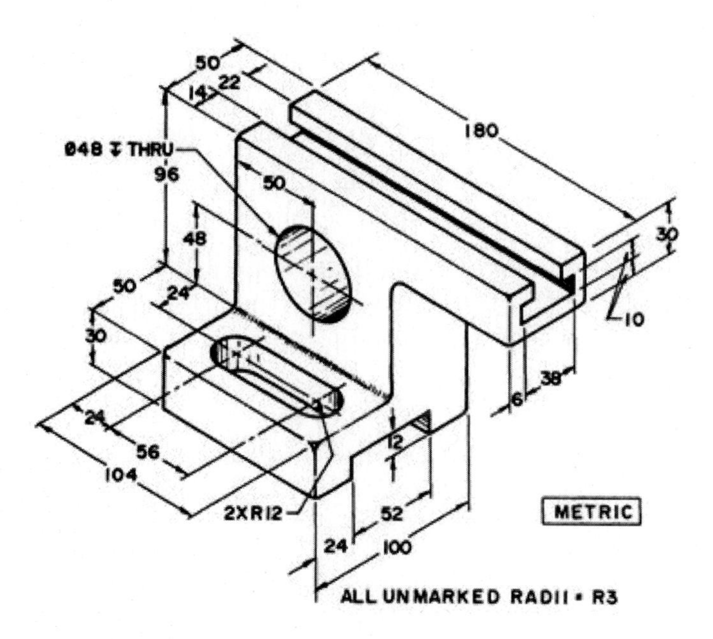

5.5

Create a drawing for English units assignment 4.1. Select a drawing size as assigned by the instructor. Choose the appropriate scale with the assistance of SolidWorks' automatic scaling. Create a Detail view of the pry end as shown. Make sure to add all necessary centerlines and centermarks. Completely fill in the titleblock and fully dimension the part. Make the main shaft (4.00 inch section) Datum A. Using GD&T, dimension the two angled surfaces to be angled within .005" from datum A.

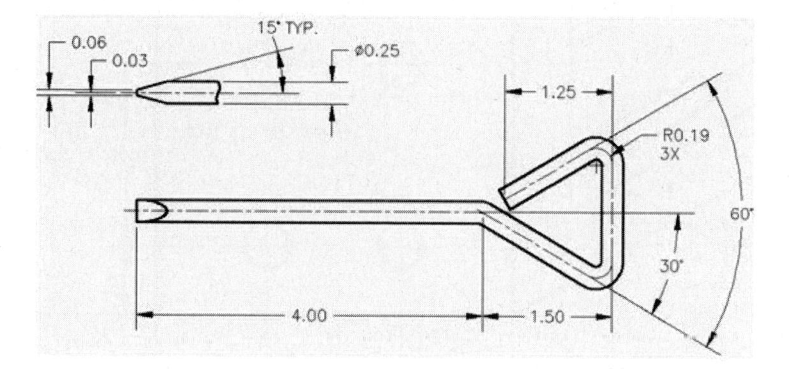

5.6

Create a drawing for Metric units assignment 4.3. Select a drawing size as assigned by the instructor. Choose the appropriate scale with the assistance of SolidWorks' automatic scaling. Create a Detail view of the tip details as shown. Make sure to add all necessary centerlines and centermarks. Completely fill in the titleblock and fully dimension the part.

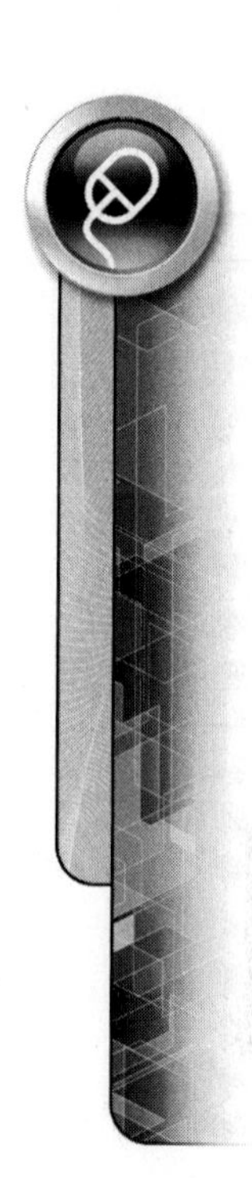

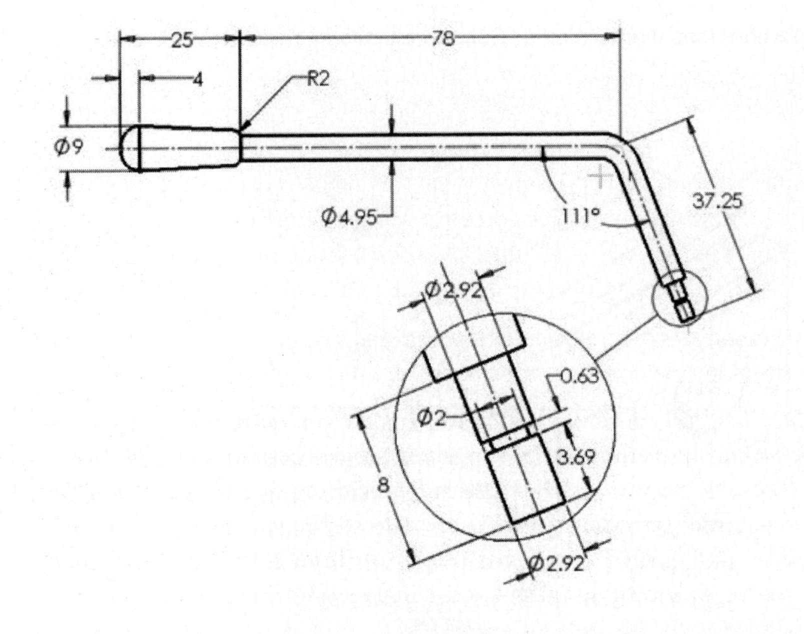

5.7

Create a drawing for English units assignment 4.4. Select a drawing size as assigned by the instructor. Choose the appropriate scale with the assistance of SolidWorks' automatic scaling. Create an auxiliary view (note: this may change the scale of the drawing). Make sure to add all necessary centerlines and centermarks. Completely fill in the titleblock and fully dimension the part. The whole part should have a general finish of 250. Also indicate that the bottom surface as well as the inclined face surface has a surface finish of 16.

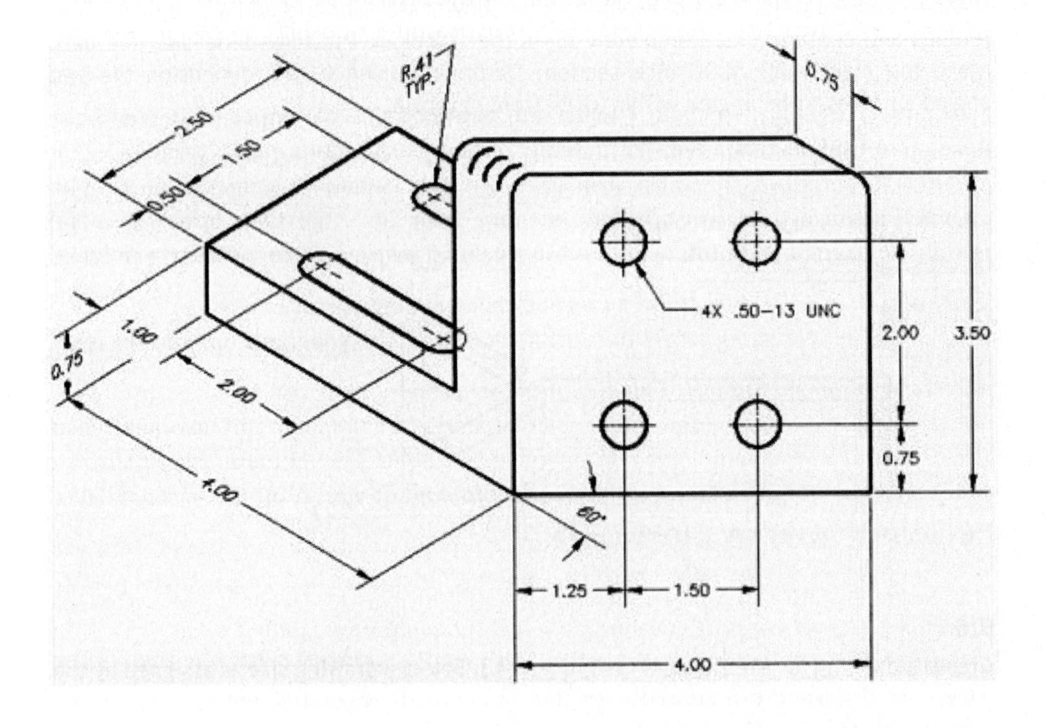

5.8

Create a drawing for English units assignment 4.5. Select a drawing size as assigned by the instructor. Choose the appropriate scale with the assistance of SolidWorks' automatic scaling. Create an auxiliary view (note: this may change the scale of the drawing). Make sure to add all necessary centerlines and centermarks. Completely fill in the titleblock and fully dimension the part.

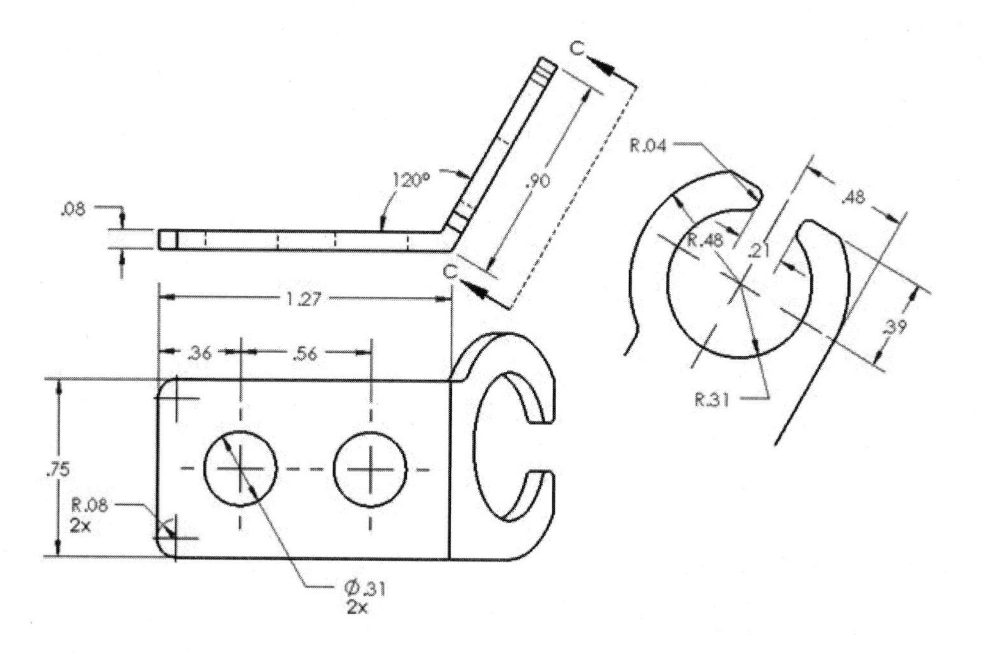

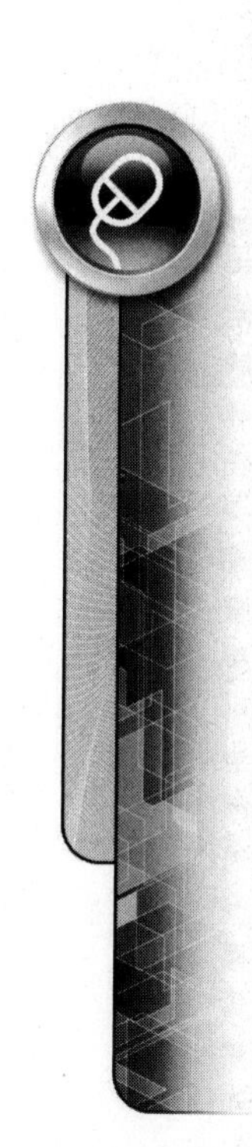

5.9

Create a drawing for English units assignment 4.6. Select a drawing size as assigned by the instructor. Choose the appropriate scale with the assistance of SolidWorks' automatic scaling. Make sure to add all necessary centerlines and centermarks. Completely fill in the titleblock and fully dimension the part. Call out all holes using the hole style and depth symbols. Make the back surface Datum A. Make the horizontal surface of the slot on the top of the part Datum B. Also make the left edge of the left notch Datum C. All holes should have a GD&T positioning tolerance of .003 from datums A, B, and C.

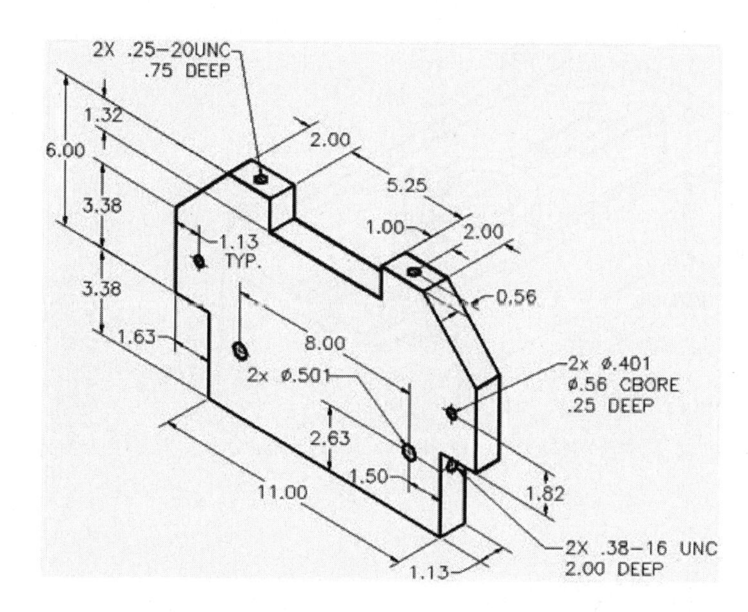

5.10

Create a drawing for English units assignment 4.7. Select a drawing size as assigned by the instructor. Choose the appropriate scale with the assistance of SolidWorks' automatic scaling. Make sure to add all necessary centerlines and centermarks. Completely fill in the titleblock and fully dimension the part. Create the appropriate section view. The whole part should have a general finish of 125. Also indicate that the top and bottom surfaces have a surface finish of 8.

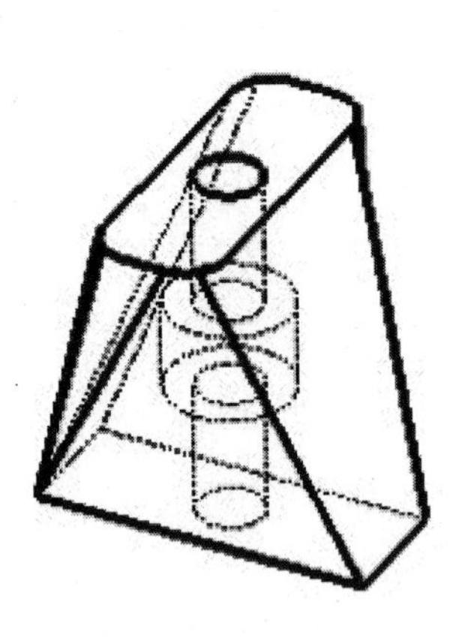

5.11

Create a drawing for English units assignment 4.9. Select a drawing size as assigned by the instructor. Choose the appropriate scale with the assistance of SolidWorks' automatic scaling. Make sure to add all necessary centerlines and centermarks. Completely fill in the titleblock and fully dimension the part. Create a full section view. The whole part should have a general finish of 250. Also indicate that the top surface as well as all holes has a surface finish of 16. Make the top surface Datum A. Indicate that the center hole has a perpendicular GD&T tolerance of .002 from datum A.

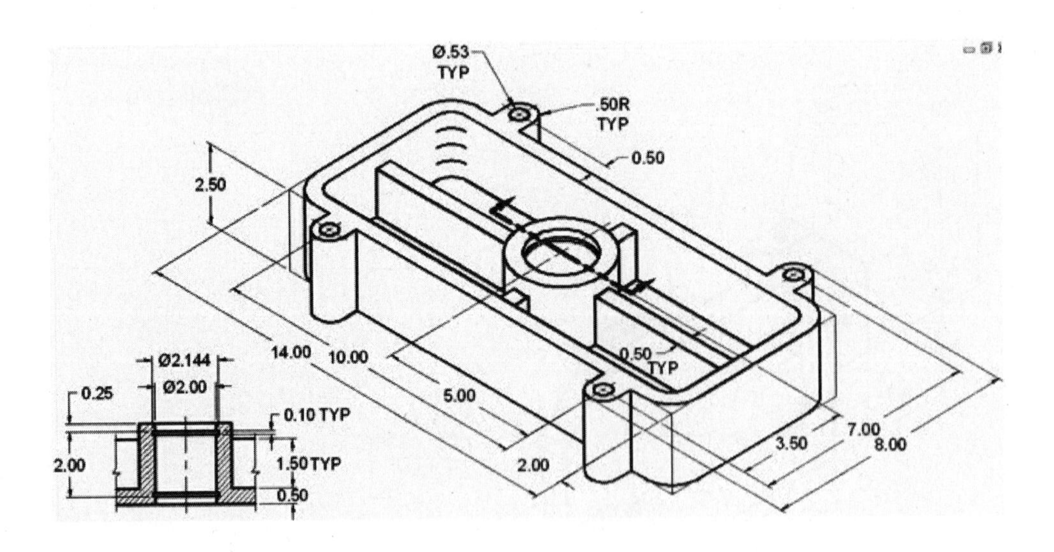

5.12

Create a drawing for Metric units assignment 3.9.a. Select a drawing size as assigned by the instructor. Choose the appropriate scale with the assistance of SolidWorks' automatic scaling. Make sure to add all necessary centerlines and centermarks. Completely fill in the titleblock and fully dimension the part. Create a half section view. Call out the hole using the hole style and depth symbols.

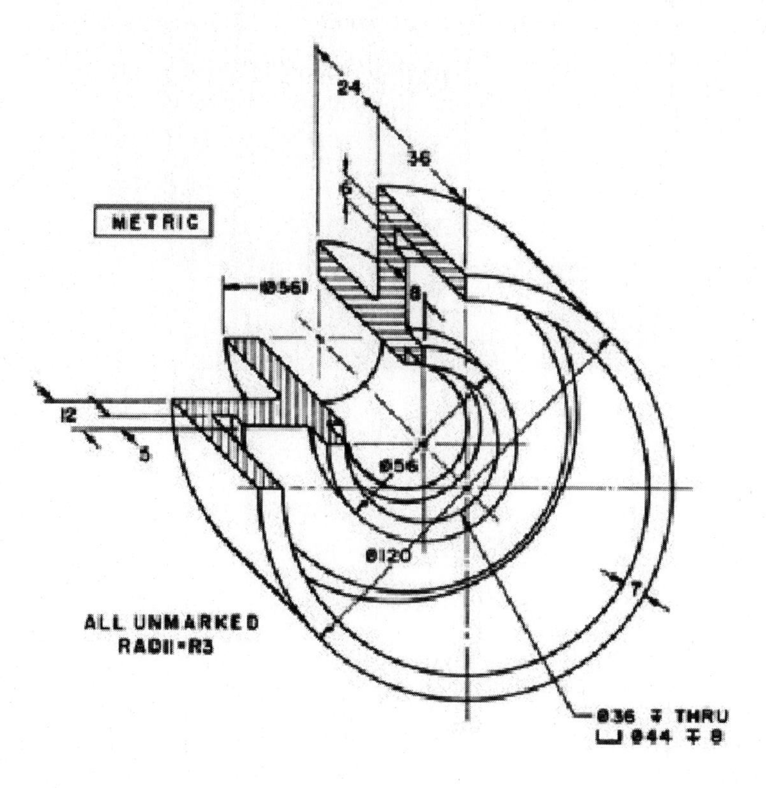

5.13

Create a drawing for Metric units assignment 3.9.e. Select a drawing size as assigned by the instructor. Choose the appropriate scale with the assistance of SolidWorks' automatic scaling. Make sure to add all necessary centerlines and centermarks. Completely fill in the titleblock and fully dimension the part. Create a half section view. Call out the holes using the hole style and depth symbols.

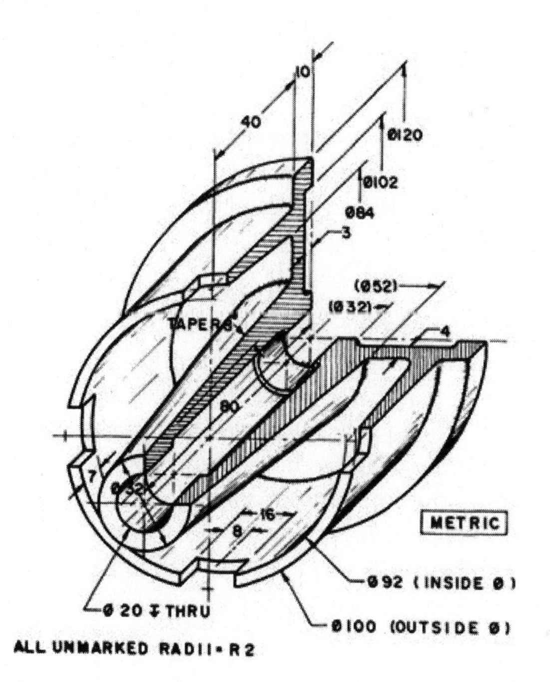

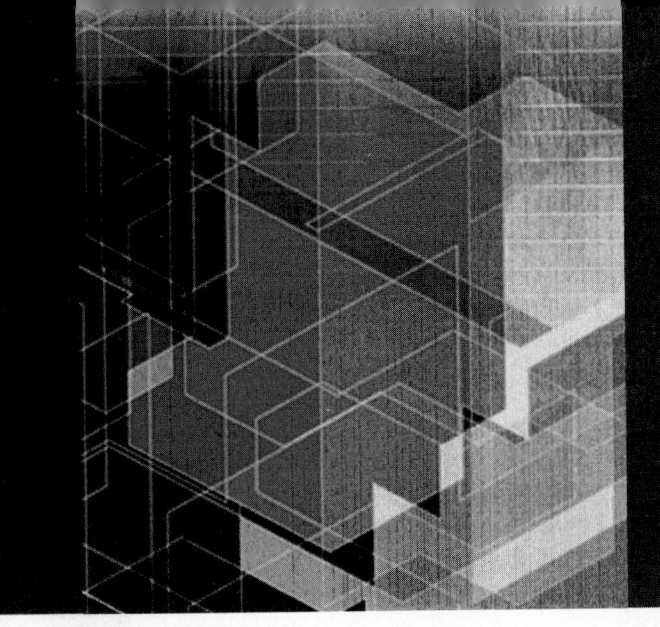

CHAPTER 6

Creating Assembly Models

This chapter discusses various techniques and tools used to assemble part files as well as imported parts to complete an assembly file:

- Bottom-up designing
- Mating parts in an assembly
- Using the SolidWorks Toolbox
- Starting an assembly file and inserting parts
- Moving and revolving a part in an assembly
- Mating using reference geometry
- Defining stationary and moving parts

CREATING ASSEMBLIES

The purpose of detail drawings is so fabricators can manufacture the component that is detailed. In order to assemble components into the designed arrangement, assembly files are required.

Assembly files start as empty files and no geometry is ever created in the file. The assembly is built by bringing in copies of parts onto the assembly drawing. The parts as they are brought in are to be arranged into the desired physical state.

BOTTOM-UP DESIGN

Bottom-up design is the process of creating the part files for an assembly prior to creating an assembly file. This would be a common method in reverse engineering where each part can be fully created because all of the final dimensions can be measured prior to modeling it. Once all of the part files have been created, an assembly file can be created so each component can be brought in to the assembly one part at a time. When designing an assembly from concept, bottom-up design is not the method to use if you want to remain efficient in the design process; for a new conceptual design, the top-down method is recommended as explained in Chapter 8.

ARRANGING PARTS IN AN ASSEMBLY USING MATES

When parts are physically assembled in an industrial setting such as a production line, components are located where surfaces of mating parts touch each other, holes and hole patterns are aligned to each other, surfaces are spaced from each other, and edges are aligned parallel with each other. The Mate command matches these assembling location methods and more in the virtual environment.

The Mate command has eight basic mating methods to mimic industrial assembly methods, and they are:

Mate Type	Description	Icon
Coincident	Takes two planar faces and makes then coplanar.	
Distance	Takes two features (planes, lines, or points) and holds them to a specified distance from each other.	
Parallel	Takes two planes or lines and holds them parallel to each other.	
Perpendicular	Takes two planes or lines and holds them perpendicular to each other.	
Angle	Takes two planes or lines and holds them at a specified angle to each other.	
Tangent	Takes two features and holds them tangent to each other.	
Concentric	Takes two cylindrical faces and holds them concentric to each other.	
Lock	Takes two parts and maintains the position and orientation to each other.	

USING THE TOOLBOX

While it is said that "once a part is modeled it should never have to be modeled again" but because there is no universal method of collaborating and sharing files, this is often untrue. SolidWorks has started to establish positive collaboration methods that are unparalleled and constantly expanding through the Design Library. There are countless components available through many companies through the 3D Content Central.

In addition, SolidWorks offers a Toolbox of Standardized parts such as fasteners that can also be loaded into the Design Library.

Because of its vast size, in order to use the Toolbox it must first be loaded. To load the Toolbox first select the Tools pull-down menu and select Add-ins.... When the Add-ins menu appears, select both the SolidWorks Toolbox and the SolidWorks Toolbox Browser as shown in Figure 6.1. Because the Toolbox is so large, it may take a moment to load.

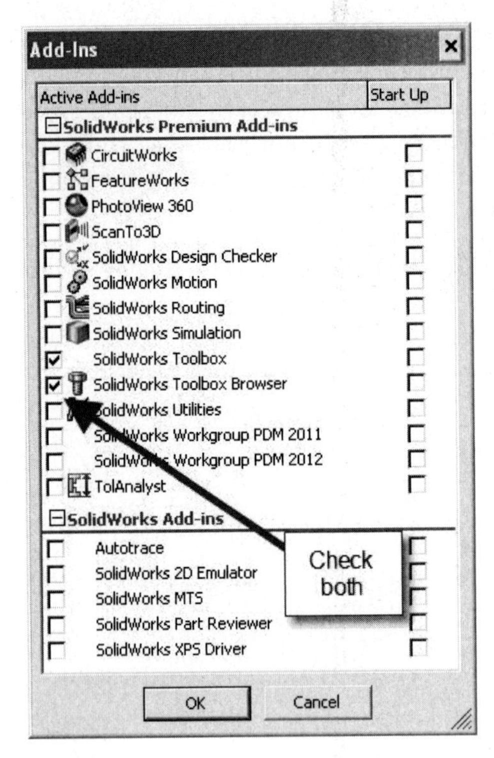

FIGURE 6.1

Once the Toolbox is loaded, the contents are available in the Design Library found to the right of the model area. The fasteners are available in various international standard formats as shown in Figure 6.2.

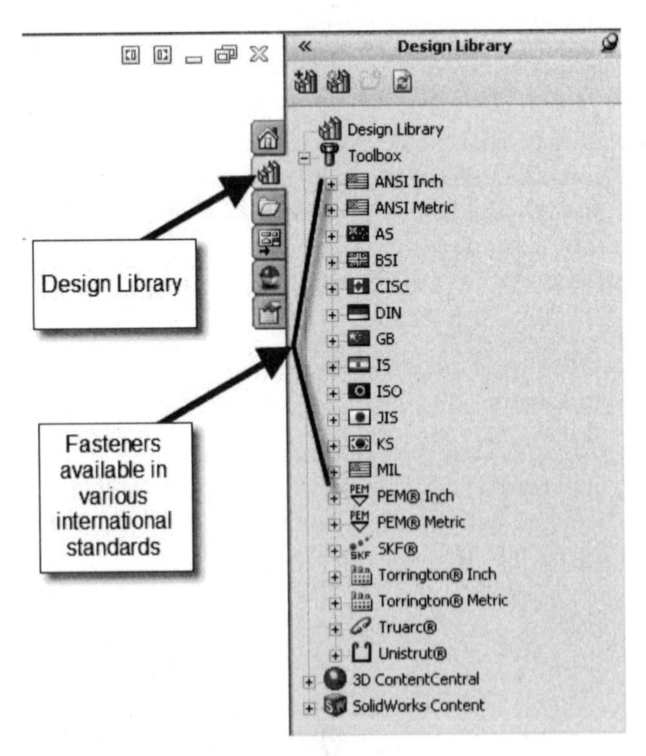

FIGURE 6.2

Within each standard, there are bearings, bolts and screws, jig bushings, keys, nuts, O-rings, pins, power transmission components (sprockets, gears, and pulleys), retaining rings, structural members, and washers. Each of these categories there is further refined into additional subgroups (Figure 6.3).

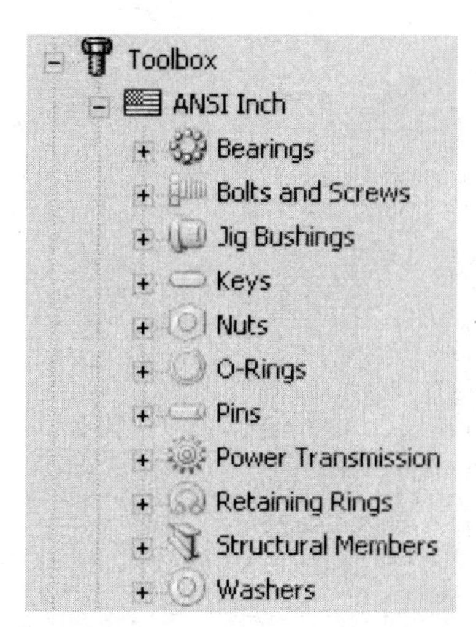

FIGURE 6.3

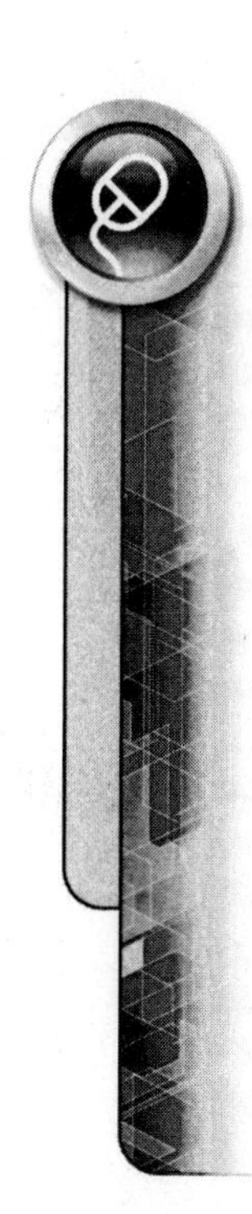

TUTORIAL EXERCISE: 06_ASSEMBLY_01.SLDPRT

Assembly Number: SWT-CH06-A1-Assembly.Sldasm

Part Numbers: SWT-CH06-A1-Axle.sldprt,

SWT-CH06-A1-BasePlate.sldprt

SWT-CH06-A1-Bracket.sldprt

SWT-CH06-A1-Bushing.sldprt

SWT-CH06-A1-Cam.sldprt

SWT-CH06-A1-Collar.sldprt

SWT-CH06-A1-Housing.sldprt

SWT-CH06-A1-Piston.sldprt

SWT-CH06-A1-Wheel.sldprt

SWT-CH06-A1-PistonArm.sldprt

SWT-CH06-A1-PistonHead.sldprt

SWT-CH06-A1-PistonPin.sldprt

Description: Assembly 1

Units: English

This exercise will fully assemble the following manual piston. Twelve components will be brought into the assembly one at a time, and as each part will be inserted into the assembly, it will be arranged to the existing parts using mates. Each part will be fully located before the next part is brought in. Once the parts are assembled, fasteners will be brought in using the Smart Fastener which uses premodeled fasteners to finalize the assembly (Figure 6.4).

FIGURE 6.4

STEP 1

Create a new file and select the Assembly file type (Figure 6.5).

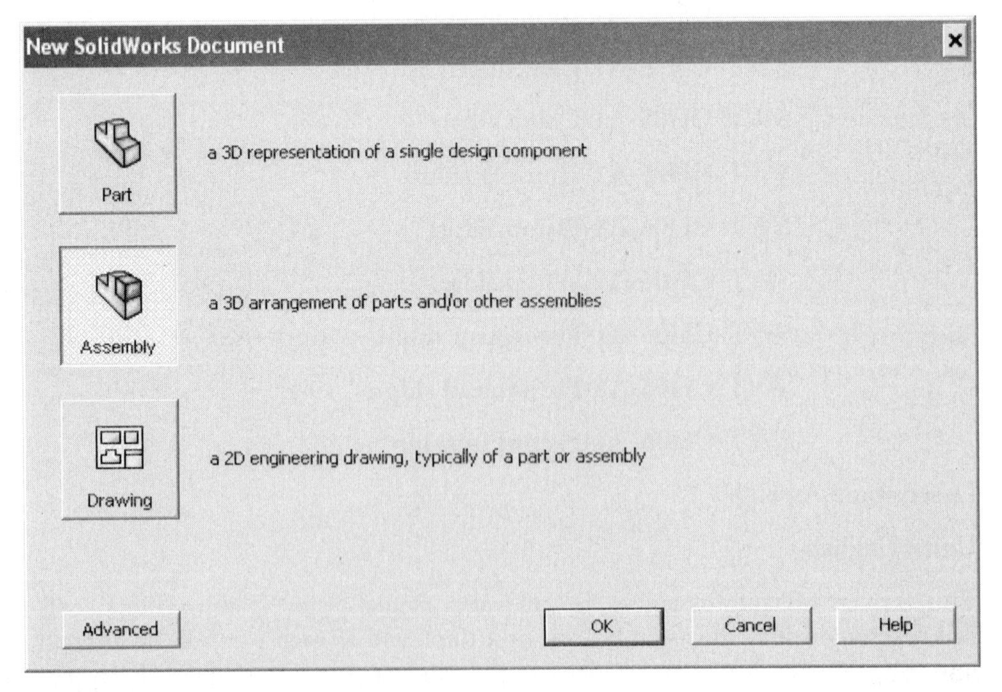

FIGURE 6.5

STEP 2

At the Begin Assembly prompts in the Command Manager, the first step is to insert the first part of the assembly. If there is any part or assembly files currently open, they will be listed in the Open Documents field and can be selected to insert into the assembly. If there are no other files open, then the Browse button will need to be selected in order to find the first file to insert. The selection of the first component is an important decision; it should be a stationary component, oftentimes the base of the assembly. The first component will not have the ability to dynamically move within the assembly once it is placed.

Select the Browse button and search for the Base Plate part file. When placing the Base Plate, it can be located anywhere within the assembly file; instead of placing it at an arbitrary location in the model area, simply accept the part by selecting the green checkmark (OK). This will locate the Base Plate in the same place with respect to the origin that it is in the part file. The origin is not used to locate any other parts (Figure 6.6).

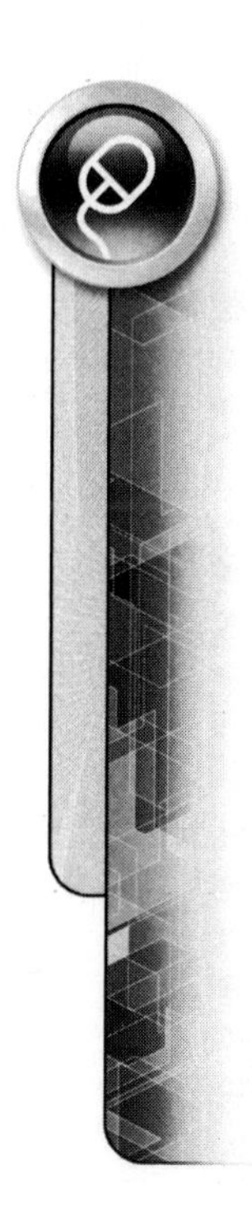

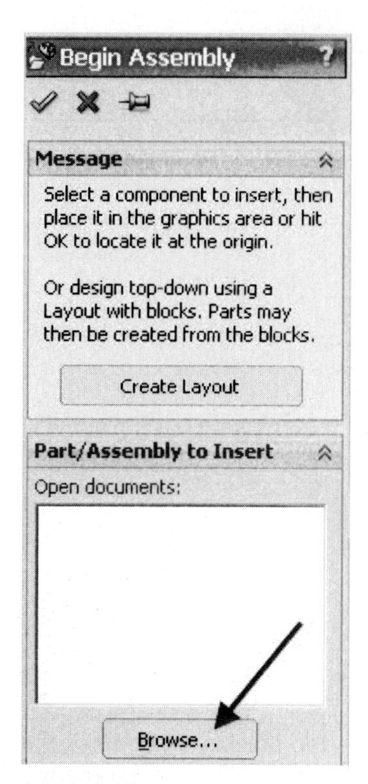

FIGURE 6.6

STEP 3

The next logical part to insert is the bracket because the bracket is the part that is assembled to the base plate. To add this part, select the Insert Components command from the Assembly command manager. Select browse from the Feature Manager, and find and select the Bracket part file (Figure 6.7).

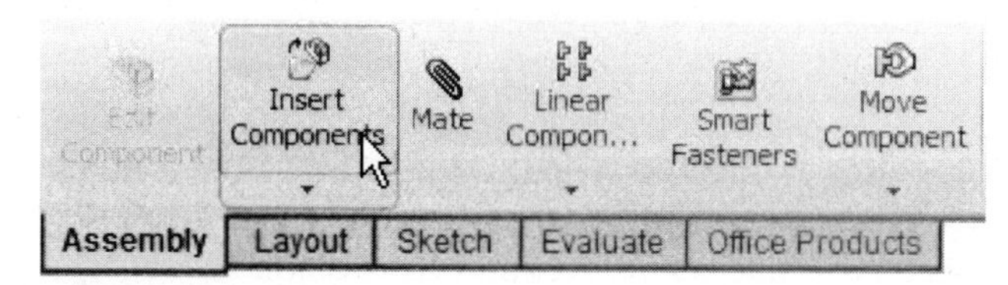

FIGURE 6.7

Instead of selecting OK, place the bracket in the model area away from the base plate so they are not touching each other, offset to one side as seen in Figure 6.8. In the mating process, features on each part may need to be selected, such as surfaces and holes. Placing parts in a manner that all of the features are easily visible and selectable will assist in the mating process.

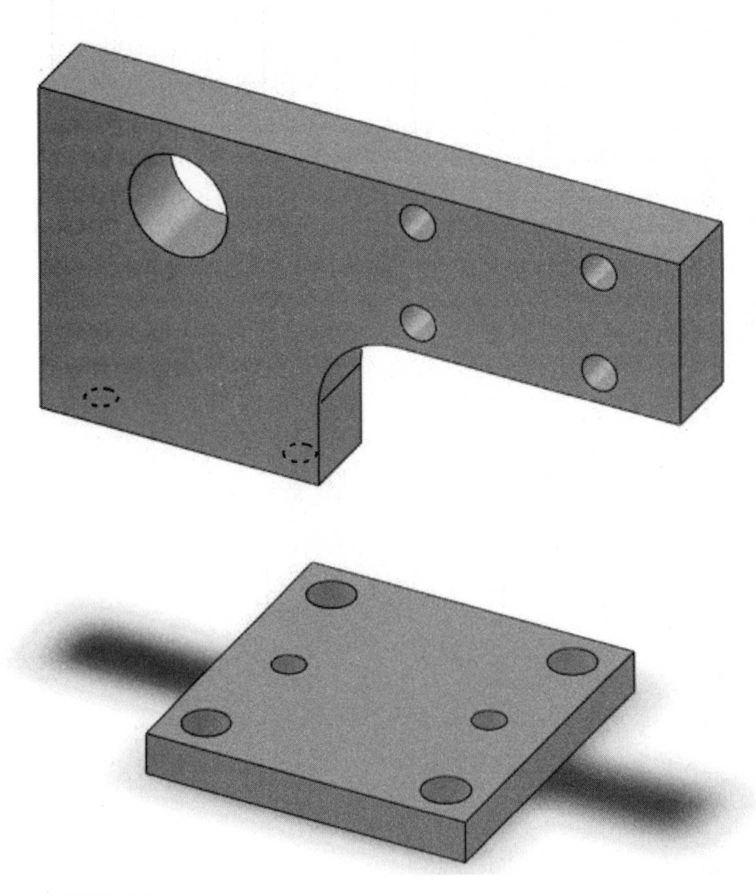

FIGURE 6.8

STEP 4

Next is to mate the bracket to the base plate by locating the bracket using all of the known methods of contact and location. First, the bottom of the bracket sits on the top of the base plate and then the holes of the bracket are aligned with the countersink holes of the base plate.

Select the Mate command from the Assembly command manager (Figure 6.9).

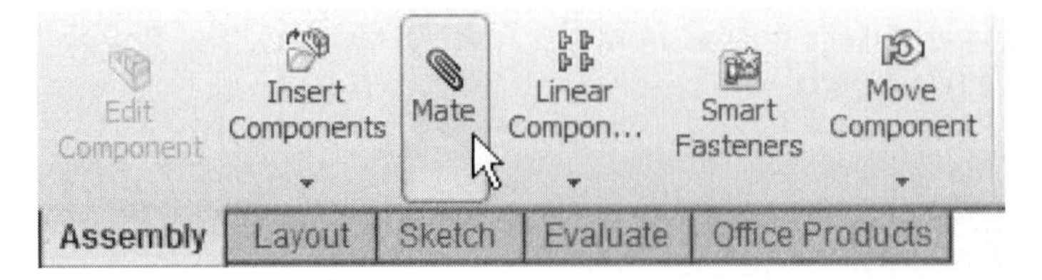

FIGURE 6.9

Once in the Mate command, select the Coincident mate from the Command Manager. Now the mating surfaces need to be selected, so first select the top of the base plate.

Then rotate the model to see the bottom of the bracket. Select the bottom face. Accept the mate by selecting OK (green checkmark) or by hitting the return button on the keyboard. When selecting a surface make sure not to select edges or points or else an undesired result may happen (Figure 6.10).

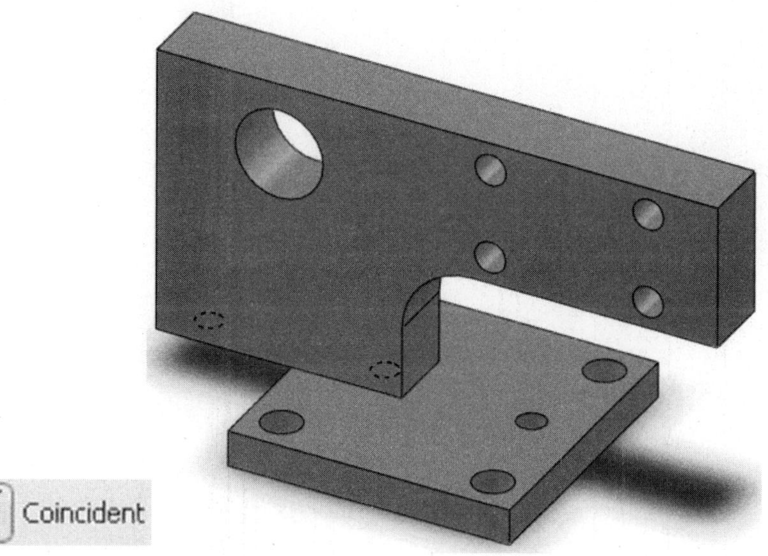

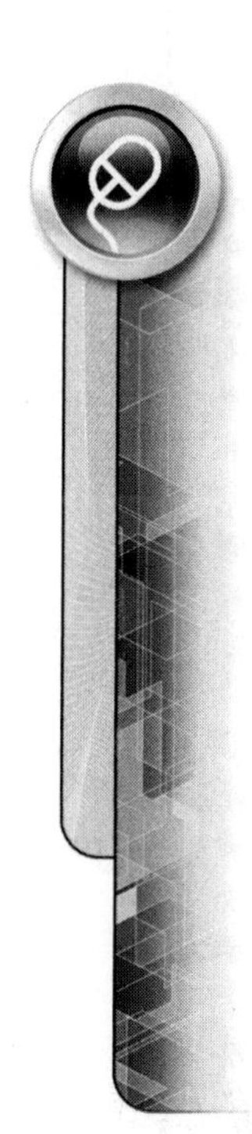

FIGURE 6.10

STEP 5

When mating features to each other, it is easiest if the features are exposed so they can easily be selected. If a feature is hidden, such as the front bottom hole in the bracket, a part can be moved to expose the feature. When moving a part, any current mates will not be ignored. For example, when moving the bracket, the bottom surface will always be aligned with the top of the top plate but the bracket can be moved around the surface of the plate and even off the plate.

Select and hold the bracket to move it away from the base plate as shown in Figure 6.11.

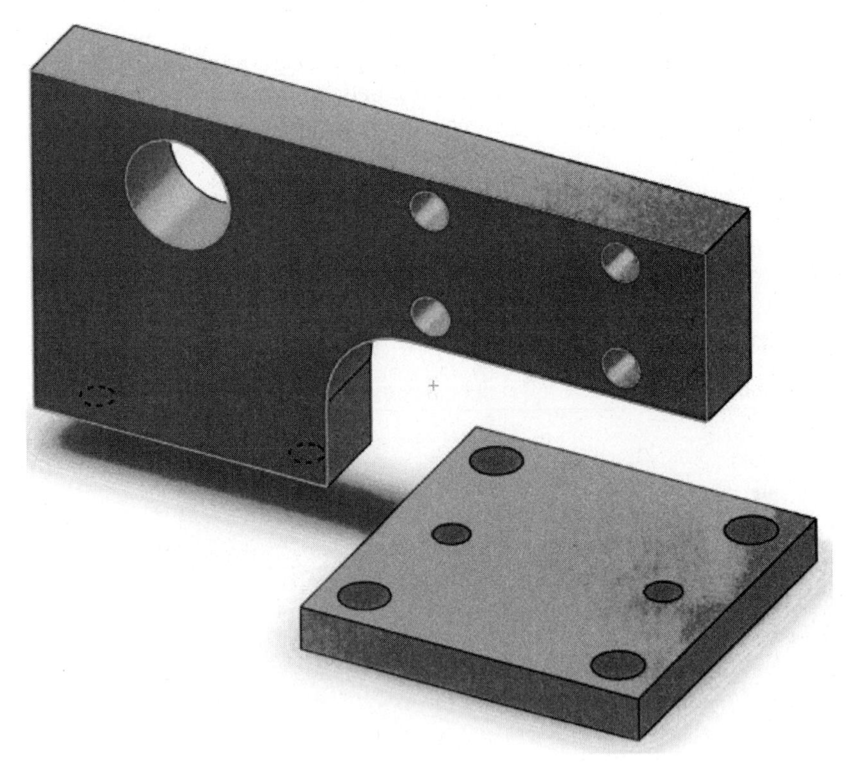

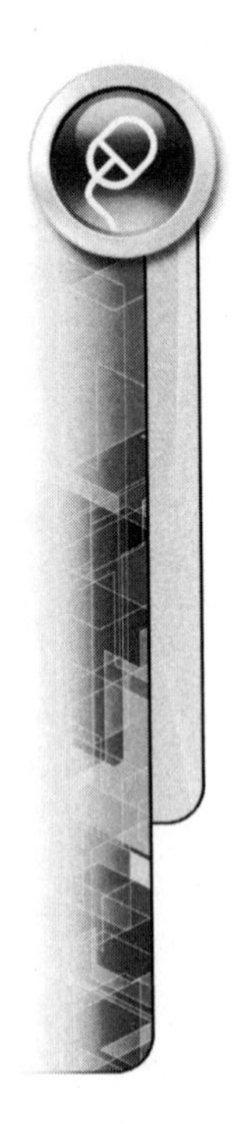

FIGURE 6.11

STEP 6

Next is to select the mate: Concentric. Select the front countersink hole in the base plate, make sure to select in the hole, and do not select an edge. Select the middle mouse button and rotate the view in order to see the underside of the bracket. Select front tapped hole in the bracket. The bracket should move over the base plate and align the two holes concentrically. Accept the mate by selecting the green checkmark (Figure 6.12).

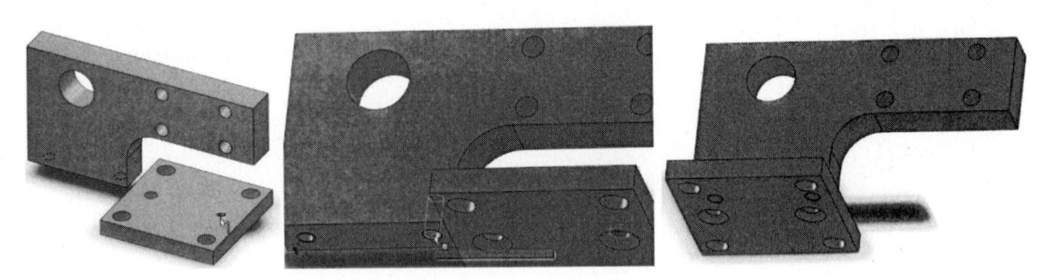

FIGURE 6.12

NOTE

Now that the bracket has two mates constraining it, the bracket can still be rotated around the constrained concentric mate axis. The bracket can be clicked and held to rotate. Rotating it will not affect the next mate.

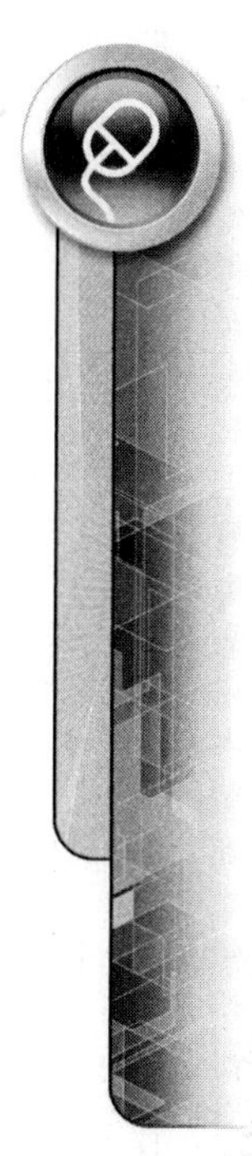

STEP 7

Although the parts appear to be assembled, the bracket is not fully defined (located). With the surfaces aligned and one set of holes concentric, the bracket could still spin around the axis of the mated holes. To fully define the bracket so it has no freedom to move, we need to concentric mate the other countersink hole in the base plate with the other tapped hole in the bracket.

Rotate the view as to allow viewing into the other countersink hole at an angle where the tapped hole can also be seen through the countersink hole. Select the Mate command and select the concentric mate. Select the countersink cylindrical surface, then select the tapped hole cylindrical surface. Accept the mate.

The bracket is now fully defined. Note in the lower right corner the status indicator stating "Fully Defined." Exit the Mate command by selecting OK (Figure 6.13).

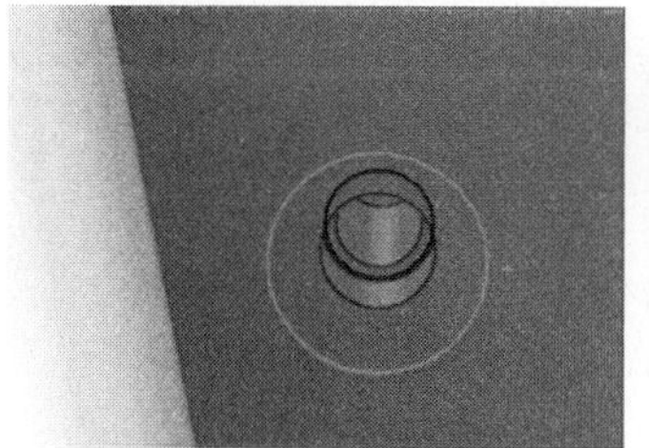

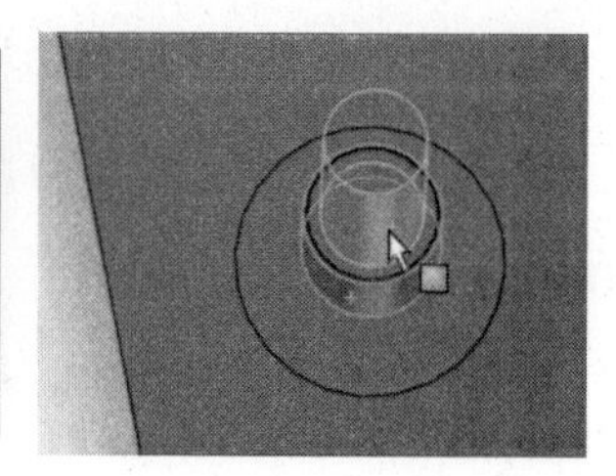

FIGURE 6.13

STEP 8

Next is to insert the collar. Select the Insert Component command and browse for the collar part file. Place the collar in the view off to the side so selecting features on it is easier. Select the Mate command and the coincident mate. Select the side face of the bracket as seen in Figure 6.14. Rotate the view and select the face of the collar that is on the underside of the head. The collar should rotate into alignment; accept the mate.

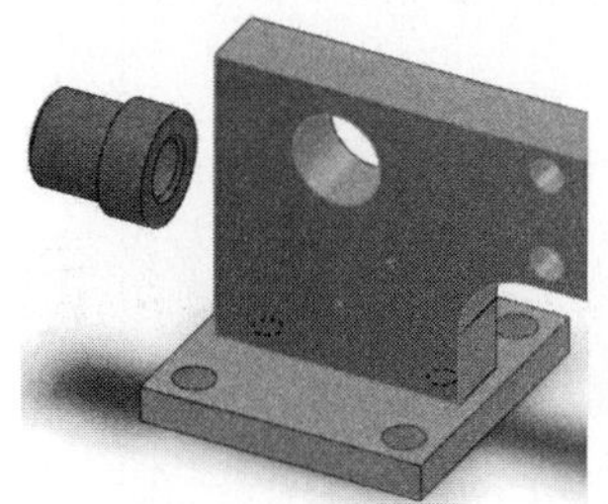
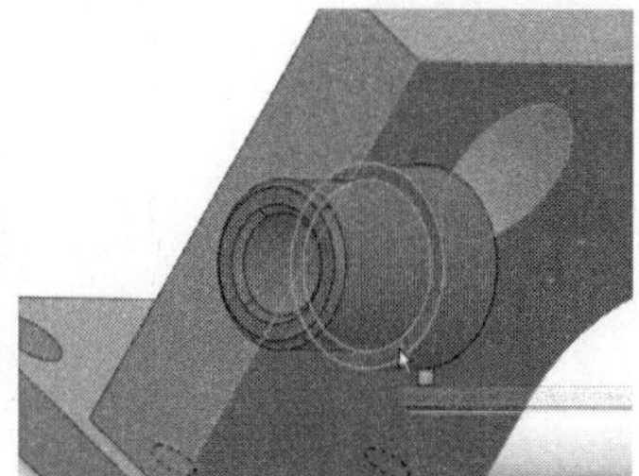

FIGURE 6.14

STEP 9

Select the Mate command and select the concentric mate. Select the inside of the large hole in the bracket, then select the outside diameter of the collar. Accept the mate. Notice that the collar is not fully defined. This is OK if the part is a moving part which in this case the collar can freely rotate in the bracket hole. There is no need to fully define this part. Exit the Mate command by selecting OK (Figure 6.15).

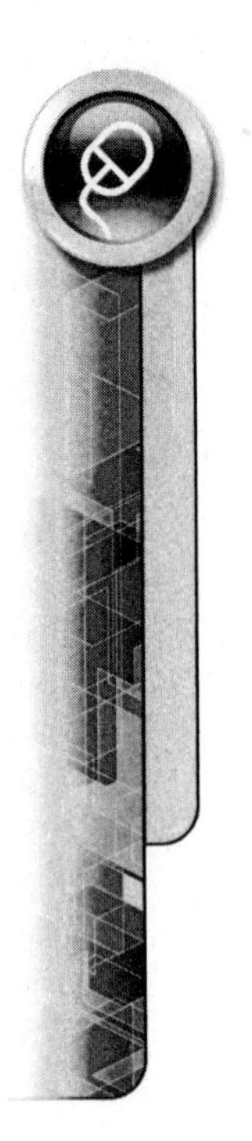

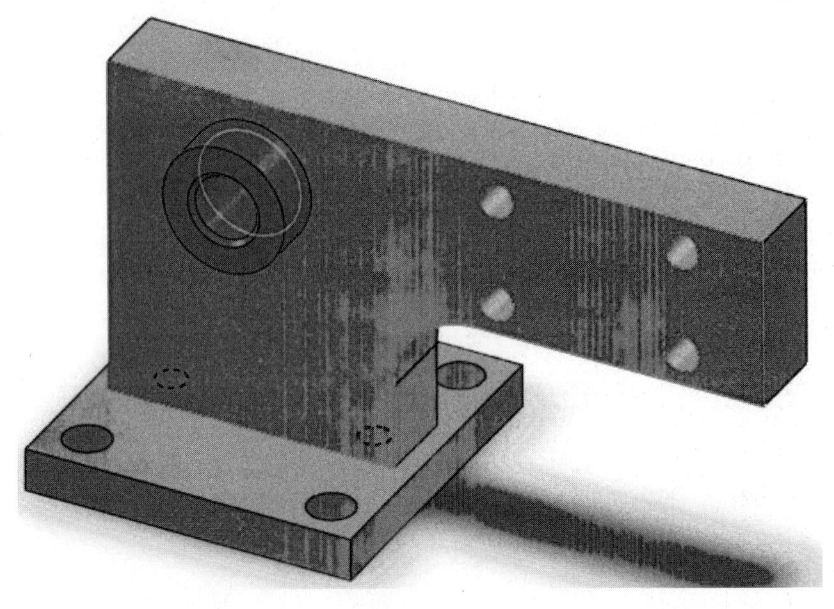

FIGURE 6.15

STEP 10

Next is to locate the wheel against the collar. Select the Insert Component command and browse for the wheel part file. Place the wheel away from the assembly in the model area. While in the Mate command, select the coincident mate, then select the face of the boss of the wheel. Next, select the end face of the collar. Accept the mate (Figure 6.16).

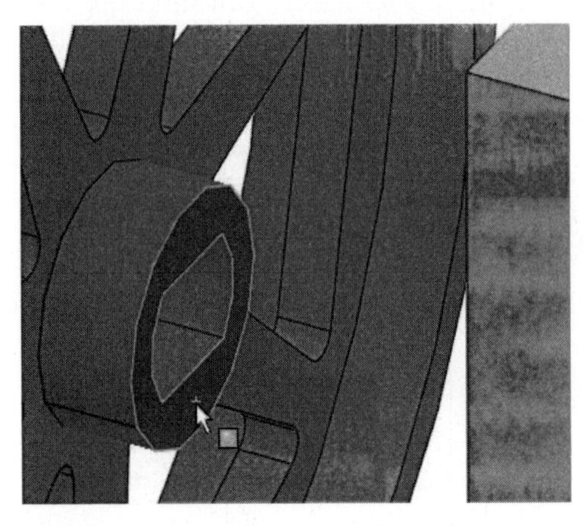

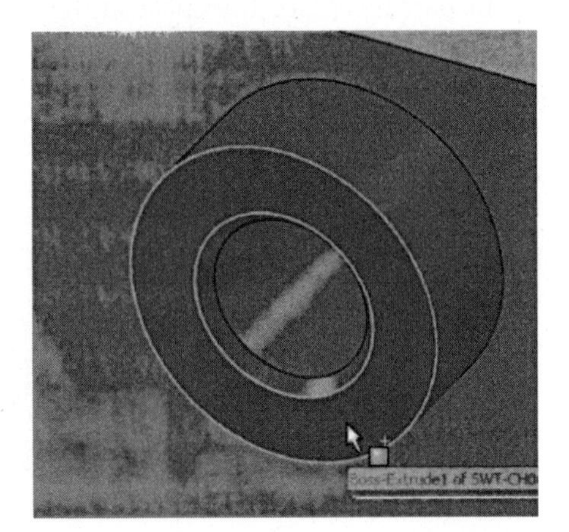

FIGURE 6.16

STEP 11

Select the concentric mate. Select the inside *cylindrical* surface of the wheel bore, then select the inside bore of the collar. Accept the mate. The wheel can also be left underdefined to allow it to spin freely. Exit the Mate command (Figure 6.17).

FIGURE 6.17

STEP 12

Next is to bring in the axle. Select the Insert Component command and browse for the axle. Place the axle in the assembly but away from the other parts. Select the Mate command and the coincident mate. Select the outer face of the wheel boss, then select the end face of the axle (Figure 6.18).

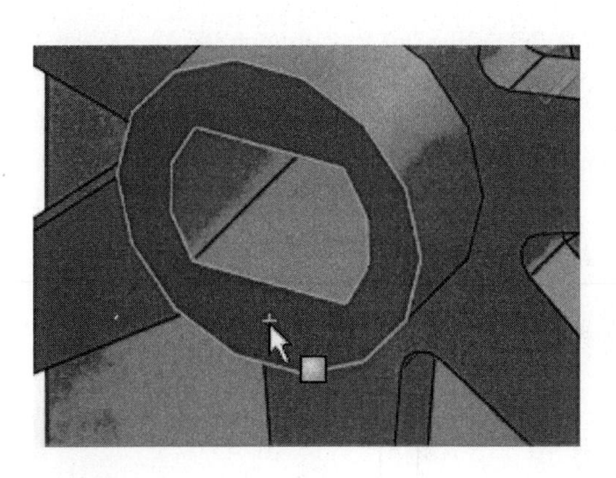

FIGURE 6.18

If the axle appears to be facing in the wrong direction, which may happen when adding a mate, the part may need to be flipped. On the Mate pop-up toolbar, select the "Flip Mate Alignment" button. The axle will flip around into the proper alignment. Accept the mate (Figure 6.19).

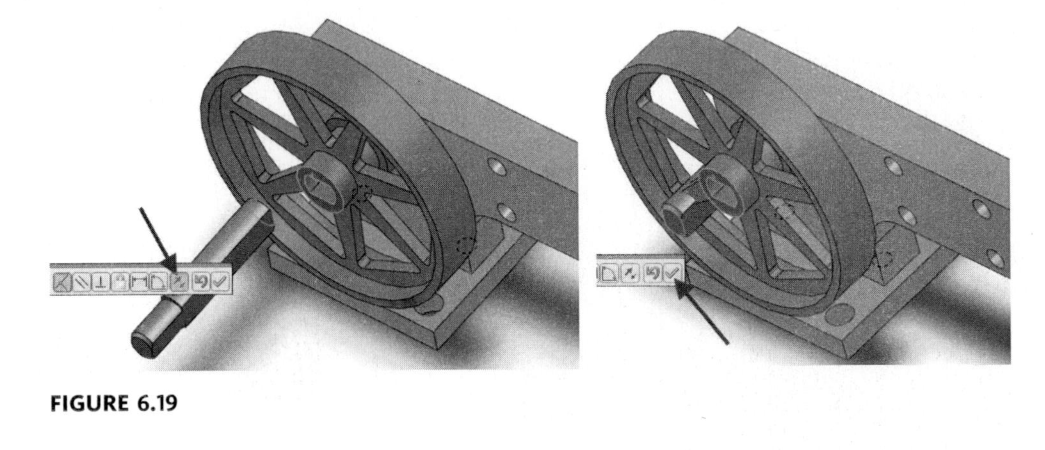

FIGURE 6.19

STEP 13

Select the concentric mate, then select the *cylindrical* part of inner bore of the wheel, then select the outer *cylindrical* surface of the axle. Accept the mate (Figure 6.20).

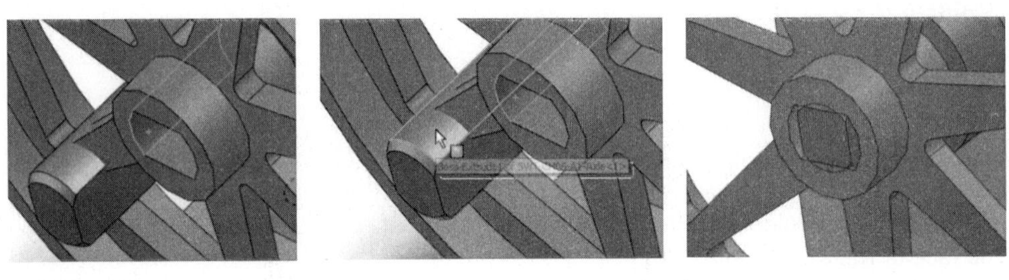

FIGURE 6.20

Notice how the flats of the axle do not align with the flats of the wheel bore. The parallel mate can be used to align these. While in the Mate command select a flat side of the axle, then carefully select a flat of the wheel bore; zooming in may be necessary. Make sure that the parallel mate is selected (not coincident), then accept the Mate command (Figure 6.21).

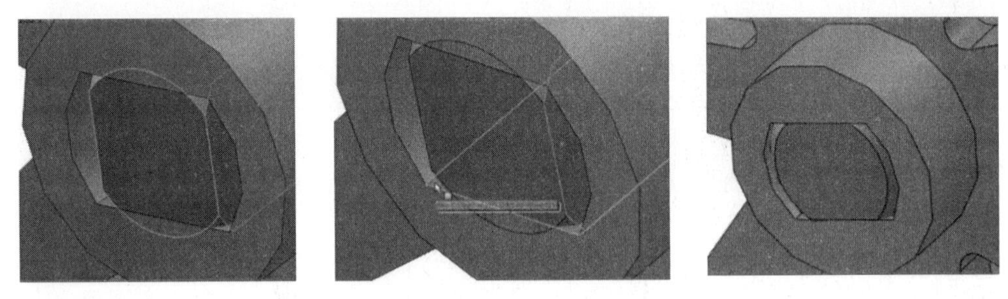

FIGURE 6.21

STEP 14

Rotate the assembly to the back side . Select the Insert Components command and browse for the cam part. Place the cam near the assembly. Select the Mate command and select the Coincident mate option. Select the end face of the collar. Next, without rotating the view, select the cam with the *right mouse button* where the cursor is located over where the back face of the cam is (even though it is hidden) as shown in the middle of Figure 6.22. Select the "Select Other" command. This is a method of selecting features without having to rotate the view. A Select Other option window will appear that gives options of all of the other features other than the front face that the cursor could possibly be selecting. As the cursor scrolls through the list, each feature will highlight. Select the "Face@[SWT-ch06-cam]." Accept the mate. Use this method whenever you see fit.

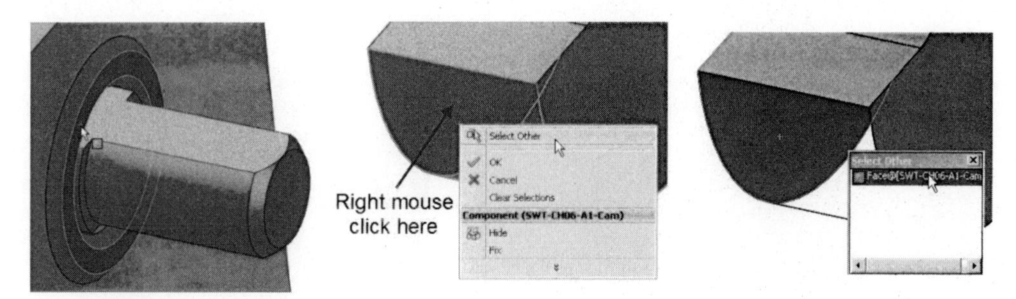

FIGURE 6.22

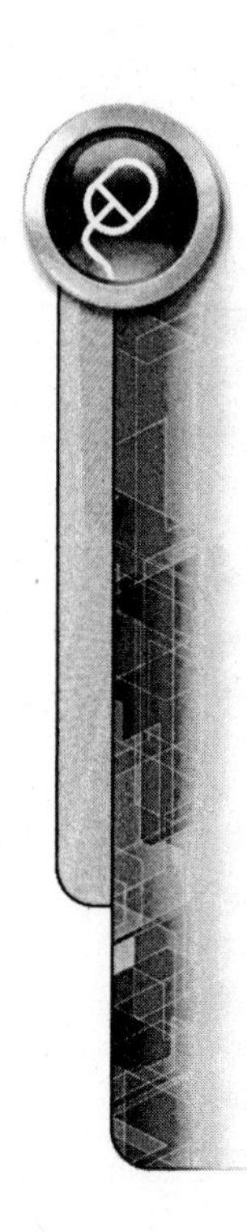

STEP 15

Next is to move the cam onto the axle. Select the Concentric Mate command. Select the inside cylindrical surface of the larger hole on the cam, then select the cylindrical surface of the outer diameter of the axle. Accept the mate.

Next, select the flat surface of the cam bore as well as the flat surface of the axle (the Select Other method may need to be used to accomplish this). Accept the mate. The cam should now be located on the axle. Exit the Mate command (Figure 6.23).

FIGURE 6.23

STEP 16

Next is to add the bushing. Open the Insert Components command and browse for the bushing part file. Insert the part near the assembly but off to the side so faces can be easily selected.

Zoom way into the end of the bushing and select the end face as shown in Figure 6.24. Then select the backside surface of the cam. Select the "Flip Mate Alignment" button. Accept the mate.

Select the concentric mate, then select the hole surface in the small cam hole, then select the outside diameter of the bushing. Accept the mate. Exit the Mate command.

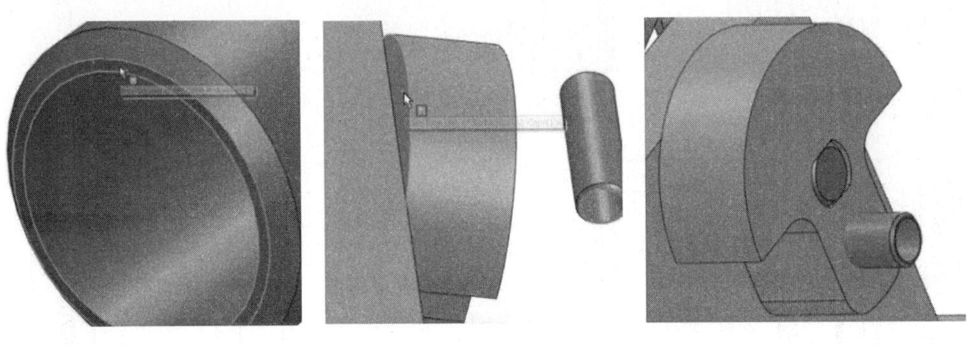

FIGURE 6.24

STEP 17

Next is to add the housing. Select the Insert Components command and browse for the housing. Locate it near the assembly but far enough away to select features on it. Select the coincident mate. Select the side surface of the housing that has the holes. Select the cam side of the bracket as shown in Figure 6.25. Accept the mate.

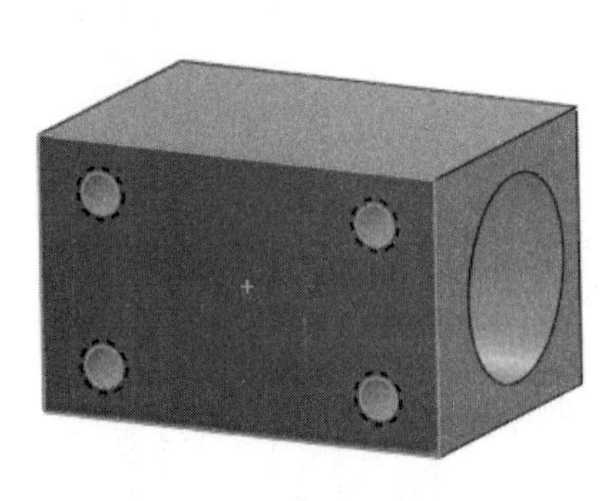

FIGURE 6.25

STEP 18

Next is to align the hole patterns. If the tapped holes in the housing are not visible, select and hold on the housing and drag it away from the other parts. Then using the concentric mate, align one tapped hole in the housing with the matching through hole in the bracket. Align a second pair of holes and the housing is fully defined (Figure 6.26).

FIGURE 6.26

STEP 19

Next is to add the Piston Arm. Select the Insert Components command and browse for the Piston Arm. Insert the arm near the assembly. Select the Mate command and select the concentric mate. Select the bore of the piston arm and the outside cylindrical surface of the bushing. The parts should be concentric (Figure 6.27).

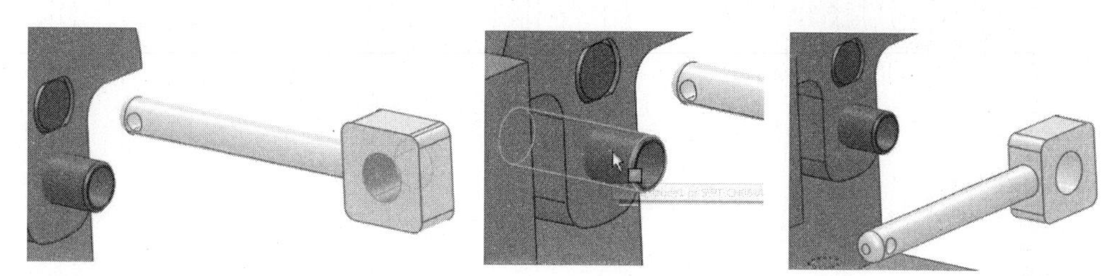

FIGURE 6.27

Then select the coincident mate. Select the back face of the piston arm, then select the face of the cam. Accept the mate. The arm is underconstrained as is a couple of the other past components. Because the parts are underconstrained, they can still move in the unconstrained direction, in this case, rotation. Rotate the view so the arm can be clearly seen. Grab the bar of the arm and rotate it as shown in Figure 6.28. Notice how other components (axle and wheel) rotate with the arm due to mates that lock them together.

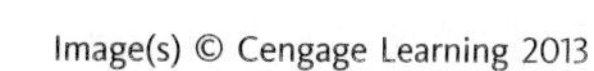

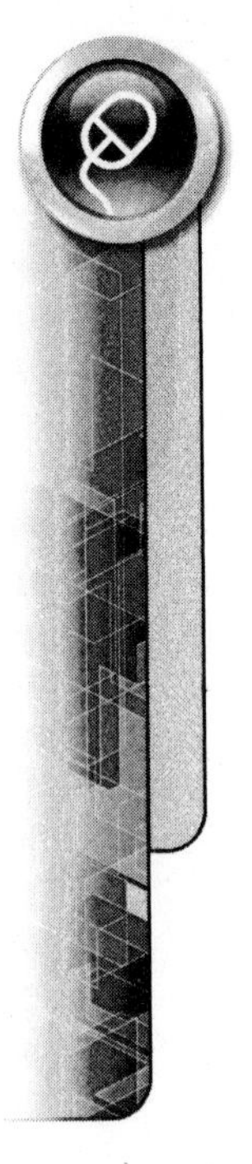

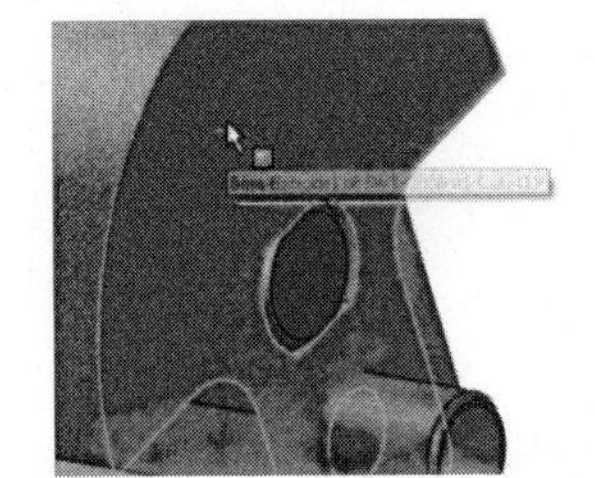
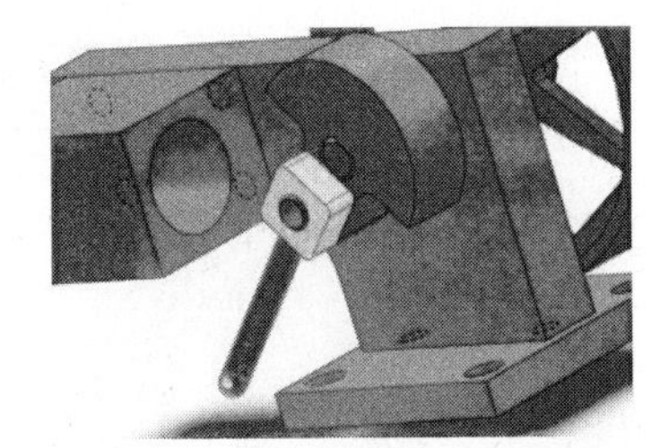

FIGURE 6.28

STEP 20

Select the Insert Components command and browse for the Piston Head. Locate it near the assembly. Select the Mate command and the concentric mate. Select the outside diameter of the head. Then select the inside diameter of the housing bore. The head should now be located in the bore. If the end of the piston head is not partially exposed as shown in Figure 6.29, grab the piston with a click-and-hold of the mouse selection button and partially pull the head out of the hole.

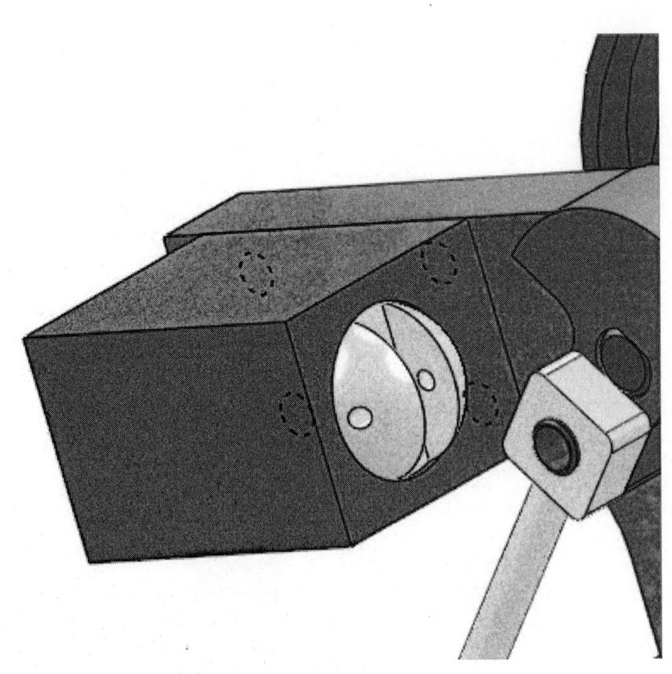

FIGURE 6.29

STEP 21

Using the Insert Components command, bring in the Piston Head Pin and locate it near the assembly. Using the Mate command, select concentric mates, select the inside diameter of the piston head and the outside diameter of the pin. Accept the mate (Figure 6.30).

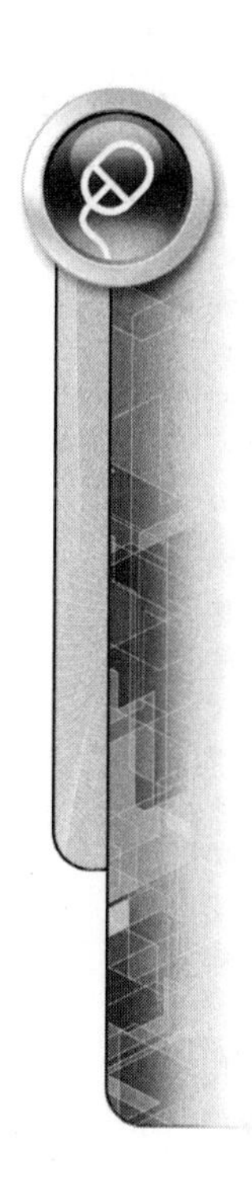

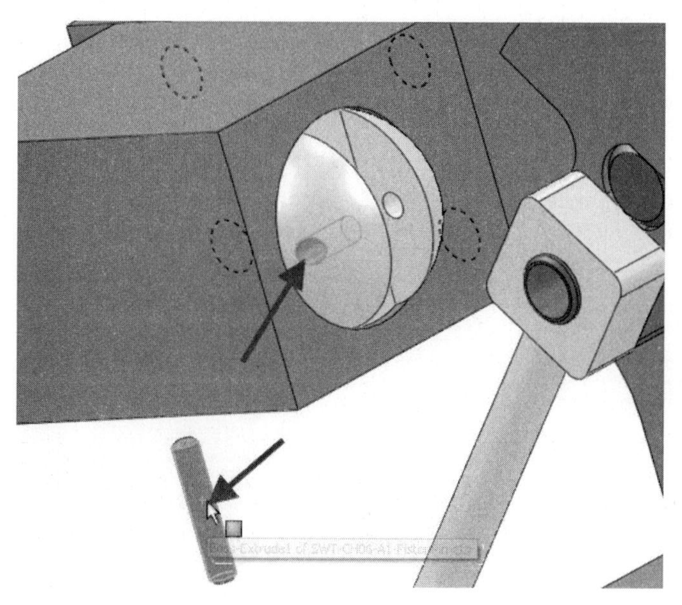

FIGURE 6.30

STEP 22

The pin is a total length of 0.50 inches and the groove in the piston head is 0.22 inches. That means in order to center the pin in the hole, the pin must be inserted into the hole of 0.14 inches a side.

While still in the Mate command, select the end face of the pin that is facing the piston. Then select the groove face to the right. Then select the Distance mate option. Key in 0.14. Then select the Flip Dimension button from the Mates menu bar as shown in Figure 6.31 (not the Flip Mate Alignment button). The pin should be equally into both sides of the hole. Accept the mate.

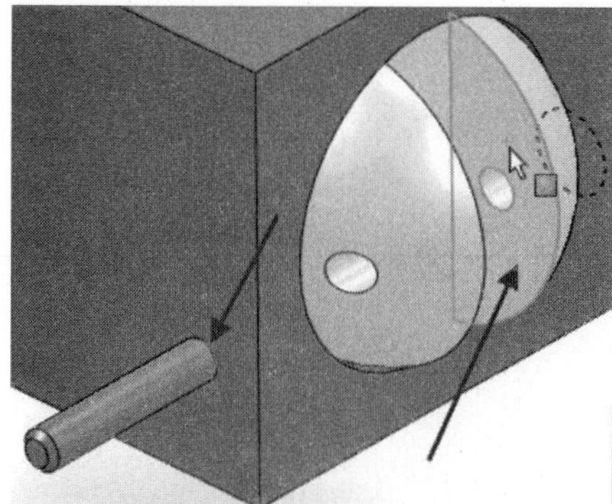

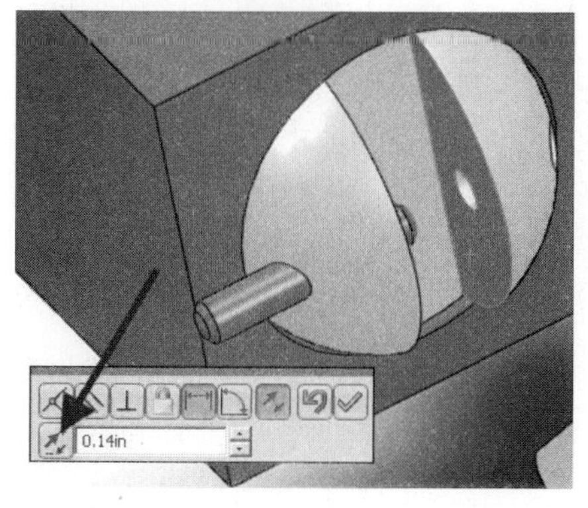

FIGURE 6.31

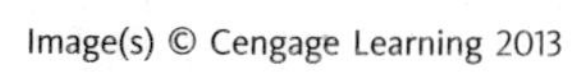

STEP 23

While still in the Mate command, select the hole surface on the piston arm, then select the outer surface of the pin. The Concentric command is automatically chosen. Accept the mate by selecting OK (Figure 6.32).

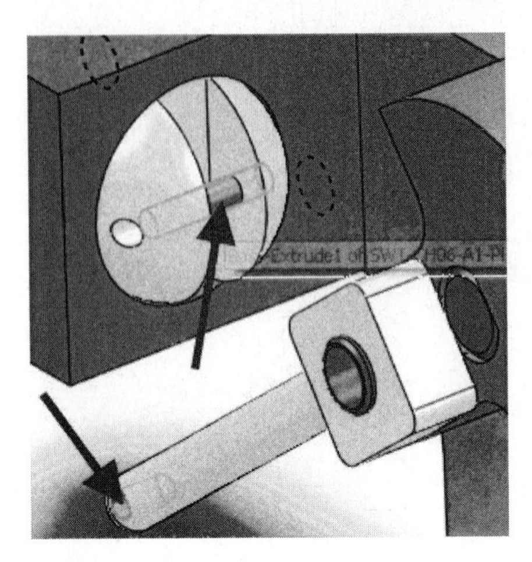

FIGURE 6.32

The piston head may be hidden inside the bore. The assembly is fully mated. Exit the Mate command by selecting OK in the command manager.

STEP 24

Take the cursor and place it over the wheel. Click and hold on the wheel, then move the cursor to rotate the wheel in a circle. Because the appropriate parts were not fully mated to allow dynamics, the wheel should rotate and the piston should slide back and forth in the housing. This is a good way to check the validity of all of the assembly mates (Figure 6.33).

FIGURE 6.33

STEP 25

Load the SolidWorks Toolbox by selecting Add-Ins from the Tools pull-down menu and selecting SolidWorks Toolbox and SolidWorks Toolbox Browser. In the Design Library select the ANSI Inch folder, then select the Bolts and Screws folder (Figure 6.34).

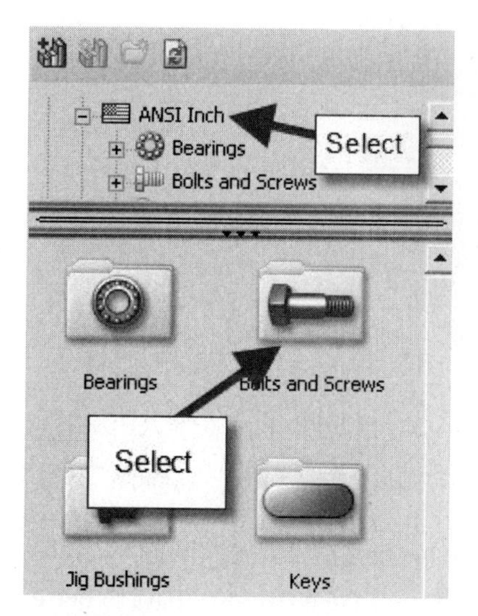

FIGURE 6.34

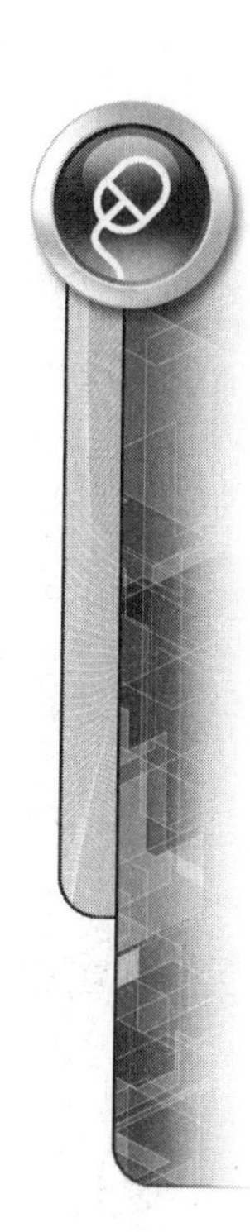

Then select the Socket Head Screws folder, then click and drag the Socket Button Head Cap Screw icon into the model view (Figure 6.35).

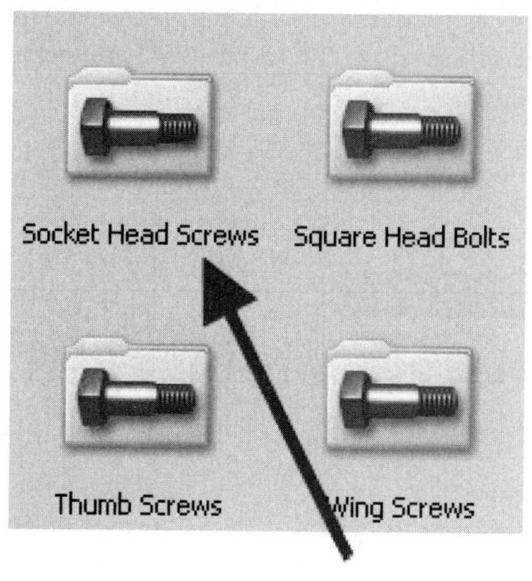

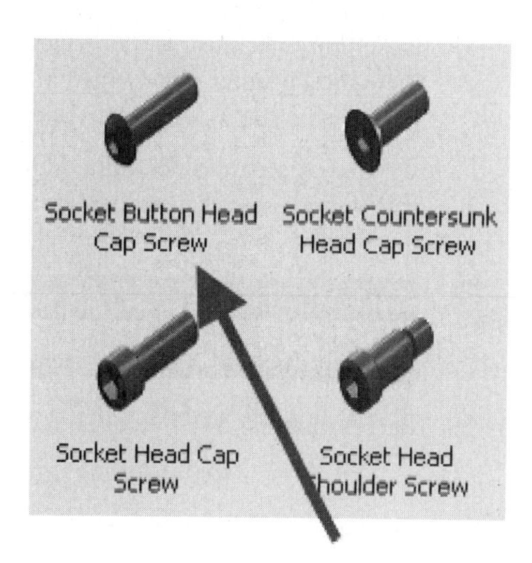

FIGURE 6.35

A #10-24 UNC × 0.75 long fastener is desired. Select these sizes either in the command manager or using the on-screen menu. While the fastener is dynamic and attached to the cursor, move the cursor over one of the bracket holes. Note that the fastener will visually and automatically locate itself centered in the hole and flush to the bracket face. Place four screws into the four bracket holes (Figure 6.36).

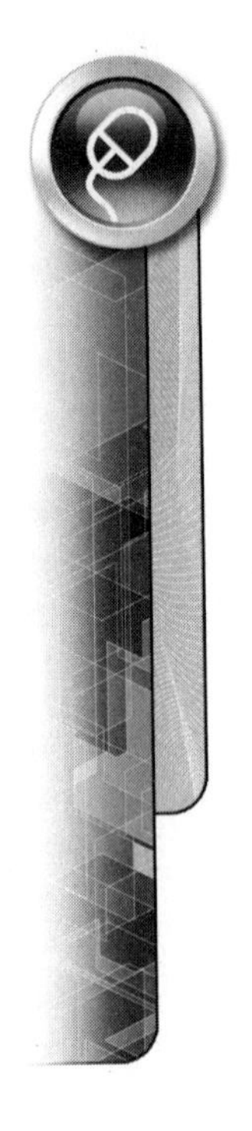

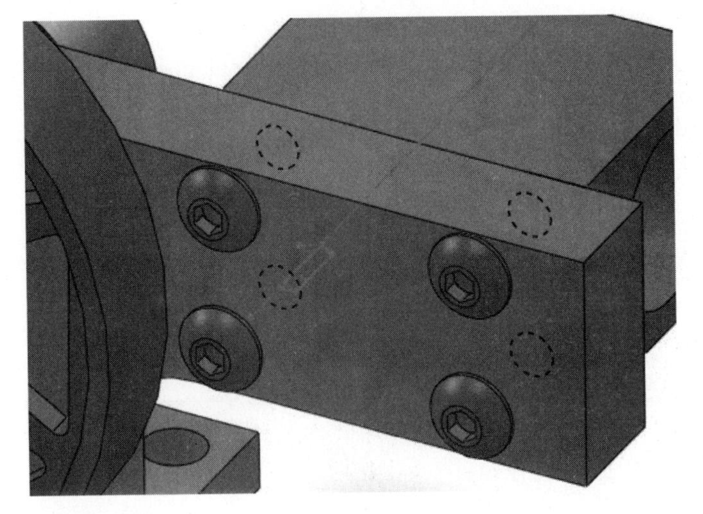

FIGURE 6.36

The assembly is complete.

MOVING AND REVOLVING A PART IN AN ASSEMBLY

While parts are being mated, oftentimes it is helpful to move or rotate a part to better expose a feature for selecting. As demonstrated in the tutorial (Step 4 and Step 5), if a part is constrained with a coincident mate, one part can be moved around the other in any 2D direction but the parts will remain coincident (aligned) in the third dimension.

If two parts are mated with a concentric mate, the two cylindrical features will remain locked on their axes, and one part can spin with relation to the other on the axis. If there are no other mates other than the concentric mate, the parts can move away from each other as long as the axes remain aligned.

As long as a part is not fully defined, it will have the ability to move and/or rotate within the assembly based on what freedoms of movement still remain.

TUTORIAL EXERCISE: 06_ASSEMBLY_02.SLDASM

Assembly Number: SWT-CH06-A2-Assembly.Sldasm

Part Numbers: SWT-CH06-A2-Plate.sldprt

SWT-CH06-A2-Knife.sldprt

Description: Assembly 2

Units: English

This is a short tutorial used to demonstrate moving and rotating parts in an assembly by assembling two parts together.

STEP 1

Create a new assembly file, then using the Insert Components command, bring file SWT-CH06-A2-Plate.sldprt (Figure 6.37).

FIGURE 6.37

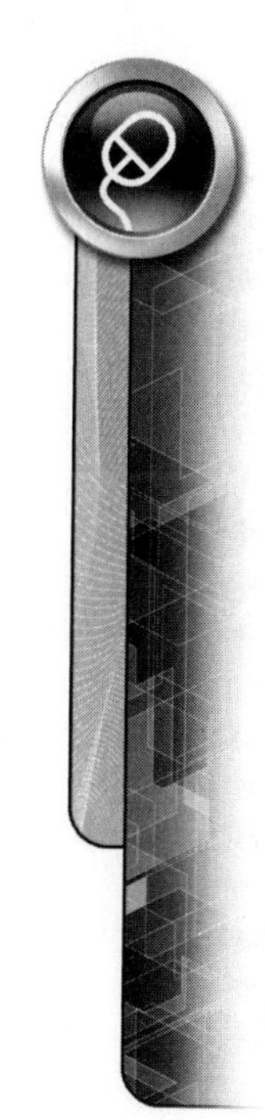

STEP 2

Using the Insert Components command, insert the SWT-CH06-A2-Knife.sldprt file and locate it near the plate without touching it (Figure 6.38).

FIGURE 6.38

STEP 3

Select the Mate command and select the concentric mate. Select the hole in the knife and then select the boss projecting from the plate. Select the Flip Mate Alignment option, then accept the mate (Figure 6.39).

NOTE

The concentric mate does not need to be selected before selecting the holes. SolidWorks will automatically choose the most logical mate, concentric. If a different mate is chosen, it can be changed to the concentric mate before accepting.

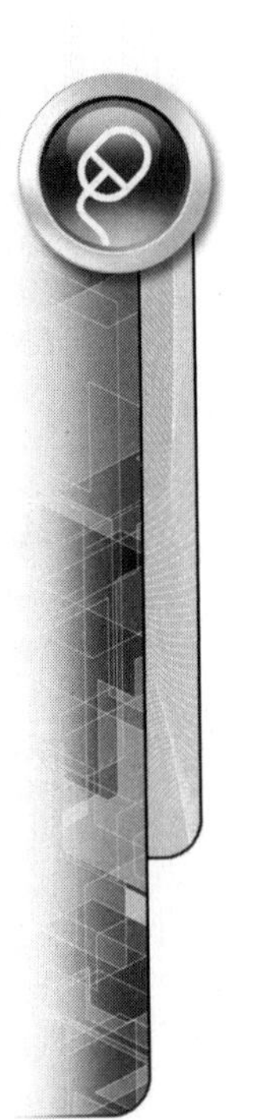

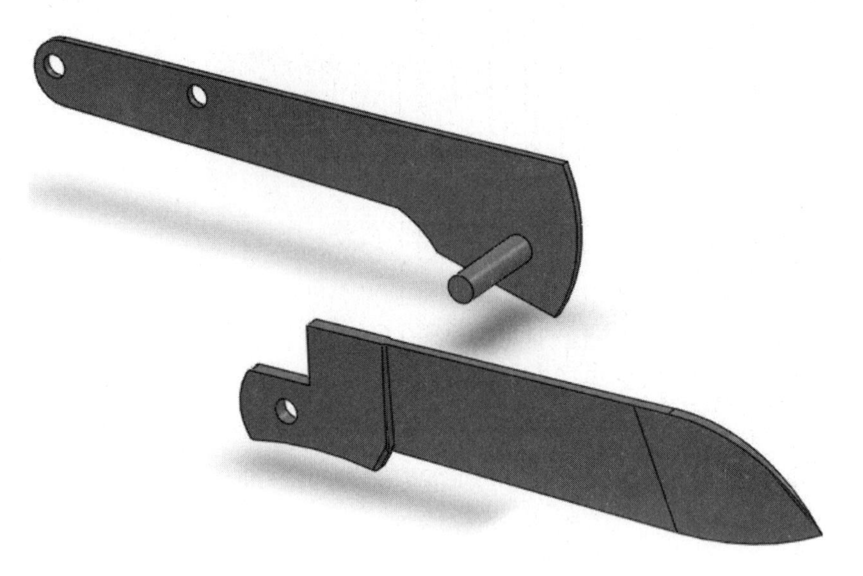

FIGURE 6.39

STEP 4

While no command is selected, click and grab the knife near the hole. Move the knife back and forth. The concentric constraint remains valid and the knife can be moved back and forth along the pin, including past the pin in either direction (Figure 6.40).

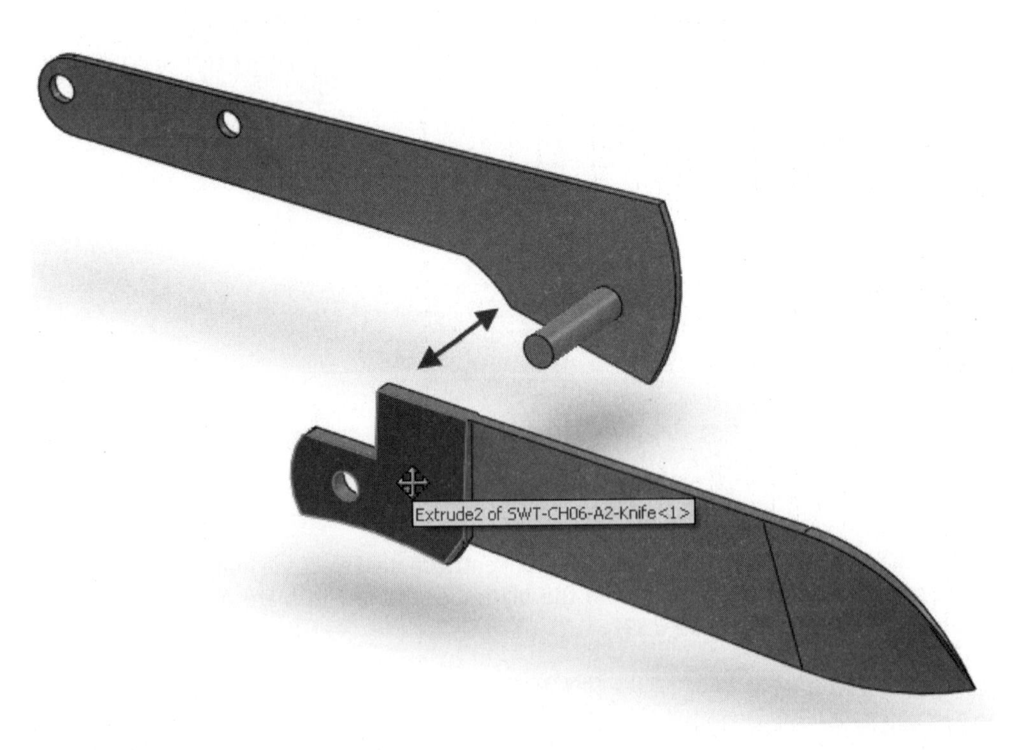

FIGURE 6.40

STEP 5

While no command is selected, click and grab the knife near the tip. Although the knife can be grabbed anywhere, it is easier to rotate the knife from this end. The knife will rotate around the pin because the concentric mate remains (Figure 6.41).

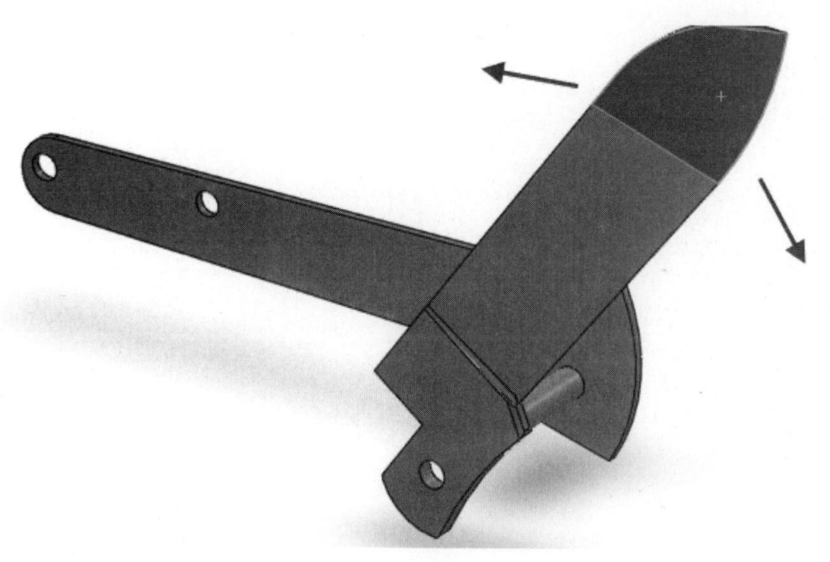

FIGURE 6.41

The ability to regularly move and rotate parts while in an assembly assists in exposing features for visibility and selection while selecting them for mating.

MATING USING REFERENCE GEOMETRY

Whenever possible, mates should be applied to features such as surfaces and holes but oftentimes though the design intent may be a more advanced relationship such as a part being centered on another part. This can easily be done by using reference geometry.

Reference planes as well as features can be used when mating parts to each other. The most common is to relate the coincident mate of two Reference Planes to each other to center parts to each other.

TUTORIAL EXERCISE: 06_ASSEMBLY_03.SLDASM

Assembly Number: SWT-CH06-A3-Assembly.Sldasm

Part Numbers: SWT-CH06-A3-Collar.sldprt,

SWT-CH06-A3-Holder.sldprt

Description: Assembly 3

Units: English

STEP 1

Create a new assembly file, then using the Insert Components command, bring file SWT-CH06-A3-Holder.sldprt (Figure 6.42).

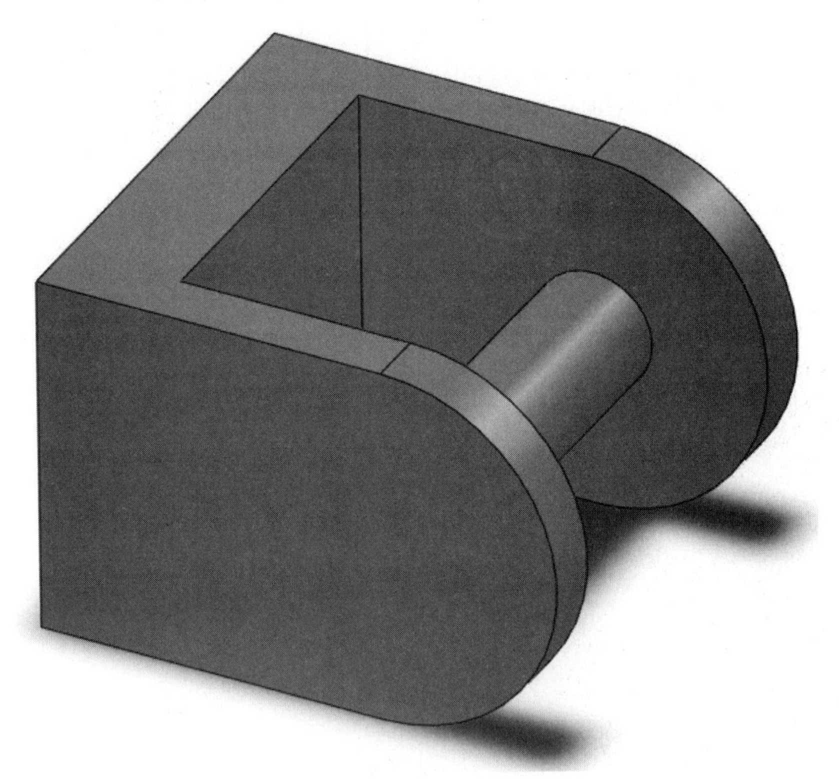

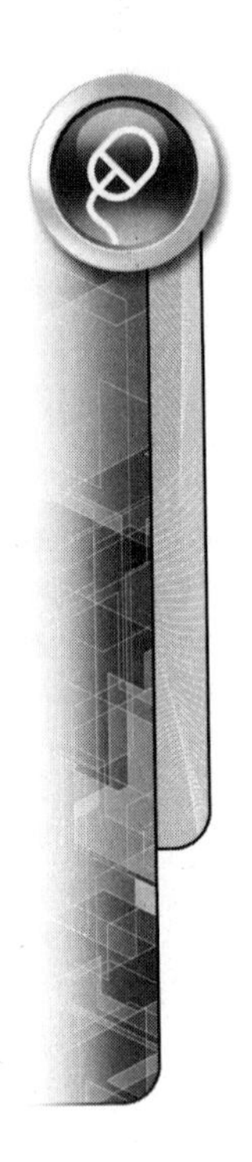

FIGURE 6.42

STEP 2

Using the Insert Component command, insert the file SWT-CH06-A3-Collar.sldprt and locate it near the holder without touching it (Figure 6.43).

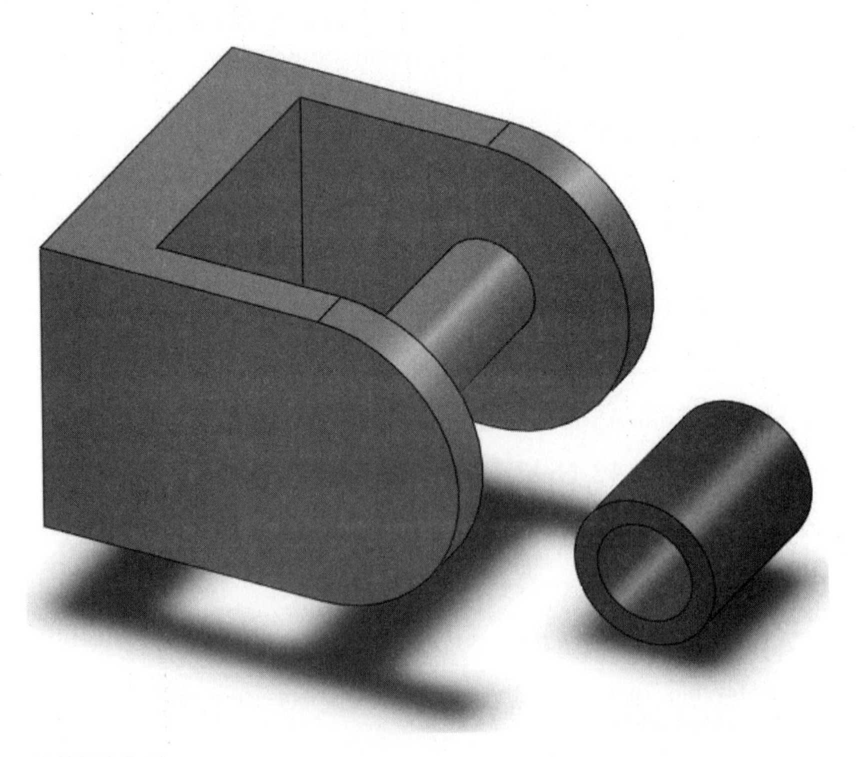

FIGURE 6.43

STEP 3

Using the Mate command, and connect the inside diameter of the collar to the axle on the holder using the concentric command (Figure 6.44).

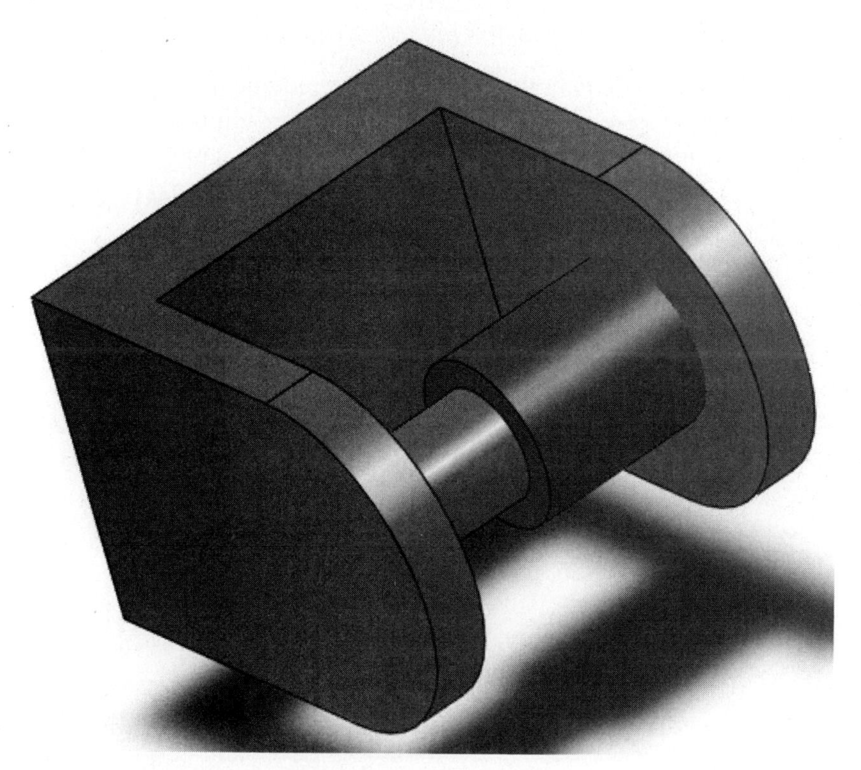

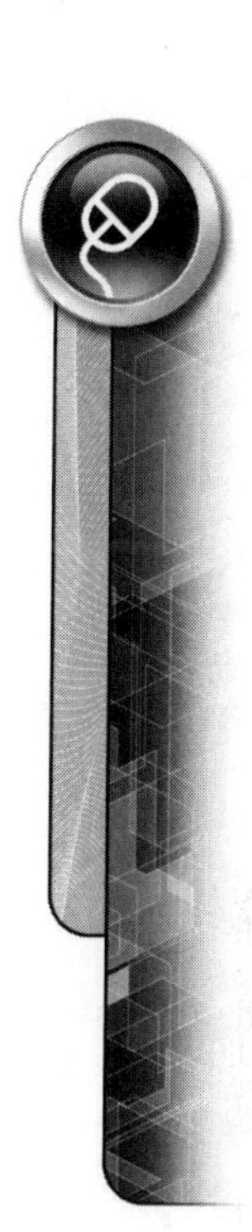

FIGURE 6.44

STEP 4

Both the Holder and the Collar have midplane reference planes and in this design the collar is 1.00 inch long and the space on the holder axle is 1.50 inches. The design intent of these two parts is to have the collar centered on the holder axle. The two midplanes will be used to center the collar on the axle.

Select the Mate command. Notice that when the Feature Command area changes to the Mate command options, on the left side of the model area the assembly Design Tree is shown (Figure 6.45). Select the "+" box next to the Assembly to open the tree. Also select the "+" next to each part. The Design Tree should look like the figure.

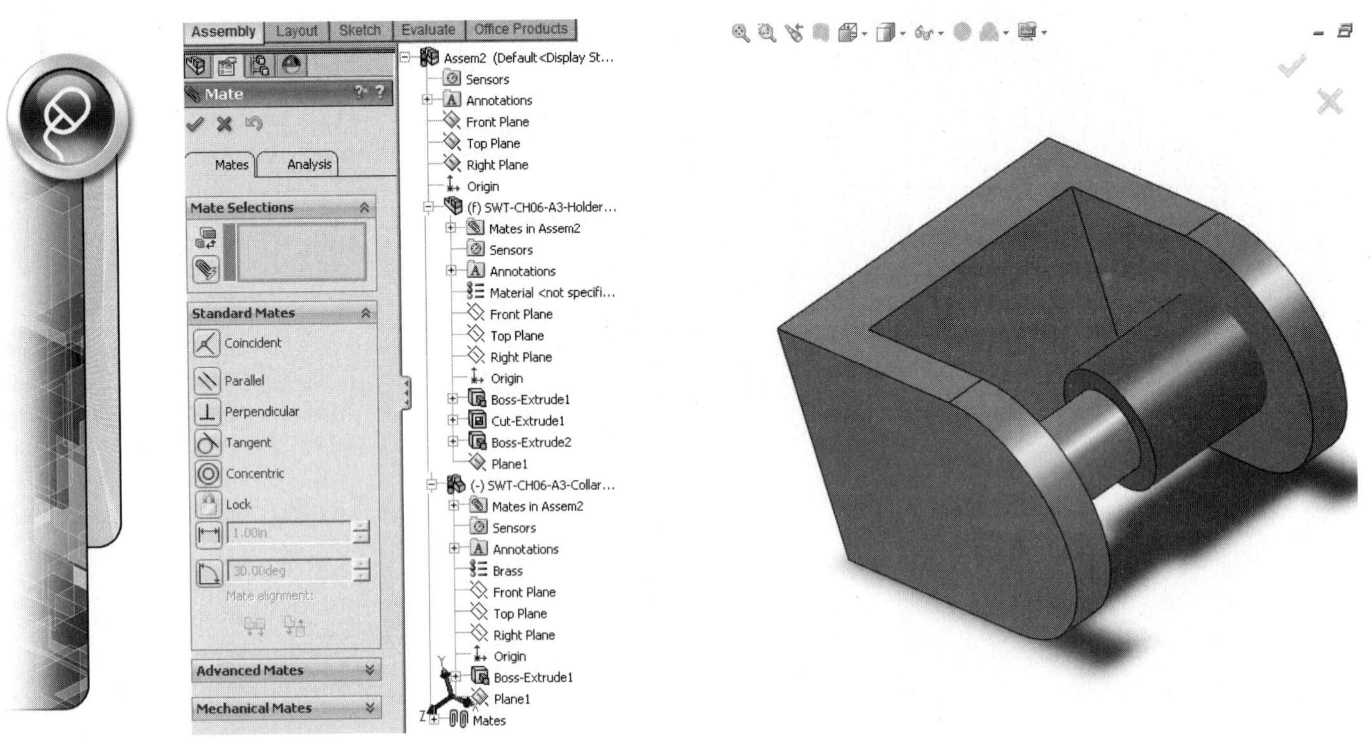

FIGURE 6.45

STEP 5

Notice in the Design Tree that the midplane for the holder is called "Plane 1." The midplane for the collar is also called "Plane 1." Select both of these planes while in the Mate command. The coincident mate is automatically selected, so select OK. The planes will align and the collar will be centered on the holder axle (Figure 6.46).

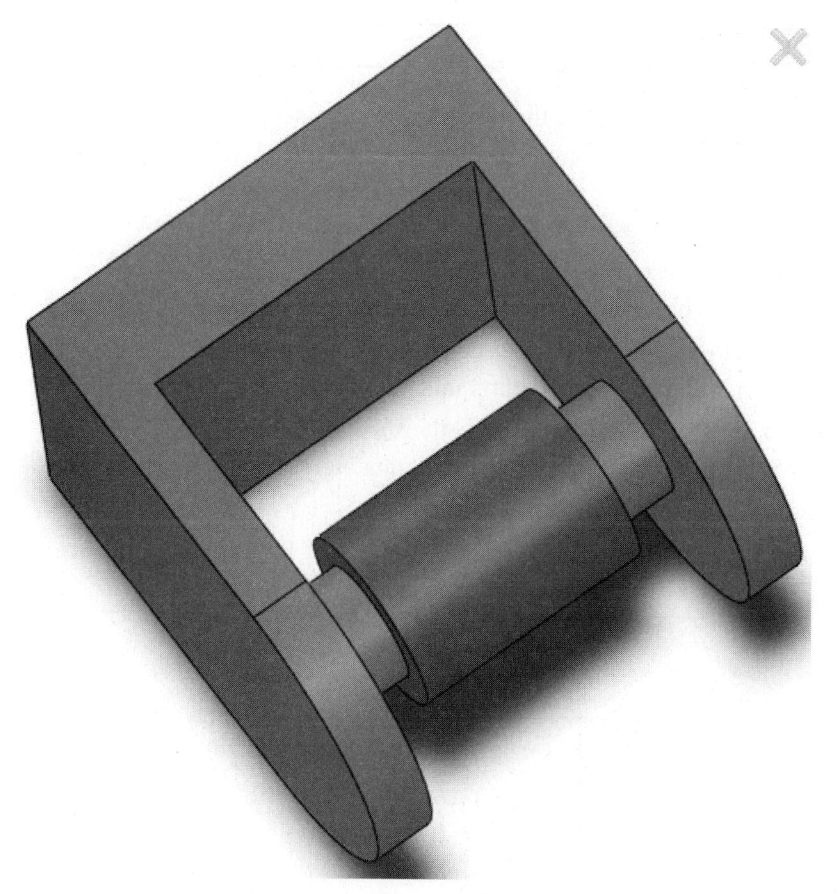

FIGURE 6.46

DEFINING STATIONARY AND MOVING PARTS

When mating parts in an assembly, they should be fully mated to match the same constraints realized in a physical assembly. Fully mated does not mean fully defined though.

Parts that are stationary in the assembly and will have no freedom of movement in any direction or any ability to rotate should be fully defined. Parts that are dynamic in any way such as the ability to rotate or slide, as a function of the assembly, should not to be fully defined. The freedoms that a part may have in a physical assembly are the same freedoms that should not be defined in the model. This assists in the motion of the assembly when completed.

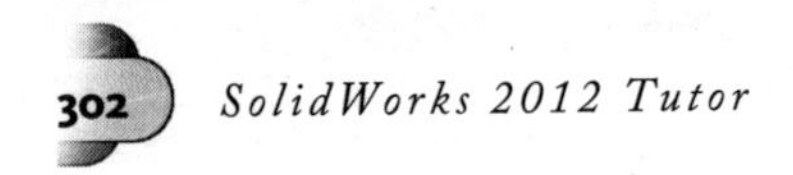

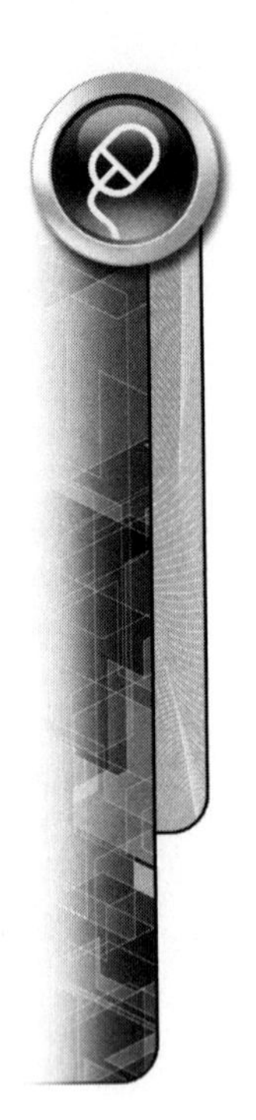

REVIEW QUESTIONS

1. Which part is best when selecting the first part for an assembly?
2. When placing additional new components in an assembly, where should they initially be located?
3. Can parts be moved in an assembly once a mate is defined to the part?
4. How can a surface be selected without rotating the view to visually see it?
5. Describe why a part may need to be flipped while mating.
6. Describe why a dimension may need to be flipped while mating.
7. What is the main reason for leaving parts underdefined while mating?
8. Why is there a need to move and rotate parts while mating?
9. When would reference geometry be needed for mating?
10. How can parts show motion within its constraints?

EXERCISES

6.1

Create the following weldment assembly with MMGS units. The assembly contains three individual pieces of steel stock. The gusset support is to be centered onto the back of the face plate.

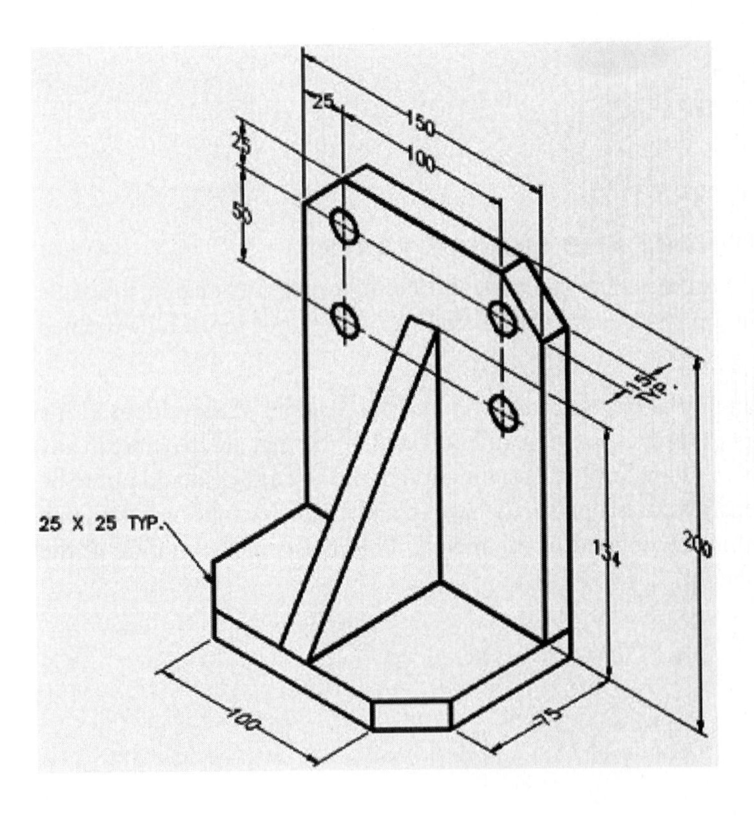

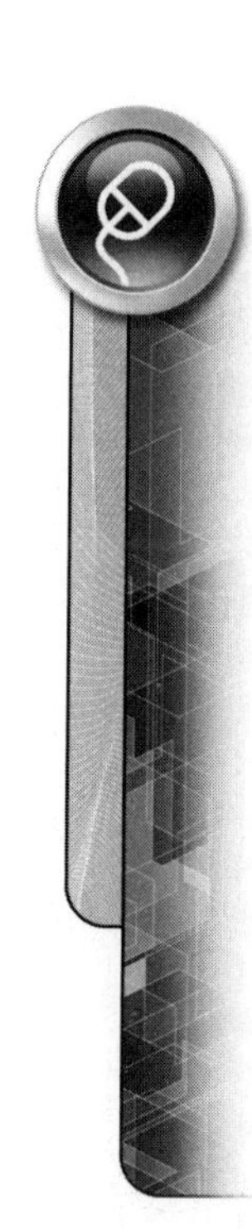

6.2

Create the following weldment assembly with IPS units. The assembly contains four individual pieces of steel stock.

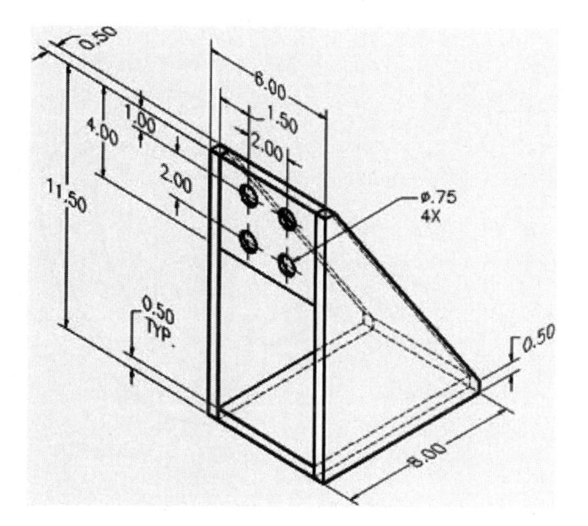

6.3

Using the precreated part files found on the textbook CD, assemble them in an assembly file using the appropriate mates.

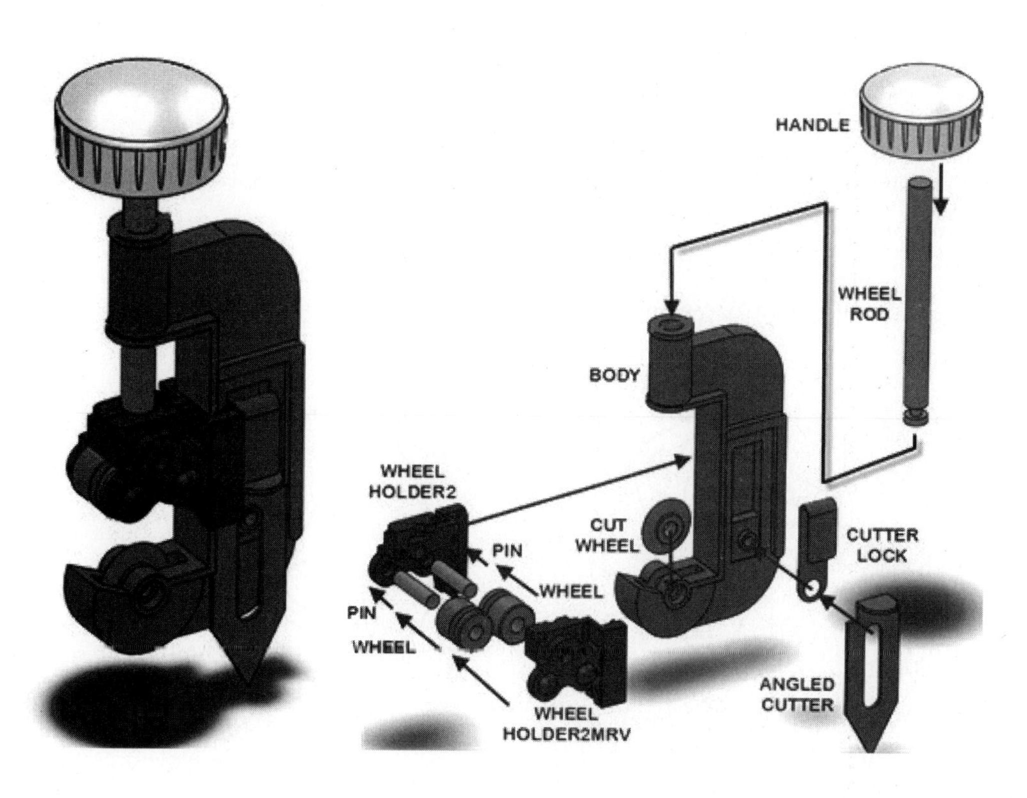

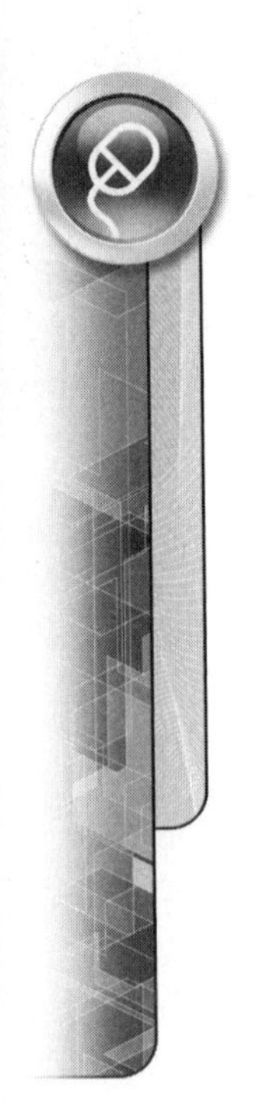

6.4

Create the following five parts as individual part files. Assemble the components in an assembly file. The shaft is held in place on the outside of the bracket with two #11-100-MSH15 retaining rings.

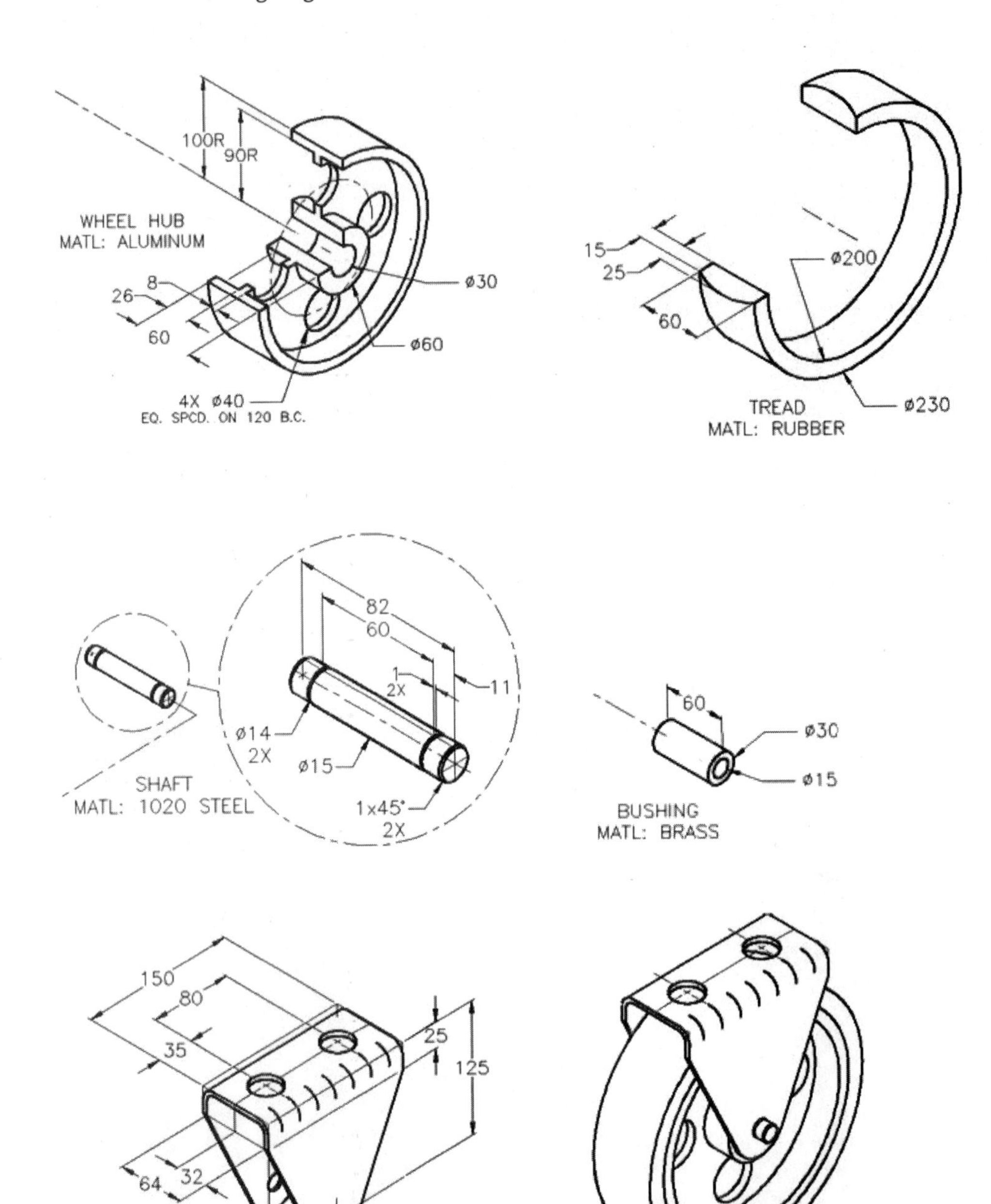

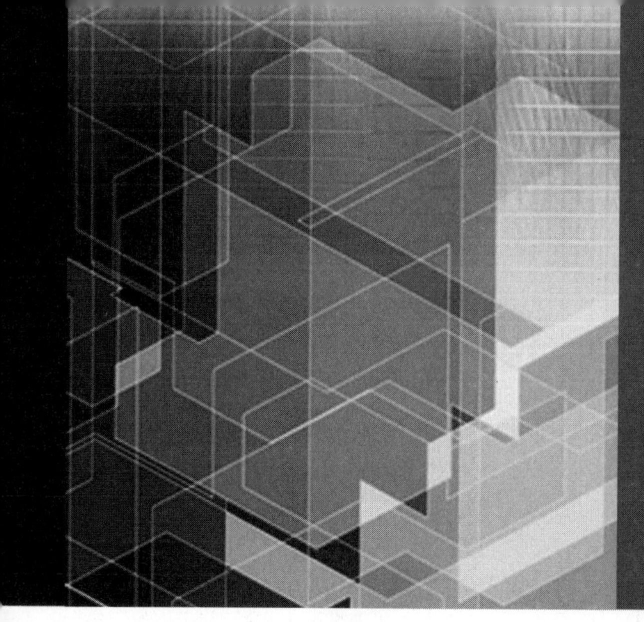

CHAPTER 7

Generating 2D Assembly Drawings

This chapter discusses various techniques and tools used to assemble part files as well as imported parts to generate 2D assembly drawings:

- Importing an assembly file into a drawing
- Manipulating section views in an assembly drawing
- Creating a bill of materials
- Adding balloons
- Adding welding symbols

Initially, assembly drawings are created similarly to detail drawings as discussed in Chapter 5—Generating 2D Part Drawings; the difference is instead of importing a part file into the drawing file, an assembly file is imported. All of the topics, such as editing the titleblock, adding dimensions, adding text and other annotations, and creating section views, covered in Chapter 5 are also relevant.

There are additional techniques needed to complete assembly drawings, such as adding an Items List (Bill of Material), adding item balloons, and adding weld symbols (Figure 7.1).

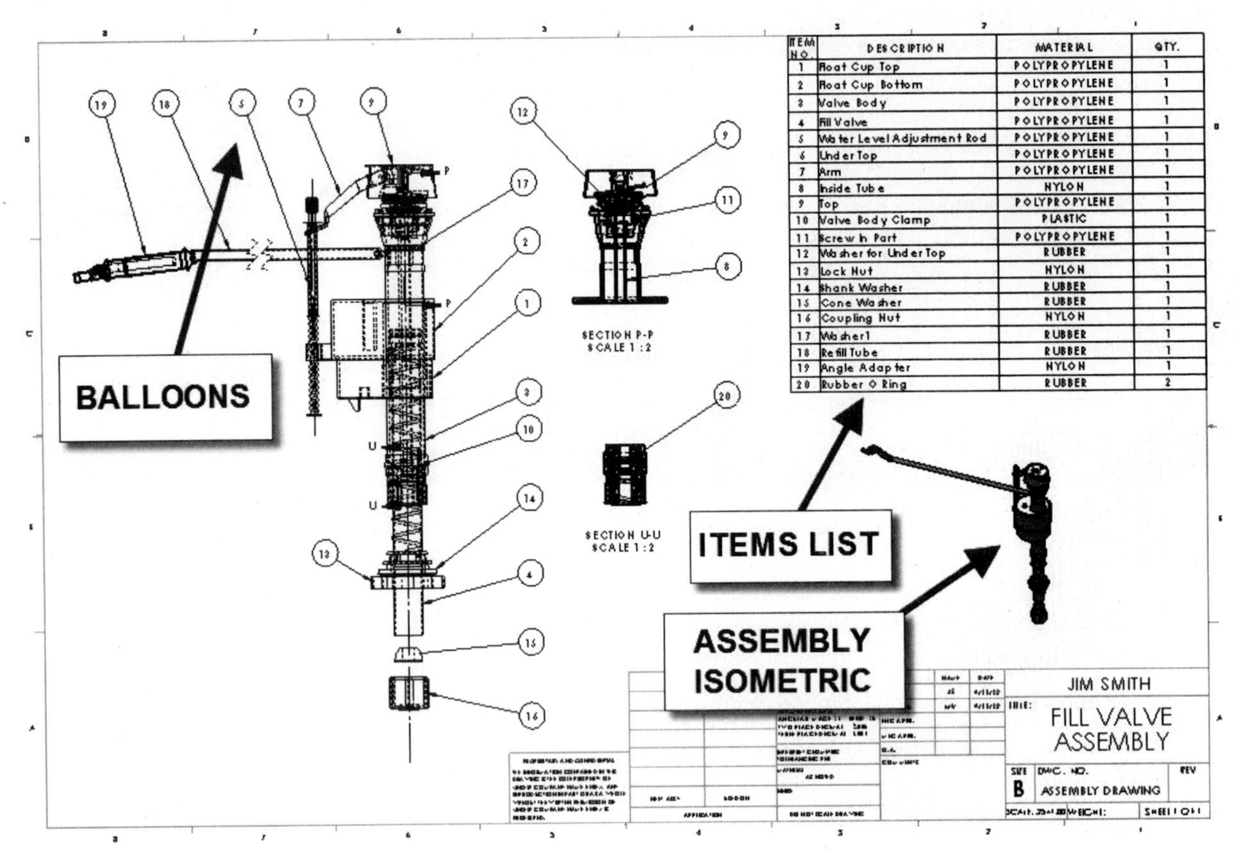

FIGURE 7.1

IMPORTING AN ASSEMBLY FILE INTO A DRAWING

Creating an assembly drawing is the same process as creating a detail drawing as discussed in Chapter 5; the only difference is when selecting a file to view on a drawing, an assembly file is selected instead of a part file. The first step is to create a new drawing file and while in the Model View command, browse for the desired file and accept it, then add any additional views desired. The views will reflect the multiple parts as they are assembled in the assembly file.

MANIPULATING SECTION VIEWS IN AN ASSEMBLY DRAWING

The Section View command can be used in an assembly drawing and more likely used on most assembly drawings. Since most assemblies have internal components, at least one section view is typically needed to expose the internal components for ballooning.

Once a section view is created on a drawing, the hatch patterns can be modified to change the style, scale, and angle of the hatch patterns.

TUTORIAL EXERCISE: 07_ASSEMBLY_SECTION.SLDDRW

Assembly Drawing: SWT-CH07-Assembly_Section.SLDDRW

Assembly Number: SWT-CH07-Pepper-Mill-Assembly.SLDASM

Description: Pepper Mill Assembly Drawing

Units: English

This exercise will demonstrate creating an assembly drawing, adding views, creating an assembly section view, and manipulating the hatching in the section view.

STEP 1

Create new drawing file making it *B (ANSI) Landscape* size, then change the units to English (IPS) and change the drafting standard to ANSI in Options. In the Command Manager, right-click on Sheet and select Properties. Change the Type of Projection to Third angle.

Select Model View and browse for the file SWT-CH07-Pepper-Mill-Assembly.SLDASM. Since hidden lines are not typically shown on assembly drawings, use the display style of hidden lines removed and bring in the bottom view to place it on the drawing sheet as shown in Figure 7.2. Accept the view and exit the Model View command.

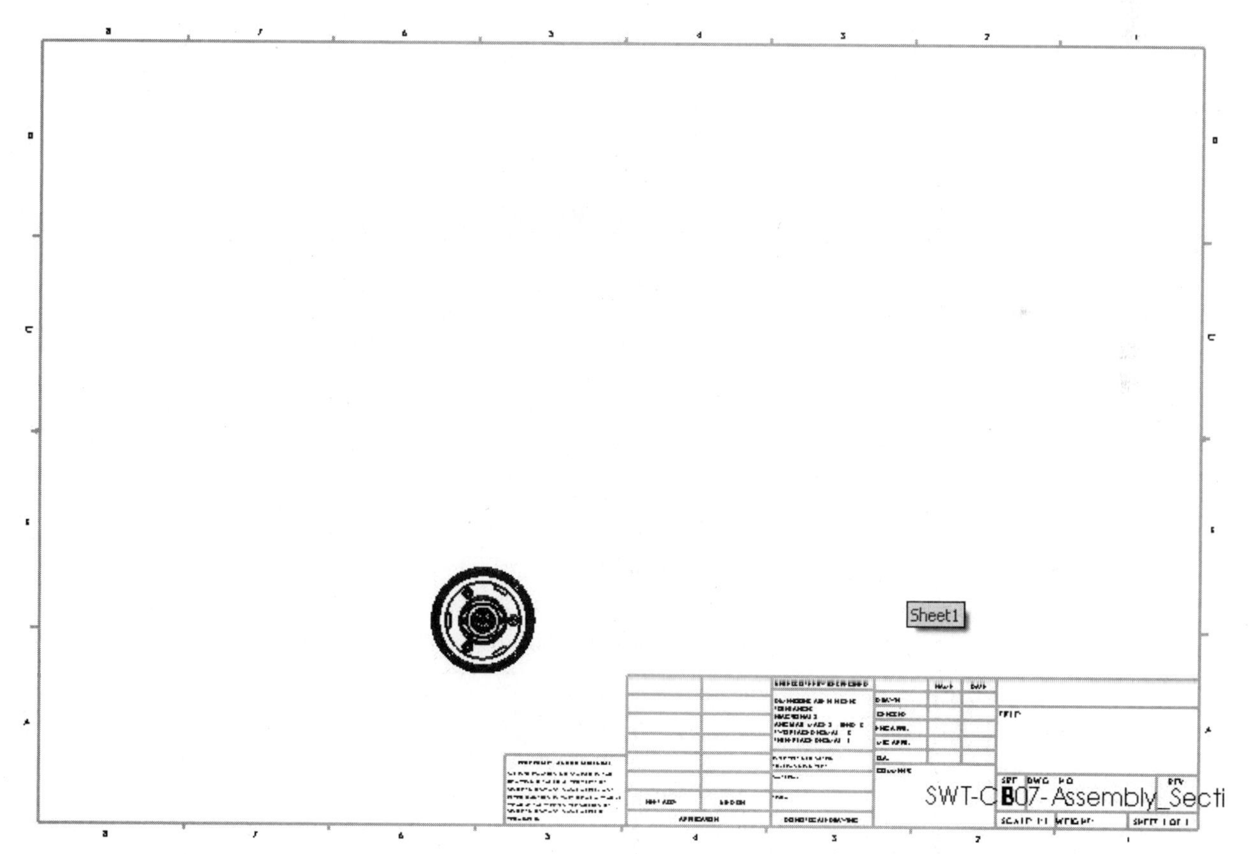

FIGURE 7.2

STEP 2

Select the Section View command and sketch a section line as shown in Figure 7.3.

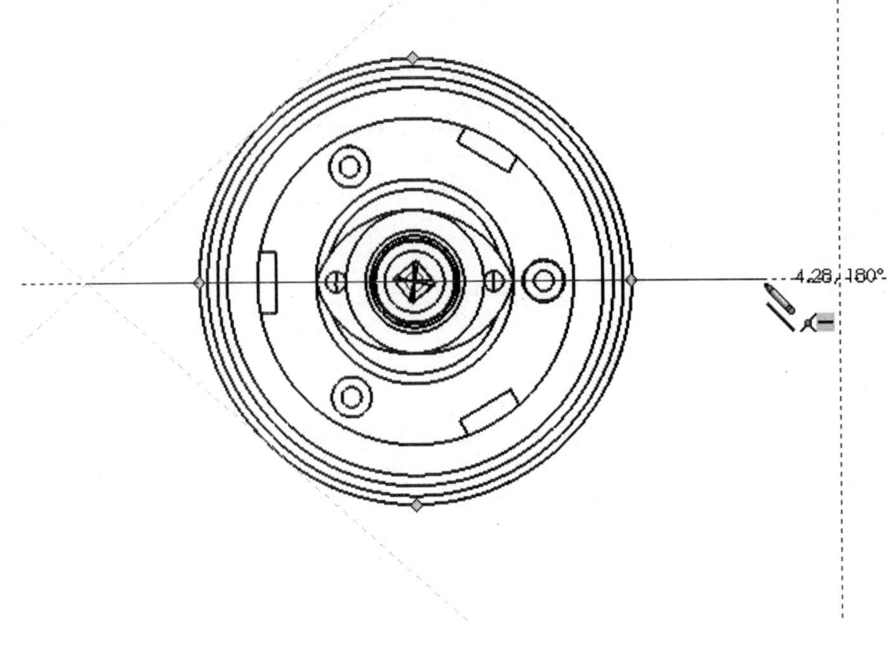

FIGURE 7.3

STEP 3

When the Section View window appears, ensure that Autohatching is selected and that Exclude fasteners is not selected. The various parts will be automatically assigned unique hatch styles and settings to each part (Figure 7.4).

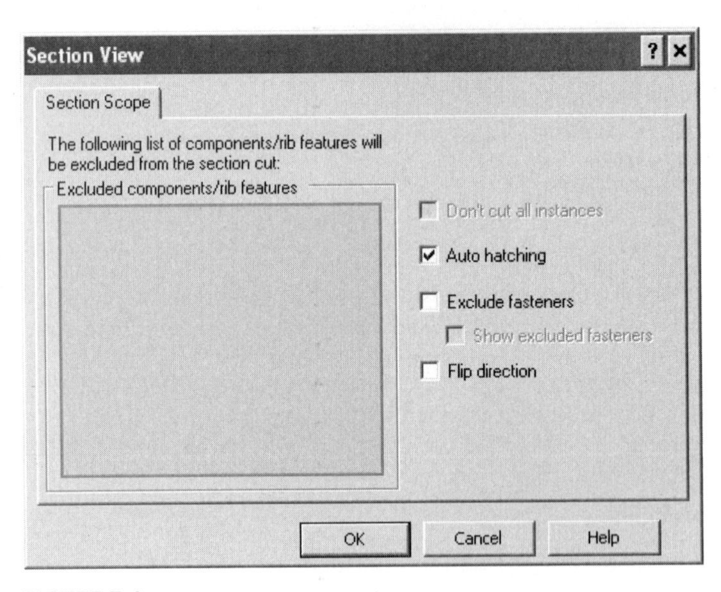

FIGURE 7.4

STEP 4

Locate the section view above the bottom view as shown in Figure 7.5. Notice the various hatch patterns defined for the various parts.

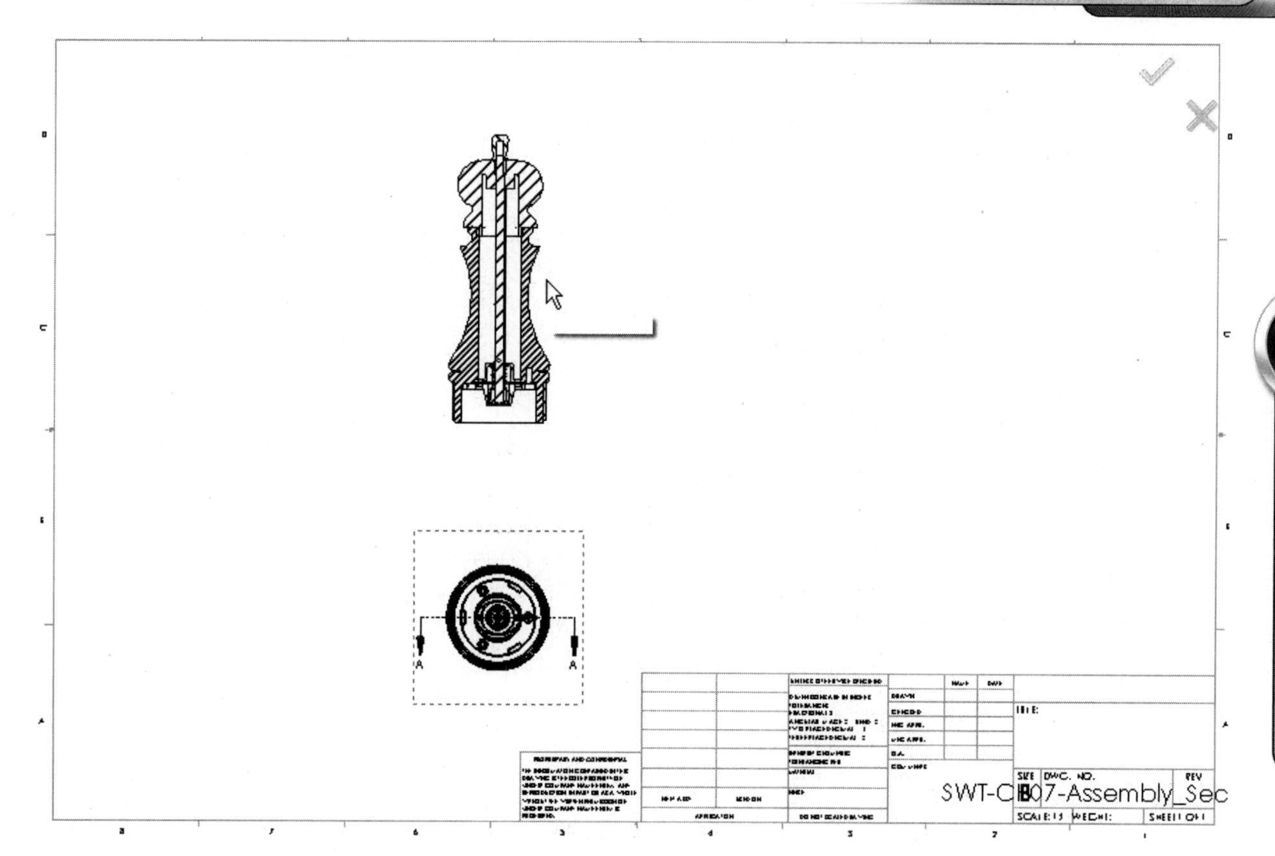

FIGURE 7.5

STEP 5

Zoom in to see the front view more clearly. Although different hatch patterns were utilized using the autohatching method, some are too similar to accept, such as the center post and the mill head. Click on the hatch pattern that is in the head of the mill as shown in Figure 7.6. The feature manager provides the properties that can be edited with respect to the hatch pattern.

Uncheck the Material crosshatch which will allow the settings to be changed. Change the hatch pattern to ANSI32 (Steel), change the scale to 2, and change the angle to 15. Accept the new hatch pattern settings.

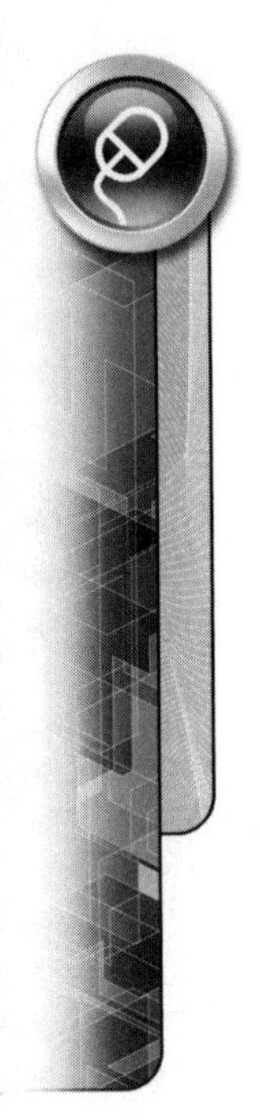

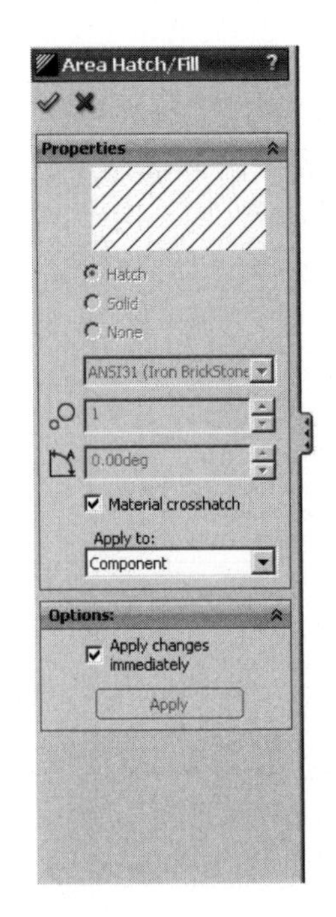

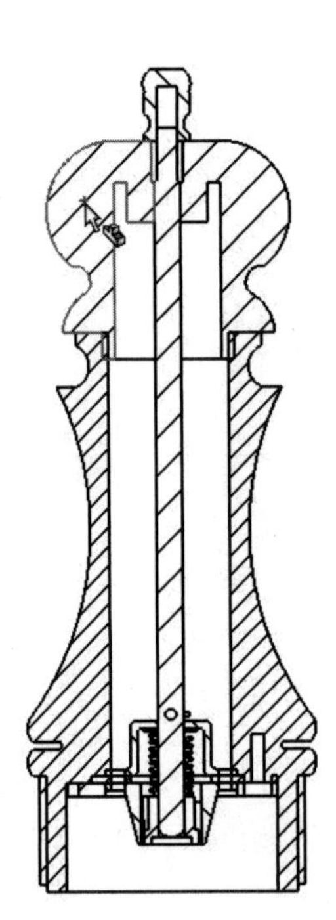

FIGURE 7.6

STEP 6

In the same manner, select the mill body and change the hatch pattern settings to a scale of 2 and a hatch angle of –15. Also change the hatch pattern of the bottom ring (part 2) to a scale of 4 and an angle of 90. Located at the bottom of the post, change the hatch pattern of the nut (part 10) to the settings of pattern ANSI33 (Bronze Brass), a scale of 8, and an angle of –20. Also at the bottom of the post, edit the hatch settings for the plug (part 9) to the settings of a scale of 8 and an angle of 120. The various parts now have distinctly different hatch patterns and can easily be identified visually from each other as shown in Figure 7.7.

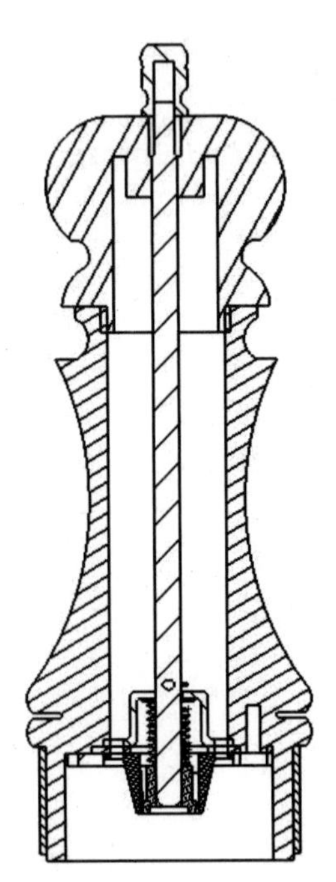

FIGURE 7.7

CREATING A BILL OF MATERIAL

Once the desired views are added to an assembly drawing, a Bill of Material can be added to identify all of the parts used in the assembly. The Bill of Material is created as a table and the process is semi-automated by using the Bill of Material command found on the Annotation toolbar in the Table pull-down menu as shown in Figure 7.8.

The format of the table is predefined with four columns: Item, Part Number, Description, and Quantity. These columns can be modified after the fact with respect to width, and the fields can also be modified. The table reacts similarly to a spreadsheet.

The part number and description cell values are linked to properties of the part files. The part number is linked to the part file name and the description is linked to a custom property that can be set in the part file options. These links can be broken by modifying the BOM by typing in overriding values and can be restored by clearing the cell of the overriding values.

When placing the BOM, the corners will snap to common points, such as corners of the border, as the BOM is moved around the drawing until a point is selected.

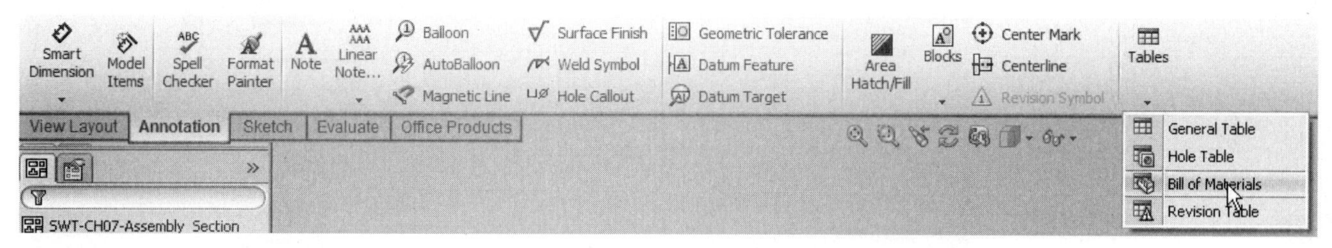

FIGURE 7.8

ADDING BALLOONS

It is recommended to place a Bill of Material prior to ballooning an assembly drawing because the balloons can be attributed to the BOM. Ballooning can be accomplished in two ways, autoballooning or manually ballooning. Both methods of ballooning can be found on the Annotation menu (Ballooning & Autoballooning).

The Autoballoon command will automatically identify all of the parts in the assembly and place one balloon for each part with a leader going from the balloon and pointing to the part. If no view is selected, the balloons will be arbitrarily assigned views; typically the majority of balloons will go on the first drawing view that was inserted and if any parts are not visible in that view, they will go to views that do visually represent the parts. If one view is selected prior to selecting the Autoballoon command, then all of the balloons will be attached to that view.

There are several settings that can be changed when autoballooning, such as the numbering sequence of the balloons, the style of the balloons, and the pattern type (where the balloons are located). Manipulation of these settings will be demonstrated in the tutorial (Figure 7.9).

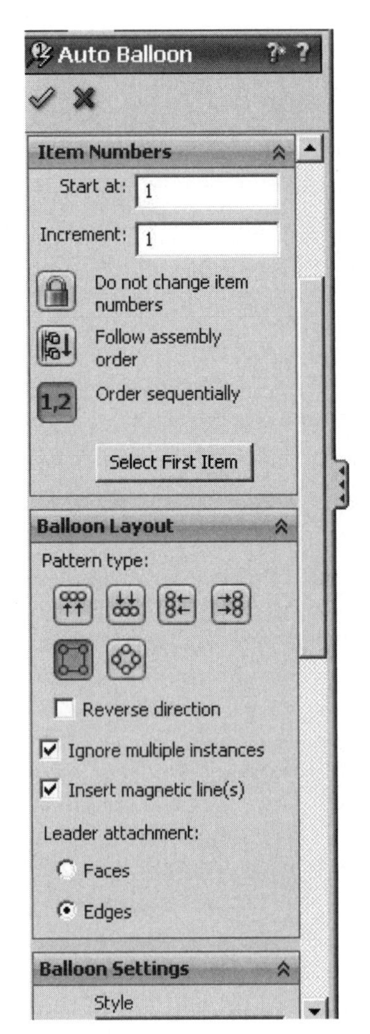

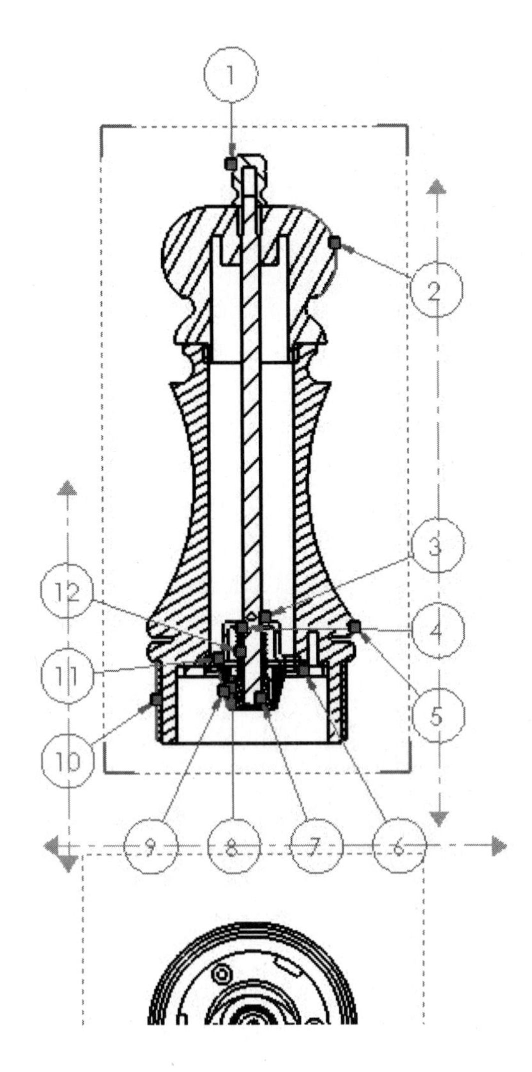

FIGURE 7.9

Manual ballooning although more tedious will give you a better opportunity to locate the balloon positions exactly where they are desired such as aligning the balloons in multiple sets (left and right side of a view) while still maintaining the association of the balloon numbers and their matching part on the Bill of Material. When manually ballooning, a good knowledge of the location of each component is required to ensure each part is properly identified as shown in Figure 7.10.

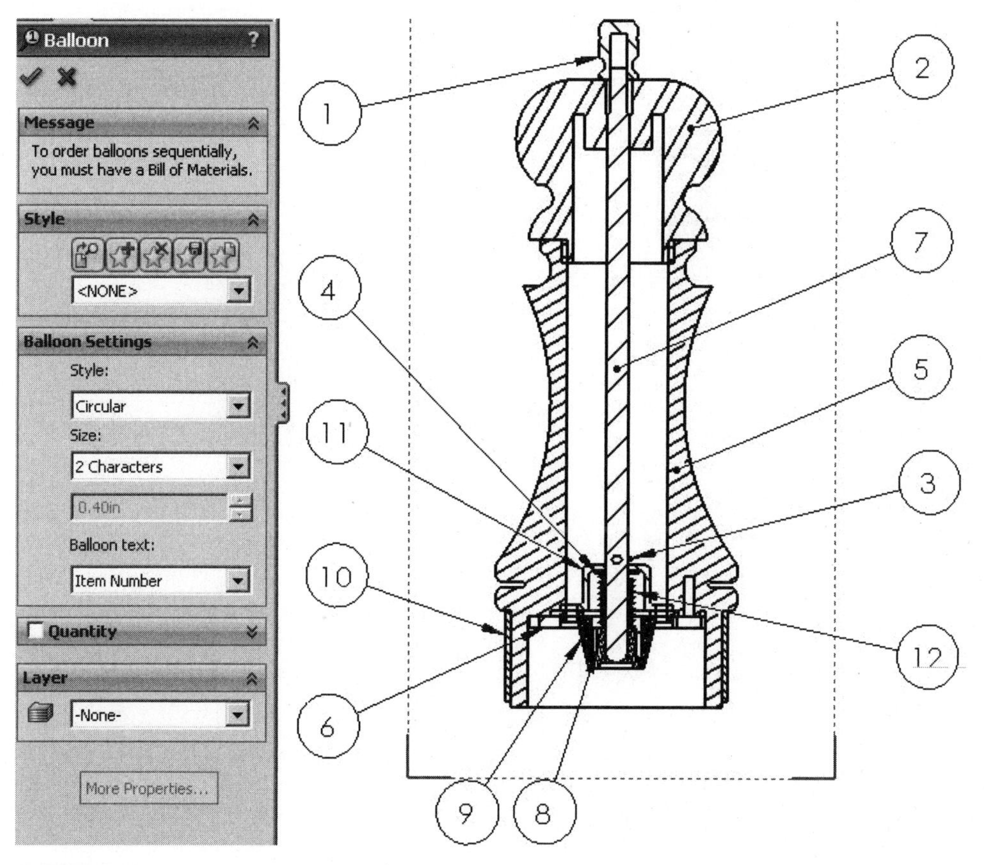

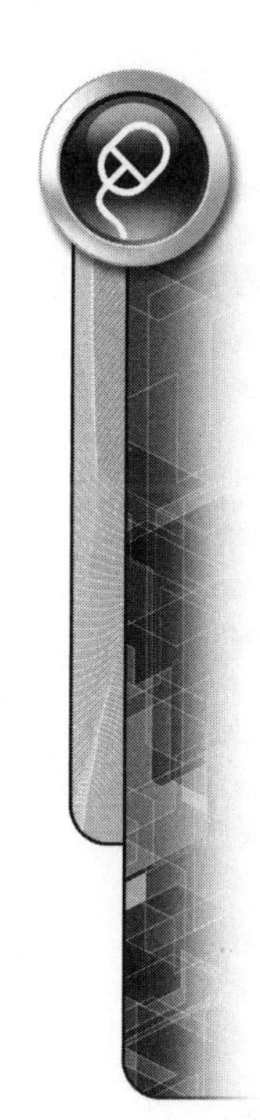

FIGURE 7.10

TUTORIAL EXERCISE: 07_ASSEMBLY_01.SLDPRT

Assembly Drawing: SWT-CH07-Assembly_Valve.SLDDRW

Assembly Number: SWT-CH07-Valve_Assy.SLDASM

Description: Assembly Drawing

Units: English

This exercise will demonstrate how to add a Bill of Material to an assembly drawing as well as how to add balloons using the manual method of ballooning.

STEP 1

Create new drawing file making it *B (ANSI) Landscape* size. Then change the units to English (IPS) and change the drafting standard to ANSI in Options. In the command manager, right-click on Sheet and select Properties. Change the Type of Projection to Third angle.

Select Model View and browse for the file SWT-CH07-Valve_Assy.SLDASM. Using the display style of hidden lines removed, bring in the top view and place it on the drawing sheet as shown in Figure 7.11. Accept the view and exit the Model View command.

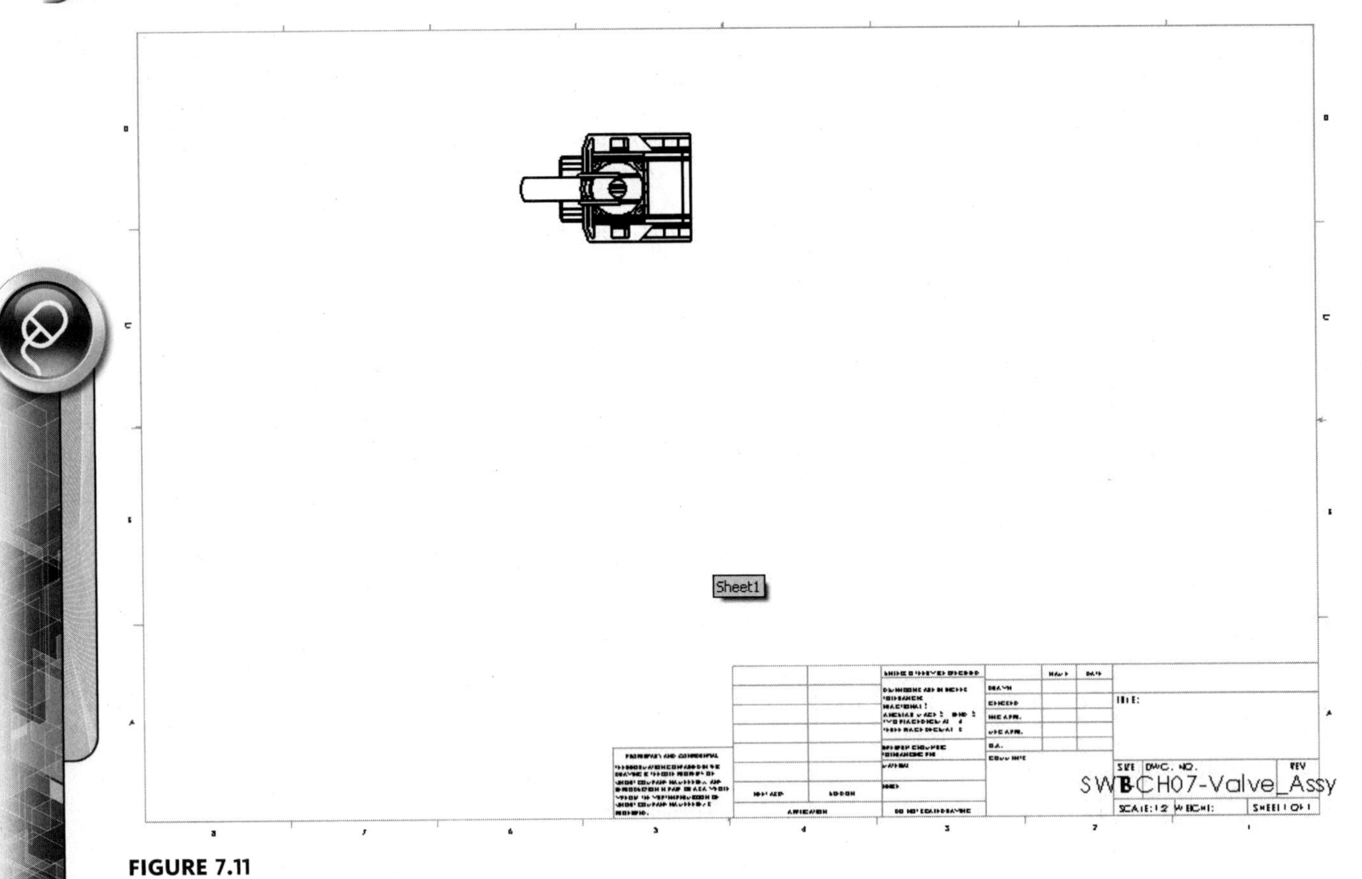

FIGURE 7.11

STEP 2

On the View Layout toolbar select the Section View command. Create a cutting plane line as represented by the blue line in Figure 7.12. When the Section View menu pops up, select the Flip direction and accept the settings by selecting OK.

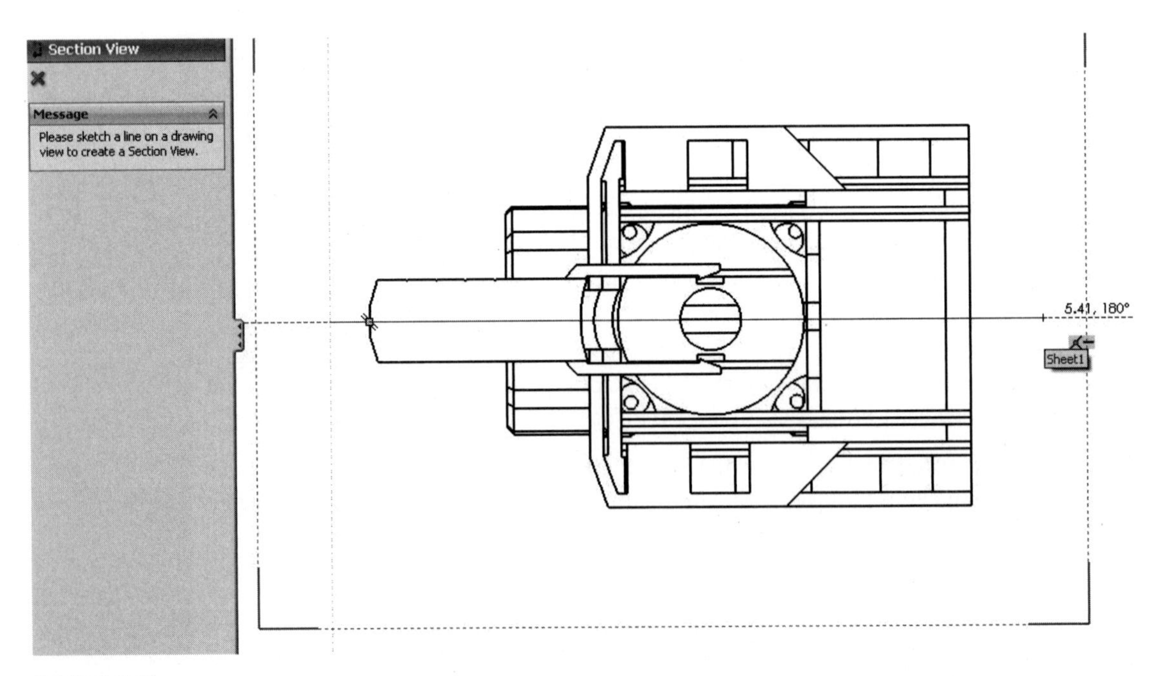

FIGURE 7.12

STEP 3

Place the section view as a front view located below the top view as shown in Figure 7.13. In the command manager on the left, remove the section letter callout by removing the "A" from the field, then accept the command. Select the "Section A-A" text that is identifying the front view and hit the delete key on the keyboard. If a section view is in projection as it is here, labeling of the cutting plane line and the view itself are not necessary.

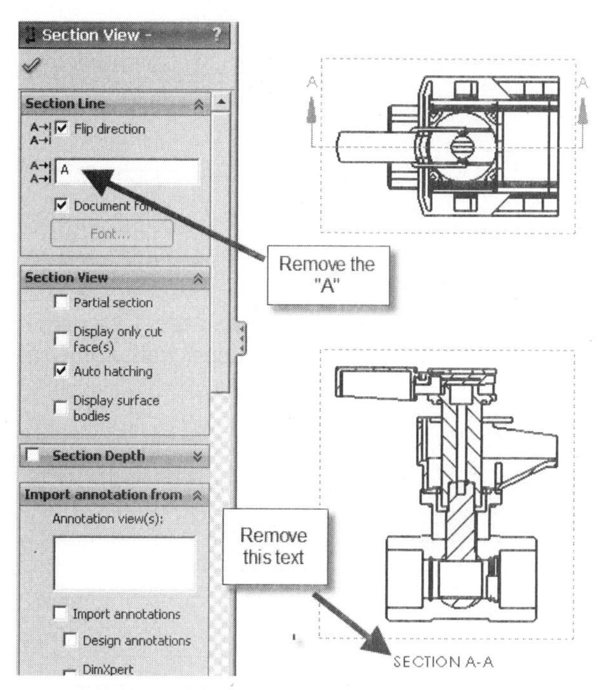

FIGURE 7.13

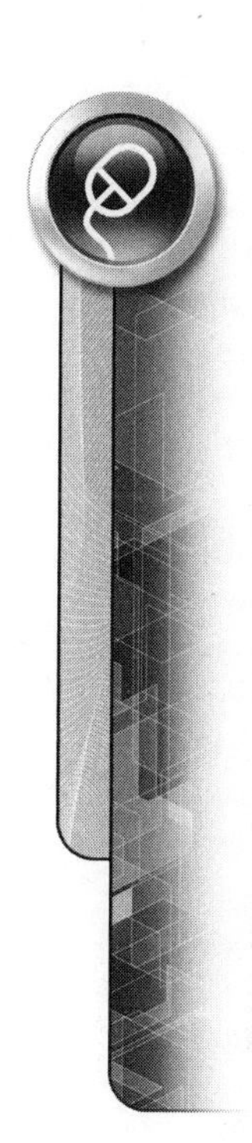

(Note that the valve body is made up of two halves that are assembled and they are parted by the cutting plane line and therefore the half that is exposed is not hatched.)

STEP 4

Add the appropriate center lines and center marks. Modify any hatching patterns to ensure each is visually distinct. The O-rings in the front view can have a solid hatch added to them. Zoom in on the front view. Ensure no component is selected when entering the Area Hatch/Fill command on the annotation toolbar. Select the Solid fill property in the command manager at the left, then select inside the four circles representing the sectioning of the two O-rings as shown in Figure 7.14. Accept the Hatch command.

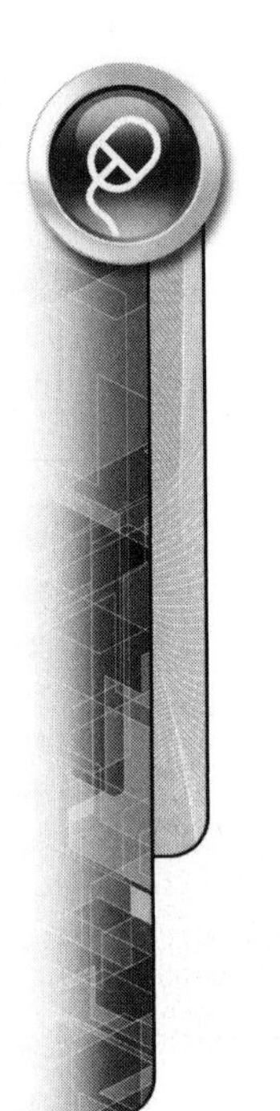

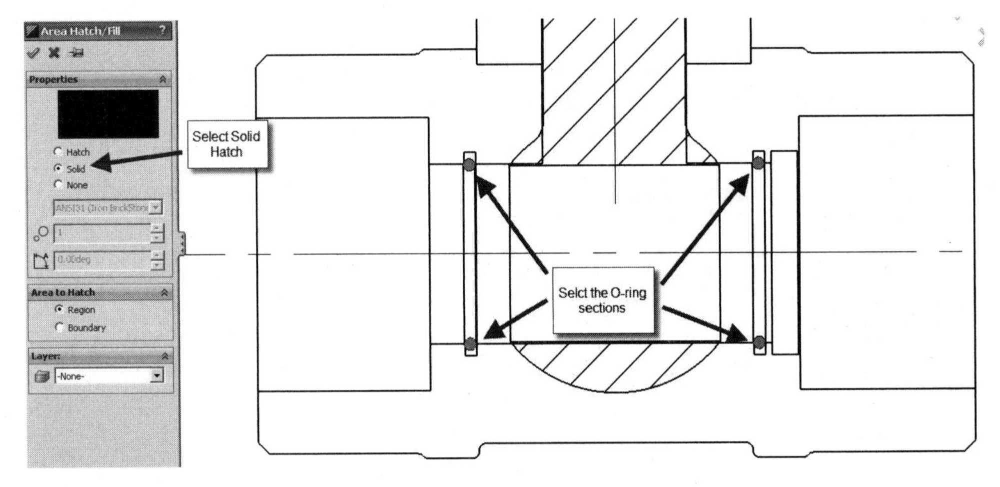

FIGURE 7.14

STEP 5

On the Annotation toolbar, select the Bill of Material command from the Tables pull-down menu. Accept the default settings and place the table in the upper right corner of the drawing (Figure 7.15).

ITEM NO.	PART NUMBER	DESCRIPTION	QTY.
1	NC10050		1
2	NC10052		1
3	NC10056		2
4	NC10051		1
5	NC10053		1
6	NC10054		1
7	NC10055		1

FIGURE 7.15

STEP 6

Click on the first description field (for item 1) and type in the drawing title name for that part. When the field is initially selected, a warning window appears that controls the link from the assembly drawing and the part file. It is desired to keep the link between the files, so select the "Keep Link" button and continue filling out the description field (in all capitals) (Figure 7.16).

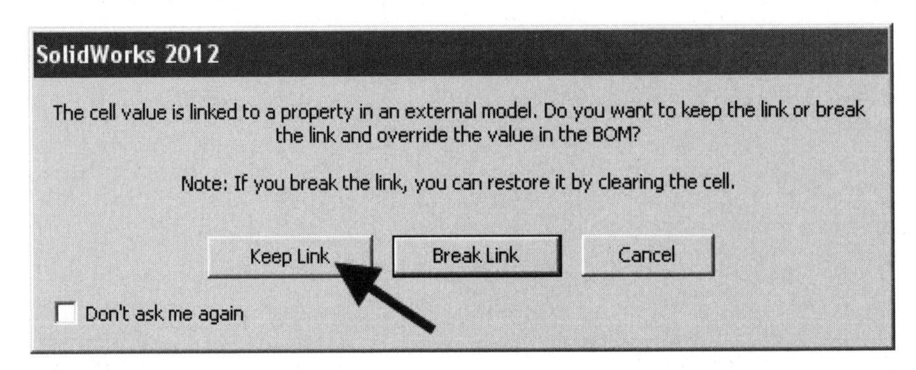

FIGURE 7.16

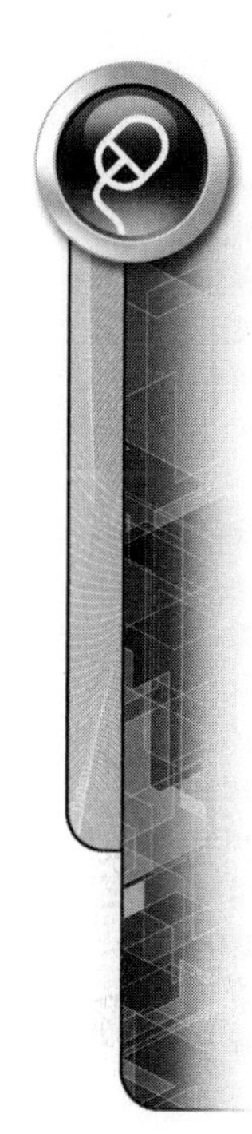

STEP 7

Continue filling out all of the description fields while continuing keeping the cell value links. Ensure that all text is all capital letters (Figure 7.17).

ITEM NO.	PART NUMBER	DESCRIPTION	QTY.
1	NC10050	L VALVE DETAIL	1
2	NC10052	BALL VALVE DETAIL	1
3	NC10056	O-RING DETAIL	2
4	NC10051	R VALVE DETAIL	1
5	NC10053	HANDLE HOLDER DETAIL	1
6	NC10054	HANDLE DETAIL	1
7	NC10055	HOUSING DETAIL	1

FIGURE 7.17

STEP 8

Move the cursor over the table. Note that the table visually changes to an editing mode where the size of columns and rows can be modified. As shown in Figure 7.18, grab one column line controlling the width and make each column narrower to minimize the size of the table. As the table is being shortened, it will pull away from the corner of the border. Grab the last column by the top cell labeled "D" and move this column to the left and place it as the second column.

	A	B	C	D
1	ITEM NO.	PART NUMBER	DESCRIPTION	QTY.
2	1	NC10050	L VALVE DETAIL	1
3	2	NC10052	BALL VALVE DETAIL	1
4	3	NC10056	O-RING DETAIL	2
5	4	NC10051	R VALVE DETAIL	1
6	5	NC10053	HANDLE HOLDER DETAIL	1
7	6	NC10054	HANDLE DETAIL	1
8	7	NC10055	HOUSING DETAIL	1

FIGURE 7.18

STEP 9

Once the columns are narrowed, grab the table by the move table icon as shown in Figure 7.19 and move the table back into the drawing corner.

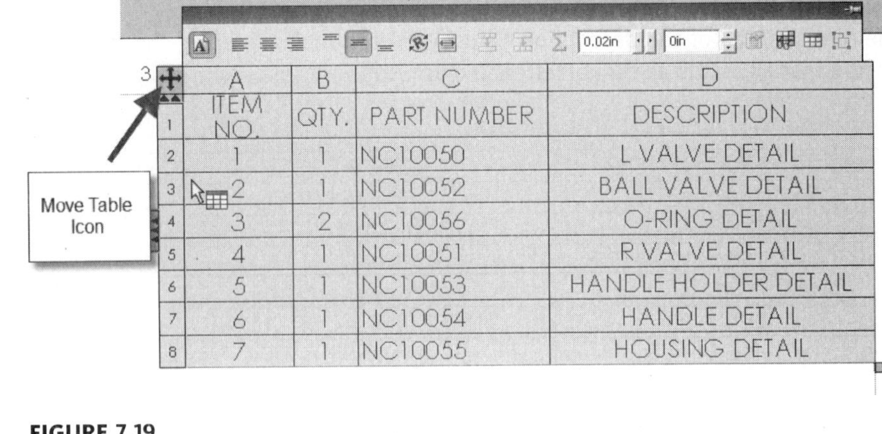

ITEM NO.	QTY.	PART NUMBER	DESCRIPTION
1	1	NC10050	L VALVE DETAIL
2	1	NC10052	BALL VALVE DETAIL
3	2	NC10056	O-RING DETAIL
4	1	NC10051	R VALVE DETAIL
5	1	NC10053	HANDLE HOLDER DETAIL
6	1	NC10054	HANDLE DETAIL
7	1	NC10055	HOUSING DETAIL

FIGURE 7.19

STEP 10

Zoom in on the front section view. On the Annotation toolbar select the Balloon command. Move the cursor over one of the parts and a leader will automatically be added to the balloon and the leader will touch the part. Select the right side of the ball valve as shown in the left side of Figure 7.20. Then place the balloon off the right side of the front view as shown in the right side of Figure 7.20. The appropriate item number will automatically be put in the balloon.

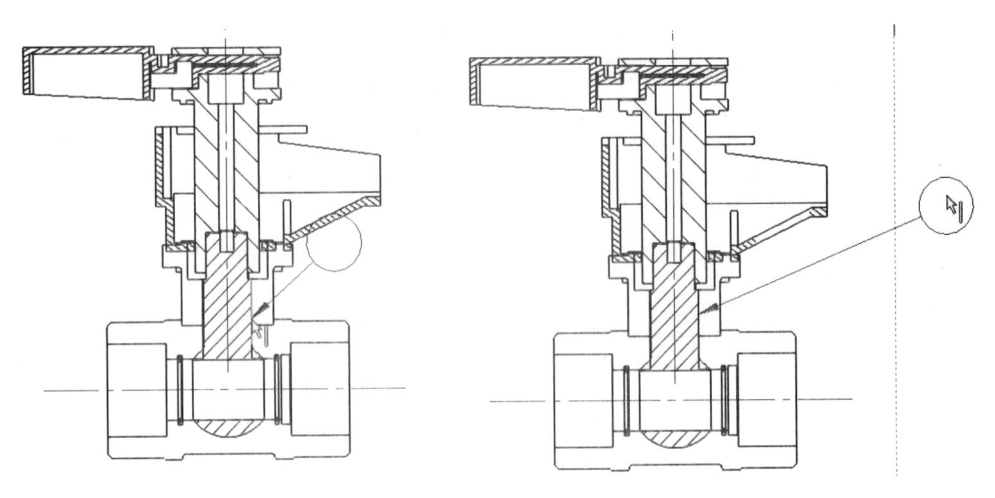

FIGURE 7.20

STEP 11

While still in the Balloon command, select the handle and move the cursor to the right slightly above the first balloon. Notice a yellow vertical line that appears as shown in Figure 7.21. This line ensures that the balloons are aligned vertically (also will indicate horizontal alignments). Select the location of the balloon while the vertical alignment condition is available.

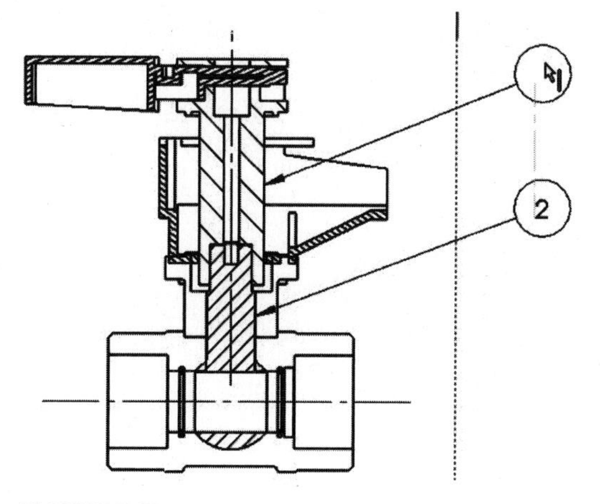

FIGURE 7.21

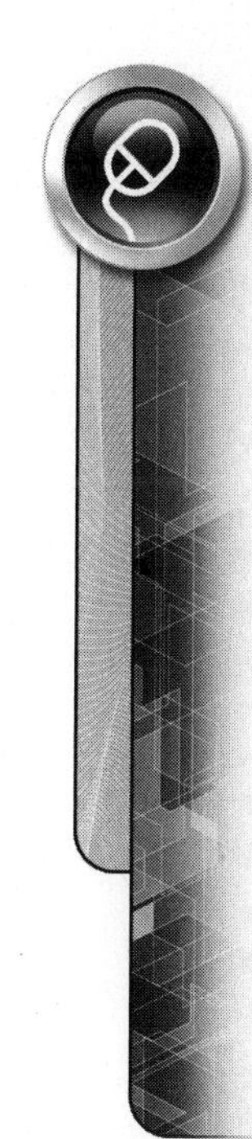

STEP 12

Continue to place all necessary balloons as shown in Figure 7.22. When complete, accept the Balloon command by selecting OK.

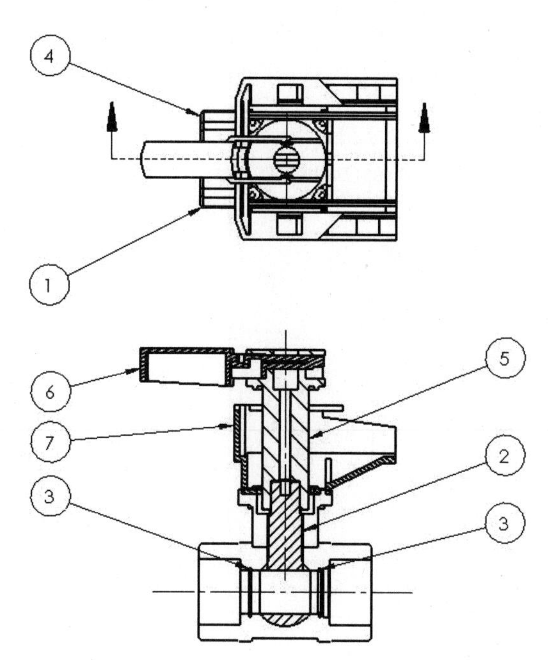

FIGURE 7.22

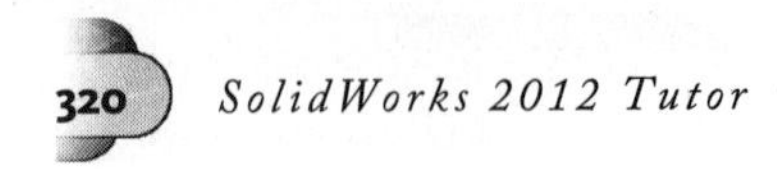

ADDING WELDING SYMBOLS

Welding symbols can be placed on drawings whether they are assembly, detail, or weldment drawings. A drawing must reflect the appropriate welding symbol as defined by the American Welding Society (AWS). The Weld Symbol command is found on the Annotation toolbar (Figure 7.23).

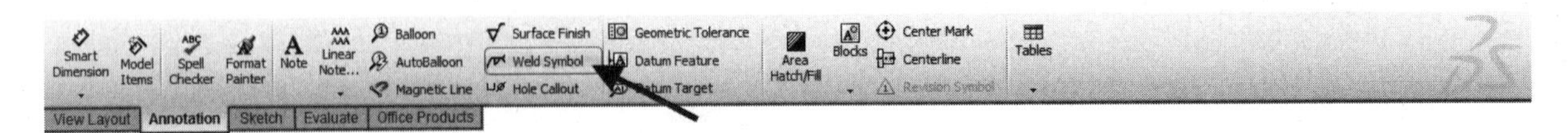

FIGURE 7.23

When the Weld Symbol command is selected, the weld symbol property menu appears. This menu controls all of the possible settings and symbols that can be put on a weld symbol. The two most common settings to be placed around the weld reference line are the weld symbol and the size of the weld. When the weld symbol button is selected, all possible weld types can be selected. The weld size is typed into the field located to the left of the symbols (Figure 7.24).

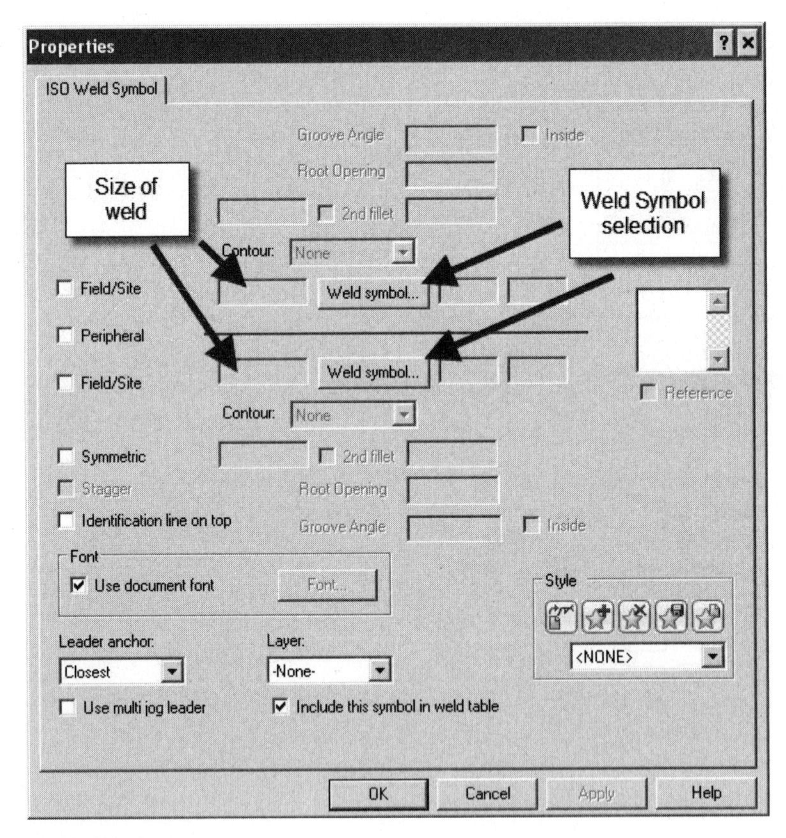

FIGURE 7.24

If the weld joint is to be welded on the arrow side, the weld symbol is to be placed below the weld line and if the weld joint is to be welded on the other side of the joint (opposite where the arrow is pointing), the symbol is placed above the line. If both sides of the joint are to be welded, then the symbol is put on both sides of the reference line as shown in Figure 7.25.

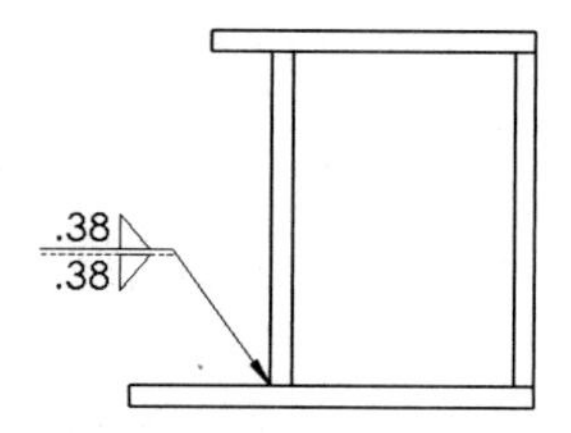

FIGURE 7.25

Additional features can be added to the weld symbol such as field weld, weld all around, and weld contours as shown in Figure 7.26.

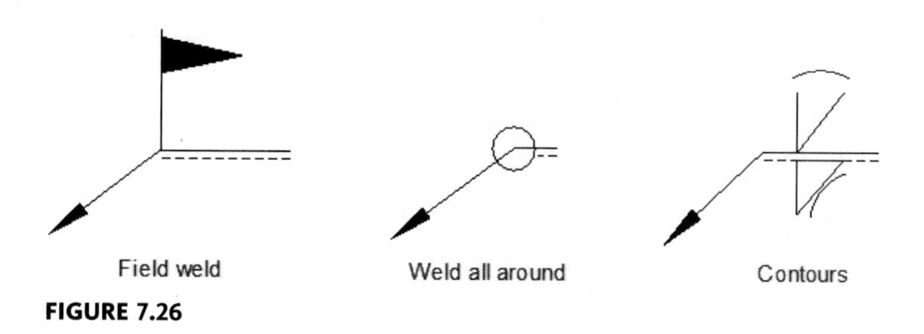

FIGURE 7.26

ADDING AN ASSEMBLY ISOMETRIC

A fairly new concept is adding an assembly isometric view to assembly drawings. In past 2D CAD systems which used manual drawing methods, isometrics were not added to assembly drawings because they were time consuming to create. Due to the nature of adding views fairly effortlessly, isometric views are now commonly added to the drawing. An isometric view further enhances the drawing by adding a visual of the assembly beyond the typical orthographic views.

When bringing in an assembly isometric view, two questions need to be answered: "Where is there space on the drawing to place the isometric view?" and "What scale does the isometric view need to be in order to fit in that space?" Once an area on the drawing has been identified, and oftentimes slight movement of existing views to make room for the isometric, the view can be brought in.

Adding the view is completed by using the Model View command found on the View Layout menu. The assembly drawing may need to be browsed for if it is not available in the Open Documents field. Once the assembly file is selected, one of the isometric views should be selected (isometric, trimetric, or diametric) from the Orientation options. Of the Display Style options, Shaded With Edges gives the most complete view, although any of the display styles can be used. For the Scale options, it is easiest to use the "Use Custom Scale" option to allow control of the view. While placing the view in the desired area, the overall border of the view is shown. If the view appears to be at an undesirable size, the custom scale can be changed before the view is placed. Once the size appears to be appropriate, place the view in the desired area and accept the view by selecting OK (Figure 7.27).

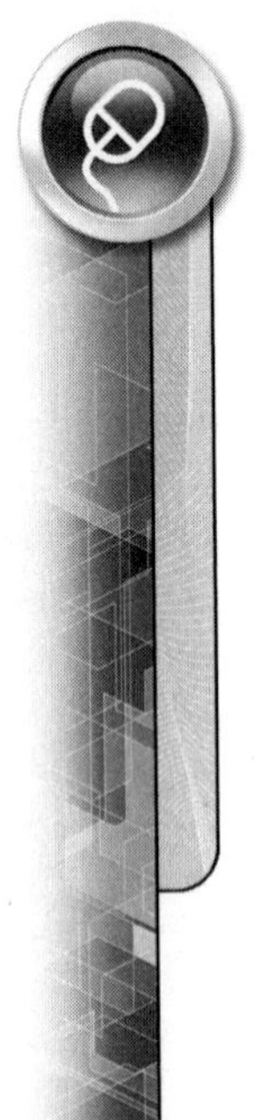

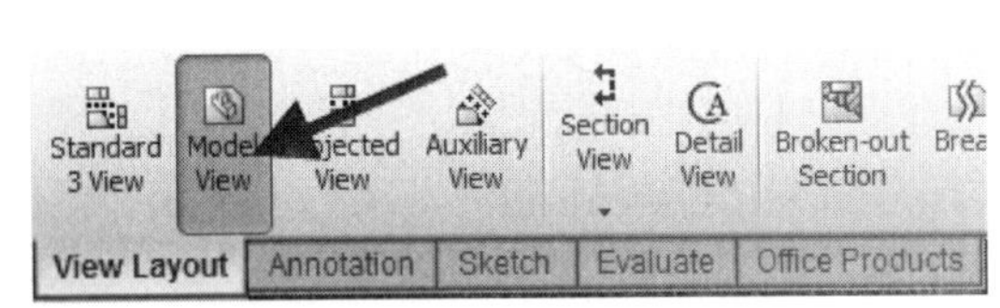

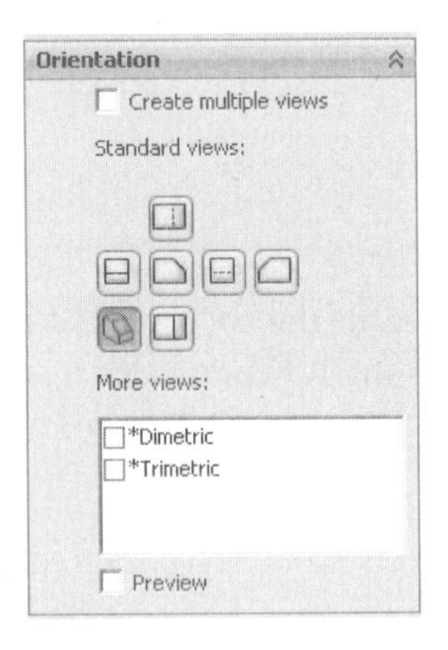

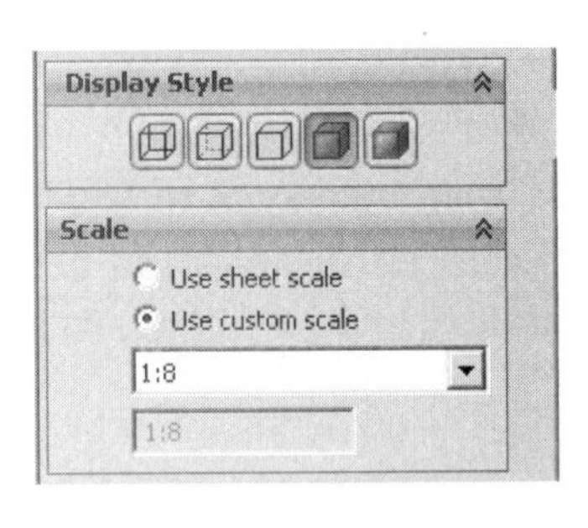

FIGURE 7.27

REVIEW QUESTIONS

1. Which drawing features are used specifically for assembly drawings?
2. Why are section views so common on assembly drawings?
3. Why should hatch patterns be modified from the autohatch settings?
4. How is a Bill of Material associated to the original part files?
5. Are hidden lines typically shown on assembly views?
6. How can a Bill of Material be dimensionally modified?
7. How can balloons be aligned with each other using the manual method of ballooning?
8. Why should balloons be aligned in one or two areas of the drawing?
9. What additional features can be added to a weld symbol?
10. Why are assembly isometric views more commonly being added to assembly drawings?

EXERCISES

For all drawing assignments, the paper size and scale are to be selected by the student.

7.1

Create a Weldment drawing for Exercise 6.1. The weldment drawing is partly an assembly drawing where each part (generally raw stock sizes) is itemized in a Bill of Material and ballooned. The weldment drawing should also have detail information such as locations and sizes of holes and other machined features. Weld the base plate to the vertical plate with a 12mm fillet weld/bevel weld. Weld the rib with a 12mm fillet weld. Show weld information by using weld symbols. Name the drawing: Base Weldment.

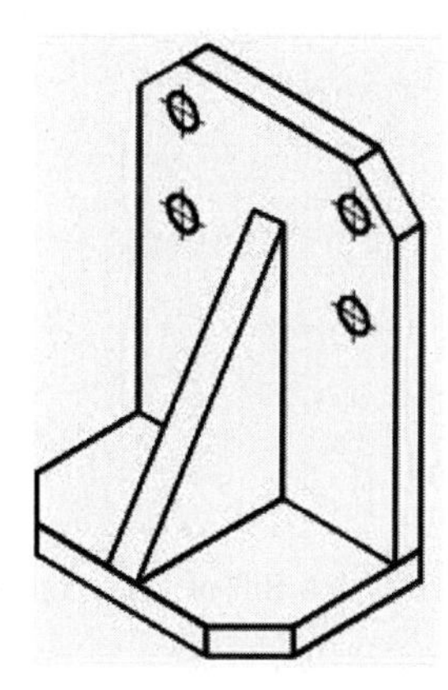

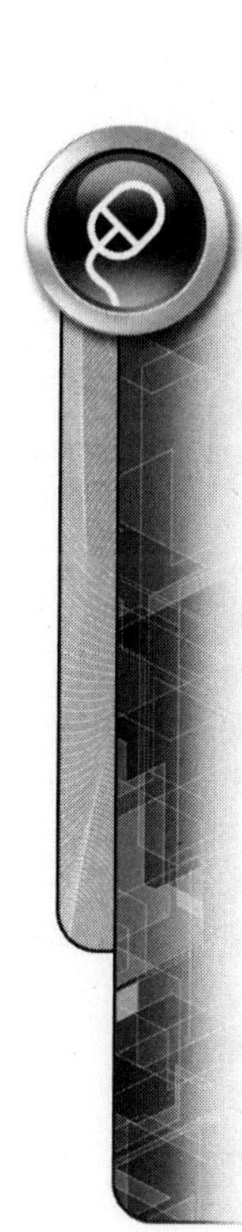

7.2

Create a Weldment drawing for Exercise 6.2. Fully balloon in conjunction with a Bill of Material. Show all welding information with weld symbols. Fully dimension and note the drawing.

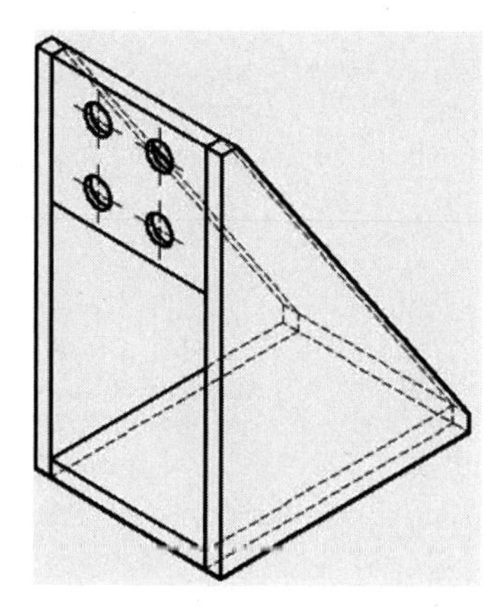

7.3

Create an assembly drawing for Exercise 6.3. Fully itemize that part with a Bill of Material and balloon.

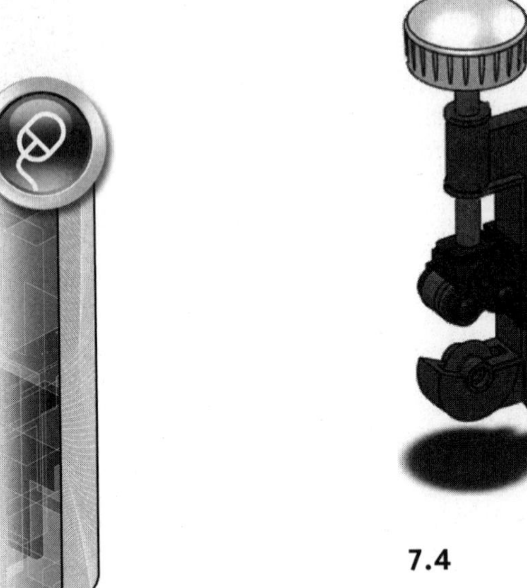

7.4

Create an assembly drawing for Exercise 6.4. Fully itemize that part with a Bill of Material and balloon.

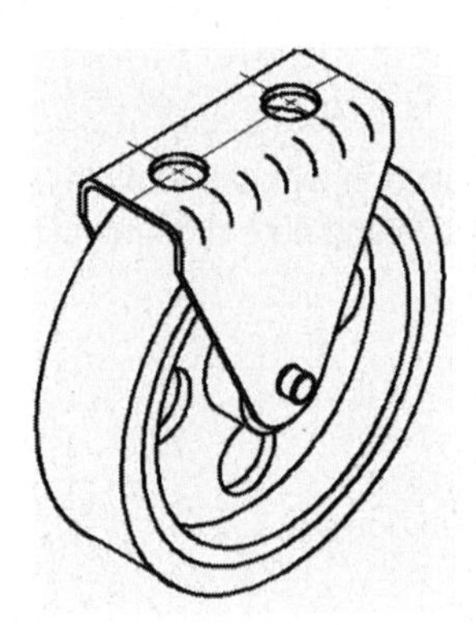

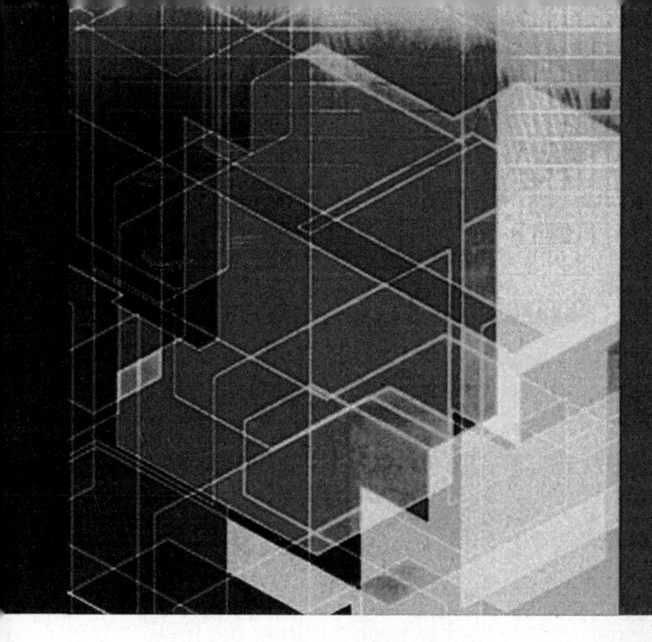

CHAPTER 8

Advanced Assembly Modeling

This chapter discusses various techniques and tools used to assemble part files as well as imported parts to complete an advanced assembly file:

- Top-down design methodology
- Using assembly layouts
- Creating new parts in an assembly
- Editing part files in an assembly
- Creating a belt in an assembly
- Interference checking
- Creating a subassembly

TOP-DOWN DESIGN METHODOLOGY

Up until now the design process has been to first create parts then insert and combine them into a new assembly; this method is called *bottom-up* design. Although bottom-up design is common in reverse engineering and is also helpful when learning the basic commands of solid modeling, it is not the preferred method for efficiently designing a new design. The more common method of designing is called top-down design where the initial design is created in a new assembly and only when the design is refined are parts broken out into part files. Now that the basic functions of solid modeling have been learned, the top-down method of design is recommended for use in your future design endeavors.

Parametric Modeling

Top-down design relies on *Parametric Modeling*, in which the parts in assembly interact with each other. For example, a new hole that is created for a part is defined by a sketch circle and that sketch circle diameter is defined by linking it to the diameter of a hole of a mating part. So when designing top-down, one should use these parametric relationships to assist in designing. By understanding the parametric relationships between features, as one feature is changed there may be features on other parts that will change automatically along with it (as long as all files are open and the locations of the files have not changed).

External References

The links in Parametric Modeling when using top-down design techniques are the creation of *External References*. That is if one file is using data from another file to define itself, the files are then linked.

Initially, when parts are being created in an assembly, these links between features on different parts are called *in-context* features. An in-context feature is a feature that is defined by geometry of another part (or assembly) by an external reference. Any parts that have in-context mates because they are defined in an assembly must maintain their association with the assembly file; if not the external references are broken and the parts will not be able to update appropriately.

USING ASSEMBLY LAYOUTS

One method of starting a design using the top-down method is to create a new assembly file and create an Assembly Layout. An Assembly Layout allows the basis of the design to be started. On the layout, several parts can be represented by creating the geometry in the layout which is very similar to a sketch. Once the sketch geometry of one part is created, then it should be grouped into a block. A layout block can then be converted into a part file.

The layout can be used to create as many sketches to convert blocks to part files as desired (Figure 8.1).

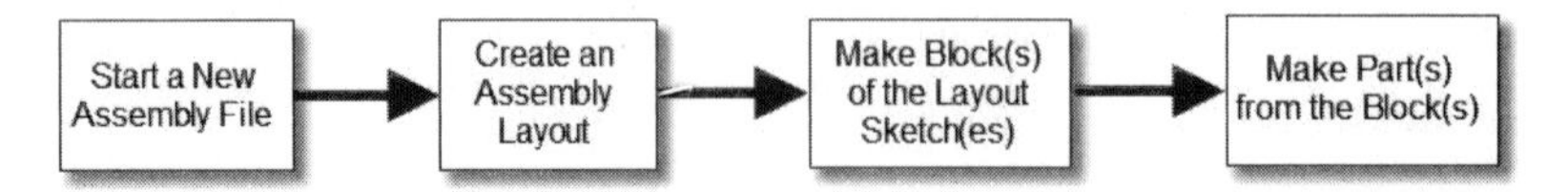

FIGURE 8.1

Creating an Assembly Layout

When a new assembly file is started, the option to Create a Layout is immediately available in the command manager menu. If it is not chosen immediately, the Create Layout command can always be launched from the Layout Toolbar (Figure 8.2).

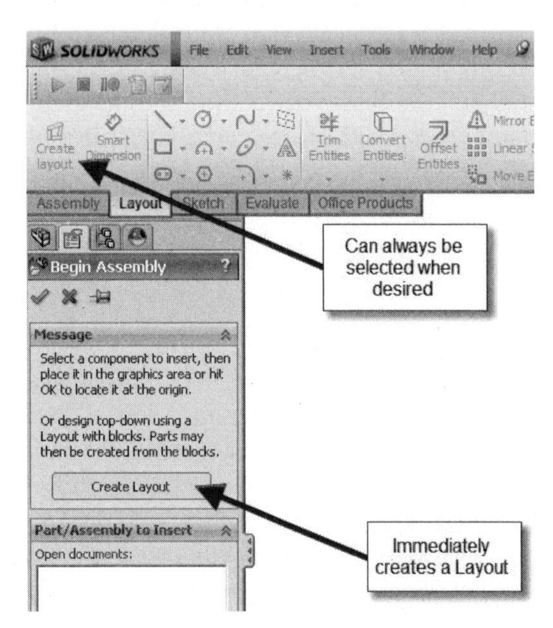

FIGURE 8.2

The layout is automatically created on the front plane. It is easier to work on the Layout sketch plane by selecting the View Display Menu for View Orientation and Front View (Figure 8.3).

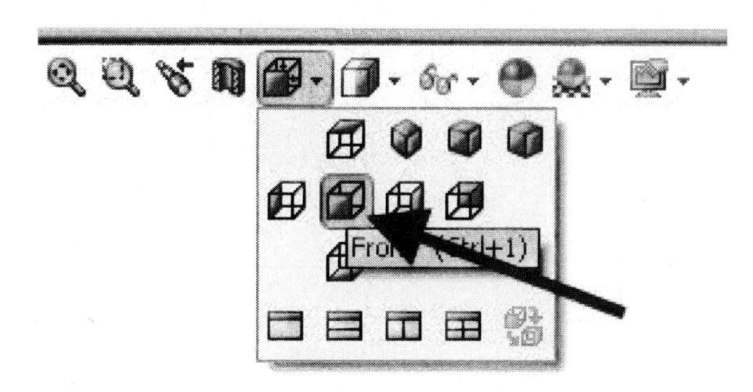

FIGURE 8.3

Creating a Block from an Assembly Layout Sketch

While in the Assembly Layout, create a sketch that will represent the first part in the design. Creating the block is done while still in the Layout command by first selecting all of the geometry desired in the block, then right-clicking and selecting the Make Block icon on the pop-up menu as shown in Figure 8.4. Then accept the block in the Command Manager.

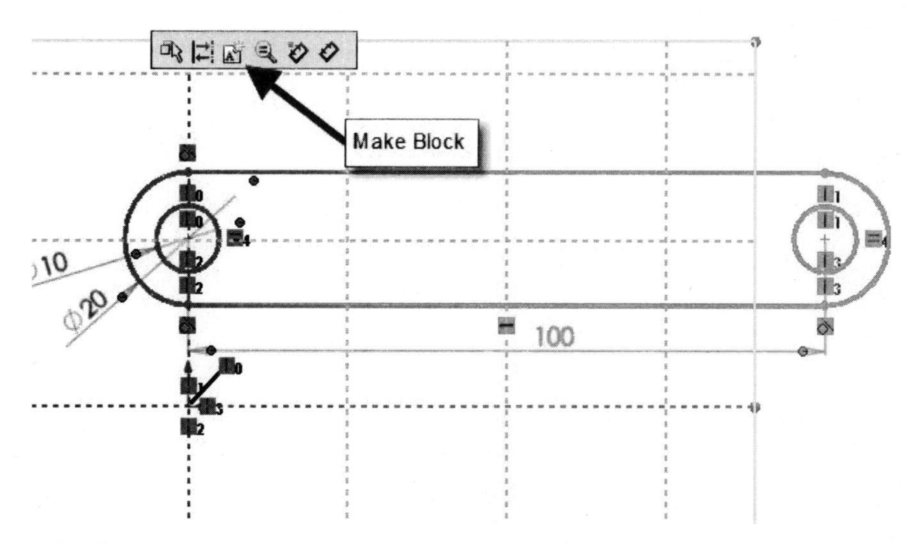

FIGURE 8.4

At this point, multiple sketches can be created and also converted to blocks. Whether one block or multiple blocks are created, the Layout command should be exited before part files are created.

Creating a Part from a Layout Block

Once the Layout command is exited, part files can be created from blocks. This is done by right-clicking on any part of the geometry that represents the block and selecting the Make Part from Block command (Figure 8.5).

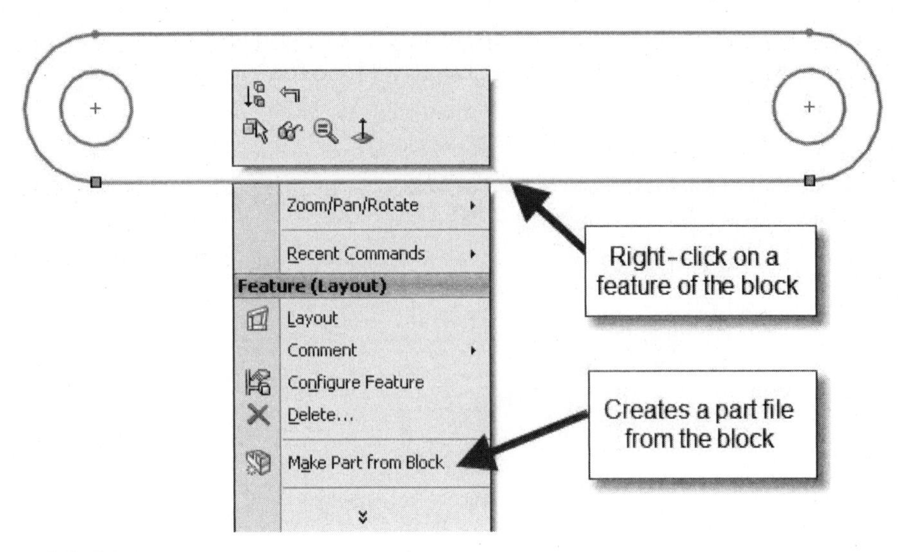

FIGURE 8.5

The part can be given an initial constraint where the part is made to be co-planar to the plane it was created on (Front Plane) by selecting the On Block option, or the new part can remain free for the time being by selecting the Project option where mates will be defined later (Figure 8.6).

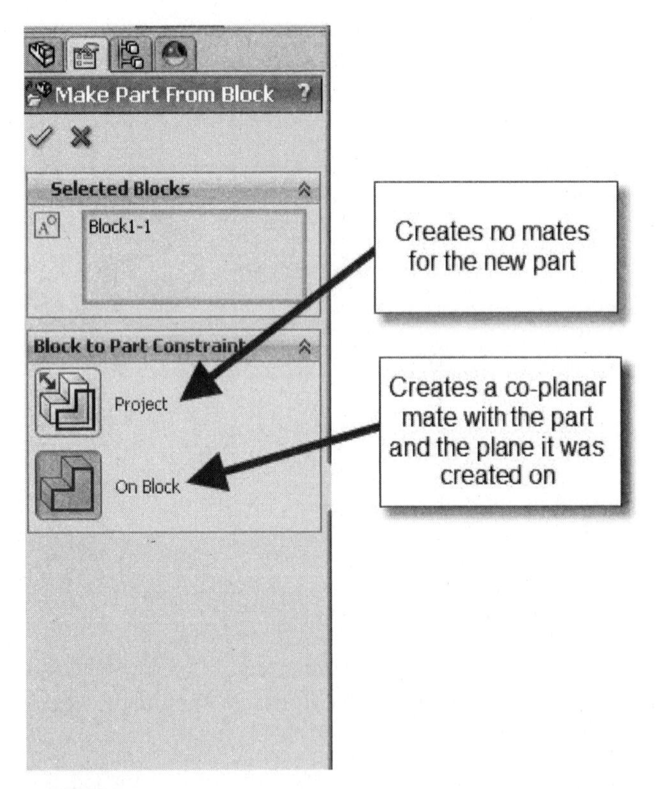

FIGURE 8.6

Once the command is accepted, the Part icon should be selected on the New SolidWorks Document menu. This will allow the part to be created. That part at this point IS NOT SAVED. The part is created but will need to be saved at some point before exiting the assembly. When the assembly file is saved and exited upon completion, a prompt will appear to save any part files that have not been saved yet.

CREATING NEW PARTS IN AN ASSEMBLY

Top-down design can be accomplished without ever using Assembly Layouts. Although using Assembly Layouts is a good method for creating parts on an assembly file, there is a more standard method of creating parts on an assembly file. This is done using the New Part command found on the Assembly toolbar in the Insert Part pull-down menu (Figure 8.7).

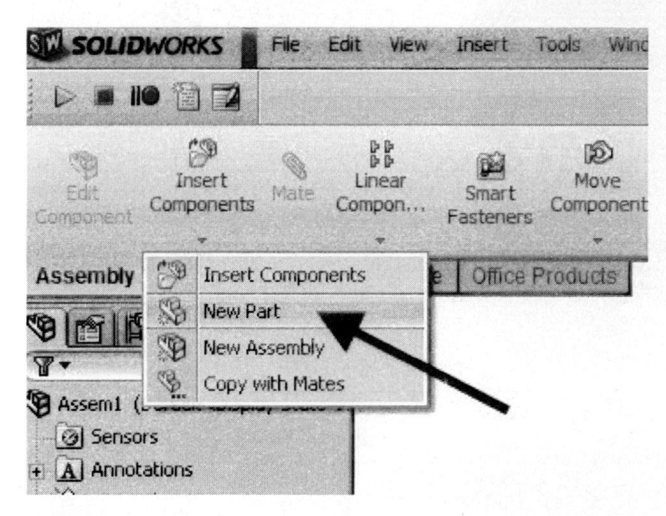

FIGURE 8.7

An initial plane needs to be selected to start the new part. It should be noted that whatever plane is selected on the assembly file, a coincident mate between the selected plane and the FRONT plane of the new part will be created. This mate can be used or deleted later.

While designing using the top-down method, there is the need to regularly switch back and forth between the assembly and the part files. The archaic way of accomplishing this is to simply use the Window pull-down menu to go from file to file, but this method is less efficient than using the top-down assembly drawing Feature Manager functions.

When an assembly file is open either the assembly file or any of the part files can be the active file. The file that is currently activated is indicated in the top tool bar. Figure 8.8 shows the assembly file name to indicate that the assembly file is the active file. When a part file is the active file, the top of the drawing will indicate the part file name.

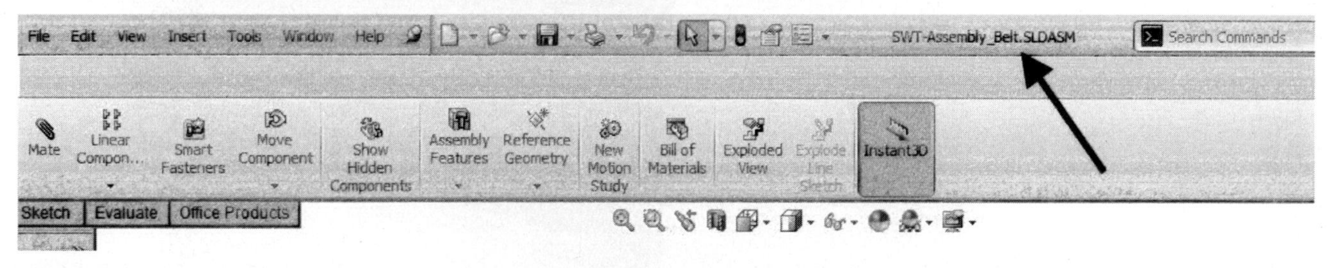

FIGURE 8.8

EDITING PART FILES IN AN ASSEMBLY

Once the new part is created, the part can immediately be edited, and since the part is new, this means creating the first sketch desired. Although the assembly is still open, the part file can be modified. This is indicated a couple of ways. First, in the Feature Manager Design Tree, the assembly file features are shown, *but the new part file is highlighted in blue.* Secondly, in the Command Manager along the top, the left half is for the assembly and the right half is sketch commands for the part file. At this point the part file should be developed to a point where it can be used in the assembly. And lastly, along the top of the screen the part file is named (Part6^Assem6 in Assem6) (Figure 8.9).

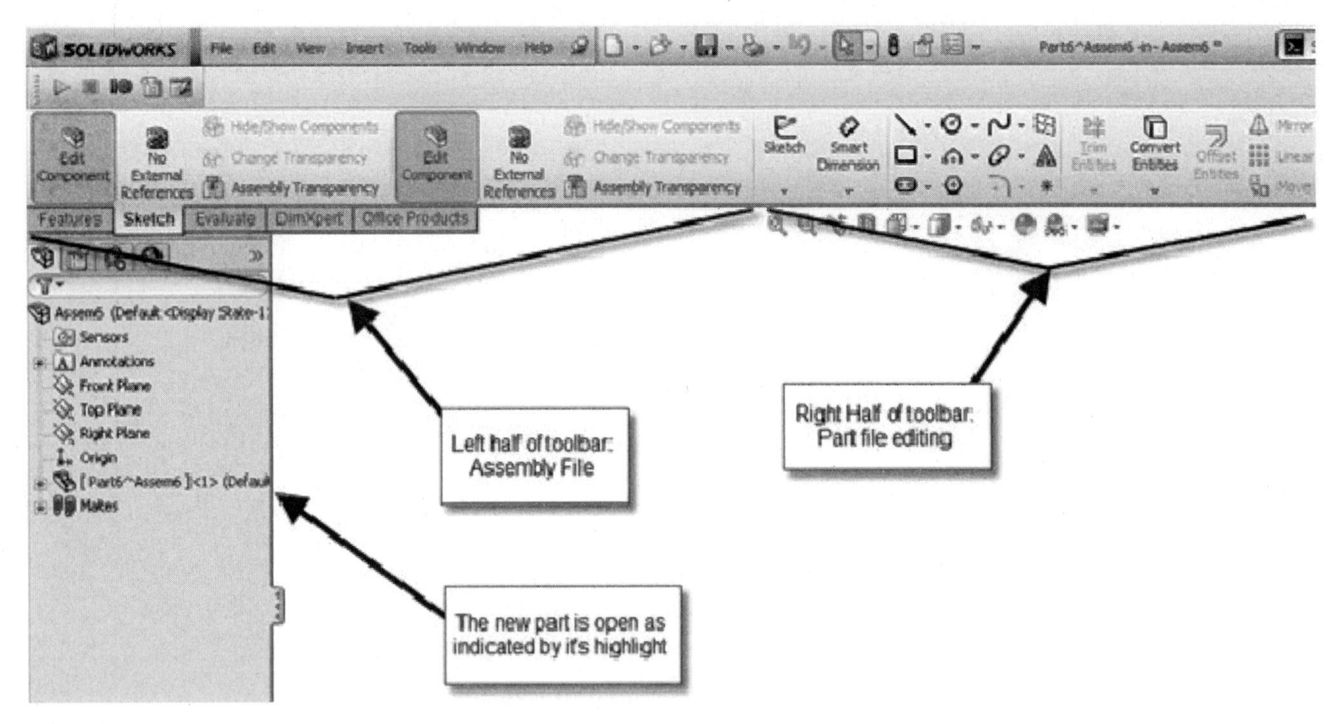

FIGURE 8.9

Once the part is developed, it is recommended to save it and locate the file appropriately on the computer.

CREATING A BELT/CHAIN

Within an assembly file, flexible parts such as a belt and a chain can be designed. Given the simple assembly in the figure there are two pulleys that are mounted on a mounting plate with two shafts. Select the Belt/Chain command found in the Insert pull-down menu and the Assembly Feature pull-out menu (Figure 8.10).

FIGURE 8.10

When the faces of each pulley are selected, a belt size is calculated and inserted into the assembly as shown in Figure 8.11. A part file for the belt can be created if desired. Also, if a specific belt length is desired, the Driving option can be selected and the belt length would drive the design or move pulleys if it can.

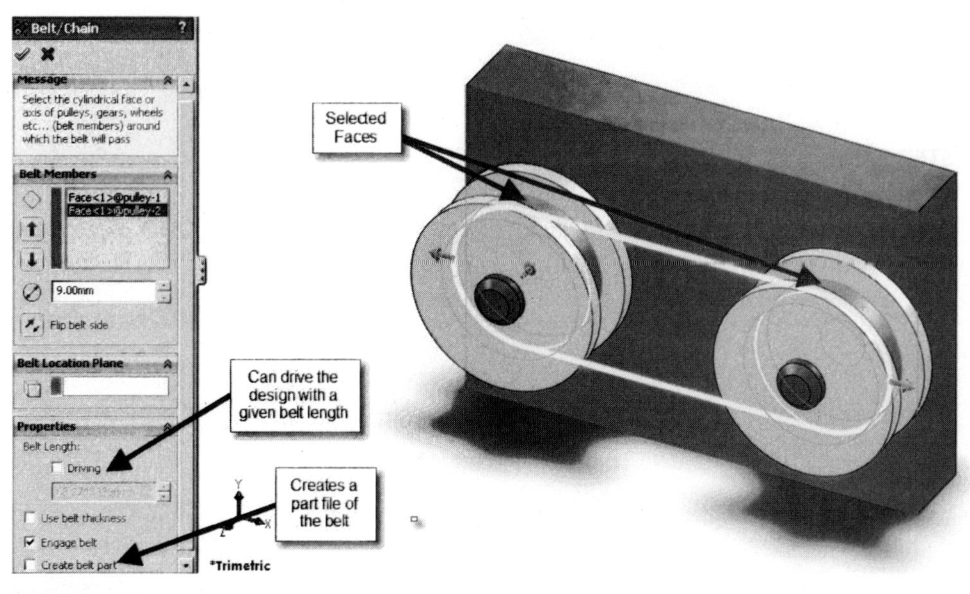

FIGURE 8.11

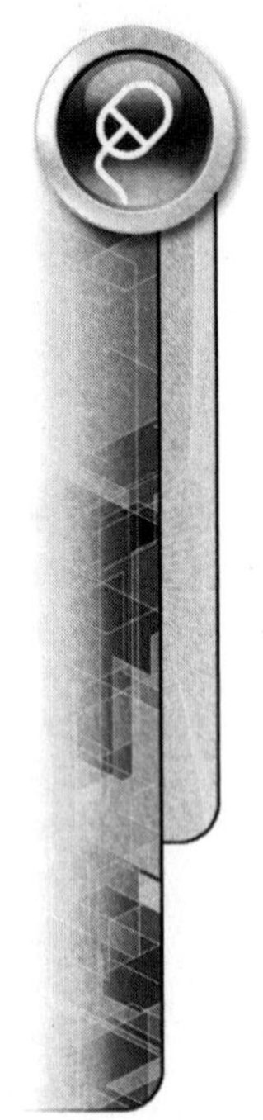

TUTORIAL EXERCISE: SWT_CH08_PULLEYS.SLDASM

Part Number: SWT-CH08-pulleys.SLDASM

Description: Pulley and Belt Assembly

Units: Metric

This exercise will create a pulley assembly that includes a base plate, three shafts, three pulleys of different sizes, and a belt, all using the standard method of top-down design.

STEP 1

Start a new assembly file and exit out of the Create Layout option. Select New Part from the Insert Component pull-down menu and select Part File. Select the Front Plane from the Feature Manager Design Tree on the left, then select the front view orientation from the display menu.

Using the sketch commands, draw the following sketch shown in Figure 8.12. Add the *equal* relation to the three circles. This sketch represents the mounting plate of the assembly.

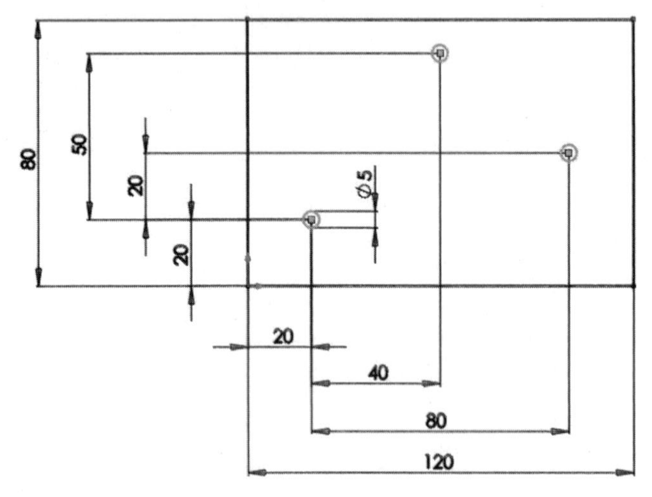

FIGURE 8.12

STEP 2

Exit the sketch. Change the view orientation to trimetric. On the Features toolbar select the Extrude Boss/Base command. Select the sketch, then extrude it by 10mm. Accept the Extrude command (Figure 8.13).

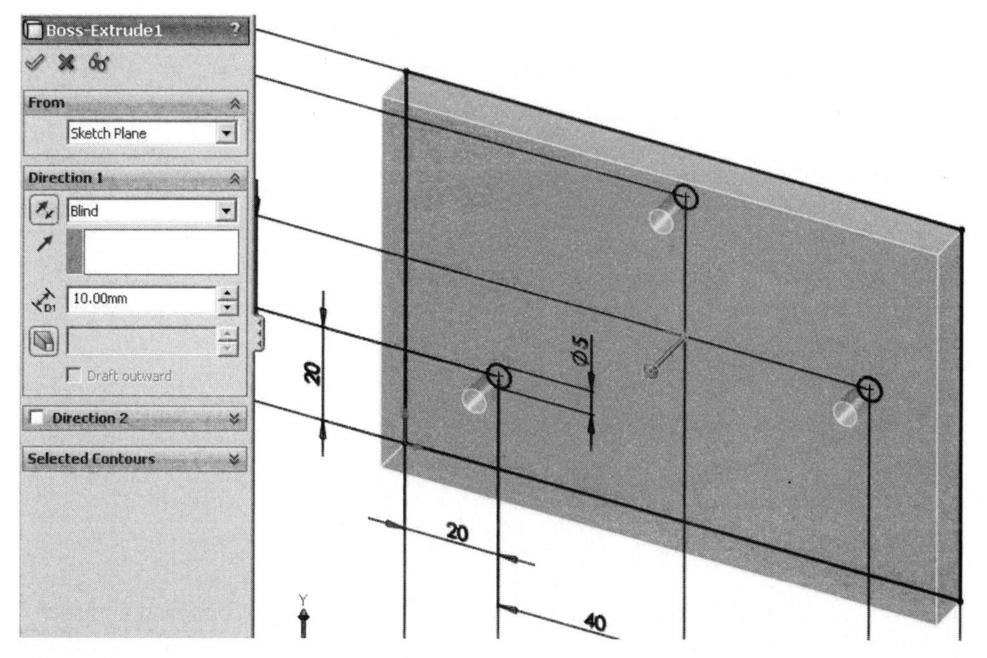

FIGURE 8.13

STEP 3

Leave the part file by right-clicking on the assembly in the Feature Manager Design Tree and selecting Edit Assembly. This reactivates the assembly file (Figure 8.14).

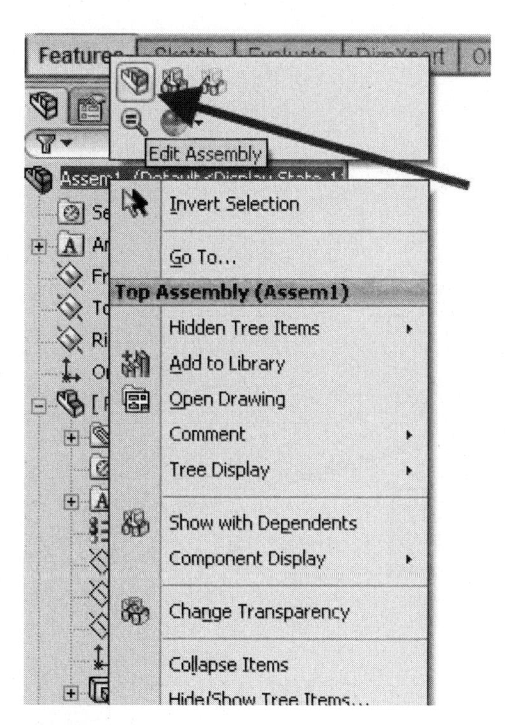

FIGURE 8.14

STEP 4

Select the New Part command and select and accept the Part File option to create the first axle. Select the front face of the mounting plate to start a new sketch. Create a circle

by selecting the circle command, then the center of the lower left hole, and then a point on the hole radius (Figure 8.15).

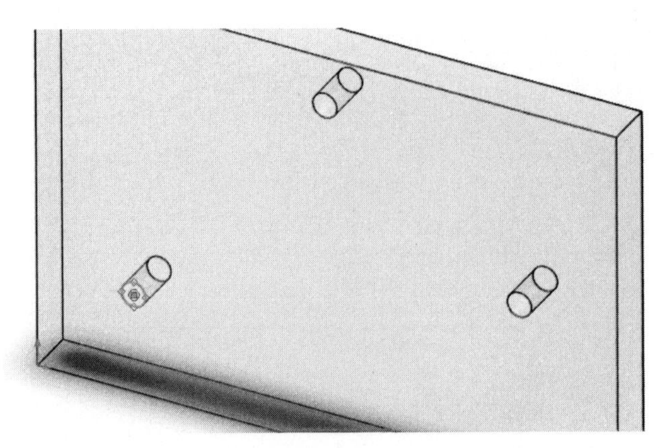

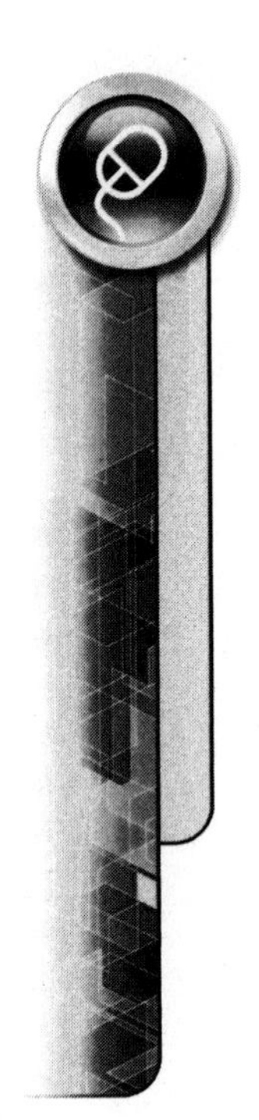

FIGURE 8.15

Accept the circle and exit the sketch. Enter the Extrude Boss/Base command and select the new circle. Set direction 1 for 12mm and set direction 2 for 10mm, then accept the Extrude command. Add a 1mm chamfer on both ends of the shaft. Right-click on the Material for the part and select Brass (Figure 8.16).

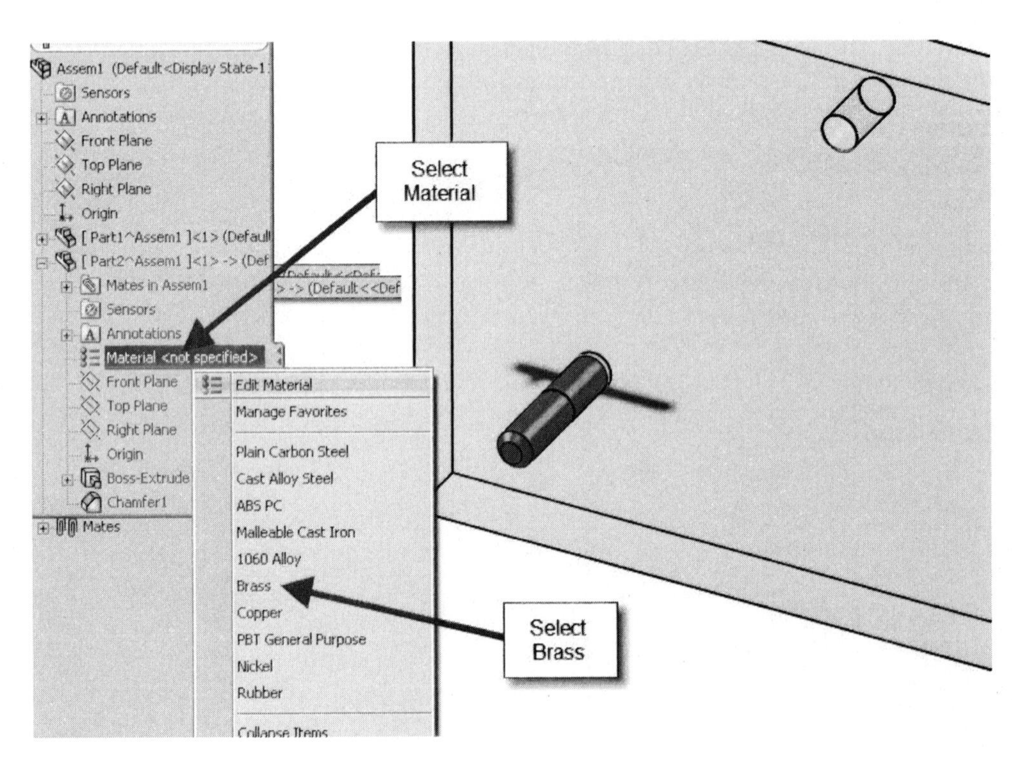

FIGURE 8.16

STEP 5

Enter back into the assembly by right-clicking on the assembly in the design tree and selecting Edit Assembly. Save the file by selecting the Save File command. This allows naming and locating of the assembly file as well as naming and locating of the part files. This is required so the part files can be reused such as inserting multiple copies of the axle.

When the Save Modified Documents window appears, select Save All to save the parts as well as the assembly. Locate and name the assembly file (Figure 8.17).

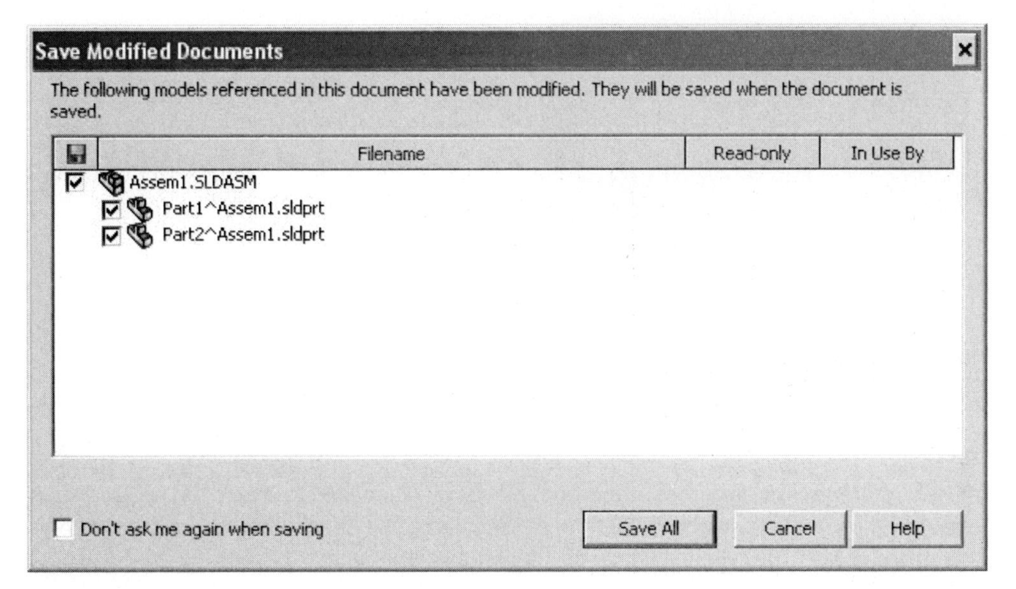

FIGURE 8.17

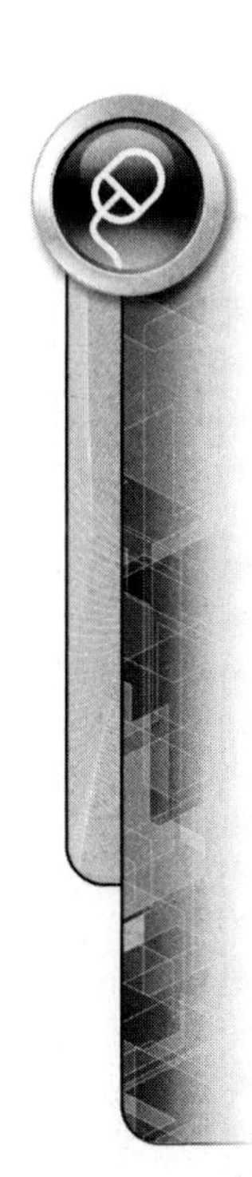

When the Save As window appears, select Save externally. Locate and name each part. This allows the parts to be recalled when inserting new components (Figure 8.18).

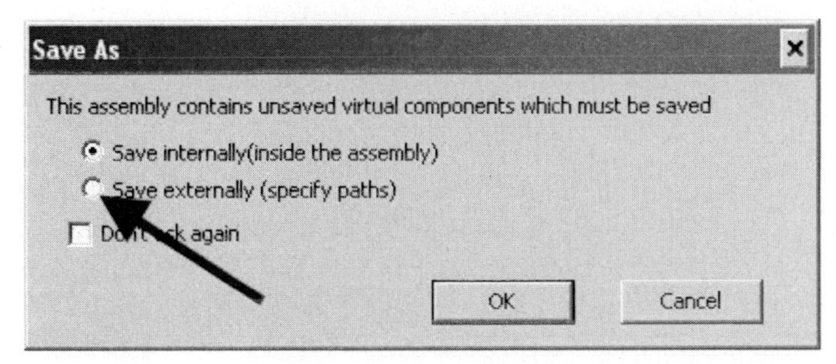

FIGURE 8.18

STEP 6

Select the Insert Component command and insert a second axle. Locate the axle away from the mounting plate temporarily. Create a concentric mate between the cylindrical surface of the top hole and the outside cylindrical surface of the axle. Next create a mate between the back end face of the axle and the back face of the mounting plate.

Insert another pin and mate it to the third hole accordingly (Figure 8.19).

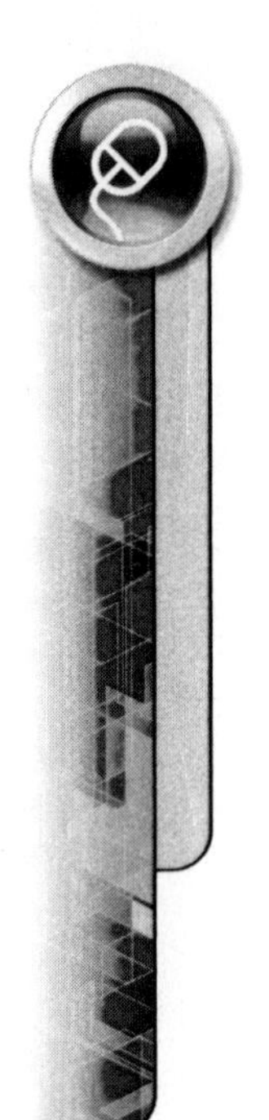

FIGURE 8.19

STEP 7

A new reference axis as well as a reference plane will be needed to create the next part. On the Assembly toolbar, select the Reference Geometry pull-down menu and select the Axis command. Select the cylindrical face of the first axle and accept the axis (Figure 8.20).

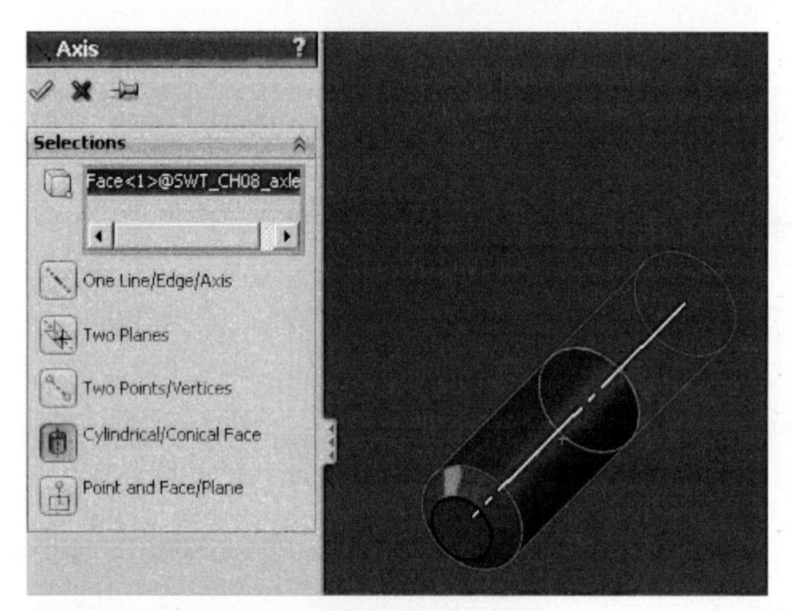

FIGURE 8.20

Next, on the Assembly toolbar, select the Reference Geometry pull-down menu and select the Plane command. To define the plane, first select the new axis, then select the Top Plane, then select Parallel to the second reference. The Design Tree in the model area may need to be expanded to select the Top Plane of the assembly. Accept the plane. This sequence places a plane directly through the center of the axle (Figure 8.21).

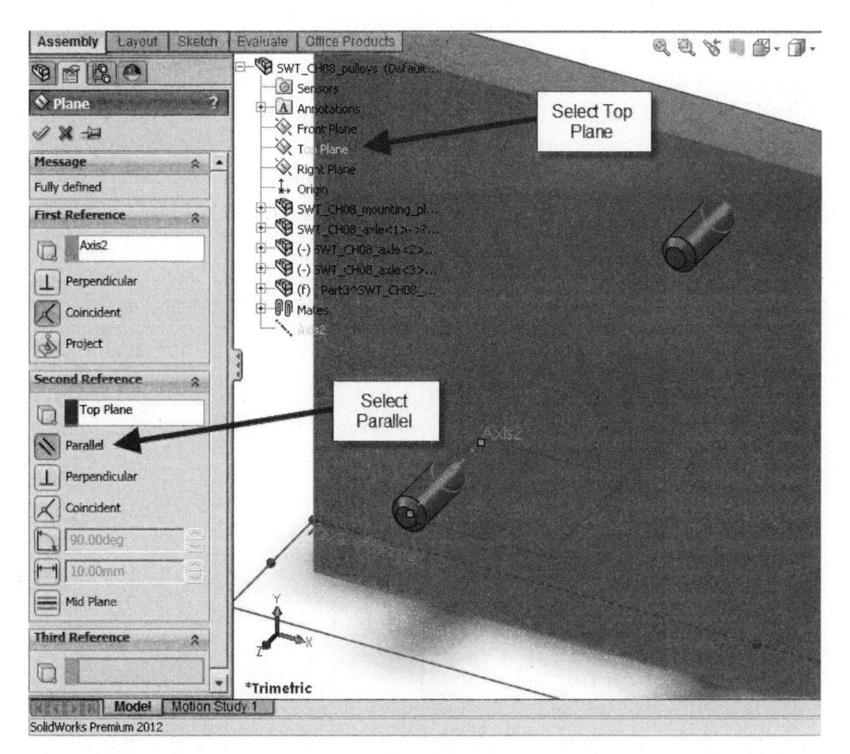

FIGURE 8.21

STEP 8

Select the New Part command from the Insert Component pull-down, then select the newly created plane. Change the view orientation to a top view and zoom in around the first pin. Create and define the sketch shown in Figure 8.22 on this plane. Start the sketch in the lower right by selecting the corner of the chamfer. Continue clockwise to the left, and when creating the top line select the front face line of the mounting plate first, then the intersection of the plate and axle. Close the shape with the last line, then place the given dimensions as seen in Figure 8.22. Because of the relationships between the sketch and the existing geometry, any additional dimensions would cause the sketch to overconstrain. Accept and exit the sketch.

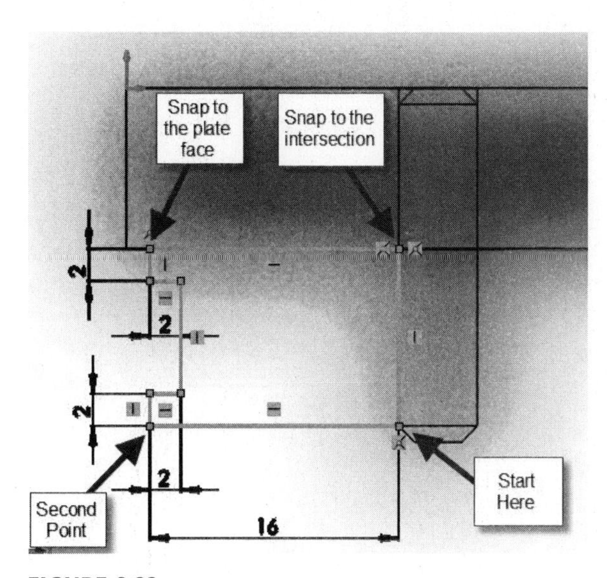

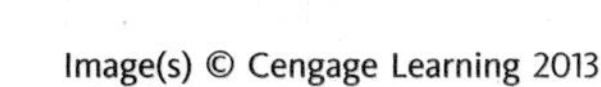

FIGURE 8.22

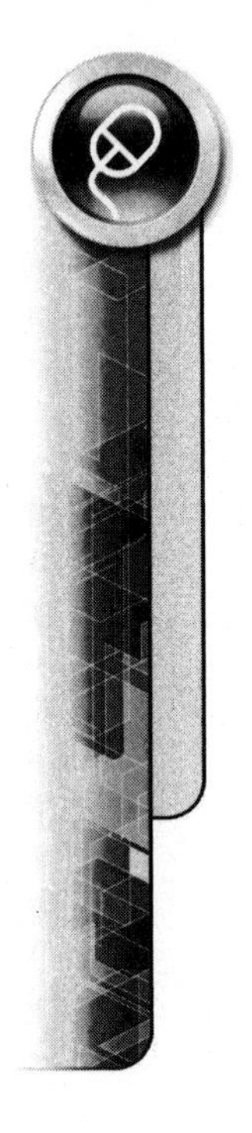

STEP 9

Select the Features toolbar and select the Reference Geometry Axis command. Select the cylindrical face of the axle and accept the axis; this will create an axis on the part file. Then select the Revolved Boss/Base command and select the sketch to be revolved. Select the newly created axis as the Axis of Revolution. The Model View Design Tree may need to be expanded to select the axis. Accept the revolve. Save the part by using the Save As command (Figure 8.23).

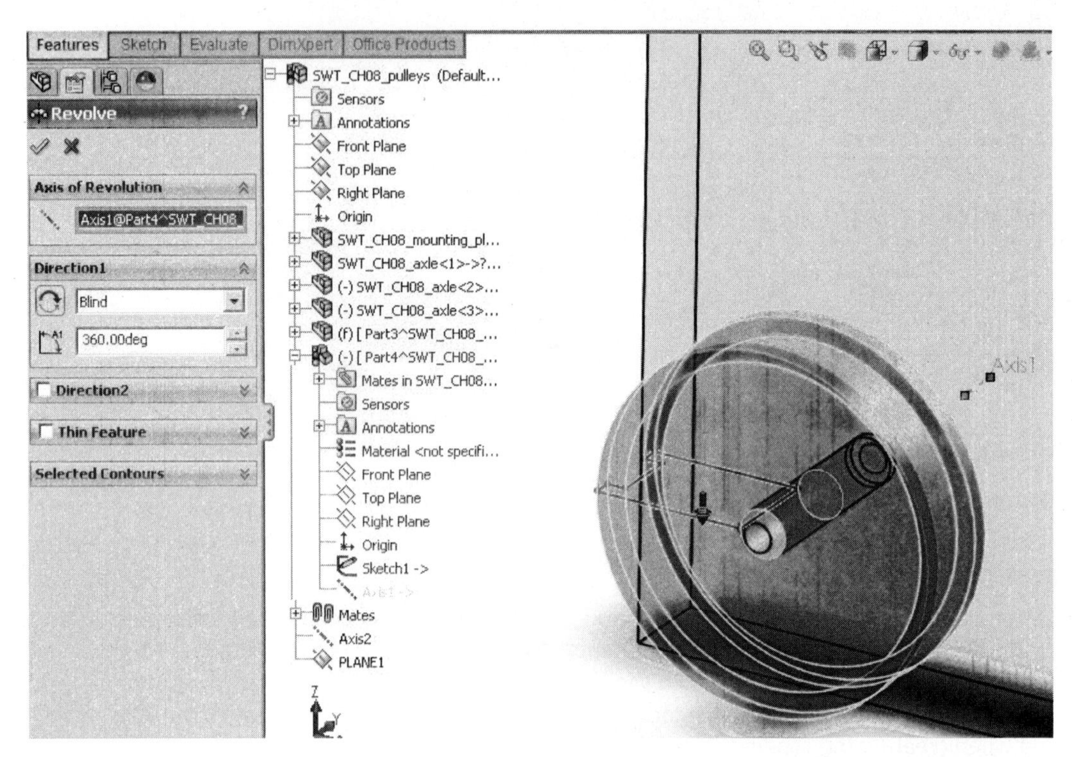

FIGURE 8.23

Right-click on the assembly in the Design Tree and select Edit Assembly.

STEP 10

Create a second pulley about the second axle using the same process as the first by repeating steps 7 through 9. Refer to the sketch dimensions shown in Figure 8.24. Save the second pulley using the Save As command.

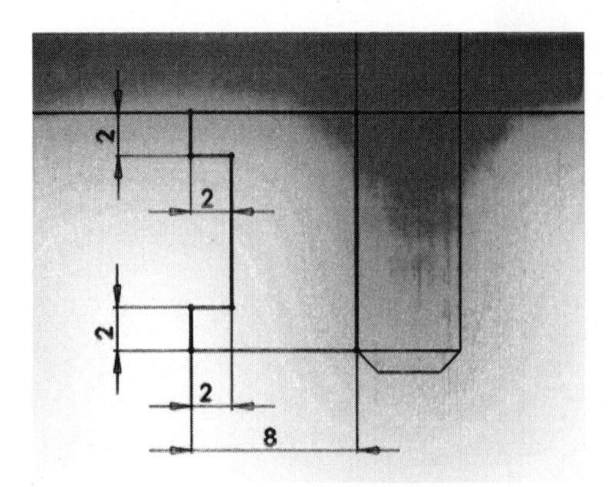

FIGURE 8.24

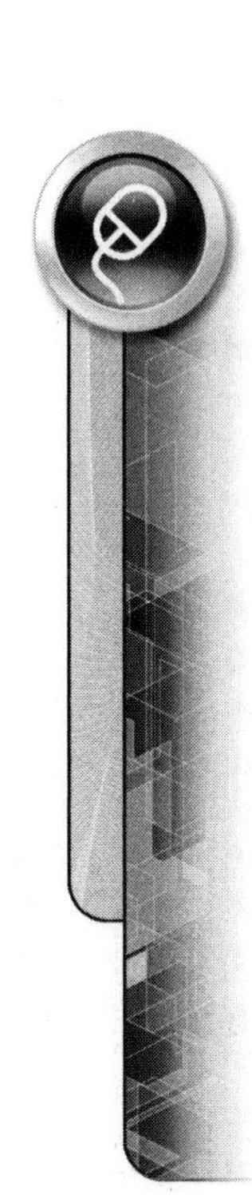

STEP 11

Create a third pulley about the third axle using the same process as the first by repeating steps 7 through 9. Refer to the sketch dimensions shown in Figure 8.25. Save the third pulley using the Save As command.

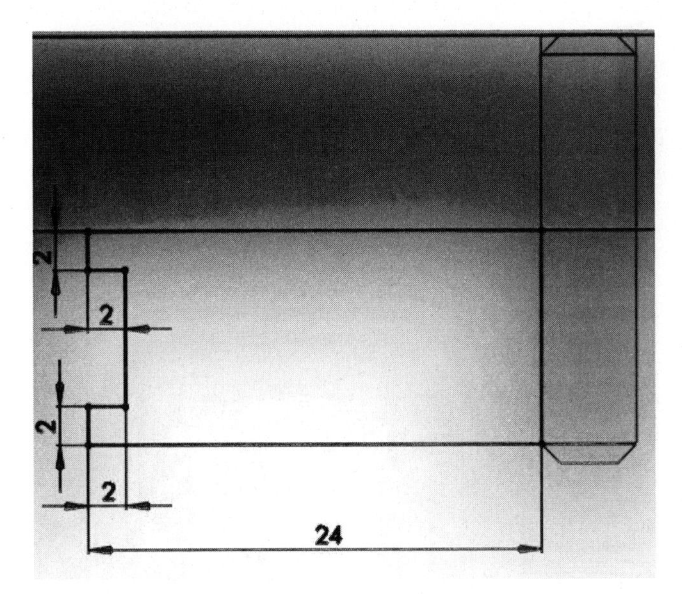

FIGURE 8.25

STEP 12

Create a belt around the three pulleys by selecting the Belt/Chain command from the Insert pull-down menu and the Assembly Feature pull-out menu (Figure 8.26).

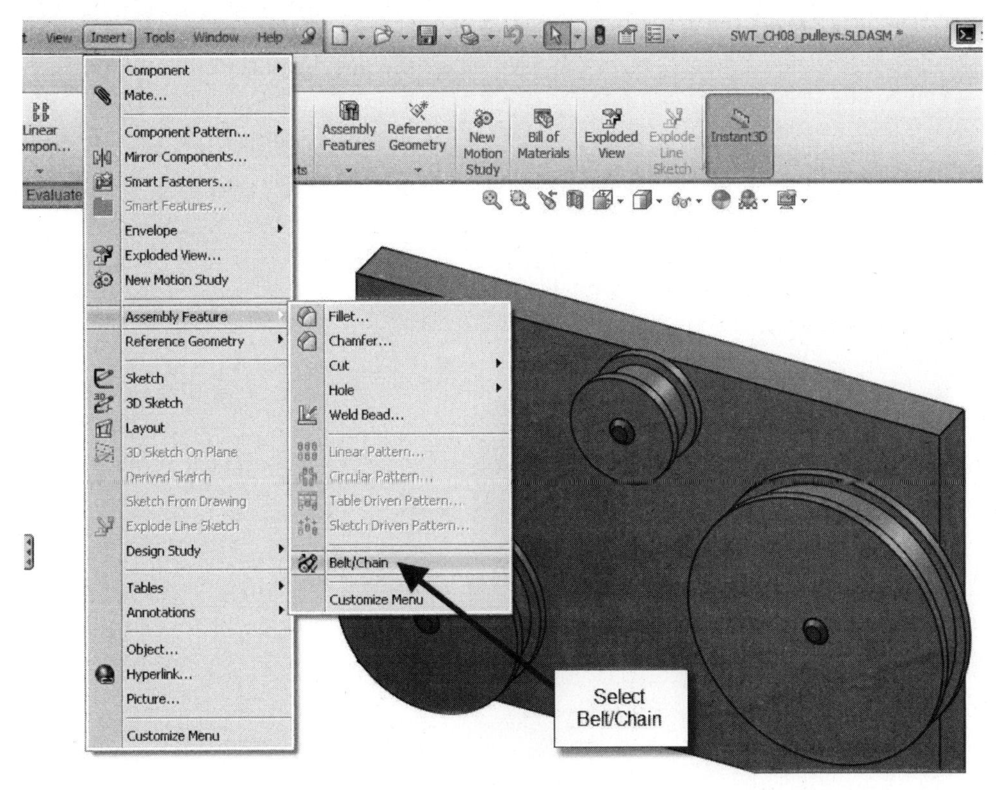

FIGURE 8.26

STEP 13

Select the three pulley faces as shown in Figure 8.27. Then select the Create belt part from the properties in the feature manager. Accept the command.

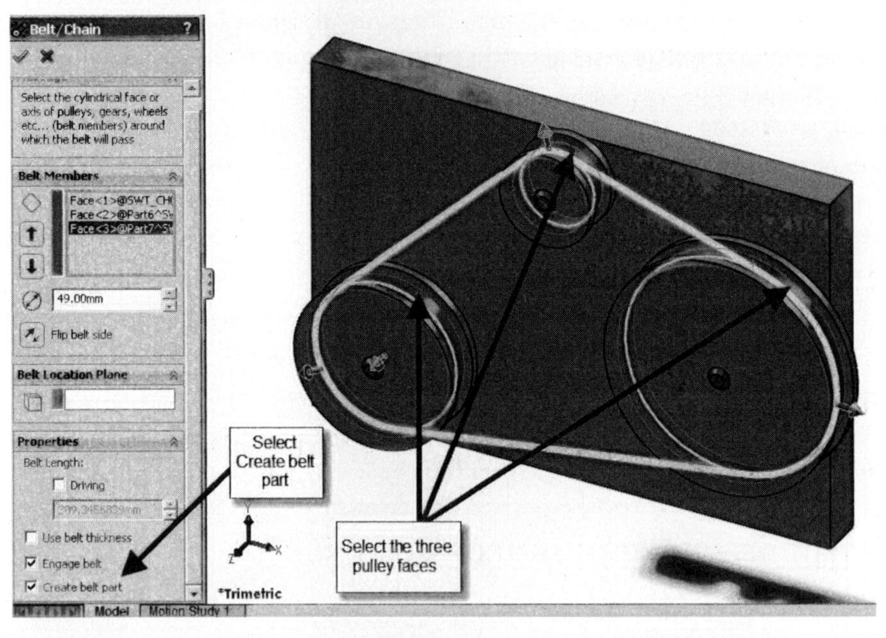

FIGURE 8.27

Select the Part File option on the New SolidWorks Document menu and select OK.

STEP 14

Save all of the files by first saving the assembly. Select the Save All button on the Save Modified Documents menu. Select the Save externally per figure selection on the Save As menu and select OK (Figure 8.28).

For each part file, open in the Design Tree and Edit Part. Individually complete the Save As command for each, and name and locate each file. The assembly is now completed and fully saved.

FIGURE 8.28

At this point the design could be continued by adding more features, refining the parts with more detail, or adding additional mates between components. The top-down method can continue until the parts are designed to the level of detail that can be manufactured.

INTERFERENCE CHECKING

While in an assembly, parts can be mated and located in a manner where they share space, or in other words the volumes of the parts may interfere with each other. SolidWorks does not restrain the designer from doing this. At times this may be intentional, such as when an interference fit has been designed, and sometimes it may be an unintentional error. Interference checks can be made at any time and one should always be completed at the end of every assembly design.

Interference Detection is found on the Evaluate toolbar. While in an assembly select the Interference Detection command and then select Calculate button. All interferences are listed in the results area. As each result is selected, a visual representation of the interference is shown in the model view (Figure 8.29). If the result is expanded, a list of the parts that are interfering are also listed.

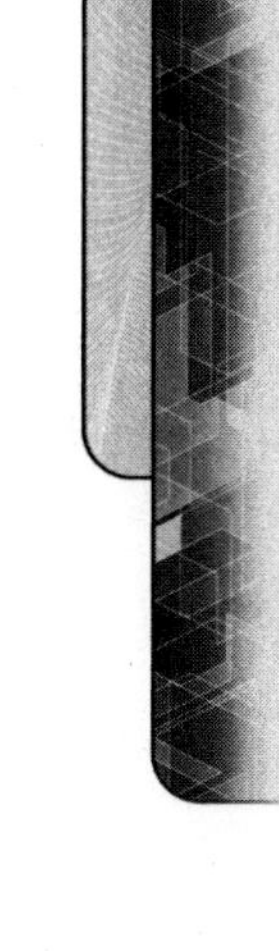

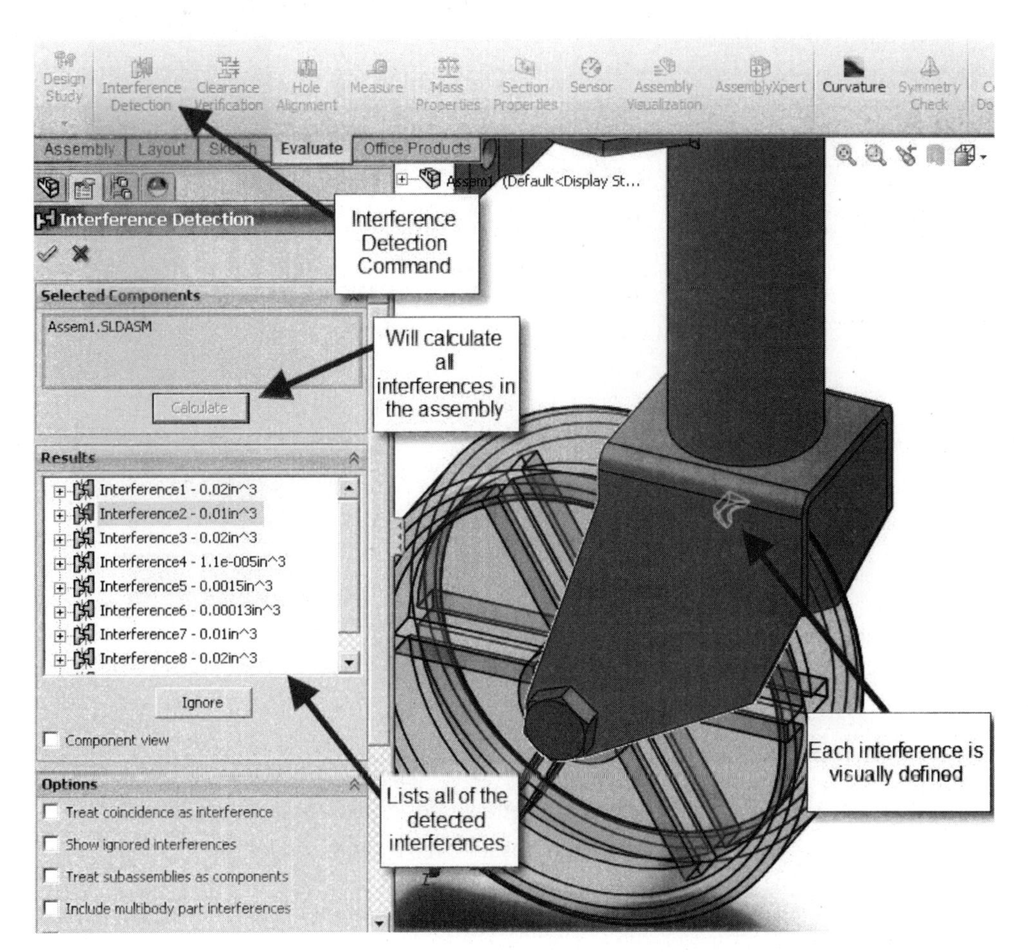

FIGURE 8.29

CREATING A SUBASSEMBLY

While in a large assembly, subassemblies can be quickly created and saved. This would simplify the top assembly by breaking out many components neatly into subassemblies.

To accomplish this select several components while holding the Control key down on the keyboard; this is most easily done in the design tree. When all components are selected for the subassembly, right-click in the graphics area and select the Form New Subassembly Here command. When the New SolidWorks Document appears, select the Assembly file option and accept it. The parts are now grouped in the design tree as an assembly (Figure 8.30).

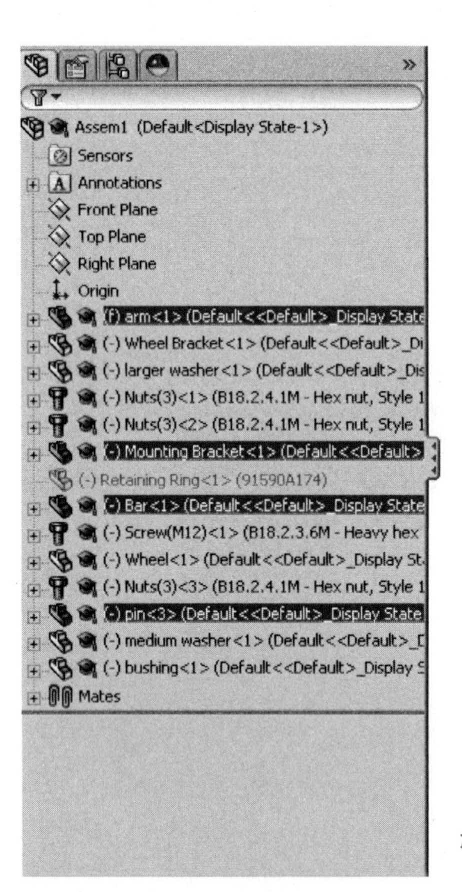

FIGURE 8.30

REVIEW QUESTIONS

1. What are the differences between top-down design and bottom-up design?
2. What is the advantage of top-down design over bottom-up design?
3. When should bottom-up design be used?
4. What is the purpose of using Assembly Layouts?
5. Is it required to use Assembly Layouts when designing top-down?
6. How can parts be edited when in an assembly?

7. What is the advantage of not having to exit the assembly file when creating a new part?
8. What is the advantage of using reference geometry in an assembly?
9. When interferences are detected, are they all design faults? Explain.
10. What is the main advantage of creating subassemblies in an assembly file?

EXERCISES

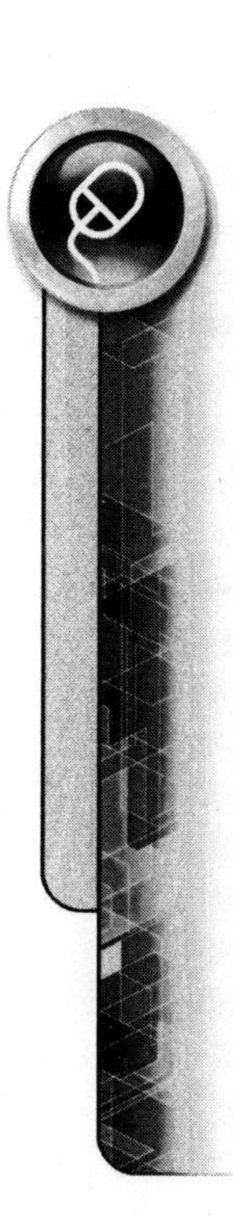

8.1

Using top-down design methodology design a caster assembly. A series of these castors will be used to guide and move large pressure vessels from a welding station into a shipping container. The wheel should be 8.00" diameter and 2.50" wide. The bore in the wheel has two bearings with an inside diameter of 1.00". This wheel needs to be mounted to a ½" mounting plate that is located below the wheel and has a hole pattern of .53" diameter holes that are 8 inches and 6 inches apart. The wheel should ride on a shaft that goes through the wheel bearings and into wheel supports. The wheel supports are to be designed to hold the wheel in place and are fastened to the mounting plate. Refer to the figure for general design ideas.

Add additional features to the assembly parts for functionality. For instance, design the shaft to have retaining rings to hold the shaft in place. Fasten the brackets to the mounting plate. Design the axle supports to be light by minimizing their volume wherever possible and removing all sharp corners. Design the wheel to be lightened in volume where possible. Design the axle to be locked in axially. Add chamfers to the axle. Add any other appropriate details.

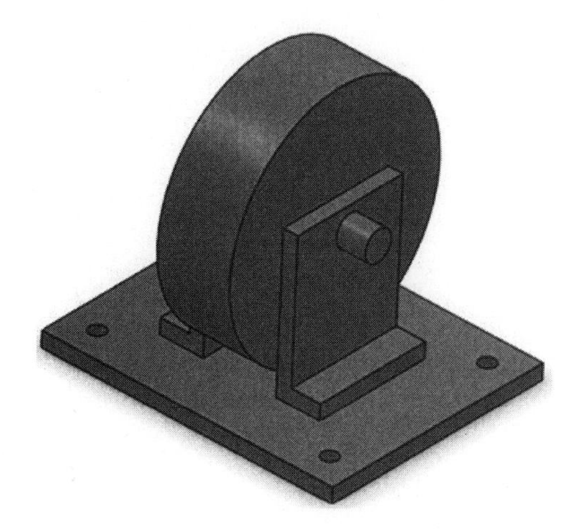

8.2

Using top-down design methodology design a C-clamp. The clamp is to have two clamping faces of 0.75" in diameter. The clamp is to have one face stationary on the C-clamp body. The other end of the clamp is to have a 0.50 flat thread that is 0.75 deep (thru). When fully extended, the clamp should have a 5.00" gap opening. The threaded post is to be turned by a T-bar. Refer to the figure for basic design layout.

Add additional features to the design, such as reducing the body's volume. Make the T-bar to be able to slide in the axle hole yet not be able to slide completely out.

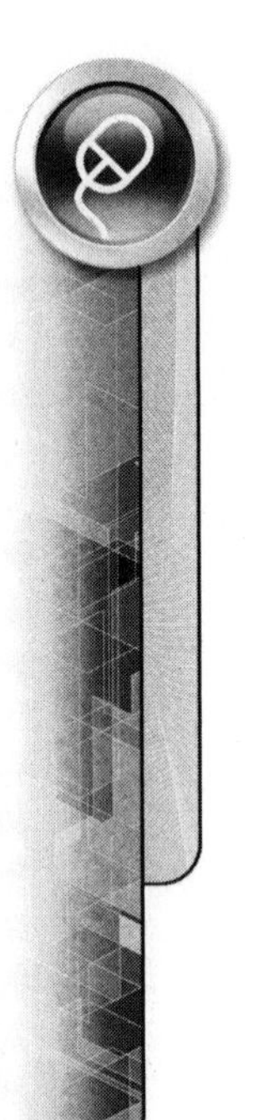

INDEX

G

H

I

L

M

N

O

P

Q

R

S

T